Contents

STUDENT'S SOLUTIONS MANUAL

CINDY TRIMBLE & ASSOCIATES

BASIC COLLEGE MATHEMATICS
FOURTH EDITION

Elayn Martin-Gay
University of New Orleans

Prentice Hall
is an imprint of

The author and publisher of this book have used their best efforts in preparing this book. These efforts include the development, research, and testing of the theories and programs to determine their effectiveness. The author and publisher make no warranty of any kind, expressed or implied, with regard to these programs or the documentation contained in this book. The author and publisher shall not be liable in any event for incidental or consequential damages in connection with, or arising out of, the furnishing, performance, or use of these programs.

Reproduced by Pearson Prentice Hall from electronic files supplied by the author.

ISBN-13: 978-0-321-64665-1
ISBN-10: 0-321-64665-7

2 3 4 5 6 BRR 14 13 12 11

Prentice Hall
is an imprint of

www.pearsonhighered.com

Chapter 1

Section 1.2 Practice

1. The place value of the 8 in 38,760,005 is millions.

2. The place value of the 8 in 67,890 is hundreds.

3. The place value of the 8 in 481,922 is ten-thousands.

4. 67 is written as sixty-seven.

5. 395 is written as three hundred ninety-five.

6. 12,804 is written as twelve thousand, eight hundred four.

7. 321,670,200 is written as three hundred twenty-one million, six hundred seventy thousand, two hundred.

8. Twenty-nine in standard form is 29.

9. Seven hundred ten in standard form is 710.

10. Twenty-six thousand, seventy-one in standard form is 26,071.

11. Six million, five hundred seven in standard form is 6,000,507.

12. 1,047,608
    = 1,000,000 + 40,000 + 7000 + 600 + 8

13. a. Find "France" in the left column. Then read from left to right until the "Literature" column is reached. We find that France has 14 Nobel Prize winners in Literature.

    b. Look at the "Total" column. Three countries have more than 60 Nobel Prize winners. The United States has 320, the United Kingdom has 110, and Germany has 82.

Vocabulary and Readiness Check

1. The numbers 0, 1, 2, 3, 4, 5, 6, 7, 8, 9, 10, 11, 12, ... are called whole numbers.

2. The number 1,286 is written in standard form.

3. The number "twenty-one" is written in words.

4. The number 900 + 60 + 5 is written in expanded form.

5. In a whole number, each group of 3 digits is called a period.

6. The place value of the digit 4 in the whole number 264 is ones.

Exercise Set 1.2

1. The place value of the 5 in 657 is tens.

3. The place value of the 5 in 5423 is thousands.

5. The place value of the 5 in 43,526,000 is hundred-thousands.

7. The place value of the 5 in 5,408,092 is millions.

9. 354 is written as three hundred fifty-four.

11. 8279 is written as eight thousand, two hundred seventy-nine.

13. 26,990 is written as twenty-six thousand, nine hundred ninety.

15. 2,388,000 is written as two million, three hundred eighty-eight thousand.

17. 24,350,185 is written as twenty-four million, three hundred fifty thousand, one hundred eighty-five.

19. 304,367 is written as three hundred four thousand, three hundred sixty-seven.

21. 2600 is written as two thousand, six hundred.

23. 15,800,000 is written as fifteen million, eight hundred thousand.

25. 14,433 is written as fourteen thousand, four hundred thirty-three.

27. 13,000,000 is written as thirteen million.

29. Six thousand, five hundred eighty-seven in standard form is 6587.

31. Fifty-nine thousand, eight hundred in standard form is 59,800.

33. Thirteen million, six hundred one thousand, eleven in standard form is 13,601,011.

1

35. Seven million, seventeen in standard form is 7,000,017.

37. Two hundred sixty thousand, nine hundred ninety-seven in standard form is 260,997.

39. Three hundred ninety-five in standard form is 395.

41. Sixteen thousand, seven hundred thirty-two in standard form is 16,732.

43. Seventy-two million, seven hundred four thousand in standard form is 72,704,000.

45. One thousand, three hundred seventeen in standard form is 1317.

47. $406 = 400 + 6$

49. $3470 = 3000 + 400 + 70$

51. $80,774 = 80,000 + 700 + 70 + 4$

53. $66,049 = 60,000 + 6000 + 40 + 9$

55. 39,680,000
$= 30,000,000 + 9,000,000 + 600,000 + 80,000$

57. The elevation of Mt. Clay in standard form is 5532. 5532 is written as five thousand, five hundred thirty-two.

59. $5492 = 5000 + 400 + 90 + 2$

61. The tallest mountain in New England is Mt. Washington.

63. Boxer has fewer dogs registered than Dachshund.

65. Labrador Retrievers have the most registrations; 123,760 is written as one hundred twenty-three thousand, seven hundred sixty.

67. The maximum weight of an average-size Dachshund is 25 pounds.

69. The largest number is 9861.

71. No; 105.00 should be written as one hundred five.

73. answers may vary

75. 1000 trillion in standard form is 1,000,000,000,000,000.

77. The Pro-Football Hall of Fame is in the town of Canton.

Section 1.3 Practice

1.
$$
\begin{array}{r}
7235 \\
+\ 542 \\
\hline
7777
\end{array}
$$

2.
$$
\begin{array}{r}
{\scriptstyle 11\ 11} \\
27,364 \\
+\ 92,977 \\
\hline
120,341
\end{array}
$$

3.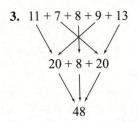

4.
$$
\begin{array}{r}
{\scriptstyle 11\ 2} \\
19 \\
5042 \\
638 \\
+\ 526 \\
\hline
6225
\end{array}
$$

5. $2\text{ cm} + 8\text{ cm} + 15\text{ cm} + 5\text{ cm} = 30\text{ cm}$
The perimeter is 30 centimeters.

6. $647 + 647 + 647 = 1941$
The perimeter is 1941 feet.

7.
$$
\begin{array}{r}
70 \\
+\ 50 \\
\hline
120
\end{array}
$$
Georgia produces 120 million pounds of freestone peaches.

8. a. The country with the fewest endangered species corresponds to the shortest bar, which is Australia.

  b. To find the total number of endangered species for Brazil, India, and Mexico, we add.
$$
\begin{array}{r}
73 \\
89 \\
+\ 72 \\
\hline
234
\end{array}
$$
The total number of endangered species for Brazil, India, and Mexico is 234.

Calculator Explorations

1. $89 + 45 = 134$

2. $76 + 91 = 173$

3. $285 + 55 = 340$

4. $8773 + 652 = 9425$

5.
```
   985
  1210
   562
+  77
------
  2834
```

6.
```
    465
   9888
    620
+ 1550
-------
 12,523
```

Vocabulary and Readiness Check

1. The sum of 0 and any number is the same <u>number</u>.

2. The sum of any number and 0 is the same <u>number</u>.

3. In $35 + 20 = 55$, the number 55 is called the <u>sum</u> and 35 and 20 are each called an <u>addend</u>.

4. The distance around a polygon is called its <u>perimeter</u>.

5. Since $(3 + 1) + 20 = 3 + (1 + 20)$, we say that changing the <u>grouping</u> in addition does not change the sum. This property is called the <u>associative</u> property of addition.

6. Since $7 + 10 = 10 + 7$, we say that changing the <u>order</u> in addition does not change the sum. This property is called the <u>commutative</u> property of addition.

Exercise Set 1.3

1.
```
   14
+ 22
-----
   36
```

3.
```
    62
+ 230
------
   292
```

5.
```
   12
   13
+ 24
-----
   49
```

7.
```
  5267
+ 132
------
  5399
```

9.
```
    1
   53
+ 64
-----
  117
```

11.
```
     1
    22
+ 490
------
   512
```

13.
```
    1  1 1
   22,781
+ 186,297
---------
  209,078
```

15.
```
   2
   8
   9
   2
   5
+ 1
----
  25
```

17.
```
    2
    6
   21
   14
    9
+ 12
-----
   62
```

19.
```
   2 2
   81
   17
   23
   79
+ 12
-----
  212
```

21.
```
    1
   62
   18
+ 14
-----
   94
```

23.
$$
\begin{array}{r}
\overset{1}{}40 \\
800 \\
+\;70 \\
\hline
910
\end{array}
$$

25.
$$
\begin{array}{r}
\overset{111}{7542} \\
49 \\
+\;682 \\
\hline
8273
\end{array}
$$

27.
$$
\begin{array}{r}
\overset{1\;2}{}24 \\
9\;006 \\
489 \\
+\;2\;407 \\
\hline
11,926
\end{array}
$$

29.
$$
\begin{array}{r}
\overset{1\;\;2}{627} \\
628 \\
+\;629 \\
\hline
1884
\end{array}
$$

31.
$$
\begin{array}{r}
\overset{1\;1\;1}{6\;820} \\
4\;271 \\
+\;5\;626 \\
\hline
16,717
\end{array}
$$

33.
$$
\begin{array}{r}
\overset{11}{507} \\
593 \\
+\;10 \\
\hline
1110
\end{array}
$$

35.
$$
\begin{array}{r}
4200 \\
2107 \\
+\;2692 \\
\hline
8999
\end{array}
$$

37.
$$
\begin{array}{r}
\overset{11\;2\;2}{}49 \\
628 \\
5\;762 \\
+\;29,462 \\
\hline
35,901
\end{array}
$$

39.
$$
\begin{array}{r}
\overset{1\;22\;\;21}{121,742} \\
57,279 \\
26,586 \\
+\;426,782 \\
\hline
632,389
\end{array}
$$

41.
$$
\begin{array}{r}
\overset{2}{}9 \\
12 \\
9 \\
+\;12 \\
\hline
42
\end{array}
$$
The perimeter is 42 inches.

43.
$$
\begin{array}{r}
\overset{1}{}7 \\
8 \\
+\;10 \\
\hline
25
\end{array}
$$
The perimeter is 25 feet.

45. Opposite sides of a rectangle have the same length.
$$
\begin{array}{r}
\overset{2}{}4 \\
8 \\
4 \\
+\;8 \\
\hline
24
\end{array}
$$
The perimeter is 24 inches.

47.
$$
\begin{array}{r}
2 \\
2 \\
2 \\
+\;2 \\
\hline
8
\end{array}
$$
The perimeter is 8 yards.

49. $8 + 3 + 5 + 7 + 5 + 1 = 29$
The perimeter is 29 inches.

51. The unknown vertical side has length
$12 - 5 = 7$ meters. The unknown horizontal side has length $10 - 5 = 5$ meters.
$10 + 12 + 5 + 7 + 5 + 5 = 44$
The perimeter is 44 meters.

53. "Find the sum" indicates addition.
$$
\begin{array}{r}
\overset{111}{297} \\
+\;1796 \\
\hline
2093
\end{array}
$$
The sum of 297 and 1796 is 2093.

55. "Find the total" indicates addition.

$$
\begin{array}{r}
\overset{1\,3}{} \\
76 \\
39 \\
8 \\
17 \\
+\ 126 \\
\hline
266
\end{array}
$$

The total of 76, 39, 8, 17, and 126 is 266.

57. "Increased by" indicates addition.

$$
\begin{array}{r}
\overset{1}{} \\
452 \\
+\ 92 \\
\hline
544
\end{array}
$$

452 increased by 92 is 544.

59. "Plus" indicates addition.

$$
\begin{array}{r}
\overset{1\,2\,1}{} \\
2686 \\
686 \\
+\ \ 80 \\
\hline
3452
\end{array}
$$

2686 plus 686 plus 80 is 3452.

61. Add 3170 to 19,308.

$$
\begin{array}{r}
\overset{1}{} \\
19,308 \\
+\ 3\ 170 \\
\hline
22,478
\end{array}
$$

Florida's projected population in 2020 is 22,478 thousand.

63.
$$
\begin{array}{r}
3124 \\
+\ 3560 \\
\hline
6684
\end{array}
$$

The height of Mt. Mitchell is 6684 feet.

65.
$$
\begin{array}{r}
\overset{2\,1}{} \\
78 \\
90 \\
102 \\
+\ 70 \\
\hline
340
\end{array}
$$

He needs 340 feet of wiring.

67.
$$
\begin{array}{r}
\overset{1\,1}{} \\
1430 \\
675 \\
+\ 320 \\
\hline
2425
\end{array}
$$

The total height of Yosemite Falls is 2425 feet.

69. Add 78,328 to 235,441.

$$
\begin{array}{r}
\overset{1\,1}{} \\
235,441 \\
+\ \ 78,328 \\
\hline
313,769
\end{array}
$$

The total number of Harley-Davidson motorcycles sold in 2008 was 313,769.

71. Find the sum of 41,382 and 44,064.

$$
\begin{array}{r}
\overset{1}{} \\
41,382 \\
+\ 44,064 \\
\hline
85,446
\end{array}
$$

The total number of Accords and Camrys sold was 85,446.

73. The sides of a square are all the same length.

$$
\begin{array}{r}
\overset{1}{} \\
31 \\
31 \\
31 \\
+\ 31 \\
\hline
124
\end{array}
$$

The perimeter of the board is 124 feet.

75. Find the sum of 2677 and 493.

$$
\begin{array}{r}
\overset{1\,1\,1}{} \\
2677 \\
+\ \ 493 \\
\hline
3170
\end{array}
$$

There were 3170 Gap Inc. stores located worldwide.

77. California has the most Target stores.

79.
$$
\begin{array}{r}
236 \\
124 \\
+\ 146 \\
\hline
506
\end{array}
$$

The total number of Target stores in California, Florida, and Texas is 506 stores.

81. Florida and Georgia:
$$
\begin{array}{r}
124 \\
+\ 56 \\
\hline
180
\end{array}
$$

Michigan and Ohio:
$$
\begin{array}{r}
60 \\
+\ 64 \\
\hline
124
\end{array}
$$

Florida and Georgia have more Target stores.

83.
$$\begin{array}{r} \overset{1}{2029} \\ + 3865 \\ \hline 5894 \end{array}$$

The total highway mileage in Delaware is 5894 miles.

85. answers may vary

87. answers may vary

89.
$$\begin{array}{r} \overset{111\ 121\ 1}{56,468,980} \\ 1,236,785 \\ + 986,768,000 \\ \hline 1,044,473,765 \end{array}$$

91.
$$\begin{array}{r} \overset{21}{566} \\ 932 \\ + 871 \\ \hline 2369 \end{array}$$

The given answer is correct.

93.
$$\begin{array}{r} \overset{22}{14} \\ 173 \\ 86 \\ + 257 \\ \hline 530 \end{array}$$

The given answer is incorrect.

Section 1.4 Practice

1. a. $14 - 6 = 8$ because $8 + 6 = 14$.

b. $20 - 8 = 12$ because $12 + 8 = 20$.

c. $93 - 93 = 0$ because $0 + 93 = 93$.

d. $42 - 0 = 42$ because $42 + 0 = 42$.

2. a.
$$\begin{array}{r} 9143 \\ - 122 \\ \hline 9021 \end{array} \qquad Check: \begin{array}{r} 9021 \\ + 122 \\ \hline 9143 \end{array}$$

b.
$$\begin{array}{r} 978 \\ - 851 \\ \hline 127 \end{array} \qquad Check: \begin{array}{r} 127 \\ + 851 \\ \hline 978 \end{array}$$

3. a.
$$\begin{array}{r} \overset{8\ 17}{69\ 7} \\ - 4\ 9 \\ \hline 64\ 8 \end{array} \qquad Check: \begin{array}{r} 648 \\ + 49 \\ \hline 697 \end{array}$$

b.
$$\begin{array}{r} \overset{2\ 12}{326} \\ - 245 \\ \hline 81 \end{array} \qquad Check: \begin{array}{r} 81 \\ + 245 \\ \hline 326 \end{array}$$

c.
$$\begin{array}{r} 1234 \\ - 822 \\ \hline 412 \end{array} \qquad Check: \begin{array}{r} 412 \\ + 822 \\ \hline 1234 \end{array}$$

4. a.
$$\begin{array}{r} \overset{9}{\underset{3\ 10}{\ \ 10}} \\ 4\ 0\ 0 \\ - 1\ 6\ 4 \\ \hline 2\ 3\ 6 \end{array} \qquad Check: \begin{array}{r} 236 \\ + 164 \\ \hline 400 \end{array}$$

b.
$$\begin{array}{r} \overset{9}{\underset{9\ 10}{\ \ 10}} \\ 1\ 0\ 0\ 0 \\ - 7\ 6\ 2 \\ \hline 2\ 3\ 8 \end{array} \qquad Check: \begin{array}{r} 238 \\ + 762 \\ \hline 1000 \end{array}$$

5.
$$\begin{array}{r} 15,759 \\ - \ \ \ 458 \\ \hline 15,301 \end{array}$$

The radius of Neptune is 15,301 miles.

6.
$$\begin{array}{r} 92 \\ - 47 \\ \hline 45 \end{array}$$

The sale price of the suit is $45.

Calculator Explorations

1. $865 - 95 = 770$

2. $76 - 27 = 49$

3. $147 - 38 = 109$

4. $366 - 87 = 279$

5. $9625 - 647 = 8978$

6. $10,711 - 8925 = 1786$

Vocabulary and Readiness Check

1. The difference of any number and that same number is <u>0</u>.

2. The difference of any number and 0 is the same <u>number</u>.

3. In $37 - 19 = 18$, the number 37 is the <u>minuend</u>, and the number 19 is the <u>subtrahend</u>.

4. In $37 - 19 = 18$, the number 18 is called the <u>difference</u>.

5. $6 - 6 = 0$

6. $93 - 93 = 0$

7. $600 - 0 = 600$

8. $5 - 0 = 5$

Exercise Set 1.4

1.
$$
\begin{array}{r}
67 \\
-23 \\
\hline
44
\end{array}
$$
Check:
$$
\begin{array}{r}
44 \\
+23 \\
\hline
67
\end{array}
$$

3.
$$
\begin{array}{r}
389 \\
-124 \\
\hline
265
\end{array}
$$
Check:
$$
\begin{array}{r}
265 \\
+124 \\
\hline
389
\end{array}
$$

5.
$$
\begin{array}{r}
167 \\
-32 \\
\hline
135
\end{array}
$$
Check:
$$
\begin{array}{r}
135 \\
+32 \\
\hline
167
\end{array}
$$

7.
$$
\begin{array}{r}
2677 \\
-423 \\
\hline
2254
\end{array}
$$
Check:
$$
\begin{array}{r}
2254 \\
+423 \\
\hline
2677
\end{array}
$$

9.
$$
\begin{array}{r}
6998 \\
-1453 \\
\hline
5545
\end{array}
$$
Check:
$$
\begin{array}{r}
5545 \\
+1453 \\
\hline
6998
\end{array}
$$

11.
$$
\begin{array}{r}
749 \\
-149 \\
\hline
600
\end{array}
$$
Check:
$$
\begin{array}{r}
600 \\
+149 \\
\hline
749
\end{array}
$$

13.
$$
\begin{array}{r}
62 \\
-37 \\
\hline
25
\end{array}
$$
Check:
$$
\begin{array}{r}
1 \\
25 \\
+37 \\
\hline
62
\end{array}
$$

15.
$$
\begin{array}{r}
70 \\
-25 \\
\hline
45
\end{array}
$$
Check:
$$
\begin{array}{r}
1 \\
45 \\
+25 \\
\hline
70
\end{array}
$$

17.
$$
\begin{array}{r}
938 \\
-792 \\
\hline
146
\end{array}
$$
Check:
$$
\begin{array}{r}
1 \\
146 \\
+792 \\
\hline
938
\end{array}
$$

19.
$$
\begin{array}{r}
922 \\
-634 \\
\hline
288
\end{array}
$$
Check:
$$
\begin{array}{r}
11 \\
288 \\
+634 \\
\hline
922
\end{array}
$$

21.
$$
\begin{array}{r}
600 \\
-432 \\
\hline
168
\end{array}
$$
Check:
$$
\begin{array}{r}
11 \\
168 \\
+432 \\
\hline
600
\end{array}
$$

23. $\begin{array}{r} 142 \\ -\ 36 \\ \hline 106 \end{array}$

   Check:

   $\begin{array}{r} 1 \\ 106 \\ +\ 36 \\ \hline 142 \end{array}$

25. $\begin{array}{r} 923 \\ -\ 476 \\ \hline 447 \end{array}$

   Check:

   $\begin{array}{r} 11 \\ 447 \\ +\ 476 \\ \hline 923 \end{array}$

27. $\begin{array}{r} 6283 \\ -\ 560 \\ \hline 5723 \end{array}$

   Check:

   $\begin{array}{r} 1 \\ 5723 \\ +\ 560 \\ \hline 6283 \end{array}$

29. $\begin{array}{r} 533 \\ -\ 29 \\ \hline 504 \end{array}$

   Check:

   $\begin{array}{r} 1 \\ 504 \\ +\ 29 \\ \hline 533 \end{array}$

31. $\begin{array}{r} 200 \\ -111 \\ \hline 89 \end{array}$

   Check:

   $\begin{array}{r} 11 \\ 89 \\ +\ 111 \\ \hline 200 \end{array}$

33. $\begin{array}{r} 1983 \\ -\ 1904 \\ \hline 79 \end{array}$

   Check:

   $\begin{array}{r} 1 \\ 79 \\ +\ 1904 \\ \hline 1983 \end{array}$

35. $\begin{array}{r} 56,422 \\ -\ 16,508 \\ \hline 39,914 \end{array}$

   Check:

   $\begin{array}{r} 11\ \ 1 \\ 39,914 \\ +\ 16,508 \\ \hline 56,422 \end{array}$

37. $\begin{array}{r} 50,000 \\ -\ 17,289 \\ \hline 32,711 \end{array}$

   Check:

   $\begin{array}{r} 11\ \ 11 \\ 32,711 \\ +\ 17,289 \\ \hline 50,000 \end{array}$

39. $\begin{array}{r} 7020 \\ -\ 1979 \\ \hline 5041 \end{array}$

   Check:

   $\begin{array}{r} 111 \\ 5041 \\ +\ 1979 \\ \hline 7020 \end{array}$

41. $\begin{array}{r} 51,111 \\ -\ 19,898 \\ \hline 31,213 \end{array}$

   Check:

   $\begin{array}{r} 11\ 11 \\ 31,213 \\ +\ 19,898 \\ \hline 51,111 \end{array}$

43. $\begin{array}{r} 9 \\ -\ 5 \\ \hline 4 \end{array}$

   9 subtract 5 is 4.

45. $\begin{array}{r} 41 \\ -\ 21 \\ \hline 20 \end{array}$

   The difference of 41 and 21 is 20.

47. $\begin{array}{r} 63 \\ -\ 56 \\ \hline 7 \end{array}$

   63 subtract 56 is 7.

49.   108
          − 36
          ─────
            72
108 less 36 is 72.

51.   100
          − 12
          ─────
            88
12 subtracted from 100 is 88.

53.   503
          − 239
          ─────
            264
She must read 264 more pages.

55.   25,000,000
        − 22,000,000
        ────────────
          3,000,000
The hole in the ozone layer has grown 3,000,000 square kilometers from 2002 to 2008.

57.   189,000
          + 75,000
          ─────────
            264,000
The total U.S. land area drained by the Upper Mississippi and Lower Mississippi sub-basins is 264,000 square miles.

59.   530,000
          − 247,000
          ──────────
            283,000
The Missouri sub-basin drains 283,000 square miles more than the Arkansas Red-White sub-basin.

61.   20,320
          − 14,255
          ─────────
            6 065
Mt. McKinley's peak is 6065 feet higher than Long's Peak.

63.   754
          − 726
          ─────
            28
The Oroville Dam is 28 feet taller than the Hoover Dam.

65.   645
          − 287
          ─────
            358
The distance between Hays and Denver is 358 miles.

67.   914
          − 295
          ─────
            619
She will have $619 left in her savings account after buying the DVD player.

69.   4100
          − 3648
          ──────
            452
Oklahoma's projected population increase is 452 thousand.

71. Live rock music has a decibel level of 100 dB.

73.   88
          − 30
          ────
            58
The sound of snoring is 58 dB louder than normal conversation.

75.   539
          − 219
          ─────
            320
320 members were never Boy Scouts.

77.   100,000
          − 94,080
          ─────────
            5 920
The Dole Plantation maze is 5920 square feet larger than the Ruurlo maze.

79. The tallest bar corresponds to Hartsfield-Jackson Atlanta International, so that is the busiest airport.

81.   76,000,000
        − 50,000,000
        ────────────
          26,000,000
Chicago O'Hare International Airport has 26 million more passengers per year than the Denver International Airport.

83. Votes for Jo:
        1 2 1
          276
          362
          201
        + 179
        ──────
          1018

Votes for Trudy:
$\overset{1\;1}{}$
295
122
312
+ 18
―――
747

Since Jo received more votes, she won the election.

1018
− 747
―――
271

Jo won the election by 271 votes.

85. $\overset{1\;1}{986}$
   + 48
   ―――
   1034

87. 76
   −67
   ――
   9

89. 9000
   − 482
   ―――
   8518

91. 10,962
   4 851
   + 7 063
   ―――
   22,876

93. In 48 , 48 is the minuend and 1 is the
   − 1
   ――
subtrahend.

95. In subtract 7 from 70, 70 is the minuend and 7 is the subtrahend.

97. 741
   − 56
   ―――
   685

The given answer is incorrect.
Check:
$\overset{1\;1}{685}$
+ 56
――
741

99. 1029
   − 888
   ―――
   141

The given answer is correct. Check:
$\overset{1\;1}{141}$
+ 888
―――
1029

101. 5269
   − 2385
   ―――
   2884

103. answers may vary.

105. $\overset{1\,2\,1\;3\,2}{289,462}$
   369,477
   218,287
   + 121,685
   ―――――
   998,911

The college students have not reached a goal of one million pages.

1,000,000
− 998,911
―――――
1 089

The college students need to read 1089 more pages to reach a goal of one million pages.

Section 1.5 Practice

1. a. To round 57 to the nearest ten, observe that the digit in the ones place is 7. Since the digit is at least 5, we add 1 to the digit in the tens place. The number 57 rounded to the nearest ten is 60.

b. To round 641 to the nearest ten, observe that the digit in the ones place is 1. Since the digit is less than 5, we do not add 1 to the digit in the tens place. The number 641 rounded to the nearest ten is 640.

c. To round 325 to the nearest ten observe that the digit in the ones place is 5. Since the digit is at least 5, we add 1 to the digit in the tens place. The number 325 rounded to the nearest ten is 330.

2. a. To round 72,304 to the nearest thousand, observe that the digit in the hundreds place is 3. Since the digit is less than 5, we do not add 1 to the digit in the thousands place. The number 72,304 rounded to the nearest thousand is 72,000.

b. To round 9222 to the nearest thousand, observe that the digit in the hundreds place is 2. Since the digit is less than 5, we do not add 1 to the digit in the thousands place. The number 9222 rounded to the nearest thousand is 9000.

c. To round 671,800 to the nearest thousand, observe that the digit in the hundreds place is 8. Since this digit is at least 5, we add 1 to the digit in the thousands place. The number 671,800 rounded to the nearest thousand is 672,000.

3. a. To round 3474 to the nearest hundred, observe that the digit in the tens place is 7. Since this digit is at least 5, we add 1 to the digit in the hundreds place. The number 3474 rounded to the nearest hundred is 3500.

b. To round 76,243 to the nearest hundred, observe that the digit in the tens place is 4. Since this digit is less than 5, we do not add 1 to the digit in the hundreds place. The number 76,243 rounded to the nearest hundred is 76,200.

c. To round 978,965 to the nearest hundred, observe that the digit in the tens place is 6. Since this digit is at least 5, we add 1 to the digit in the hundreds place. The number 978,865 rounded to the nearest hundred is 979,000.

4.
49	rounds to	50
25	rounds to	30
32	rounds to	30
51	rounds to	50
98	rounds to	+ 100
		260

5.
3785	rounds to	4000
− 2479	rounds to	− 2000
		2000

6.
11	rounds to	10
16	rounds to	20
19	rounds to	20
+ 31	rounds to	+ 30
		80

The total distance is approximately 80 miles.

7.
48,445	rounds to	48,000
6,584	rounds to	7,000
+ 15,632	rounds to	+ 16,000
		71,000

The total number of cases is approximately 71,000.

Vocabulary and Readiness Check

1. To graph a number on a number line, darken the point representing the location of the number.

2. Another word for approximating a whole number is rounding.

3. The number 65 rounded to the nearest ten is 70 but the number 61 rounded to the nearest ten is 60.

4. An exact number of products is 1265, but an estimate is 1000.

Exercise Set 1.5

1. To round 423 to the nearest ten, observe that the digit in the ones place is 3. Since this digit is less than 5, we do not add 1 to the digit in the tens place. The number 423 rounded to the nearest ten is 420.

3. To round 635 to the nearest ten, observe that the digit in the ones place is 5. Since this digit is at least 5, we add 1 to the digit in the tens place. The number 635 rounded to the nearest ten is 640.

5. To round 2791 to the nearest hundred, observe that the digit in the tens place is 9. Since this digit is at least 5, we add 1 to the digit in the hundreds place. The number 2791 rounded to the nearest hundred is 2800.

7. To round 495 to the nearest ten, observe that the digit in the ones place is 5. Since this digit is at least 5, we add 1 to the digit in the tens place. The number 495 rounded to the nearest ten is 500.

9. To round 21,094 to the nearest thousand, observe that the digit in the hundreds place is 0. Since this digit is less than 5, we do not add 1 to the digit in the thousands place. The number 21,094 rounded to the nearest thousand is 21,000.

11. To round 33,762 to the nearest thousand, observe that the digit in the hundreds place is 7. Since this digit is at least 5, we add 1 to the digit in the thousands place. The number 33,762 rounded to the nearest thousand is 34,000.

13. To round 328,495 to the nearest hundred, observe that the digit in the tens place is 9. Since this digit is at least 5, we add 1 to the digit in the hundreds place. The number 328,495 rounded to the nearest hundred is 328,500.

15. To round 36,499 to the nearest thousand, observe that the digit in the hundreds place is 4. Since this digit is less than 5, we do not add 1 to the digit in the thousands place. The number 36,499 rounded to the nearest thousand is 36,000.

17. To round 39,994 to the nearest ten, observe that the digit in the ones place is 4. Since this digit is less than 5, we do not add 1 to the digit in the tens place. The number 39,994 rounded to the nearest ten is 39,990.

19. To round 29,834,235 to the nearest ten-million, observe that the digit in the millions place is 9. Since this digit is at least 5, we add 1 to the digit in the ten-millions place. The number 29,834,235 rounded to the nearest ten-million is 30,000,000.

21. Estimate 5281 to a given place value by rounding it to that place value. 5281 rounded to the tens place is 5280, to the hundreds place is 5300, and to the thousands place is 5000.

23. Estimate 9444 to a given place value by rounding it to that place value. 9444 rounded to the tens place is 9440, to the hundreds place is 9400, and to the thousands place is 9000.

25. Estimate 14,876 to a given place value by rounding it to that place value. 14,876 rounded to the tens place is 14,880, to the hundreds place is 14,900, and to the thousands place is 15,000.

27. To round 83,659 to the nearest thousand, observe that the digit in the hundreds place is 6. Since this digit is at least 5, we add 1 to the digit in the thousands place. Therefore, 83,659 members rounded to the nearest thousand is 84,000 members.

29. To round 38,387 to the nearest thousand, observe that the digit in the hundreds place is 3. Since this digit is less than 5, we do not add 1 to the digit in the thousands place. Therefore, 38,387 points rounded to the nearest thousand is 38,000 points.

31. To round 42,570,000,000 to the nearest billion, observe that the digit in the hundred-millions place is 5. Since this digit is at least 5, we add 1 to the digit in the billions place. Therefore, $42,570,000,000 rounded to the nearest billion is $43,000,000,000.

33. To round 4,089,867 to the nearest hundred-thousand, observe that the digit in the ten-thousands place is 8. Since this digit is at least 5, we add 1 to the digit in the hundred thousands place. Therefore, $4,089,867 rounded to the nearest hundred-thousand is $4,100,000.

35. U.S.: To round 262,700,000 to the nearest million, observe that the digit in the hundred-thousands place is 7. Since this digit is at least 5, we add 1 to the digit in the millions place. The number 262,700,000 rounded to the nearest million is 263,000,000.
India: To round 296,886,000 to the nearest million, observe that the digit in the hundred-thousands place is 8. Since this digit is at least 5, we add 1 to the digit in the millions place. The number 296,886,000 rounded to the nearest million is 297,000,000.

37.

39	rounds to	40
45	rounds to	50
22	rounds to	20
+ 17	rounds to	+ 20
		130

39.

449	rounds to	450
− 373	rounds to	− 370
		80

41.

1913	rounds to	1900
1886	rounds to	1900
+ 1925	rounds to	+ 1900
		5700

43.

1774	rounds to	1800
− 1492	rounds to	− 1500
		300

45.
3995	rounds to	4000
2549	rounds to	2500
+ 4944	rounds to	+ 4900
		11,400

47. 463 + 219 is approximately 460 + 220 = 680.
The answer of 602 is incorrect.

49. 229 + 443 + 606 is approximately
230 + 440 + 610 = 1280.
The answer of 1278 is correct.

51. 7806 + 5150 is approximately
7800 + 5200 = 13,000.
The answer of 12,956 is correct.

53.
899	rounds to	900
1499	rounds to	1500
+ 999	rounds to	+ 1000
		3400

The total cost is approximately $3400.

55.
1429	rounds to	1400
− 530	rounds to	− 500
		900

Boston in approximately 900 miles farther from Kansas City than Chicago is.

57.
20,320	rounds to	20,000
− 14,410	rounds to	− 14,000
		6 000

The difference in elevation is approximately 6000 feet.

59.
142,702	rounds to	140,000
− 75,543	rounds to	− 80,000
		60,000

Joliet was approximately 60,000 larger than Evanston.

61.
908,412	rounds to	908,000
− 905,851	rounds to	− 906,000
		2 000

The increase in enrollment is approximately 2000 children.

63. 391 million dollars written in standard form is $391,000,000. $391,000,000 rounded to the nearest ten-million is $390,000,000. $391,000,000 rounded to the nearest hundred-million is $400,000,000.

65. 349 million dollars written in standard form is $349,000,000. $349,000,000 rounded to the nearest ten-million is $350,000,000. $349,000,000 rounded to the nearest hundred-million is $300,000,000.

67. 5723, for example, rounded to the nearest hundred is 5700.

69. a. The smallest possible number that rounds to 8600 is 8550.

   b. The largest possible number that rounds to 8600 is 8649.

71. answers may vary

73. 54 rounds to 50
17 rounds to 20
50 + 20 + 50 + 20 = 140
The perimeter is approximately 140 meters.

Section 1.6 Practice

1. a. $3 \times 0 = 0$

   b. $4(1) = 4$

   c. $(0)(34) = 0$

   d. $1 \cdot 76 = 76$

2. a. $5(2 + 3) = 5 \cdot 2 + 5 \cdot 3$

   b. $9(8 + 7) = 9 \cdot 8 + 9 \cdot 7$

   c. $3(6 + 1) = 3 \cdot 6 = 3 \cdot 1$

3. a.
$$\begin{array}{r} \overset{2}{36} \\ \times 4 \\ \hline 144 \end{array}$$

   b.
$$\begin{array}{r} \overset{2\,1}{132} \\ \times \ 9 \\ \hline 1188 \end{array}$$

4. a.
$$\begin{array}{r} 594 \\ \times \ 72 \\ \hline 1\ 188 \\ 41\ 580 \\ \hline 42,768 \end{array}$$

b.
$$\begin{array}{r} 306 \\ \times\ \ 81 \\ \hline 306 \\ 24\ 480 \\ \hline 24,786 \end{array}$$

5. a.
$$\begin{array}{r} 726 \\ \times\ \ 142 \\ \hline 1\ 452 \\ 29\ 040 \\ 72\ 600 \\ \hline 103,092 \end{array}$$

b.
$$\begin{array}{r} 288 \\ \times\ \ 4 \\ \hline 1152 \end{array}$$

6. $75 \cdot 100 = 7500$

7. $808 \cdot 1000 = 808,000$

8.
$$\begin{array}{r} 35 \\ \times\ 3 \\ \hline 105 \end{array}$$
$35 \cdot 3000 = 105,000$
Attach 3 zeros.

9. $600 \cdot 600 = 360,000$

10. Area $=$ length $\cdot$ width
$= (360 \text{ miles})(280 \text{ miles})$
$= 100,800$ square miles
The area of Wyoming is 100,800 square miles.

11.
$$\begin{array}{r} 16 \\ \times 45 \\ \hline 80 \\ 640 \\ \hline 720 \end{array}$$
The printer can print 720 pages in 45 minutes.

12. $8 \times 11 = 88$
$5 \times 9 = 45$
$$\begin{array}{r} {\scriptstyle 1} \\ 88 \\ +\ 45 \\ \hline 133 \end{array}$$
The total cost is $133.

13.
163	rounds to	200
$\times\,391$	rounds to	$\times\ 400$
		$\overline{80,000}$

There are approximately 80,000 words on 391 pages.

Calculator Explorations

1. $72 \times 48 = 3456$

2. $81 \times 92 = 7452$

3. $163 \cdot 94 = 15,322$

4. $285 \cdot 144 = 41,040$

5. $983(277) = 272,291$

6. $1562(843) = 1,316,766$

Vocabulary and Readiness Check

1. The product of 0 and any number is <u>0</u>.

2. The product of 1 and any number is the <u>number</u>.

3. In $8 \cdot 12 = 96$, the 96 is called the <u>product</u> and 8 and 12 are each called a <u>factor</u>.

4. Since $9 \cdot 10 = 10 \cdot 9$, we say that changing the <u>order</u> in multiplication does not change the product. This property is called the <u>commutative</u> property of multiplication.

5. Since $(3 \cdot 4) \cdot 6 = 3 \cdot (4 \cdot 6)$, we say that changing the <u>grouping</u> in multiplication does not change the product. This property is called the <u>associative</u> property of multiplication.

6. <u>Area</u> measures the amount of surface of a region.

7. Area of a rectangle $=$ <u>length</u> $\cdot$ width.

8. We know $9(10 + 8) = 9 \cdot 10 + 9 \cdot 8$ by the <u>distributive</u> property.

Exercise Set 1.6

1. $1 \cdot 24 = 24$

3. $0 \cdot 19 = 0$

5. $8 \cdot 0 \cdot 9 = 0$

7. $87 \cdot 1 = 87$

9. $6(3 + 8) = 6 \cdot 3 + 6 \cdot 8$

11. $4(3 + 9) = 4 \cdot 3 + 4 \cdot 9$

13. $20(14 + 6) = 20 \cdot 14 + 20 \cdot 6$

15.
$$\begin{array}{r} 64 \\ \times\ 8 \\ \hline 512 \end{array}$$

17.
$$\begin{array}{r} 613 \\ \times\ 6 \\ \hline 3678 \end{array}$$

19.
$$\begin{array}{r} 277 \\ \times\ 6 \\ \hline 1662 \end{array}$$

21.
$$\begin{array}{r} 1074 \\ \times\ 6 \\ \hline 6444 \end{array}$$

23.
$$\begin{array}{r} 89 \\ \times\ 13 \\ \hline 267 \\ 890 \\ \hline 1157 \end{array}$$

25.
$$\begin{array}{r} 421 \\ \times\ 58 \\ \hline 3\ 368 \\ 21\ 050 \\ \hline 24,418 \end{array}$$

27.
$$\begin{array}{r} 306 \\ \times\ 81 \\ \hline 306 \\ 24\ 480 \\ \hline 24,786 \end{array}$$

29.
$$\begin{array}{r} 780 \\ \times\ 20 \\ \hline 15,600 \end{array}$$

31. $(495)(13)(0) = 0$

33. $(640)(1)(10) = (640)(10) = 6400$

35.
$$\begin{array}{r} 1234 \\ \times\ 39 \\ \hline 11\ 106 \\ 37\ 020 \\ \hline 48,126 \end{array}$$

37.
$$\begin{array}{r} 609 \\ \times\ 234 \\ \hline 2\ 436 \\ 18\ 270 \\ 121\ 800 \\ \hline 142,506 \end{array}$$

39.
$$\begin{array}{r} 8649 \\ \times\ 274 \\ \hline 34\ 596 \\ 605\ 430 \\ 1\ 729\ 800 \\ \hline 2,369,826 \end{array}$$

41.
$$\begin{array}{r} 589 \\ \times\ 110 \\ \hline 5\ 890 \\ 58\ 900 \\ \hline 64,790 \end{array}$$

43.
$$\begin{array}{r} 1941 \\ \times\ 2035 \\ \hline 9\ 705 \\ 58\ 230 \\ 3\ 882\ 000 \\ \hline 3,949,935 \end{array}$$

45. $8 \times 100 = 800$

47. $11 \times 1000 = 11,000$

49. $7406 \cdot 10 = 74,060$

51. $6 \cdot 4 = 24$
$6 \cdot 4000 = 24,000$
(attach 3 zeros)

53. $5 \cdot 9 = 45$
$50 \cdot 900 = 45,000$
(attach 3 zeros)

55. $41 \cdot 8 = 328$
$41 \cdot 80,000 = 3,280,000$
(attach 4 zeros)

57. Area = (length)(width)
= (9 meters)(7 meters)
= 63 square meters

Perimeter = length + width + length + width
= $9 + 7 + 9 + 7$
= 32 meters

59. Area = (length)(width)
$$= (40 \text{ feet})(17 \text{ feet})$$
$$= 680 \text{ square feet}$$

Perimeter = length + width + length + width
$$= 40 + 17 + 40 + 17$$
$$= 114 \text{ feet}$$

61.
$$\begin{array}{r} 576 \\ \times\, 354 \\ \hline \end{array} \quad \begin{array}{l} \text{rounds to} \\ \text{rounds to} \end{array} \quad \begin{array}{r} 600 \\ \times\, 400 \\ \hline 240,000 \end{array}$$

63.
$$\begin{array}{r} 604 \\ \times\, 451 \\ \hline \end{array} \quad \begin{array}{l} \text{rounds to} \\ \text{rounds to} \end{array} \quad \begin{array}{r} 600 \\ \times\, 500 \\ \hline 300,000 \end{array}$$

65. 38×42 is approximately 40×40, which is 1600. The best estimate is c.

67. 612×29 is approximately 600×30, which is 18,000.
The best estimate is c.

69. $80 \times 11 = (8 \times 10) \times 11$
$$= 8 \times (10 \times 11)$$
$$= 8 \times 110$$
$$= 880$$

71. $6 \times 700 = 4200$

73.
$$\begin{array}{r} 2240 \\ \times\quad 2 \\ \hline 4480 \end{array}$$

75.
$$\begin{array}{r} 125 \\ \times\quad 3 \\ \hline 375 \end{array}$$
There are 375 calories in 3 tablespoons of olive oil.

77.
$$\begin{array}{r} 94 \\ \times\, 35 \\ \hline 470 \\ 2820 \\ \hline 3290 \end{array}$$
The total cost is $3290.

79. a. $4 \times 5 = 20$
There are 20 boxes in one layer.

b.
$$\begin{array}{r} 20 \\ \times\quad 5 \\ \hline 100 \end{array}$$
There are 100 boxes on the pallet.

c.
$$\begin{array}{r} 100 \\ \times\quad 20 \\ \hline 2000 \end{array}$$
The weight of the cheese on the pallet is 2000 pounds.

81. Area = (length)(width)
$$= (110 \text{ feet})(80 \text{ feet})$$
$$= 8800 \text{ square feet}$$
The area is 8800 square feet.

83. Area = (length)(width)
$$= (350 \text{ feet})(160 \text{ feet})$$
$$= 56,000 \text{ square feet}$$
The area is 56,000 square feet.

85.
$$\begin{array}{r} 94 \\ \times\, 62 \\ \hline 188 \\ 5640 \\ \hline 5828 \end{array}$$
There are 5828 pixels on the screen.

87.
$$\begin{array}{r} 60 \\ \times\, 35 \\ \hline 300 \\ 1\,800 \\ \hline 2\,100 \end{array}$$
There are 2100 characters in 35 lines.

89.
$$\begin{array}{r} 160 \\ \times\quad 8 \\ \hline 1280 \end{array}$$
There are 1280 calories in 8 ounces.

91.

T-Shirt Size	Number of Shirts Ordered	Cost per Shirt	Cost per Size Ordered
S	4	$10	$40
M	6	$10	$60
L	20	$10	$200
XL	3	$12	$36
XXL	3	$12	$36
Total Cost			$372

93. There are 60 minutes in one hour.
$24 \times 60 \times 1000 = 1440 \times 1000 = 1,440,000$
1,440,000 tea bags are produced in one day.

95.
$$\begin{array}{r} 128 \\ +\ \ 7 \\ \hline 135 \end{array}$$

97.
$$\begin{array}{r} 134 \\ \times 16 \\ \hline 804 \\ 1340 \\ \hline 2144 \end{array}$$

99.
$$\begin{array}{r} 19 \\ +\ 4 \\ \hline 23 \end{array}$$
The sum of 19 and 4 is 23.

101.
$$\begin{array}{r} 19 \\ -\ 4 \\ \hline 15 \end{array}$$
The difference of 19 and 4 is 15.

103. $6 + 6 + 6 + 6 + 6 = 5 \cdot 6$ or $6 \cdot 5$

105. a. $3 \cdot 5 = 5 + 5 + 5$ or $3 + 3 + 3 + 3 + 3$

b. answers may vary

107.
$$\begin{array}{r} 203 \\ \times\ 14 \\ \hline 812 \\ 2030 \\ \hline 2842 \end{array}$$

109. $42 \times 3 = 126$
$42 \times 9 = 378$
The problem is
$$\begin{array}{r} 42 \\ \times 93 \\ \hline \end{array}$$

111. answers may vary

113. On a side with 7 windows per row, there are $7 \times 23 = 161$ windows. On a side with 4 windows per row, there are $4 \times 23 = 92$ windows.
$161 + 161 + 92 + 92 = 506$
There are 506 windows on the building.

Section 1.7 Practice

1. a. $9\overline{)72}$ $\overset{8}{}$ because $8 \cdot 9 = 72$.

b. $40 \div 5 = 8$ because $8 \cdot 5 = 40$.

c. $\dfrac{24}{6} = 4$ because $4 \cdot 6 = 24$.

2. a. $\dfrac{7}{7} = 1$ because $1 \cdot 7 = 7$.

b. $5 \div 1 = 5$ because $5 \cdot 1 = 5$.

c. $1\overline{)11}$ $\overset{11}{}$ because $11 \cdot 1 = 11$.

d. $4 \div 1 = 4$ because $4 \cdot 1 = 4$.

e. $\dfrac{10}{1} = 10$ because $10 \cdot 1 = 10$.

f. $21 \div 21 = 1$ because $1 \cdot 21 = 21$.

3. a. $\dfrac{0}{7} = 0$ because $0 \cdot 7 = 0$.

b. $8\overline{)0}$ $\overset{0}{}$ because $0 \cdot 8 = 0$.

c. $7 \div 0$ is undefined because if $7 \div 0$ is a number, then the number times 0 would be 7.

d. $0 \div 14 = 0$ because $0 \cdot 14 = 0$.

4. a.

$$
\begin{array}{r}
818 \\
6\overline{)4908} \\
\underline{-48} \\
10 \\
\underline{-6} \\
48 \\
\underline{-48} \\
0
\end{array}
$$

Check:
$$
\begin{array}{r}
818 \\
\times\ \ 6 \\
\hline
4908
\end{array}
$$

b.

$$
\begin{array}{r}
553 \\
4\overline{)2212} \\
\underline{-20} \\
21 \\
\underline{-20} \\
12 \\
\underline{-12} \\
0
\end{array}
$$

Check:
$$
\begin{array}{r}
553 \\
\times\ \ 4 \\
\hline
2212
\end{array}
$$

c.

$$
\begin{array}{r}
251 \\
3\overline{)753} \\
\underline{-6} \\
15 \\
\underline{-15} \\
03 \\
\underline{-3} \\
0
\end{array}
$$

Check:
$$
\begin{array}{r}
251 \\
\times\ \ 3 \\
\hline
753
\end{array}
$$

5. a.

$$
\begin{array}{r}
304 \\
7\overline{)2128} \\
\underline{-21} \\
02 \\
\underline{-0} \\
28 \\
\underline{-28} \\
0
\end{array}
$$

Check: $304 \times 7 = 2128$

b.

$$
\begin{array}{r}
5\,100 \\
9\overline{)45,900} \\
\underline{-45} \\
0\ 9 \\
\underline{-9} \\
000
\end{array}
$$

Check: $5100 \times 9 = 45,900$

6. a.

$$
\begin{array}{r}
234\ \text{R}\ 3 \\
4\overline{)939} \\
\underline{-8} \\
13 \\
\underline{-12} \\
19 \\
\underline{-16} \\
3
\end{array}
$$

Check: $234 \cdot 4 + 3 = 939$

b.

$$
\begin{array}{r}
657\ \text{R}\ 2 \\
5\overline{)3287} \\
\underline{-30} \\
28 \\
\underline{-25} \\
37 \\
\underline{-35} \\
2
\end{array}
$$

Check: $657 \cdot 5 + 2 = 3287$

7. a.

$$
\begin{array}{r}
9067\ \text{R}\ 2 \\
9\overline{)81,605} \\
\underline{-81} \\
0\ 6 \\
\underline{-0} \\
60 \\
\underline{-54} \\
65 \\
\underline{-63} \\
2
\end{array}
$$

Check: $9067 \cdot 9 + 2 = 81,605$

b.

$$
\begin{array}{r}
5827 \text{ R } 2 \\
4\overline{)23{,}310} \\
\underline{-20} \\
3\ 3 \\
\underline{-3\ 2} \\
11 \\
\underline{-8} \\
30 \\
\underline{-28} \\
2
\end{array}
$$

Check: $5827 \cdot 4 + 2 = 23{,}310$

8.
$$
\begin{array}{r}
524 \text{ R } 12 \\
17\overline{)8920} \\
\underline{-85} \\
42 \\
\underline{-34} \\
80 \\
\underline{-68} \\
12
\end{array}
$$

9.
$$
\begin{array}{r}
49 \text{ R } 60 \\
678\overline{)33{,}282} \\
\underline{-27\ 12} \\
6\ 162 \\
\underline{-6\ 102} \\
60
\end{array}
$$

10.
$$
\begin{array}{r}
57 \\
3\overline{)171} \\
\underline{-15} \\
21 \\
\underline{-21} \\
0
\end{array}
$$

Each student got 57 CDs.

11.
$$
\begin{array}{r}
44 \\
12\overline{)532} \\
\underline{-48} \\
52 \\
\underline{-48} \\
4
\end{array}
$$

There will be 44 full boxes and 4 printers left over.

12. Find the sum and divide by 7.

$$
\begin{array}{r}
4 \\
7 \\
35 \\
16 \\
9 \\
3 \\
\underline{+\ 52} \\
126
\end{array}
\qquad
\begin{array}{r}
18 \\
7\overline{)126} \\
\underline{-7} \\
56 \\
\underline{-56} \\
0
\end{array}
$$

The average time is 18 minutes.

Calculator Explorations

1. $848 \div 16 = 53$

2. $564 \div 12 = 47$

3. $5890 \div 95 = 62$

4. $1053 \div 27 = 39$

5. $\dfrac{32{,}886}{126} = 261$

6. $\dfrac{143{,}088}{264} = 542$

7. $0 \div 315 = 0$

8. $315 \div 0$ is an error.

Vocabulary and Readiness Check

1. In $90 \div 2 = 45$, the answer 45 is called the <u>quotient</u>, 90 is called the <u>dividend</u>, and 2 is called the <u>divisor</u>.

2. The quotient of any number and 1 is the same <u>number</u>.

3. The quotient of any number (except 0) and the same number is <u>1</u>.

4. The quotient of 0 and any number (except 0) is <u>0</u>.

5. The quotient of any number and 0 is <u>undefined</u>.

6. The <u>average</u> of a list of numbers is the sum of the numbers divided by the <u>number</u> of numbers.

Exercise Set 1.7

1. $54 \div 9 = 6$

3. $36 \div 3 = 12$

5. $0 \div 8 = 0$

7. $31 \div 1 = 31$

9. $\dfrac{18}{18} = 1$

11. $\dfrac{24}{3} = 8$

13. $26 \div 0$ is undefined

15. $26 \div 26 = 1$

17. $0 \div 14 = 0$

19. $18 \div 2 = 9$

21.
$$
\begin{array}{r}
29 \\
3\overline{)\,87} \\
-6 \\
\hline
27 \\
-27 \\
\hline
0
\end{array}
$$
Check: $3 \cdot 29 = 87$

23.
$$
\begin{array}{r}
74 \\
3\overline{)\,222} \\
-21 \\
\hline
12 \\
-12 \\
\hline
0
\end{array}
$$
Check: $74 \cdot 3 = 222$

25.
$$
\begin{array}{r}
338 \\
3\overline{)\,1014} \\
-9 \\
\hline
11 \\
-9 \\
\hline
24 \\
-24 \\
\hline
0
\end{array}
$$
Check: $3 \cdot 338 = 1014$

27. $\dfrac{30}{0}$ is undefined.

29.
$$
\begin{array}{r}
9 \\
7\overline{)\,63} \\
-63 \\
\hline
0
\end{array}
$$
Check: $7 \cdot 9 = 63$

31.
$$
\begin{array}{r}
25 \\
6\overline{)\,150} \\
-12 \\
\hline
30 \\
-30 \\
\hline
0
\end{array}
$$
Check: $25 \cdot 6 = 150$

33.
$$
\begin{array}{r}
68 \text{ R } 3 \\
7\overline{)\,479} \\
-42 \\
\hline
59 \\
-56 \\
\hline
3
\end{array}
$$
Check: $7 \cdot 68 + 3 = 479$

35.
$$
\begin{array}{r}
236 \text{ R } 5 \\
6\overline{)\,1421} \\
-12 \\
\hline
22 \\
-18 \\
\hline
41 \\
-36 \\
\hline
5
\end{array}
$$
Check: $236 \cdot 6 + 5 = 1421$

37.
$$
\begin{array}{r}
38 \text{ R } 1 \\
8\overline{)\,305} \\
-24 \\
\hline
65 \\
-64 \\
\hline
1
\end{array}
$$
Check: $8 \cdot 38 + 1 = 305$

39.

$$
\begin{array}{r}
326 \text{ R } 4 \\
7\overline{)2286} \\
\underline{-21} \\
18 \\
\underline{-14} \\
46 \\
\underline{-42} \\
4
\end{array}
$$

Check: $326 \cdot 7 + 4 = 2286$

41.

$$
\begin{array}{r}
13 \\
55\overline{)715} \\
\underline{-55} \\
165 \\
\underline{-165} \\
0
\end{array}
$$

Check: $55 \cdot 13 = 715$

43.

$$
\begin{array}{r}
49 \\
23\overline{)1127} \\
\underline{-92} \\
207 \\
\underline{-207} \\
0
\end{array}
$$

Check: $49 \cdot 23 = 1127$

45.

$$
\begin{array}{r}
97 \text{ R } 8 \\
97\overline{)9417} \\
\underline{-873} \\
687 \\
\underline{-679} \\
8
\end{array}
$$

Check: $97 \cdot 97 + 8 = 9417$

47.

$$
\begin{array}{r}
209 \text{ R } 11 \\
15\overline{)3146} \\
\underline{-30} \\
14 \\
\underline{-0} \\
146 \\
\underline{-135} \\
11
\end{array}
$$

Check: $209 \cdot 15 + 11 = 3146$

49.

$$
\begin{array}{r}
506 \\
13\overline{)6578} \\
\underline{-65} \\
07 \\
\underline{-0} \\
78 \\
\underline{-78} \\
0
\end{array}
$$

Check: $13 \cdot 506 = 6578$

51.

$$
\begin{array}{r}
202 \text{ R } 7 \\
46\overline{)9299} \\
\underline{-92} \\
09 \\
\underline{-0} \\
99 \\
\underline{-92} \\
7
\end{array}
$$

Check: $202 \cdot 46 + 7 = 9299$

53.

$$
\begin{array}{r}
54 \\
236\overline{)12744} \\
\underline{-1180} \\
944 \\
\underline{-944} \\
0
\end{array}
$$

Check: $236 \cdot 54 = 12,744$

55.

$$
\begin{array}{r}
99 \text{ R } 100 \\
103\overline{)10,297} \\
\underline{-9\ 27} \\
1\ 027 \\
\underline{-927} \\
100
\end{array}
$$

Check: $99 \cdot 103 + 100 = 10,297$

57.

$$
\begin{array}{r}
202 \text{ R } 15 \\
102\overline{)20619} \\
\underline{-204} \\
21 \\
\underline{-0} \\
219 \\
\underline{-204} \\
15
\end{array}
$$

Check: $102 \cdot 202 + 15 = 20,619$

59.
$$
\begin{array}{r}
579 \text{ R } 72 \\
423\overline{)244{,}989} \\
-211\ 5 \\
\hline
33\ 48 \\
-29\ 61 \\
\hline
3\ 879 \\
-3\ 807 \\
\hline
72
\end{array}
$$

Check: $579 \cdot 423 + 72 = 244{,}989$

61.
$$
\begin{array}{r}
17 \\
7\overline{)119} \\
-7 \\
\hline
49 \\
-49 \\
\hline
0
\end{array}
$$

63.
$$
\begin{array}{r}
511 \text{ R } 3 \\
7\overline{)3580} \\
-35 \\
\hline
08 \\
-7 \\
\hline
10 \\
-7 \\
\hline
3
\end{array}
$$

65.
$$
\begin{array}{r}
2132 \text{ R } 32 \\
40\overline{)85312} \\
-80 \\
\hline
53 \\
-40 \\
\hline
131 \\
-120 \\
\hline
112 \\
-80 \\
\hline
32
\end{array}
$$

67.
$$
\begin{array}{r}
6\ 080 \\
142\overline{)863{,}360} \\
-852 \\
\hline
11\ 3 \\
-0 \\
\hline
11\ 36 \\
-11\ 36 \\
\hline
00 \\
-0 \\
\hline
0
\end{array}
$$

69.
$$
\begin{array}{r}
23 \text{ R } 2 \\
5\overline{)117} \\
-10 \\
\hline
17 \\
-15 \\
\hline
2
\end{array}
$$

The quotient is 23 R 2.

71.
$$
\begin{array}{r}
5 \text{ R } 25 \\
35\overline{)200} \\
-175 \\
\hline
25
\end{array}
$$

200 divided by 35 is 5 R 25.

73.
$$
\begin{array}{r}
20 \text{ R } 2 \\
3\overline{)62} \\
-6 \\
\hline
02 \\
-0 \\
\hline
2
\end{array}
$$

The quotient is 20 R 2.

75.
$$
\begin{array}{r}
33 \\
65\overline{)2145} \\
-195 \\
\hline
195 \\
-195 \\
\hline
0
\end{array}
$$

There are 33 students in the group.

77.
$$
\begin{array}{r}
165 \\
318\overline{)52470} \\
-318 \\
\hline
2067 \\
-1908 \\
\hline
1590 \\
-1590 \\
\hline
0
\end{array}
$$

The person weighs 165 pounds on Earth.

79.
$$
\begin{array}{r}
310 \\
18\overline{)5580} \\
-54 \\
\hline
18 \\
-18 \\
\hline
0
\end{array}
$$

The distance is 310 yards.

81.

$$\begin{array}{r} 88 \text{ R } 1 \\ 3\overline{)265} \\ -24 \\ \hline 25 \\ -24 \\ \hline 1 \end{array}$$

There are 88 bridges every 3 miles over the 265 miles, plus the first bridge, for a total of 89 bridges.

83.

$$\begin{array}{r} 10 \\ 492\overline{)5280} \\ -492 \\ \hline 360 \end{array}$$

There should be 10 poles, plus the first pole for a total of 11 light poles.

85.

$$\begin{array}{r} 5 \\ 5280\overline{)26400} \\ -26400 \\ \hline 0 \end{array}$$

Broad Peak is 5 miles tall.

87.

$$\begin{array}{r} 1760 \\ 3\overline{)5280} \\ -3 \\ \hline 22 \\ -21 \\ \hline 18 \\ -18 \\ \hline 0 \end{array}$$

There are 1760 yards in 1 mile.

89.

$$\begin{array}{r} 2 \\ 10 \\ 24 \\ 35 \\ 22 \\ 17 \\ + 12 \\ \hline 120 \end{array} \qquad \begin{array}{r} 20 \\ 6\overline{)120} \\ -12 \\ \hline 00 \end{array}$$

$$\text{Average} = \frac{120}{6} = 20$$

91.

$$\begin{array}{r} 1 \\ 205 \\ 972 \\ 210 \\ + 161 \\ \hline 1548 \end{array} \qquad \begin{array}{r} 387 \\ 4\overline{)1548} \\ -12 \\ \hline 34 \\ -32 \\ \hline 28 \\ -28 \\ \hline 0 \end{array}$$

$$\text{Average} = \frac{1548}{4} = 387$$

93.

$$\begin{array}{r} 2 \\ 86 \\ 79 \\ 81 \\ 69 \\ + 80 \\ \hline 395 \end{array} \qquad \begin{array}{r} 79 \\ 5\overline{)395} \\ -35 \\ \hline 45 \\ -45 \\ \hline 0 \end{array}$$

$$\text{Average} = \frac{395}{5} = 79$$

95.

$$\begin{array}{r} 2 \\ 69 \\ 77 \\ + 76 \\ \hline 222 \end{array} \qquad \begin{array}{r} 74 \\ 3\overline{)222} \\ -21 \\ \hline 12 \\ -12 \\ \hline 0 \end{array}$$

The average temperature is 74°.

97.

$$\begin{array}{r} 1\,1\,1 \\ 82 \\ 463 \\ 29 \\ + 8704 \\ \hline 9278 \end{array}$$

99.

$$\begin{array}{r} 546 \\ \times \quad 28 \\ \hline 4\,368 \\ 10\,920 \\ \hline 15,288 \end{array}$$

101.

$$\begin{array}{r} 722 \\ - 43 \\ \hline 679 \end{array}$$

103. $\dfrac{45}{0}$ is undefined.

105.
$$
\begin{array}{r}
9 \text{ R } 12 \\
24\overline{)228} \\
-216 \\
\hline
12
\end{array}
$$

107. The quotient of 40 and 8 is $40 \div 8$, which is choice c.

109. 200 divided by 20 is $200 \div 20$, which is choice b.

111.
$$
\begin{array}{r}
320 \\
110 \\
82 \\
58 \\
+\ 30 \\
\hline
600
\end{array}
\qquad
\begin{array}{r}
120 \\
5\overline{)600} \\
-5 \\
\hline
10 \\
-10 \\
\hline
00 \\
-0 \\
\hline
0
\end{array}
$$

The average number of Nobel Prize winners is 120.

113. The average will increase; answers may vary.

115. No; answers may vary
Possible answer: The average cannot be less than each of the four numbers.

117.
$$
\begin{array}{r}
12 \\
5\overline{)60} \\
-5 \\
\hline
10 \\
-10 \\
\hline
0
\end{array}
$$

The length is 12 feet.
Notice that Area = length × width = $12 \times 5 = 60$.

119. answers may vary

121.
$$
\begin{array}{r}
26 \\
-5 \\
\hline
21 \\
-5 \\
\hline
16 \\
-5 \\
\hline
11 \\
-5 \\
\hline
6 \\
-5 \\
\hline
1
\end{array}
$$

Therefore, $26 \div 5 = 5$ R 1

Integrated Review

1.
$$
\begin{array}{r}
\overset{1\ 1}{23} \\
46 \\
+\ 79 \\
\hline
148
\end{array}
$$

2.
$$
\begin{array}{r}
7006 \\
-\ 451 \\
\hline
6555
\end{array}
$$

3.
$$
\begin{array}{r}
36 \\
\times 45 \\
\hline
180 \\
1440 \\
\hline
1620
\end{array}
$$

4.
$$
\begin{array}{r}
562 \\
8\overline{)4496} \\
-40 \\
\hline
49 \\
-48 \\
\hline
16 \\
-16 \\
\hline
0
\end{array}
$$

5. $1 \cdot 79 = 79$

6. $\dfrac{36}{0}$ is undefined.

7. $9 \div 1 = 9$

8. $9 \div 9 = 1$

9. $0 \cdot 13 = 0$

10. $7 \cdot 0 \cdot 8 = 0 \cdot 8 = 0$

11. $0 \div 2 = 0$

12. $12 \div 4 = 3$

13.
$$
\begin{array}{r}
4219 \\
-1786 \\
\hline
2433
\end{array}
$$

14.
$$
\begin{array}{r}
\overset{1\ 1}{1861} \\
+\ 7965 \\
\hline
9826
\end{array}
$$

15.
$$
\begin{array}{r}
213\ \text{R }3 \\
5{\overline{\smash{\big)}\,1068}} \\
\underline{-10} \\
06 \\
\underline{-5} \\
18 \\
\underline{-15} \\
3
\end{array}
$$

16.
$$
\begin{array}{r}
1259 \\
\times\ \ 63 \\
\hline
3\,777 \\
75\,540 \\
\hline
79{,}317
\end{array}
$$

17. $3 \cdot 9 = 27$

18. $45 \div 5 = 9$

19.
$$
\begin{array}{r}
207 \\
-\ 69 \\
\hline
138
\end{array}
$$

20.
$$
\begin{array}{r}
207 \\
+\ 69 \\
\hline
276
\end{array}
$$

21.
$$
\begin{array}{r}
1099\ \text{R }2 \\
7{\overline{\smash{\big)}\,7695}} \\
\underline{-7} \\
06 \\
\underline{-0} \\
69 \\
\underline{-63} \\
65 \\
\underline{-63} \\
2
\end{array}
$$

22.
$$
\begin{array}{r}
111\ \text{R }1 \\
9{\overline{\smash{\big)}\,1000}} \\
\underline{-9} \\
10 \\
\underline{-9} \\
10 \\
\underline{-9} \\
1
\end{array}
$$

23.
$$
\begin{array}{r}
663\ \text{R }6 \\
32{\overline{\smash{\big)}\,21222}} \\
\underline{-192} \\
202 \\
\underline{-192} \\
102 \\
\underline{-96} \\
6
\end{array}
$$

24.
$$
\begin{array}{r}
1076\ \text{R }60 \\
65{\overline{\smash{\big)}\,70000}} \\
\underline{-65} \\
50 \\
\underline{-0} \\
500 \\
\underline{-455} \\
450 \\
\underline{-390} \\
60
\end{array}
$$

25.
$$
\begin{array}{r}
4000 \\
-\ 2976 \\
\hline
1024
\end{array}
$$

26.
$$
\begin{array}{r}
10{,}000 \\
-\ \ \ \ 101 \\
\hline
9{,}899
\end{array}
$$

27.
$$
\begin{array}{r}
303 \\
\times\ \ 101 \\
\hline
303 \\
0 \\
30\,300 \\
\hline
30{,}603
\end{array}
$$

28. $(475)(100) = 47{,}500$

29.
$$
\begin{array}{r}
\overset{1}{} \\
57 \\
+\ 8 \\
\hline
65
\end{array}
$$
The total of 57 and 8 is 65.

30.
$$
\begin{array}{r}
57 \\
\times\ 8 \\
\hline
456
\end{array}
$$
The product of 57 and 8 is 456.

31.
$$
\begin{array}{r}
6\ \text{R }8 \\
9{\overline{\smash{\big)}\,62}} \\
\underline{-54} \\
8
\end{array}
$$
The quotient of 62 and 9 is 6 R 8.

32.
$$\begin{array}{r} 62 \\ -\,9 \\ \hline 53 \end{array}$$
The difference of 62 and 9 is 53.

33.
$$\begin{array}{r} 200 \\ -\,17 \\ \hline 183 \end{array}$$
17 subtracted from 200 is 183.

34.
$$\begin{array}{r} 432 \\ -201 \\ \hline 231 \end{array}$$
The difference of 432 and 201 is 231.

		Tens	Hundreds	Thousands
35.	9735	9740	9700	10,000
36.	1429	1430	1400	1000
37.	20,801	20,800	20,800	21,000
38.	432,198	432,200	432,200	432,000

39.
$$\begin{array}{r} \overset{2}{6} \\ 6 \\ 6 \\ +\,6 \\ \hline 24 \end{array}$$
The perimeter is 24 feet.
$$\begin{aligned} \text{Area} &= \text{side} \times \text{side} \\ &= 6\text{ feet} \times 6\text{ feet} \\ &= 36\text{ square feet} \end{aligned}$$
The area is 36 square feet.

40.
$$\begin{array}{r} \overset{2}{14} \\ 7 \\ 14 \\ +\,7 \\ \hline 42 \end{array}$$
The perimeter is 42 inches.
Area = length · width = 14 · 7 = 98
The area is 98 square inches.

41.
$$\begin{array}{r} \overset{1}{13} \\ 9 \\ +\,6 \\ \hline 28 \end{array}$$
The perimeter is 28 miles.

42.
$$
\begin{array}{r}
\overset{2}{3} \\
7 \\
6 \\
3 \\
3 \\
+\,4 \\
\hline
26
\end{array}
$$

The perimeter is 26 meters.

43.
$$
\begin{array}{r}
\overset{3}{19} \\
15 \\
25 \\
37 \\
+\,24 \\
\hline
120
\end{array}
\qquad
\begin{array}{r}
24 \\
5\overline{)120} \\
-10 \\
\hline
20 \\
-20 \\
\hline
0
\end{array}
$$

The average is 24.

44.
$$
\begin{array}{r}
\overset{1\,2}{108} \\
131 \\
98 \\
+\,159 \\
\hline
496
\end{array}
\qquad
\begin{array}{r}
124 \\
4\overline{)496} \\
-4 \\
\hline
09 \\
-8 \\
\hline
16 \\
-16 \\
\hline
0
\end{array}
$$

The average is 124.

45.
$$
\begin{array}{r}
28,547 \\
-\,26,372 \\
\hline
2\;175
\end{array}
$$

Lake Pontchartrain Bridge is 2175 feet longer than the Mackinac Bridge.

46.
$$
\begin{array}{r}
365 \\
\times\quad 2 \\
\hline
730
\end{array}
$$

On average, 730 quarts of carbonated soft drinks would be consumed in a year.

Section 1.8 Practice

1. | Transamerica Pyramid | is | 74 feet | taller than | the Bank of America Building |
$$\downarrow\qquad\qquad\downarrow\quad\;\;\downarrow\qquad\downarrow\qquad\qquad\qquad\downarrow$$
Transamerica Pyramid = 74 + 779

$$
\begin{array}{r}
\overset{1\;1}{74} \\
+\,779 \\
\hline
853
\end{array}
$$

The Transamerica Pyramid is 853 feet tall.

2. | Amount of money | is | $65,000 | divided by | four friends |
$$\downarrow\qquad\quad\downarrow\quad\;\;\downarrow\qquad\;\downarrow\qquad\;\;\downarrow$$
Amount of money = 65,000 ÷ 4

```
      16250
4) 65000
    −4
    25
   −24
    10
    −8
    20
   −20
    00
    −0
     0
```

Each person receives $16,250.

3.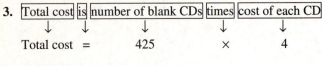

Total cost =   425   ×   4

```
    425
  ×  4
  1700
```

The total cost for the blank CDs is $1700.

4.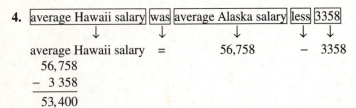

average Hawaii salary   =   56,758   −   3358

```
  56,758
 − 3 358
  53,400
```

The average public school teacher's salary in Hawaii was $53,400.

5. Area of the lot = length × width = 120 feet × 90 feet = 10,800 square feet
   Area of the house = length × width = 65 feet × 45 feet = 2925 square feet

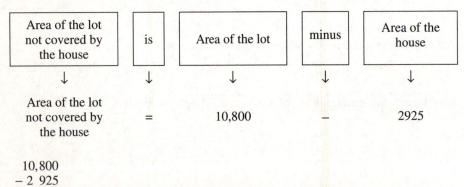

```
  10,800
 − 2 925
   7 875
```

The area of the lot not covered by the house is 7875 square feet.

Exercise Set 1.8

1. 41 increased by 8 is what number

41   +   8 = what number

$$\begin{array}{r} 41 \\ +\ 8 \\ \hline 49 \end{array}$$

3. The quotient of 1185 and 5 is some number

$$\downarrow \qquad \downarrow \quad\ \downarrow\ \downarrow \qquad\quad \downarrow$$

$$1185\ \div\ 5\ =\ \text{some number}$$

$$\begin{array}{r} 237 \\ 5\overline{)1185} \\ \underline{-10} \\ 18 \\ \underline{-15} \\ 35 \\ \underline{-35} \\ 0 \end{array}$$

5. The total of 35 and 7 is some number

$$\downarrow \qquad \downarrow \quad\ \downarrow\ \downarrow \qquad \downarrow$$

$$35\ +\ 7\ =\ \text{some number}$$

$$\begin{array}{r} 35 \\ +\ 7 \\ \hline 42 \end{array}$$

7. 60 times 10 is some number

$$\downarrow \quad\ \downarrow \quad\ \downarrow\ \downarrow \qquad \downarrow$$

$$60\quad\cdot\quad 10\ =\ \text{some number}$$

$$\begin{array}{r} 60 \\ \times 10 \\ \hline 0 \\ 600 \\ \hline 600 \end{array}$$

9. a. Perimeter is two times length plus two times width

$$\downarrow \qquad \downarrow\ \ \downarrow\quad \downarrow \qquad \downarrow \qquad \downarrow\ \ \downarrow \quad \downarrow \qquad \downarrow$$

$$\begin{aligned} \text{Perimeter} &=\ 2\ \quad\cdot\quad\ 120\ \ +\ \ 2\ \ \cdot\quad\ 80 \\ &=2\cdot 120 + 2\cdot 80 \\ &=240+160 \\ &=400 \end{aligned}$$

The perimeter is 400 feet.

b. Area is length times width

$$\downarrow \quad \downarrow \quad\ \downarrow \qquad \downarrow \qquad \downarrow$$

$$\text{Area}\ =\ 120\ \quad\times\quad\ 80$$

$$\begin{array}{r} 120 \\ \times\ 80 \\ \hline 9600 \end{array}$$

The area is 9600 square feet.

11.

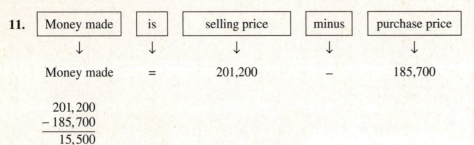

Money made	is	selling price	minus	purchase price
↓	↓	↓	↓	↓
Money made	=	201,200	–	185,700

$$\begin{array}{r} 201,200 \\ -\ 185,700 \\ \hline 15,500 \end{array}$$

The family made $15,500 by selling the house.

13.

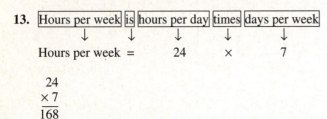

Hours per week	is	hours per day	times	days per week
↓	↓	↓	↓	↓
Hours per week	=	24	×	7

$$\begin{array}{r} 24 \\ \times\ 7 \\ \hline 168 \end{array}$$

There are 168 hours in a week.

15.

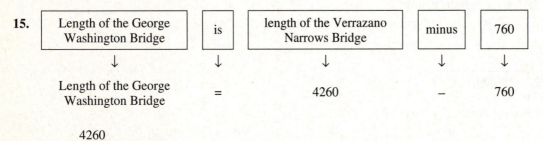

Length of the George Washington Bridge	is	length of the Verrazano Narrows Bridge	minus	760
↓	↓	↓	↓	↓
Length of the George Washington Bridge	=	4260	–	760

$$\begin{array}{r} 4260 \\ -\ 760 \\ \hline 3500 \end{array}$$

The length of the George Washington Bridge is 3500 feet.

17.

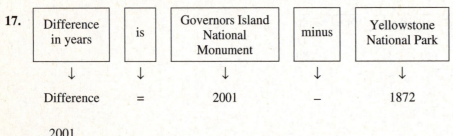

Difference in years	is	Governors Island National Monument	minus	Yellowstone National Park
↓	↓	↓	↓	↓
Difference	=	2001	–	1872

$$\begin{array}{r} 2001 \\ -\ 1872 \\ \hline 129 \end{array}$$

Yellowstone is 129 years older than Governors Island.

19.

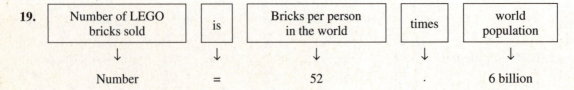

Number of LEGO bricks sold	is	Bricks per person in the world	times	world population
↓	↓	↓	↓	↓
Number	=	52	·	6 billion

$$\begin{array}{r} 52 \\ \times\ 6 \\ \hline 312 \end{array}$$

312 billion LEGO bricks have been sold since their introduction.

21.

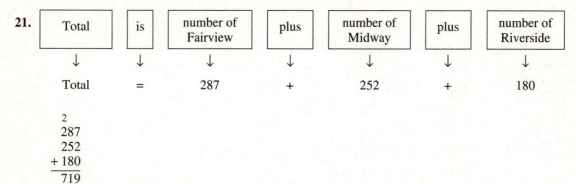

Total	is	number of Fairview	plus	number of Midway	plus	number of Riverside
↓	↓	↓	↓	↓	↓	↓
Total	=	287	+	252	+	180

$$\begin{array}{r} 2 \\ 287 \\ 252 \\ +\ 180 \\ \hline 719 \end{array}$$

There are a total of 719 towns named Fairview, Midway, or Riverside.

23.

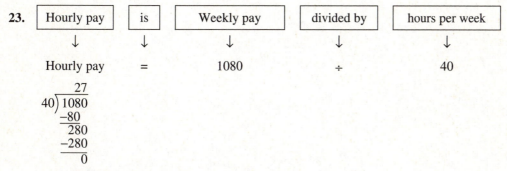

Hourly pay	is	Weekly pay	divided by	hours per week
↓	↓	↓	↓	↓
Hourly pay	=	1080	÷	40

$$\begin{array}{r} 27 \\ 40\overline{)1080} \\ \underline{-80} \\ 280 \\ \underline{-280} \\ 0 \end{array}$$

The hourly pay of the supervisor is $27.

25.

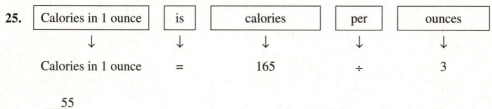

Calories in 1 ounce	is	calories	per	ounces
↓	↓	↓	↓	↓
Calories in 1 ounce	=	165	÷	3

$$\begin{array}{r} 55 \\ 3\overline{)165} \\ \underline{-15} \\ 15 \\ \underline{-15} \\ 0 \end{array}$$

There are 55 calories in 1 ounce of canned tuna.

27.

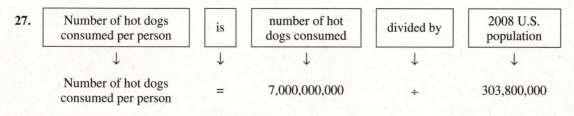

Number of hot dogs consumed per person	is	number of hot dogs consumed	divided by	2008 U.S. population
↓	↓	↓	↓	↓
Number of hot dogs consumed per person	=	7,000,000,000	÷	303,800,000

$$
\begin{array}{r}
23 \\
303800000\overline{)7000000000} \\
\underline{607600000} \\
924000000 \\
\underline{911400000} \\
12600000
\end{array}
$$

Approximately 23 hot dogs were consumed per person between Memorial and Labor Day.

29.

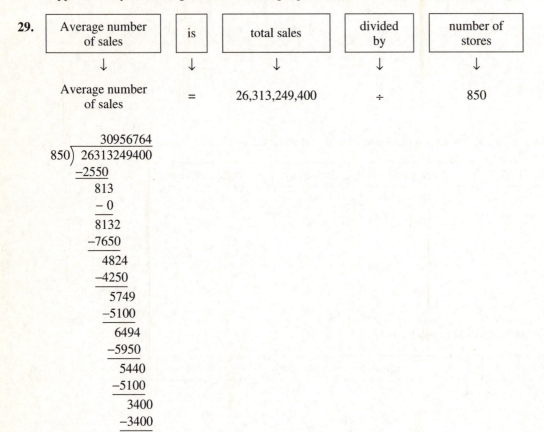

Average number of sales	is	total sales	divided by	number of stores
↓	↓	↓	↓	↓
Average number of sales	=	26,313,249,400	÷	850

$$
\begin{array}{r}
30956764 \\
850\overline{)26313249400} \\
\underline{-2550} \\
813 \\
\underline{-0} \\
8132 \\
\underline{-7650} \\
4824 \\
\underline{-4250} \\
5749 \\
\underline{-5100} \\
6494 \\
\underline{-5950} \\
5440 \\
\underline{-5100} \\
3400 \\
\underline{-3400} \\
0
\end{array}
$$

The average amount of sales made by each store $30,956,764.

31.

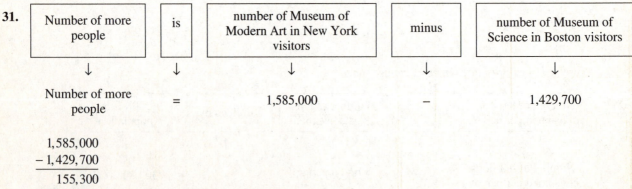

Number of more people	is	number of Museum of Modern Art in New York visitors	minus	number of Museum of Science in Boston visitors
↓	↓	↓	↓	↓
Number of more people	=	1,585,000	−	1,429,700

$$
\begin{array}{r}
1,585,000 \\
-1,429,700 \\
\hline
155,300
\end{array}
$$

155,300 more people visit the Museum of Modern Art than the science museum.

33.

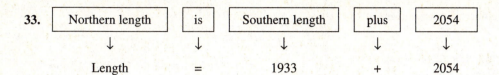

Northern length	is	Southern length	plus	2054
↓	↓	↓	↓	↓
Length	=	1933	+	2054

$$\begin{array}{r} 1933 \\ +\ 2054 \\ \hline 3987 \end{array}$$

The northern boundary of the conterminous United States is 3987 miles long.

35.

Paychecks per year	is	weeks per year	divided by	weeks per paycheck
↓	↓	↓	↓	↓
Paycheck per year	=	52	÷	4

$$\begin{array}{r} 13 \\ 4\overline{)52} \\ \underline{-4} \\ 12 \\ \underline{-12} \\ 0 \end{array}$$

He receives 13 paychecks per year.

37.

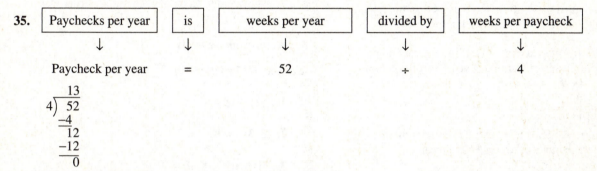

Total cost	is	number of sweaters	times	cost of sweater	plus	number of shirts	times	cost of shirt
↓	↓	↓	↓	↓	↓	↓	↓	↓
Total	=	3	·	38	+	5	·	25

$$= 3 \cdot 38 + 5 \cdot 25 = 114 + 125 = 239$$
The total cost is $239.

39.

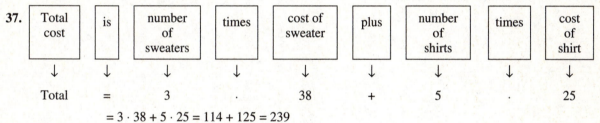

Money in account now	is	money in account	minus	money spent	plus	money deposited
↓	↓	↓	↓	↓	↓	↓
Money	=	950	–	205	+	300

$$\begin{array}{r} 950 \\ -\ 205 \\ \hline 745 \end{array}$$

$$\begin{array}{r} {\scriptstyle 1} \\ 745 \\ +\ 300 \\ \hline 1045 \end{array}$$

There is $1045 in the account now.

41. Option a: a hamburger, onion rings, a candy bar, and a soda cost
$4 + $3 + $2 + $1 = $10
Option b: a hot dog, an apple, french fries, and a soda cost
$3 + $1 + $2 + $1 = $7
Thus option b will be cheaper by $10 − $7 or $3.

43. Asia had the greatest number of Internet users in 2008.

45. The greatest number of users in a region was 579 million. The fewest number of users in a region was 20 million.

$$\begin{array}{r} 579 \\ -\ 20 \\ \hline 559 \end{array}$$

The region with the greatest number of Internet users had 559 million more users than the region with the fewest number of Internet users.

47. North America had 248 million Internet users. Latin America/Caribbean had 139 million users.

$$\begin{array}{r} 248 \\ -139 \\ \hline 109 \end{array}$$

North America had 109 million more Internet users than Latin America/Caribbean.

49. The three greatest numbers of Internet users are 579 million, 385 million, and 248 million.

$$\begin{array}{r} 579 \\ 385 \\ +\ 248 \\ \hline 1212 \end{array} \qquad \begin{array}{r} 404 \\ 3{\overline{\smash{\big)}\,1212}} \\ \underline{-12} \\ 01 \\ \underline{-0} \\ 12 \\ \underline{-12} \\ 0 \end{array}$$

The average number of Internet users is 404 million.

51. Total cost $= 22 \cdot \$615 + 3 \cdot \$408$
$\phantom{\text{Total cost }} = \$13,530 + \$1224$
$\phantom{\text{Total cost }} = \$14,754$
The total cost is $14,754.

53. There are 7 days in a week.
$7 \cdot 2400 = 16,800$
No more than 16,800 milligrams of sodium should be consumed in one week.

55. a. Area $=$ length $\cdot$ width
$\phantom{\text{Area }} = 50 \cdot 75$
$\phantom{\text{Area }} = 3750$ square feet

    b. Area $=$ length $\cdot$ width
$\phantom{\text{Area }} = 15 \cdot 25$
$\phantom{\text{Area }} = 375$ square feet

    c.
$$\begin{array}{r} 3750 \\ -\ 375 \\ \hline 3375 \end{array}$$
The area not part of the swimming pool is 3375 square feet.

57. 149,100,000 rounded to the nearest ten-million is 150,000,000. 25,649,300,000 rounded to the nearest ten-million is 25,650,000,000.
$25,650,000,000 \div 150,000,000 = 171$
The average value of each money order was about $171.

59. answers may vary

Section 1.9 Practice

1. $8 \cdot 8 \cdot 8 \cdot 8 = 8^4$

2. $3 \cdot 3 \cdot 3 = 3^3$

3. $10 \cdot 10 \cdot 10 \cdot 10 \cdot 10 = 10^5$

4. $5 \cdot 5 \cdot 4 \cdot 4 \cdot 4 \cdot 4 \cdot 4 \cdot 4 = 5^2 \cdot 4^6$

5. $4^2 = 4 \cdot 4 = 16$

6. $7^3 = 7 \cdot 7 \cdot 7 = 343$

7. $11^1 = 11$

8. $2 \cdot 3^2 = 2 \cdot 3 \cdot 3 = 18$

9. $\sqrt{100} = 10$ because $10 \cdot 10 = 100$.

10. $\sqrt{4} = 2$ because $2 \cdot 2 = 4$.

11. $\sqrt{1} = 1$ because $1 \cdot 1 = 1$.

12. $9 \cdot 3 - 8 \div 4 = 27 - 8 \div 4 = 27 - 2 = 25$

13. $48 \div 3 \cdot 2^2 = 48 \div 3 \cdot 4 = 16 \cdot 4 = 64$

14. $(10 - 7)^4 + 2 \cdot 3^2 = 3^4 + 2 \cdot 3^2$
$ = 81 + 2 \cdot 9$
$ = 81 + 18$
$ = 99$

15. $36 \div [20 - (4 \cdot 2)] + 4^3 - 6 = 36 \div [20 - 8] + 4^3 - 6$
$ = 36 \div 12 + 4^3 - 6$
$ = 36 \div 12 + 64 - 6$
$ = 3 + 64 - 6$
$ = 61$

16. $\dfrac{25+8\cdot 2-3^3}{2(3-2)} = \dfrac{25+8\cdot 2-27}{2(1)}$

$\qquad\qquad\qquad = \dfrac{25+16-27}{2}$

$\qquad\qquad\qquad = \dfrac{14}{2}$

$\qquad\qquad\qquad = 7$

17. $81\div\sqrt{81}\cdot 5+7 = 81\div 9\cdot 5+7$

$\qquad\qquad\qquad\quad = 9\cdot 5+7$

$\qquad\qquad\qquad\quad = 45+7$

$\qquad\qquad\qquad\quad = 52$

18. Area $= (\text{side})^2$

$\qquad\; = (12 \text{ centimeters})^2$

$\qquad\; = 144 \text{ square centimeters}$

The area of the square is 144 square centimeters.

Calculator Explorations

1. $4^6 = 4096$

2. $5^6 = 15,625$

3. $5^5 = 3125$

4. $7^6 = 117,649$

5. $2^{11} = 2048$

6. $6^8 = 1,679,616$

7. $7^4 + 5^3 = 2526$

8. $12^4 - 8^4 = 16,640$

9. $63 \cdot 75 - 43 \cdot 10 = 4295$

10. $8 \cdot 22 + 7 \cdot 16 = 288$

11. $4(15 \div 3 + 2) - 10 \cdot 2 = 8$

12. $155 - 2(17 + 3) + 185 = 300$

Vocabulary and Readiness Check

1. In $2^5 = 32,$ the 2 is called the base and the 5 is called the exponent.

2. To simplify $8 + 2 \cdot 6$, which operation should be performed first? multiplication

3. To simplify $(8 + 2) \cdot 6$, which operation should be performed first? addition

4. To simplify $9(3 - 2) \div 3 + 6$, which operation should be performed first? subtraction

5. To simplify $8 \div 2 \cdot 6$, which operation should be performed first? division

6. The square root of a whole number is one of two identical factors of the number.

Exercise Set 1.9

1. $4\cdot 4\cdot 4 = 4^3$

3. $7\cdot 7\cdot 7\cdot 7\cdot 7\cdot 7 = 7^6$

5. $12\cdot 12\cdot 12 = 12^3$

7. $6\cdot 6\cdot 5\cdot 5\cdot 5 = 6^2 \cdot 5^3$

9. $9\cdot 8\cdot 8 = 9\cdot 8^2$

11. $3\cdot 2\cdot 2\cdot 2\cdot 2 = 3\cdot 2^4$

13. $3\cdot 2\cdot 2\cdot 2\cdot 2\cdot 5\cdot 5\cdot 5\cdot 5\cdot 5 = 3\cdot 2^4 \cdot 5^5$

15. $8^2 = 8\cdot 8 = 64$

17. $5^3 = 5\cdot 5\cdot 5 = 125$

19. $2^5 = 2\cdot 2\cdot 2\cdot 2\cdot 2 = 32$

21. $1^{10} = 1\cdot 1\cdot 1\cdot 1\cdot 1\cdot 1\cdot 1\cdot 1\cdot 1\cdot 1 = 1$

23. $7^1 = 7$

25. $2^7 = 2\cdot 2\cdot 2\cdot 2\cdot 2\cdot 2\cdot 2 = 128$

27. $2^8 = 2\cdot 2\cdot 2\cdot 2\cdot 2\cdot 2\cdot 2\cdot 2 = 256$

29. $4^4 = 4\cdot 4\cdot 4\cdot 4 = 256$

31. $9^3 = 9\cdot 9\cdot 9 = 729$

33. $12^2 = 12\cdot 12 = 144$

35. $10^2 = 10 \cdot 10 = 100$

37. $20^1 = 20$

39. $3^6 = 3 \cdot 3 \cdot 3 \cdot 3 \cdot 3 \cdot 3 = 729$

41. $3 \cdot 2^6 = 3 \cdot 2 \cdot 2 \cdot 2 \cdot 2 \cdot 2 \cdot 2 = 192$

43. $2 \cdot 3^4 = 2 \cdot 3 \cdot 3 \cdot 3 \cdot 3 = 162$

45. $\sqrt{9} = 3$ since $3 \cdot 3 = 9$.

47. $\sqrt{64} = 8$ since $8 \cdot 8 = 64$.

49. $\sqrt{144} = 12$ since $12 \cdot 12 = 144$.

51. $\sqrt{16} = 4$ since $4 \cdot 4 = 16$.

53. $15 + 3 \cdot 2 = 15 + 6 = 21$

55. $14 \div 7 \cdot 2 + 3 = 2 \cdot 2 + 3 = 4 + 3 = 7$

57. $32 \div 4 - 3 = 8 - 3 = 5$

59. $13 + \dfrac{24}{8} = 13 + 3 = 16$

61. $6 \cdot 5 + 8 \cdot 2 = 30 + 16 = 46$

63. $\dfrac{5 + 12 \div 4}{1^7} = \dfrac{5 + 3}{1} = \dfrac{8}{1} = 8$

65. $(7 + 5^2) \div 4 \cdot 2^3 = (7 + 25) \div 4 \cdot 2^3$
$\qquad\qquad = 32 \div 4 \cdot 2^3$
$\qquad\qquad = 32 \div 4 \cdot 8$
$\qquad\qquad = 8 \cdot 8$
$\qquad\qquad = 64$

67. $5^2 \cdot (10 - 8) + 2^3 + 5^2 = 5^2 \cdot 2 + 2^3 + 5^2$
$\qquad\qquad\qquad = 25 \cdot 2 + 8 + 25$
$\qquad\qquad\qquad = 50 + 8 + 25$
$\qquad\qquad\qquad = 83$

69. $\dfrac{18 + 6}{2^4 - 2^2} = \dfrac{24}{16 - 4} = \dfrac{24}{12} = 2$

71. $(3 + 5) \cdot (9 - 3) = 8 \cdot 6 = 48$

73. $\dfrac{7(9 - 6) + 3}{3^2 - 3} = \dfrac{7(3) + 3}{9 - 3} = \dfrac{21 + 3}{6} = \dfrac{24}{6} = 4$

75. $8 \div 0 + 37 =$ undefined

77. $2^4 \cdot 4 - (25 \div 5) = 2^4 \cdot 4 - 5$
$\qquad\qquad\qquad = 16 \cdot 4 - 5$
$\qquad\qquad\qquad = 64 - 5$
$\qquad\qquad\qquad = 59$

79. $3^4 - [35 - (12 - 6)] = 3^4 - [35 - 6]$
$\qquad\qquad\qquad = 3^4 - 29$
$\qquad\qquad\qquad = 81 - 29$
$\qquad\qquad\qquad = 52$

81. $(7 \cdot 5) + [9 \div (3 \div 3)] = (7 \cdot 5) + [9 \div (1)]$
$\qquad\qquad\qquad = 35 + 9$
$\qquad\qquad\qquad = 44$

83. $8 \cdot [2^2 + (6 - 1) \cdot 2] - 50 \cdot 2 = 8 \cdot (2^2 + 5 \cdot 2) - 50 \cdot 2$
$\qquad\qquad\qquad\qquad = 8 \cdot (4 + 5 \cdot 2) - 50 \cdot 2$
$\qquad\qquad\qquad\qquad = 8 \cdot (4 + 10) - 50 \cdot 2$
$\qquad\qquad\qquad\qquad = 8 \cdot 14 - 50 \cdot 2$
$\qquad\qquad\qquad\qquad = 112 - 50 \cdot 2$
$\qquad\qquad\qquad\qquad = 112 - 100$
$\qquad\qquad\qquad\qquad = 12$

85. $\dfrac{9^2 + 2^2 - 1^2}{8 \div 2 \cdot 3 \cdot 1 \div 3} = \dfrac{81 + 4 - 1}{4 \cdot 3 \cdot 1 \div 3}$
$\qquad\qquad = \dfrac{85 - 1}{12 \cdot 1 \div 3}$
$\qquad\qquad = \dfrac{84}{12 \div 3}$
$\qquad\qquad = \dfrac{84}{4}$
$\qquad\qquad = 21$

87. $6 \cdot \sqrt{9} + 3 \cdot \sqrt{4} = 6 \cdot 3 + 3 \cdot 2 = 18 + 6 = 24$

89. $4 \cdot \sqrt{49} - 0 \div \sqrt{100} = 4 \cdot 7 - 0 \div 10$
$\qquad\qquad\qquad = 28 - 0 \div 10$
$\qquad\qquad\qquad = 28 - 0$
$\qquad\qquad\qquad = 28$

91. $\dfrac{\sqrt{4} + 4^2}{5(20 - 16) - 3^2 - 5} = \dfrac{2 + 16}{5(4) - 9 - 5}$
$\qquad\qquad = \dfrac{18}{20 - 9 - 5}$
$\qquad\qquad = \dfrac{18}{11 - 5}$
$\qquad\qquad = \dfrac{18}{6}$
$\qquad\qquad = 3$

93. $\sqrt{81} \div \sqrt{9} + 4^2 \cdot 2 - 10 = 9 \div 3 + 16 \cdot 2 - 10$
$$= 3 + 16 \cdot 2 - 10$$
$$= 3 + 32 - 10$$
$$= 35 - 10$$
$$= 25$$

95. $\left[\sqrt{225} \div (11 - 6) + 2^2\right] + \left(\sqrt{25} - \sqrt{1}\right)^2$
$$= [15 \div (5) + 4] + (5 - 1)^2$$
$$= [3 + 4] + (4)^2$$
$$= 7 + 16$$
$$= 23$$

97. $7^2 - \left\{18 - \left[40 \div (4 \cdot 2) + \sqrt{4}\right] + 5^2\right\}$
$$= 7^2 - \left\{18 - \left[40 \div 8 + \sqrt{4}\right] + 5^2\right\}$$
$$= 7^2 - \left\{18 - \left[40 \div 8 + 2\right] + 5^2\right\}$$
$$= 7^2 - \{18 - [5 + 2] + 5^2\}$$
$$= 7^2 - \{18 - 7 + 5^2\}$$
$$= 7^2 - \{18 - 7 + 25\}$$
$$= 7^2 - \{11 + 25\}$$
$$= 7^2 - 36$$
$$= 49 - 36$$
$$= 13$$

99. Area of a square $= (\text{side})^2$
$$= (7 \text{ meters})^2$$
$$= 49 \text{ square meters}$$
Perimeter $= 4(\text{side}) = 4(7 \text{ meters}) = 28 \text{ meters}$

101. Area of a square $= (\text{side})^2$
$$= (23 \text{ miles})^2$$
$$= 529 \text{ square miles}$$
Perimeter $= 4(\text{side}) = 4(23 \text{ miles}) = 92 \text{ miles}$

103. The statement is true.

105. $2^5 = 2 \cdot 2 \cdot 2 \cdot 2 \cdot 2$
The statement is false.

107. $(2 + 3) \cdot 6 - 2 = 5 \cdot 6 - 2 = 30 - 2 = 28$

109. $24 \div (3 \cdot 2) + 2 \cdot 5 = 24 \div 6 + 2 \cdot 5 = 4 + 10 = 14$

111. The unknown vertical length is
$30 - 12 = 18$ feet. The unknown horizontal
length is $60 - 40 = 20$ feet.
Perimeter $= 60 + 30 + 40 + 18 + 20 + 12 = 180$
The total perimeter of seven homes is
$7(180) = 1260$ feet.

113. $(7 + 2^4)^5 - (3^5 - 2^4)^2 = (7 + 16)^5 - (243 - 16)^2$
$$= 23^5 - 227^2$$
$$= 6,436,343 - 51,529$$
$$= 6,384,814$$

115. answers may vary; possible answer:
$(20 - 10) \cdot 5 \div 25 + 3 = 10 \cdot 5 \div 25 + 3$
$$= 50 \div 25 + 3$$
$$= 2 + 3$$
$$= 5$$

Chapter 1 Vocabulary Check

1. The <u>whole number</u> are 0, 1, 2, 3, ...

2. The <u>perimeter</u> of a polygon is its distance around or the sum of the lengths of its sides.

3. The position of each digit in a number determines its <u>place value</u>.

4. An <u>exponent</u> is a shorthand notation for repeated multiplication of the same factor.

5. To find the <u>area</u> of a rectangle, multiply length times width.

6. A <u>square root</u> of a number is one of two identical factors of the number.

7. The <u>digits</u> used to write numbers are 0, 1, 2, 3, 4, 5, 6, 7, 8, and 9.

8. The <u>average</u> of a list of numbers is their sum divided by the number of numbers.

9. The 5 above is called the <u>divisor</u>.

10. The 35 above is called the <u>dividend</u>.

11. The 7 above is called the <u>quotient</u>.

12. The 3 above is called a <u>factor</u>.

13. The 6 above is called the <u>product</u>.

14. The 20 above is called the <u>minuend</u>.

15. The 9 above is called the <u>subtrahend</u>.

16. The 11 above is called the <u>difference</u>.

17. The 4 above is called an <u>addend</u>.

18. The 21 above is called the <u>sum</u>.

Chapter 1 Review

1. The place value of 4 in 7640 is tens.

2. The place value of 4 in 46,200,120 is ten-millions.

3. 7640 is written as seven thousand, six hundred forty.

4. 46,200,120 is written as forty-six million, two hundred thousand, one hundred twenty.

5. $3158 = 3000 + 100 + 50 + 8$

6. $403,225,000 = 400,000,000 + 3,000,000$
   $+ 200,000 + 20,000 + 5000$

7. Eighty-one thousand, nine hundred in standard form is 81,900.

8. Six billion, three hundred four million in standard form is 6,304,000,000.

9. Locate Europe in the first column and read across to the number in the 2008 column. There were 384,633,765 Internet users in Europe in 2008.

10. Locate Oceania/Australia in the first column and read across to the number in the 2004 column. There were 11,805,500 Internet users in Oceania/Australia in 2004.

11. Locate the smallest number in the 2000 column. Middle East had the fewest Internet users in 2000.

12. Locate the biggest number in the 2008 column. Asia had the greatest number of Internet users in 2008.

13.
$$
\begin{array}{r}
1 \\
17 \\
+\ 46 \\
\hline
63
\end{array}
$$

14.
$$
\begin{array}{r}
1 \\
28 \\
+\ 39 \\
\hline
67
\end{array}
$$

15.
$$
\begin{array}{r}
1 \\
25 \\
8 \\
+\ 15 \\
\hline
48
\end{array}
$$

16.
$$
\begin{array}{r}
1 \\
27 \\
9 \\
+\ 41 \\
\hline
77
\end{array}
$$

17.
$$
\begin{array}{r}
932 \\
+\ 24 \\
\hline
956
\end{array}
$$

18.
$$
\begin{array}{r}
1 \\
819 \\
+\ 21 \\
\hline
840
\end{array}
$$

19.
$$
\begin{array}{r}
1\ 1 \\
567 \\
+\ 7383 \\
\hline
7950
\end{array}
$$

20.
$$
\begin{array}{r}
1\ 1\ 1 \\
463 \\
+\ 6787 \\
\hline
7250
\end{array}
$$

21.
$$
\begin{array}{r}
1\ 2\ 1 \\
91 \\
3623 \\
+\ 497 \\
\hline
4211
\end{array}
$$

22.
$$
\begin{array}{r}
1\ 1 \\
82 \\
1647 \\
+\ 238 \\
\hline
1967
\end{array}
$$

23.
$$
\begin{array}{r}
1\ 1 \\
86 \\
331 \\
+\ 909 \\
\hline
1326
\end{array}
$$
   The sum is 1326.

24.
$$
\begin{array}{r}
2 \\
49 \\
529 \\
+\ 308 \\
\hline
886
\end{array}
$$
   The sum is 886.

25.
$$
\begin{array}{r}
1\ 1 \\
26,481 \\
+\ 865 \\
\hline
27,346
\end{array}
$$
   26,481 increased by 865 is 27,346.

26.
```
  1 11
 38,556
+   744
 39,300
```
38,556 increased by 744 is 39,300.

27.
```
   1 1
 7318
+ 714
 8032
```
The total distance from Chicago to New Delhi if traveling by air through New York City is 8032 miles.

28.
```
  1 11 1
  62,589
  65,340
+ 69,770
 197,699
```
Susan Summerline's total earnings for the years 2002, 2003, and 2004 was $197,699.

29.
```
   2
   72
   72
   82
 + 50
  276
```
The perimeter is 276 feet.

30.
```
   20
   35
 + 11
   66
```
The perimeter is 66 kilometers.

31.
```
   93
 − 79
   14
```
Check:
```
    1
   14
 + 79
   93
```

32.
```
   61
 − 27
   34
```
Check:
```
    1
   34
 + 27
   61
```

33.
```
   462
 − 397
    65
```
Check:
```
   11
    65
 + 397
   462
```

34.
```
   583
 − 279
   304
```
Check:
```
    1
   304
 + 279
   583
```

35.
```
  4000
 −  86
  3914
```
Check:
```
  3914
 +  86
  4000
```

36.
```
  8000
 −  92
  7908
```
Check:
```
  7908
 +  92
  8000
```

37.
```
  25,862
 − 7 965
  17,897
```

38.
```
  39,007
 − 4 349
  34,658
```

39.
```
  1,328,984
 − 1,144,646
    184,338
```
The increase in population was 184,338 people.

40.
```
  1,517,550
 − 1,491,812
     25,738
```
The decrease in population was 25,738 people.

41.
$$\begin{array}{r} 712 \\ -\ 315 \\ \hline 397 \end{array}$$
Bob Roma has 397 pages left to proofread.

42.
$$\begin{array}{r} 28,425 \\ -\ 1\,599 \\ \hline 26,826 \end{array} \qquad \begin{array}{r} 26,826 \\ -\ 1\,200 \\ \hline 25,626 \end{array}$$
Shelly Winter paid $25,626 for the new car.

43. The balance was the least during the month of May.

44. The balance was the greatest during the month of August.

45.
$$\begin{array}{r} 280 \\ -\ 170 \\ \hline 110 \end{array}$$
The balance decrease by $110 from February to April.

46.
$$\begin{array}{r} 490 \\ -\ 250 \\ \hline 240 \end{array}$$
The balance increased by $240 from June to August.

47. To round 93 to the nearest ten, observe that the digit in the ones place is 3. Since this digit is less than 5, we do not add 1 to the digit in the tens place. The number 93 rounded to the nearest ten is 90.

48. To round 45 to the nearest ten, observe that the digit in the ones place is 5. Since this digit is at least 5, we add 1 to the digit in the tens place. The number 45 rounded to the nearest ten is 50.

49. To round 467 to the nearest ten, observe that the digit in the ones place is 5. Since this digit is at least 5, we add 1 to the digit in the tens place. The number 467 rounded to the nearest ten is 470.

50. To round 493 to the nearest hundred, observe that the digit in the tens place is 9. Since this digit is at least 5, we add 1 to the digit in the hundreds place. The number 493 rounded to the nearest hundred is 500.

51. To round 4832 to the nearest hundred, observe that the digit in the tens place is 3. Since this digit is less than 5, we do not add 1 to the digit in the hundreds place. the number 4832 rounded to the nearest hundred is 4800.

52. To round 57,534 to the nearest thousand, observe that the digit in the hundreds place is 5. Since this digit is at least 5, we add 1 to the digit in the thousands place. The number 57,534 rounded to the nearest thousand is 58,000.

53. To round 49,683,712 to the nearest million, observe that the digit in the hundred-thousands place is 6. Since this digit is at least 5, we add 1 to the digit in the millions place. $9 + 1 = 10$, so replace the digit 9 by 0 and carry 1 to the place value to the left. The number 49,683,712 rounded to the nearest million is 50,000,000.

54. To round 768,542 to the nearest hundred-thousand, observe that the digit in the ten-thousands place is 6. Since this digit is at least 5, we add 1 to the digit in the hundred-thousands place. The number 768,542 rounded to the nearest hundred-thousand is 800,000.

55. To round 65,025,901 to the nearest million, observe that the digit in the hundred-thousandths place is 0. Since this digit is less than 5, we do not add 1 to the digit in the millions place. The number 65,025,901 rounded to the nearest million is 65,000,000.

56. To round 93,295 to the nearest thousand, observe that the digit in the hundreds place is 2. Since this digit is less than 5, we do not add 1 to the digit in the thousands place. The number 93,295 rounded to the nearest thousand is 93,000.

57.
$$\begin{array}{llr} 4892 & \text{rounds to} & 4900 \\ 647 & \text{rounds to} & 600 \\ +\ 1876 & \text{rounds to} & +\ 1900 \\ & \text{rounds to} & \hline 7400 \end{array}$$
The estimated sum is 7400.

58.
$$\begin{array}{llr} 5925 & \text{rounds to} & 5900 \\ -\ 1787 & \text{rounds to} & -\ 1800 \\ & & \hline 4100 \end{array}$$
The estimated difference is 4100.

59.
$$\begin{array}{llr} 628 & \text{rounds to} & 600 \\ 290 & \text{rounds to} & 300 \\ 172 & \text{rounds to} & 200 \\ 58 & \text{rounds to} & 100 \\ 508 & \text{rounds to} & 500 \\ 445 & \text{rounds to} & 400 \\ +\ 383 & \text{rounds to} & +\ 400 \\ & & \hline 2500 \end{array}$$
The students traveled approximately 2500 miles on their week-long trip.

60. $\begin{array}{r} 2,208,180 \\ -1,336,865 \end{array}$ rounds to $\begin{array}{r} 2,200,000 \\ -1,300,000 \\ \hline 900,000 \end{array}$

Houston was approximately 900,000 people larger than San Diego in 2007.

61. $\begin{array}{r} 273 \\ \times 7 \\ \hline 1911 \end{array}$

62. $\begin{array}{r} 349 \\ \times 4 \\ \hline 1396 \end{array}$

63. $\begin{array}{r} 47 \\ \times 30 \\ \hline 0 \\ 1410 \\ \hline 1410 \end{array}$

64. $\begin{array}{r} 69 \\ \times 42 \\ \hline 138 \\ 2760 \\ \hline 2898 \end{array}$

65. $20(8)(5) = 160 \cdot 5 = 800$

66. $25(9)(4) = 225 \cdot (4) = 900$

67. $\begin{array}{r} 48 \\ \times 77 \\ \hline 336 \\ 3360 \\ \hline 3696 \end{array}$

68. $\begin{array}{r} 77 \\ \times 22 \\ \hline 154 \\ 1540 \\ \hline 1694 \end{array}$

69. $49 \cdot 49 \cdot 9 = 0$

70. $62 \cdot 88 \cdot 0 = 0$

71. $\begin{array}{r} 586 \\ \times 29 \\ \hline 5\,274 \\ 11\,720 \\ \hline 16,994 \end{array}$

72. $\begin{array}{r} 242 \\ \times 37 \\ \hline 1694 \\ 7260 \\ \hline 8954 \end{array}$

73. $\begin{array}{r} 642 \\ \times 177 \\ \hline 4\,494 \\ 44\,940 \\ 64\,200 \\ \hline 113,634 \end{array}$

74. $\begin{array}{r} 347 \\ \times 129 \\ \hline 3\,123 \\ 6\,940 \\ 34\,700 \\ \hline 44,763 \end{array}$

75. $\begin{array}{r} 1026 \\ \times 401 \\ \hline 1\,026 \\ 0 \\ 410\,400 \\ \hline 411,426 \end{array}$

76. $\begin{array}{r} 2107 \\ \times 302 \\ \hline 4\,214 \\ 0 \\ 632\,100 \\ \hline 636,314 \end{array}$

77. $375 \cdot 1000 = 375,000$
Attach 3 zeros.

78. $108 \cdot 1000 = 108,000$
Attach 3 zeros.

79. $\begin{array}{r} 30 \\ \times 4 \\ \hline 120 \end{array}$

$30 \cdot 400 = 30 \cdot 4 \cdot 100 = 120 \cdot 100 = 12,000$
Attach 2 zeros.

80. $\begin{array}{r} 50 \\ \times 7 \\ \hline 350 \end{array}$

$50 \cdot 700 = 50 \cdot 7 \cdot 100 = 350 \cdot 100 = 35,000$
Attach 2 zeros.

81.
$$\begin{array}{r} 17 \\ \times\ 3 \\ \hline 51 \end{array}$$
$1700 \cdot 3000 = 5{,}100{,}000$
Attach 5 zeros.

82.
$$\begin{array}{r} 19 \\ \times\ 4 \\ \hline 76 \end{array}$$
$1900 \cdot 4000 = 7{,}600{,}000$
Attach 5 zeros.

83.
$$\begin{array}{r} 230 \\ \times\ \ 5 \\ \hline 1150 \end{array}$$
The product of 5 and 230 is 1150.

84.
$$\begin{array}{r} 820 \\ \times\ \ 6 \\ \hline 4920 \end{array}$$
The product of 6 and 820 is 4920.

85.
$$\begin{array}{r} 12 \\ \times\ 9 \\ \hline 108 \end{array}$$

86.
$$\begin{array}{r} 14 \\ \times\ 8 \\ \hline 112 \end{array}$$

87.
$$\begin{array}{r} 8 \\ \times\ 3 \\ \hline 24 \end{array}$$
Three ounces of Swiss cheese has 24 grams of fat.

88.
$$\begin{array}{r} 6112 \\ \times\ \ \ 20 \\ \hline 122{,}240 \end{array}$$
The total cost is $122,240.

89. Area = length · width = $12 \times 5 = 60$
The area is 60 square miles.

90. Area = length · width = $25 \times 20 = 500$
The area is 500 square centimeters.

91.
$$\begin{array}{r} 3 \\ 6\overline{)\ 18} \\ -18 \\ \hline 0 \end{array}$$
Check:
$$\begin{array}{r} 3 \\ \times\ 6 \\ \hline 18 \end{array}$$

92.
$$\begin{array}{r} 4 \\ 9\overline{)\ 36} \\ -36 \\ \hline 0 \end{array}$$
Check:
$$\begin{array}{r} 4 \\ \times\ 9 \\ \hline 36 \end{array}$$

93.
$$\begin{array}{r} 6 \\ 7\overline{)\ 42} \\ -42 \\ \hline 0 \end{array}$$
Check:
$$\begin{array}{r} 6 \\ \times\ 7 \\ \hline 42 \end{array}$$

94.
$$\begin{array}{r} 7 \\ 5\overline{)\ 35} \\ -35 \\ \hline 0 \end{array}$$
Check:
$$\begin{array}{r} 7 \\ \times\ 5 \\ \hline 35 \end{array}$$

95.
$$\begin{array}{r} 5 \text{ R } 2 \\ 5\overline{)\ 27} \\ -25 \\ \hline 2 \end{array}$$
Check: $5 \cdot 5 + 2 = 25 + 2 = 27$

96.
$$\begin{array}{r} 4 \text{ R } 2 \\ 4\overline{)\ 18} \\ -16 \\ \hline 2 \end{array}$$
Check: $4 \cdot 4 + 2 = 16 + 2 = 18$

97. $16 \div 0$ is undefined.

98. $0 \div 8 = 0$

99. $9 \div 9 = 1$

100. $10 \div 1 = 10$

101. $0 \div 668 = 0$

102. $918 \div 0$ is undefined.

103.
$$\begin{array}{r} 33 \text{ R } 2 \\ 5\overline{)\ 167} \\ -15 \\ \hline 17 \\ -15 \\ \hline 2 \end{array}$$
Check: $33 \cdot 5 + 2 = 165 + 2 = 167$

104.

$$\begin{array}{r} 19 \text{ R } 7 \\ 8 \overline{)159} \\ -8 \\ \hline 79 \\ -72 \\ \hline 7 \end{array}$$

Check: $19 \cdot 8 + 7 = 152 + 7 = 159$

105.

$$\begin{array}{r} 24 \text{ R } 2 \\ 26 \overline{)626} \\ -52 \\ \hline 106 \\ -104 \\ \hline 2 \end{array}$$

Check: $24 \cdot 26 + 2 = 624 + 2 = 626$

106.

$$\begin{array}{r} 35 \text{ R } 15 \\ 19 \overline{)680} \\ -57 \\ \hline 110 \\ -95 \\ \hline 15 \end{array}$$

Check: $35 \cdot 19 + 15 = 665 + 15 = 680$

107.

$$\begin{array}{r} 506 \text{ R } 10 \\ 47 \overline{)23792} \\ -235 \\ \hline 29 \\ -0 \\ \hline 292 \\ -282 \\ \hline 10 \end{array}$$

Check: $506 \cdot 47 + 10 = 23{,}782 + 10 = 23{,}792$

108.

$$\begin{array}{r} 907 \text{ R } 40 \\ 53 \overline{)48111} \\ -477 \\ \hline 41 \\ -0 \\ \hline 411 \\ -371 \\ \hline 40 \end{array}$$

Check: $907 \cdot 53 + 40 = 48{,}071 + 40 = 48{,}111$

109.

$$\begin{array}{r} 2793 \text{ R } 140 \\ 207 \overline{)578291} \\ -414 \\ \hline 1642 \\ -1449 \\ \hline 1939 \\ -1863 \\ \hline 761 \\ -621 \\ \hline 140 \end{array}$$

Check:
$2793 \cdot 207 + 140 = 578{,}151 + 140 = 578{,}291$

110.

$$\begin{array}{r} 2012 \text{ R } 60 \\ 306 \overline{)615732} \\ -612 \\ \hline 37 \\ -0 \\ \hline 373 \\ -306 \\ \hline 672 \\ -612 \\ \hline 60 \end{array}$$

Check:
$2012 \cdot 306 + 60 = 615{,}672 + 60 = 615{,}732$

111.

$$\begin{array}{r} 18 \text{ R } 2 \\ 5 \overline{)92} \\ -5 \\ \hline 42 \\ -40 \\ \hline 2 \end{array}$$

The quotient of 92 and 5 is 18 R 2.

112.

$$\begin{array}{r} 21 \text{ R } 2 \\ 4 \overline{)86} \\ -8 \\ \hline 06 \\ -4 \\ \hline 2 \end{array}$$

The quotient of 86 and 4 is 21 R 2.

113.

$$\begin{array}{r} 458 \\ 12 \overline{)5496} \\ -48 \\ \hline 69 \\ -60 \\ \hline 96 \\ -96 \\ \hline 0 \end{array}$$

458 feet are in 5496 inches.

114.

$$1760{\overline{\smash{\big)}\,22880}} \atop {\begin{array}{r}13\\-1760\\\hline 5280\\-5280\\\hline 0\end{array}}$$

13 miles are in 22,880 yards.

115.

$$\begin{array}{r}^{2\,2}\\76\\49\\32\\+47\\\hline 204\end{array} \qquad \begin{array}{r}51\\4{\overline{\smash{\big)}\,204}}\\-20\\\hline 04\\-4\\\hline 0\end{array}$$

The average is 51.

116.

$$\begin{array}{r}^{2\,1}\\23\\85\\62\\+66\\\hline 236\end{array} \qquad \begin{array}{r}59\\4{\overline{\smash{\big)}\,236}}\\-20\\\hline 36\\-36\\\hline 0\end{array}$$

The average is 59.

117.

$$\begin{array}{r}27\\24{\overline{\smash{\big)}\,648}}\\-48\\\hline 168\\-168\\\hline 0\end{array}$$

27 boxes can be filled with 648 cans of corn.

118.

$$\begin{array}{r}32\\\times 6\\\hline 192\end{array}$$

The cost of 32 tickets is $192.

119.

$$\begin{array}{r}11,000,000,000\\-\ 4,000,000,000\\\hline 7,000,000,000\end{array}$$

People take 7 billion more aspirin tablets for heart disease prevention than for headaches.

120.

$$\begin{array}{r}27\\+45\\\hline 75\end{array}$$

The cost to banks for a person to deposit a check with a teller is 75¢.

121.

$$\begin{array}{r}32\\\times 15\\\hline 160\\320\\\hline 480\end{array} \qquad \begin{array}{r}38\\\times 11\\\hline 38\\380\\\hline 418\end{array} \qquad \begin{array}{r}480\\+418\\\hline 898\end{array}$$

The cost of 15 large and 11 extra-large shirts is $898.

122.

$$\begin{array}{r}110\\\times 65\\\hline 550\\6600\\\hline 7150\end{array} \qquad 200\times 80 = 16{,}000 \qquad \begin{array}{r}16{,}000\\+\ 7\,150\\\hline 23{,}150\end{array}$$

The total area of the land purchased is 23,150 square feet.

123. $7^2 = 7 \cdot 7 = 49$

124. $5^3 = 5 \cdot 5 \cdot 5 = 125$

125. $5 \cdot 3^2 = 5 \cdot 9 = 45$

126. $4 \cdot 10^2 = 4 \cdot 100 = 400$

127. $18 \div 3 + 7 = 6 + 7 = 13$

128. $12 - 8 \div 4 = 12 - 2 = 10$

129. $\dfrac{5(6^2 - 3)}{3^2 + 2} = \dfrac{5(36 - 3)}{9 + 2} = \dfrac{5(33)}{11} = \dfrac{165}{11} = 15$

130. $\dfrac{7(16 - 8)}{2^3} = \dfrac{7(8)}{8} = \dfrac{56}{8} = 7$

131. $48 \div 8 \cdot 2 = 6 \cdot 2 = 12$

132. $27 \div 9 \cdot 3 = 3 \cdot 3 = 9$

133. $\begin{aligned}2 + 3[1^5 + (20 - 17) \cdot 3] + 5 \cdot 2 &= 2 + 3[1 + (3) \cdot 3] + 10\\&= 2 + 3[1 + 9] + 10\\&= 2 + 3[10] + 10\\&= 2 + 30 + 10\\&= 32 + 10\\&= 42\end{aligned}$

134. $21-[2^4-(7-5)-10]+8\cdot2$
$=21-[16-(2)-10]+16$
$=21-[16-2-10]+16$
$=21-[14-10]+16$
$=21-[4]+16$
$=17+16$
$=33$

135. $\sqrt{81}=9$ since $9\cdot9=81$.

136. $\sqrt{4}=2$ since $2\cdot2=4$.

137. $\sqrt{1}=1$ since $1\cdot1=1$.

138. $\sqrt{0}=0$ since $0\cdot0=0$.

139. $4\cdot\sqrt{25}-2\cdot7=4\cdot5-2\cdot7=20-14=6$

140. $8\cdot\sqrt{49}-3\cdot9=8\cdot7-27=56-27=29$

141. $\left(\sqrt{36}-\sqrt{16}\right)^3\cdot[10^2\div(3+17)]$
$=(6-4)^3\cdot[100\div(20)]$
$=(2)^3\cdot[5]$
$=8\cdot[5]$
$=40$

142. $\left(\sqrt{49}-\sqrt{25}\right)^3\cdot[9^2\div(2+7)]=(7-5)^3\cdot[81\div(9)]$
$=(2)^3\cdot[9]$
$=8\cdot[9]$
$=72$

143. $\dfrac{5\cdot7-3\cdot\sqrt{25}}{2\left(\sqrt{121}-3^2\right)}=\dfrac{35-3\cdot5}{2(11-9)}=\dfrac{35-15}{2\cdot(2)}=\dfrac{20}{4}=5$

144. $\dfrac{4\cdot8-1\cdot\sqrt{121}}{3\left(\sqrt{81}-2^3\right)}=\dfrac{32-1\cdot11}{3(9-8)}=\dfrac{32-11}{3(1)}=\dfrac{21}{3}=7$

145. Area $=(\text{side})^2=(7\text{ meters})^2=49$ square meters
The area is 49 square meters.

146. Area $=(\text{side})^2=(3\text{ inches})^2=9$ square inches
The area is 9 square inches.

147.
$$\begin{array}{r} 375 \\ -\ 68 \\ \hline 307 \end{array}$$

148.
$$\begin{array}{r} 729 \\ -\ 47 \\ \hline 682 \end{array}$$

149.
$$\begin{array}{r} 723 \\ \times\ \ 3 \\ \hline 2169 \end{array}$$

150.
$$\begin{array}{r} 629 \\ \times\ \ 4 \\ \hline 2516 \end{array}$$

151.
$$\begin{array}{r} \overset{2\ 2}{264} \\ 39 \\ +\ 598 \\ \hline 901 \end{array}$$

152.
$$\begin{array}{r} \overset{2\ 1}{593} \\ 52 \\ +\ 766 \\ \hline 1411 \end{array}$$

153.
$$\begin{array}{r} 458 \text{ R } 8 \\ 13\overline{)\,5962} \\ -52 \\ \hline 76 \\ -65 \\ \hline 112 \\ -104 \\ \hline 8 \end{array}$$

154.
$$\begin{array}{r} 237 \text{ R } 1 \\ 18\overline{)\,4267} \\ -36 \\ \hline 66 \\ -54 \\ \hline 127 \\ -126 \\ \hline 1 \end{array}$$

155.
$$\begin{array}{r} 1968 \\ \times\ \ 36 \\ \hline 11\,808 \\ 59\,040 \\ \hline 70,848 \end{array}$$

156.
$$\begin{array}{r} 5324 \\ \times\ \ 18 \\ \hline 42\,592 \\ 53\,240 \\ \hline 95,832 \end{array}$$

157.
$$\begin{array}{r} 2000 \\ -\ 356 \\ \hline 1644 \end{array}$$

158.
$$\begin{array}{r} 9000 \\ -\ 519 \\ \hline 8481 \end{array}$$

159. To round 736 to the nearest ten, observe that the digit in the ones place is 6. Since this digit is at least 5, we add 1 to the digit in the tens place. The number 736 rounded to the nearest ten is 740.

160. To round 258,371 to the nearest thousand, observe that the digit in the hundreds place is 3. Since this digit is less than 5, we do not add 1 to the digit in the thousands place. The number 258,371 rounded to the nearest thousand is 258,000.

161. To round 1999 to the nearest hundred, observe that the digit in the tens place is 9. Since this digit is at least 5, we add 1 to the digit in the hundreds place. 9 + 1 = 10, so replace the digit 9 by 0 and carry 1 to the place value to the left. The number 1999 rounded to the nearest hundred is 2000.

162. To round 44,499 to the nearest ten-thousand, observe that the digit in the thousands place is 4. Since this digit is less than 5, we do not add 1 to the digit in the ten-thousands place. The number 44,499 rounded to the nearest ten-thousand is 40,000.

163. 36,911 written in words is thirty-six thousand nine hundred eleven.

164. 154,863 written in words is one hundred fifty-four thousand, eight hundred sixty-three.

165. Seventy thousand, nine hundred forty-three in standard form is 70,943.

166. Forty-three thousand, four hundred one in standard form is 43,401.

167. $4^3 = 4 \cdot 4 \cdot 4 = 64$

168. $5^3 = 5 \cdot 5 \cdot 5 = 125$

169. $\sqrt{144} = 12$ since $12 \cdot 12 = 144$.

170. $\sqrt{100} = 10$ since $10 \cdot 10 = 100$.

171. $24 \div 4 \cdot 2 = 6 \cdot 2 = 12$

172. $\sqrt{256} - 3 \cdot 5 = 16 - 15 = 1$

173. $\dfrac{8(7-4)-10}{4^2 - 3^2} = \dfrac{8(3)-10}{16-9} = \dfrac{24-10}{7} = \dfrac{14}{7} = 2$

174.
$$\dfrac{\left(15+\sqrt{9}\right) \cdot (8-5)}{2^3 + 1} = \dfrac{(15+3)(3)}{8+1}$$
$$= \dfrac{(18)(3)}{9}$$
$$= \dfrac{54}{9}$$
$$= 6$$

175.
$$\begin{array}{r} 4 \\ 9\overline{)\ 36} \\ -36 \\ \hline 0 \end{array}$$
36 divided by 9 is 4.

176.
$$\begin{array}{r} 12 \\ \times\ 2 \\ \hline 24 \end{array}$$
The product of 2 and 12 is 24.

177.
$$\begin{array}{r} 1 \\ 16 \\ +\ 8 \\ \hline 24 \end{array}$$
16 increased by 8 is 24.

178.
$$\begin{array}{r} 21 \\ -\ 7 \\ \hline 14 \end{array}$$
7 subtracted from 21 is 14.

179.
$$\begin{array}{r} 4,623,000 \\ -\ 4,110,000 \\ \hline 513,000 \end{array}$$
The average salary for a Detroit Tiger in 2009 was $513,000 less than in 2008.

180.
$$\begin{array}{r} 7,748,000 \\ -\ 3,260,000 \\ \hline 4,488,000 \end{array}$$
The average Yankee salary in 2009 was $4,488,000 more than the average salary for all teams in Major League Baseball for 2009.

181.
```
        53 R 18
   32) 1714
       -160
        114
        -96
         18
```
There are 53 full boxes of drinking glasses with 18 drinking glasses left over.

182.
```
   27        8        54
  × 2      × 4      + 32
   54       32        86
```
The total bill is $86.

Chapter 1 Test

1. 82,426 is written as eighty-two thousand, four hundred twenty-six.

2. Four hundred two thousand, five hundred fifty in standard form is 402,550.

3.
```
     1 1
     59
   + 82
    141
```

4.
```
     600
   - 487
     113
```

5.
```
       496
   ×    30
         0
   14 880
   14,880
```

6.
```
          766 R 42
   69) 52896
       -483
        459
       -414
        456
       -414
         42
```

7. $2^3 \cdot 5^2 = 2 \cdot 2 \cdot 2 \cdot 5 \cdot 5 = 200$

8. $\sqrt{4} \cdot \sqrt{25} = 2 \cdot 5 = 10$

9. $0 \div 49 = 0$

10. $62 \div 0$ is undefined.

11. $(2^4 - 5) \cdot 3 = (16 - 5) \cdot 3 = (11) \cdot 3 = 33$

12. $16 + 9 \div 3 \cdot 4 - 7 = 16 + 3 \cdot 4 - 7$
$= 16 + 12 - 7$
$= 28 - 7$
$= 21$

13. $\dfrac{64 \div 8 \cdot 2}{\left(\sqrt{9} - \sqrt{4}\right)^2 + 1} = \dfrac{8 \cdot 2}{(3-2)^2 + 1}$
$= \dfrac{16}{1^2 + 1}$
$= \dfrac{16}{1 + 1}$
$= \dfrac{16}{2}$
$= 8$

14. $2[(6-4)^2 + (22-19)^2] + 10 = 2[(2)^2 + (3)^2] + 10$
$= 2[4 + 9] + 10$
$= 2[13] + 10$
$= 26 + 10$
$= 36$

15. $5698 \cdot 1000 = 5,698,000$
Attach 3 zeros.

16.
```
    14
   × 8
   112
```
$8000 \cdot 1400 = 11,200,000$
Attach 5 zeros.

17. To round 52, 369 to the nearest thousand, observe that the digit in the hundreds place is 3. Since this digit is less than 5, we do not add 1 to the digit in the thousands place. The number 52,369 rounded to the nearest thousand is 52,000.

18.
```
     6289  rounds to     6300
     5403  rounds to     5400
   + 1957  rounds to  + 2000
           rounds to   13,700
```
The estimated sum is 13,700.

19.
```
     4267  rounds to     4300
   - 2738  rounds to  - 2700
                         1600
```
The estimated difference is 1600.

20.
$$\begin{array}{r} 107 \\ -\ 15 \\ \hline 92 \end{array}$$

21.
$$\begin{array}{r} 1 \\ 15 \\ +\ 107 \\ \hline 122 \end{array}$$
The sum of 15 and 107 is 122.

22.
$$\begin{array}{r} 107 \\ \times\ 15 \\ \hline 535 \\ 1070 \\ \hline 1605 \end{array}$$
The product of 15 and 107 is 1605.

23.
$$\begin{array}{r} 7\ R\ 2 \\ 15\overline{)\ 107} \\ \underline{-105} \\ 2 \end{array}$$
The quotient is 7 R 2.

24.
$$\begin{array}{r} 17 \\ 29\overline{)\ 493} \\ \underline{-29} \\ 203 \\ \underline{-203} \\ 0 \end{array}$$
Each can of paint was $17.

25.
$$\begin{array}{r} 725 \\ -\ 599 \\ \hline 126 \end{array}$$
The higher-priced refrigerator costs $126 more than the lower-priced one.

26.
$$\begin{array}{r} 45 \\ \times\ 8 \\ \hline 360 \end{array}$$
Eight tablespoons of sugar contains 360 calories.

27.
$$\begin{array}{r} 430 \\ \times\ 16 \\ \hline 2580 \\ 4300 \\ \hline 6880 \end{array} \qquad \begin{array}{r} 205 \\ \times\ 5 \\ \hline 1025 \end{array} \qquad \begin{array}{r} 6880 \\ +\ 1025 \\ \hline 7905 \end{array}$$
The total cost for these items is $7905.

28.
$$\begin{array}{r} 5 \\ 5 \\ 5 \\ +\ 5 \\ \hline 20 \end{array}$$
The perimeter is 20 centimeters.

$$\begin{aligned} \text{Area} &= (\text{side})^2 \\ &= (5 \text{ centimeters})^2 \\ &= 25 \text{ square centimeters} \end{aligned}$$
The area is 25 square centimeters.

29.
$$\begin{aligned} \text{Perimeter} &= 2(\text{length}) + 2(\text{width}) \\ &= 2(20 \text{ yards}) + 2(10 \text{ yards}) \\ &= 40 \text{ yards} + 20 \text{ yards} \\ &= 60 \text{ yards} \end{aligned}$$
The perimeter is 60 yards.
$$\begin{aligned} \text{Area} &= (\text{length}) \cdot (\text{width}) \\ &= (20 \text{ yards}) \cdot (10 \text{ yards}) \\ &= 200 \text{ square yards} \end{aligned}$$
The area is 200 square yards.

Chapter 2

Section 2.1 Practice

1. $\dfrac{9}{2}$ ← numerator ← denominator

2. $\dfrac{10}{17}$ ← numerator ← denominator

3. $\dfrac{0}{2} = 0$

4. $\dfrac{8}{8} = 1$

5. $\dfrac{4}{0}$ is undefined.

6. $\dfrac{20}{1} = 20$

7. In this figure, 3 of the 8 equal parts are shaded. Thus, the fraction is $\dfrac{3}{8}$.

8. In this figure, 1 of the 6 equal parts is shaded. Thus, the fraction is $\dfrac{1}{6}$.

9. Of the 10 parts of the syringe, 7 parts are filled. Thus, the fraction is $\dfrac{7}{10}$.

10. Of the 16 parts of one inch, 9 parts are measured. Thus, the fraction is $\dfrac{9}{16}$.

11. answers may vary; for example,

12. answers may vary; for example,

13. number of planets farther → 5
    number of planets in our solar system → 8
    $\dfrac{5}{8}$ of the planets in our solar system are farther from the Sun than Earth is.

14. a. $\dfrac{5}{8}$ is a proper fraction.

    b. $\dfrac{7}{7}$ is an improper fraction.

    c. $\dfrac{14}{13}$ is an improper fraction.

    d. $\dfrac{13}{14}$ is a proper fraction.

    e. $5\dfrac{1}{4}$ is a mixed number.

    f. $\dfrac{100}{49}$ is an improper fraction.

15. Each part is $\dfrac{1}{3}$ of a whole. There are 8 parts shaded, or 2 wholes and 2 more parts.
    improper fraction: $\dfrac{8}{3}$
    mixed number: $2\dfrac{2}{3}$

16. Each part is $\dfrac{1}{4}$ of a whole. There are 5 parts shaded, or 1 whole and 1 more part.
    improper fraction: $\dfrac{5}{4}$
    mixed number: $1\dfrac{1}{4}$

17. a. $2\dfrac{5}{7} = \dfrac{7 \cdot 2 + 5}{7} = \dfrac{14 + 5}{7} = \dfrac{19}{7}$

    b. $5\dfrac{1}{3} = \dfrac{3 \cdot 5 + 4}{3} = \dfrac{15 + 1}{3} = \dfrac{16}{3}$

    c. $9\dfrac{3}{10} = \dfrac{10 \cdot 9 + 3}{10} = \dfrac{90 + 3}{10} = \dfrac{93}{10}$

    d. $1\dfrac{1}{5} = \dfrac{5 \cdot 1 + 1}{5} = \dfrac{5 + 1}{5} = \dfrac{6}{5}$

49

18. a. $5\overline{)9}$ with quotient 1
$$\frac{5}{4}$$

$$\frac{9}{5}=1\frac{4}{5}$$

b. $9\overline{)23}$ with quotient 2
$$\frac{18}{5}$$

$$\frac{23}{9}=2\frac{5}{9}$$

c. $4\overline{)48}$ with quotient 12
$$\frac{4}{8}$$
$$\frac{8}{0}$$

$$\frac{48}{4}=12$$

d. $13\overline{)62}$ with quotient 4
$$\frac{52}{10}$$

$$\frac{62}{13}=4\frac{10}{13}$$

e. $7\overline{)51}$ with quotient 7
$$\frac{49}{2}$$

$$\frac{51}{7}=7\frac{2}{7}$$

f. $20\overline{)21}$ with quotient 1
$$\frac{20}{1}$$

$$\frac{21}{20}=1\frac{1}{20}$$

Vocabulary and Readiness Check

1. The number $\frac{17}{31}$ is called a <u>fraction</u>. The number 31 is called its <u>denominator</u> and 17 is called its <u>numerator</u>.

2. If we simplify each fraction, $\frac{9}{9}\underline{=1}$, $\frac{0}{4}\underline{=0}$, and we say $\frac{4}{0}$ <u>is undefined</u>.

3. The fraction $\frac{8}{3}$ is called an <u>improper</u> fraction, the fraction $\frac{3}{8}$ is called a <u>proper</u> fraction, and $10\frac{3}{8}$ is called a <u>mixed number</u>.

4. The value of an improper fraction is always <u>≥1</u> and the value of a proper fraction is always <u>≤1</u>.

Exercise Set 2.1

1. In the fraction $\frac{1}{2}$, the numerator is 1 and the denominator is 2. Since 1 < 2, the fraction is proper.

3. In the fraction $\frac{10}{3}$, the numerator is 10 and the denominator is 3. Since 10 > 3, the fraction is improper.

5. In the fraction $\frac{15}{15}$, the numerator is 15 and the denominator is 15. Since 15 ≥ 15, the fraction is improper.

7. $\frac{21}{21}=1$

9. $\frac{5}{0}$ is undefined.

11. $\frac{13}{1}=13$

13. $\frac{0}{20}=0$

15. $\dfrac{10}{0}$ is undefined.

17. $\dfrac{16}{1} = 16$

19. 5 of the 6 equal parts are shaded: $\dfrac{5}{6}$

21. 7 of the 12 equal parts are shaded: $\dfrac{7}{12}$

23. 3 of the 7 equal parts are shaded: $\dfrac{3}{7}$

25. 4 of the 9 equal parts are shaded: $\dfrac{4}{9}$

27. 1 of the 6 equal parts is shaded: $\dfrac{1}{6}$

29. 5 of the 8 equal parts are shaded: $\dfrac{5}{8}$

31. answers may vary; for example,

33. answers may vary; for example,

35. answers may vary; for example,

37. answers may vary; for example,

39. freshmen $\rightarrow 42$
students $\rightarrow \overline{131}$

$\dfrac{42}{131}$ of the students are freshmen.

41. a. number of students not freshmen $= 131 - 42$
$= 89$

b. not freshmen $\rightarrow 89$
students $\quad \rightarrow \overline{131}$

$\dfrac{89}{131}$ of the students are not freshmen.

43. born in Ohio $\rightarrow 7$
U.S. presidents $\rightarrow \overline{44}$

$\dfrac{7}{44}$ of U.S. presidents were born in Ohio.

45. number turned into hurricanes $\rightarrow 15$
number of tropical storms $\quad \rightarrow \overline{28}$

$\dfrac{15}{28}$ of the tropical storms turned into hurricanes.

47. 11 of 31 days of March is $\dfrac{11}{31}$ of the month.

49. number of sophomores $\quad \rightarrow 10$
number of students in class $\rightarrow \overline{31}$

$\dfrac{10}{31}$ of the class is sophomores.

51. There are 50 states total. 33 states contain federal Indian reservations.

a. $\dfrac{33}{50}$ of the states contain federal Indian reservations.

b. $50 - 33 = 17$
17 states do not contain federal Indian reservations.

c. $\dfrac{17}{50}$ of the states do not contain federal Indian reservations.

53. There are 50 marbles total. 21 marbles are blue.

a. $\dfrac{21}{50}$ of the marbles are blue.

b. $50 - 21 = 29$
29 of the marbles are red.

c. $\dfrac{29}{50}$ of the marbles are red.

55. Each part is $\dfrac{1}{4}$ of a whole and there are 11 parts shaded, or 2 wholes and 3 more parts.

    a. $\dfrac{11}{4}$ **b.** $2\dfrac{3}{4}$

57. Each part is $\dfrac{1}{6}$ of a whole and 23 parts are shaded, or 3 wholes and 5 more parts.

    a. $\dfrac{23}{6}$ **b.** $3\dfrac{5}{6}$

59. Each part is $\dfrac{1}{3}$ of a whole and there are 4 parts shaded, or 1 whole and 1 more part.

    a. $\dfrac{4}{3}$ **b.** $1\dfrac{1}{3}$

61. Each part is $\dfrac{1}{2}$ of a whole and 11 parts are shaded, or 5 wholes and 1 more part.

    a. $\dfrac{11}{2}$ **b.** $5\dfrac{1}{2}$

63. $2\dfrac{1}{3} = \dfrac{3\cdot 2 + 1}{3} = \dfrac{7}{3}$

65. $3\dfrac{3}{5} = \dfrac{5\cdot 3 + 3}{5} = \dfrac{18}{5}$

67. $6\dfrac{5}{8} = \dfrac{8\cdot 6 + 5}{8} = \dfrac{53}{8}$

69. $2\dfrac{11}{15} = \dfrac{15\cdot 2 + 11}{15} = \dfrac{41}{15}$

71. $11\dfrac{6}{7} = \dfrac{7\cdot 11 + 6}{7} = \dfrac{83}{7}$

73. $6\dfrac{6}{13} = \dfrac{13\cdot 6 + 6}{13} = \dfrac{84}{13}$

75. $4\dfrac{13}{24} = \dfrac{24\cdot 4 + 13}{24} = \dfrac{109}{24}$

77. $17\dfrac{7}{12} = \dfrac{12\cdot 17 + 7}{12} = \dfrac{211}{12}$

79. $9\dfrac{7}{20} = \dfrac{20\cdot 9 + 7}{20} = \dfrac{187}{20}$

81. $2\dfrac{51}{107} = \dfrac{107\cdot 2 + 51}{107} = \dfrac{265}{107}$

83. $166\dfrac{2}{3} = \dfrac{3\cdot 166 + 2}{3} = \dfrac{500}{3}$

85. $\begin{array}{r} 3\ \text{R}\ 2 \\ 5\overline{)\ 17} \\ \underline{-15} \\ 2 \end{array}$

    $\dfrac{17}{5} = 3\dfrac{2}{5}$

87. $\begin{array}{r} 4\ \text{R}\ 5 \\ 8\overline{)\ 37} \\ \underline{-32} \\ 5 \end{array}$

    $\dfrac{37}{8} = 4\dfrac{5}{8}$

89. $\begin{array}{r} 3\ \text{R}\ 2 \\ 15\overline{)\ 47} \\ \underline{-45} \\ 2 \end{array}$

    $\dfrac{47}{15} = 3\dfrac{2}{15}$

91. $\begin{array}{r} 2\ \text{R}\ 4 \\ 21\overline{)\ 46} \\ \underline{-42} \\ 4 \end{array}$

    $\dfrac{46}{21} = 2\dfrac{4}{21}$

93. $\begin{array}{r} 33 \\ 6\overline{)\ 198} \\ \underline{-18} \\ 18 \\ \underline{-18} \\ 0 \end{array}$

    $\dfrac{198}{6} = 33$

95. $15\overline{\smash{)}225}$

$$\begin{array}{r} 15 \\ 15\overline{\smash{)}225} \\ -15 \\ \hline 75 \\ -75 \\ \hline 0 \end{array}$$

$$\frac{225}{15} = 15$$

97. $3\overline{\smash{)}200}$

$$\begin{array}{r} 66\ \text{R}\ 2 \\ 3\overline{\smash{)}200} \\ -18 \\ \hline 20 \\ -18 \\ \hline 2 \end{array}$$

$$\frac{200}{3} = 66\frac{2}{3}$$

99. $23\overline{\smash{)}247}$

$$\begin{array}{r} 10\ \text{R}\ 17 \\ 23\overline{\smash{)}247} \\ -23 \\ \hline 17 \\ -0 \\ \hline 17 \end{array}$$

$$\frac{247}{23} = 10\frac{17}{23}$$

101. $18\overline{\smash{)}319}$

$$\begin{array}{r} 17\ \text{R}\ 13 \\ 18\overline{\smash{)}319} \\ -18 \\ \hline 139 \\ -126 \\ \hline 13 \end{array}$$

$$\frac{319}{18} = 17\frac{13}{18}$$

103. $175\overline{\smash{)}182}$

$$\begin{array}{r} 1\ \text{R}\ 7 \\ 175\overline{\smash{)}182} \\ -175 \\ \hline 7 \end{array}$$

$$\frac{182}{175} = 1\frac{7}{175}$$

105. $112\overline{\smash{)}737}$

$$\begin{array}{r} 6\ \text{R}\ 65 \\ 112\overline{\smash{)}737} \\ -672 \\ \hline 65 \end{array}$$

$$\frac{737}{112} = 6\frac{65}{112}$$

107. $3^2 = 3 \cdot 3 = 9$

109. $5^3 = 5 \cdot 5 \cdot 5 = 125$

111. $7 \cdot 7 \cdot 7 \cdot 7 \cdot 7 = 7^5$

113. $2 \cdot 2 \cdot 2 \cdot 3 = 2^3 \cdot 3$

115. answers may vary

117. $\frac{2}{3}$ is the larger fraction.

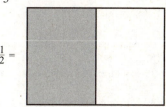

$\frac{1}{2} =$

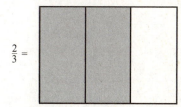

$\frac{2}{3} =$

119. ⬤⬤⬤⬤⭘⭘⭘⭘

121. $1532 + 576 + 1059 = 3167$
576 of the 3167 stores are named Banana
Republic: $\dfrac{576}{3167}$

123. $1700 + 550 = 2250$
1700 of the 2250 affiliates are located in the
United States: $\dfrac{1700}{2250}$

Section 2.2 Practice

1. a. First we write all the two-number factors of
15.
$1 \cdot 15 = 15$
$3 \cdot 5 = 15$
The factors of 15 are 1, 3, 5, and 15.

b. First we write all the two-number factors of
7.
$1 \cdot 7 = 7$
The factors of 7 are 1 and 7.

c. First we write all the two-number factors of 24.

$1 \cdot 24 = 24$

$2 \cdot 12 = 24$

$3 \cdot 8 = 24$

$4 \cdot 6 = 24$

The factors of 24 are 1, 2, 3, 4, 6, 8, 12, and 24.

2. The number 21 is composite. Its factors are 1, 3, 7, and 21.

The number 13 is prime. Its only factors are 1 and 13.

The number 18 is composite. Its factors are 1, 2, 3, 6, 9, and 18.

The number 29 is prime. Its only factors are 1 and 29.

The number 39 is composite. Its factors are 1, 3, 13, and 39.

3.
$$\begin{array}{r} 7 \\ 2\overline{)14} \\ 2\overline{)28} \end{array}$$

$28 = 2 \cdot 2 \cdot 7 = 2^2 \cdot 7$

4.
$$\begin{array}{r} 5 \\ 3\overline{)15} \\ 2\overline{)30} \\ 2\overline{)60} \\ 2\overline{)120} \end{array}$$

$120 = 2 \cdot 2 \cdot 2 \cdot 3 \cdot 5 = 2^3 \cdot 3 \cdot 5$

5.
$$\begin{array}{r} 7 \\ 3\overline{)21} \\ 3\overline{)63} \\ 3\overline{)189} \\ 3\overline{)378} \\ 2\overline{)756} \end{array}$$

$756 = 2 \cdot 2 \cdot 3 \cdot 3 \cdot 3 \cdot 7 = 2^2 \cdot 3^3 \cdot 7$

6.

```
      45
     ↙ ↘
    5    9
    ↓   ↙ ↘
    5   3   3
```

$45 = 3 \cdot 3 \cdot 5 = 3^2 \cdot 5$

7. a.

```
      30
     ↙ ↘
    2   15
    ↓   ↙ ↘
    2   3   5
```

$30 = 2 \cdot 3 \cdot 5$

b.

```
      56
     ↙ ↘
    7   8
    ↓   ↙ ↘
    7   2   4
    ↓   ↓  ↙ ↘
    7   2  2   2
```

$56 = 2 \cdot 2 \cdot 2 \cdot 7 = 2^3 \cdot 7$

c.

```
          72
        ↙    ↘
      8        9
    ↙ ↘       ↙ ↘
   2   4     3   3
   ↓  ↙ ↘    ↓   ↓
   2  2  2  2 3  3
```

$72 = 2 \cdot 2 \cdot 2 \cdot 3 \cdot 3 = 2^3 \cdot 3^2$

8.

```
      117
     ↙ ↘
    3   39
    ↓   ↙ ↘
    3   3   13
```

$117 = 3 \cdot 3 \cdot 13 = 3^2 \cdot 13$

Vocabulary and Readiness Check

1. The number 40 equals $2 \cdot 2 \cdot 2 \cdot 5$. Since each factor is prime, we call $2 \cdot 2 \cdot 2 \cdot 5$ the <u>prime factorization</u> of 40.

2. A natural number, other than 1, that is not prime is called a <u>composite</u> number.

3. A natural number that has exactly two different factors, 1 and itself, is called a <u>prime</u> number.

4. The numbers 1, 2, 3, 4, 5, ... are called the <u>natural</u> numbers.

5. Since $30 = 5 \cdot 6$, the numbers 5 and 6 are <u>factors</u> of 30.

6. True or false: $5 \cdot 6$ is the prime factorization of 30. <u>false</u>

Exercise Set 2.2

1. $1 \cdot 8 = 8$
$2 \cdot 4 = 8$
The factors of 8 are 1, 2, 4, and 8.

3. $1 \cdot 25 = 25$
$5 \cdot 5 = 25$
The factors of 25 are 1, 5, and 25.

5. $1 \cdot 4 = 4$
$2 \cdot 2 = 4$
The factors of 4 are 1, 2, and 4.

7. $1 \cdot 18 = 18$
$2 \cdot 9 = 18$
$3 \cdot 6 = 18$
The factors of 18 are 1, 2, 3, 6, 9, and 18.

9. $1 \cdot 29 = 29$
The factors of 29 are 1 and 29.

11. $1 \cdot 80 = 80$
$2 \cdot 40 = 80$
$4 \cdot 20 = 80$
$5 \cdot 16 = 80$
$8 \cdot 10 = 80$
The factors of 80 are 1, 2, 4, 5, 8, 10, 16, 20, 40, and 80.

13. $1 \cdot 12 = 12$
$2 \cdot 6 = 12$
$3 \cdot 4 = 12$
The factors of 12 are 1, 2, 3, 4, 6, and 12.

15. $1 \cdot 34 = 34$
$2 \cdot 17 = 34$
The factors of 34 are 1, 2, 17, and 34.

17. Prime, since its only factors are 1 and 7.

19. Composite, since its factors are 1, 2, and 4.

21. Prime, since its only factors are 1 and 23.

23. Composite, since its factors are 1, 7, and 49.

25. Prime, since its only factors are 1 and 67.

27. Composite, since its factors are 1, 3, 13, and 39.

29. Prime, since its only factors are 1 and 31.

31. Composite, since its factors are 1, 3, 7, 9, 21, and 63.

33. Composite, since its factors are 1, 7, 17, and 119.

35.
$$\begin{array}{r} 2 \\ 2\overline{)4} \\ 2\overline{)8} \\ 2\overline{)16} \\ 2\overline{)32} \end{array}$$
$32 = 2^5$

37.
$$\begin{array}{r} 5 \\ 3\overline{)15} \end{array}$$
$15 = 3 \cdot 5$

39.
$$\begin{array}{r} 5 \\ 2\overline{)10} \\ 2\overline{)20} \\ 2\overline{)40} \end{array}$$
$40 = 2^3 \cdot 5$

41.
$$\begin{array}{r} 3 \\ 3\overline{)9} \\ 2\overline{)18} \\ 2\overline{)36} \end{array}$$
$36 = 2^2 \cdot 3^2$

43.
$$\begin{array}{r} 13 \\ 3\overline{)39} \end{array}$$
$39 = 3 \cdot 13$

45.
$$\begin{array}{r} 5 \\ 3\overline{)15} \\ 2\overline{)30} \\ 2\overline{)60} \end{array}$$
$60 = 2^2 \cdot 3 \cdot 5$

47.
$$\begin{array}{r} 11 \\ 5\overline{)55} \\ 5\overline{)110} \end{array}$$
$110 = 2 \cdot 5 \cdot 11$

49.
$$\begin{array}{r} 17 \\ 5\overline{)85} \end{array}$$
$85 = 5 \cdot 17$

51.
$$
\begin{array}{r}
2 \\
2\overline{)4} \\
2\overline{)8} \\
2\overline{)16} \\
2\overline{)32} \\
2\overline{)64} \\
2\overline{)128}
\end{array}
$$

$128 = 2^7$

53.
$$
\begin{array}{r}
11 \\
7\overline{)77} \\
2\overline{)154}
\end{array}
$$

$154 = 2 \cdot 7 \cdot 11$

55.
$$
\begin{array}{r}
5 \\
5\overline{)25} \\
3\overline{)75} \\
2\overline{)150} \\
2\overline{)300}
\end{array}
$$

$300 = 2^2 \cdot 3 \cdot 5^2$

57.
$$
\begin{array}{r}
5 \\
3\overline{)15} \\
2\overline{)30} \\
2\overline{)60} \\
2\overline{)120} \\
2\overline{)240}
\end{array}
$$

$240 = 2^4 \cdot 3 \cdot 5$

59.
$$
\begin{array}{r}
23 \\
3\overline{)69} \\
3\overline{)207} \\
2\overline{)414} \\
2\overline{)828}
\end{array}
$$

$828 = 2^2 \cdot 3^2 \cdot 23$

61.
$$
\begin{array}{r}
7 \\
7\overline{)49} \\
3\overline{)147} \\
3\overline{)441} \\
2\overline{)882}
\end{array}
$$

$882 = 2 \cdot 3^2 \cdot 7^2$

63.
$$
\begin{array}{r}
13 \\
7\overline{)91} \\
7\overline{)637}
\end{array}
$$

$637 = 7^2 \cdot 13$

65.
$$
\begin{array}{r}
11 \\
3\overline{)33}
\end{array}
$$

$33 = 3 \cdot 11$

67.
$$
\begin{array}{r}
7 \\
7\overline{)49} \\
2\overline{)98}
\end{array}
$$

$98 = 2 \cdot 7^2$

69. 67 is prime, since its only factors are 1 and 67.

71.
$$
\begin{array}{r}
17 \\
3\overline{)51} \\
3\overline{)153} \\
3\overline{)459}
\end{array}
$$

$459 = 3^3 \cdot 17$

73. 97 is prime, since its only factors are 1 and 97.

75.
$$
\begin{array}{r}
7 \\
5\overline{)35} \\
5\overline{)175} \\
5\overline{)350} \\
5\overline{)700}
\end{array}
$$

$700 = 2^2 \cdot 5^2 \cdot 7$

77. To round 4267 to the nearest hundred, observe that the digit in the tens place is 6. Since this digit is at least 5, we need to add 1 to the digit in the hundreds place. The number 4267 rounded to the nearest hundred is 4300.

79. To round 7,658,240 to the nearest ten-thousand, observe that the digit in the thousands place is 8. Since this digit is at least 5, we add 1 to the digit in the ten-thousands place. The number 7,658,240 rounded to the nearest ten-thousand is 7,660,000.

81. To round 19,764 to the nearest thousand, observe that the digit in the hundreds place is 7. Since this digit is at least 5, we add 1 to the digit in the thousands place. The number 19,764 rounded to the nearest thousand is 20,000.

83. $27 + 49 + 43 + 40 = 159$
The total number of patents received is 159.

85. Of the 159 total number patents, 27 were granted in 2005.

$$\frac{27}{159} = \frac{9}{53}$$

$\dfrac{27}{159}$ or $\dfrac{9}{53}$ of the patents were granted in 2005.

87.
$$
\begin{array}{r}
5)\overline{35} \\[2pt]
3)\overline{105} \\[2pt]
3)\overline{315} \\[2pt]
3)\overline{945} \\[2pt]
3)\overline{2835} \\[2pt]
3)\overline{8505} \\[2pt]
2)\overline{17{,}010} \\[2pt]
2)\overline{34{,}020}
\end{array}
$$

$$34{,}020 = 2^2 \cdot 3^5 \cdot 5 \cdot 7$$

89. answers may vary

91. answers may vary

Section 2.3 Practice

1. Notice that 30 and 45 have a common factor of 15.

$$\frac{30}{45} = \frac{15 \cdot 2}{15 \cdot 3} = \frac{15}{15} \cdot \frac{2}{3} = 1 \cdot \frac{2}{3} = \frac{2}{3}$$

2. $\dfrac{39}{51} = \dfrac{3 \cdot 13}{3 \cdot 17} = \dfrac{3}{3} \cdot \dfrac{13}{17} = 1 \cdot \dfrac{13}{17} = \dfrac{13}{17}$

3. $\dfrac{9}{50} = \dfrac{3 \cdot 3}{2 \cdot 5 \cdot 5}$

Since 9 and 50 have no common factors, $\dfrac{9}{50}$ is already in simplest form.

4. $\dfrac{49}{112} = \dfrac{7 \cdot 7}{2 \cdot 2 \cdot 2 \cdot 2 \cdot 7} = \dfrac{7}{7} \cdot \dfrac{7}{2 \cdot 2 \cdot 2 \cdot 2} = 1 \cdot \dfrac{7}{16} = \dfrac{7}{16}$

5. $\dfrac{64}{20} = \dfrac{\overset{1}{\cancel{2}} \cdot \overset{1}{\cancel{2}} \cdot 2 \cdot 2 \cdot 2 \cdot 2}{\underset{1}{\cancel{2}} \cdot \underset{1}{\cancel{2}} \cdot 5} = \dfrac{1 \cdot 1 \cdot 2 \cdot 2 \cdot 2 \cdot 2}{1 \cdot 1 \cdot 5} = \dfrac{16}{5}$ or $3\dfrac{1}{5}$

6. $\dfrac{8}{56} = \dfrac{\overset{1}{\cancel{2}} \cdot \overset{1}{\cancel{2}} \cdot \overset{1}{\cancel{2}}}{\underset{1}{\cancel{2}} \cdot \underset{1}{\cancel{2}} \cdot \underset{1}{\cancel{2}} \cdot 7} = \dfrac{1 \cdot 1 \cdot 1}{1 \cdot 1 \cdot 1 \cdot 7} = \dfrac{1}{7}$

7. $\dfrac{42}{48} = \dfrac{\overset{1}{\cancel{6}} \cdot 7}{\underset{1}{\cancel{6}} \cdot 8} = \dfrac{1 \cdot 7}{1 \cdot 8} = \dfrac{7}{8}$

8. $\dfrac{7}{9}$ is in simplest form.

$$\frac{21}{27} = \frac{\overset{1}{\cancel{3}} \cdot 7}{\underset{1}{\cancel{3}} \cdot 3 \cdot 3} = \frac{1 \cdot 7}{1 \cdot 3 \cdot 3} = \frac{7}{9}$$

Since these fractions are the same, $\dfrac{7}{9} = \dfrac{21}{27}$. The fractions are equivalent.

9. Check the cross products:
$4 \cdot 18 = 72$ and $13 \cdot 5 = 65$
Since $72 \neq 65$, the fractions are not equivalent.

10. $\dfrac{6 \text{ parks in Washington}}{58 \text{ national parks}} = \dfrac{2 \cdot 3}{2 \cdot 29} = \dfrac{\overset{1}{\cancel{2}} \cdot 3}{\underset{1}{\cancel{2}} \cdot 29} = \dfrac{3}{29}$

$\dfrac{3}{29}$ of the national parks are in Washington state.

Calculator Explorations

1. $\dfrac{128}{224} = \dfrac{4}{7}$

2. $\dfrac{231}{396} = \dfrac{7}{12}$

3. $\dfrac{340}{459} = \dfrac{20}{27}$

4. $\dfrac{999}{1350} = \dfrac{37}{50}$

5. $\dfrac{810}{432} = \dfrac{15}{8}$

6. $\dfrac{315}{225} = \dfrac{7}{5}$

7. $\dfrac{243}{54} = \dfrac{9}{2}$

8. $\dfrac{689}{455} = \dfrac{53}{35}$

Vocabulary and Readiness Check

1. In $\dfrac{11}{48}$, since 11 and 48 have no common factors other than 1, $\dfrac{11}{48}$ is in <u>simplest form</u>.

2. Fractions that represent the same portion of a whole are called <u>equivalent</u> fractions.

3. In the statement $\dfrac{5}{12} = \dfrac{15}{36}$, $5 \cdot 36$ and $12 \cdot 15$ are called <u>cross products</u>.

4. The fraction $\dfrac{7}{7}$ simplifies to <u>1</u>.

5. The fraction $\dfrac{0}{7}$ simplifies to <u>0</u>.

6. The fraction $\dfrac{n}{1}$ simplifies to <u>n</u>.

Exercise Set 2.3

1. $\dfrac{3}{12} = \dfrac{3}{3 \cdot 4} = \dfrac{1 \cdot 3}{4 \cdot 3} = \dfrac{1}{4}$

3. $\dfrac{4}{42} = \dfrac{2 \cdot 2}{2 \cdot 21} = \dfrac{2}{21}$

5. $\dfrac{14}{16} = \dfrac{2 \cdot 7}{2 \cdot 8} = \dfrac{7}{8}$

7. $\dfrac{20}{30} = \dfrac{2 \cdot 10}{3 \cdot 10} = \dfrac{2}{3}$

9. $\dfrac{35}{50} = \dfrac{5 \cdot 7}{5 \cdot 10} = \dfrac{7}{10}$

11. $\dfrac{63}{81} = \dfrac{7 \cdot 9}{9 \cdot 9} = \dfrac{7}{9}$

13. $\dfrac{24}{40} = \dfrac{8 \cdot 3}{8 \cdot 5} = \dfrac{3}{5}$

15. $\dfrac{27}{64} = \dfrac{27}{64}$
27 and 64 have no common factors other than 1.

17. $\dfrac{25}{40} = \dfrac{5 \cdot 5}{5 \cdot 8} = \dfrac{5}{8}$

19. $\dfrac{40}{64} = \dfrac{5 \cdot 8}{8 \cdot 8} = \dfrac{5}{8}$

21. $\dfrac{56}{68} = \dfrac{4 \cdot 14}{4 \cdot 17} = \dfrac{14}{17}$

23. $\dfrac{36}{24} = \dfrac{3 \cdot 12}{2 \cdot 12} = \dfrac{3}{2}$ or $1\dfrac{1}{2}$

25. $\dfrac{90}{120} = \dfrac{30 \cdot 3}{30 \cdot 4} = \dfrac{3}{4}$

27. $\dfrac{70}{196} = \dfrac{5 \cdot 14}{14 \cdot 14} = \dfrac{5}{14}$

29. $\dfrac{66}{308} = \dfrac{22 \cdot 3}{22 \cdot 14} = \dfrac{3}{14}$

31. $\dfrac{55}{85} = \dfrac{5 \cdot 11}{5 \cdot 17} = \dfrac{11}{17}$

33. $\dfrac{75}{350} = \dfrac{25 \cdot 3}{25 \cdot 14} = \dfrac{3}{14}$

35. $\dfrac{189}{216} = \dfrac{27 \cdot 7}{27 \cdot 8} = \dfrac{7}{8}$

37. $\dfrac{288}{480} = \dfrac{96 \cdot 3}{96 \cdot 5} = \dfrac{3}{5}$

39. $\dfrac{224}{16} = \dfrac{14 \cdot 16}{16} = \dfrac{14 \cdot 16}{1 \cdot 16} = \dfrac{14}{1} = 14$

41. Equivalent, since the cross products are equal: $6 \cdot 4 = 24$ and $8 \cdot 3 = 24$.

43. Not equivalent, since the cross products are not equal: $11 \cdot 5 = 55$ and $8 \cdot 7 = 56$.

45. Equivalent, since the cross products are equal: $15 \cdot 6 = 90$ and $9 \cdot 10 = 90$.

47. Equivalent, since the cross products are equal: $9 \cdot 6 = 54$ and $18 \cdot 3 = 54$.

49. Not equivalent, since the cross products are not equal: $13 \cdot 12 = 156$ and $15 \cdot 10 = 150$.

51. Not equivalent, since the cross products are not equal: $18 \cdot 12 = 216$ and $24 \cdot 8 = 192$.

53. $\dfrac{2 \text{ hours}}{8 \text{ hours}} = \dfrac{2 \cdot 1}{2 \cdot 4} = \dfrac{1}{4}$

2 hours represents $\dfrac{1}{4}$ of a work shift.

55. $\dfrac{2640 \text{ feet}}{5280 \text{ feet}} = \dfrac{1 \cdot 2640}{2 \cdot 2640} = \dfrac{1}{2}$

2640 feet is $\dfrac{1}{2}$ of a mile.

57. a. $\dfrac{16 \text{ states}}{50 \text{ states}} = \dfrac{2 \cdot 8}{2 \cdot 25} = \dfrac{8}{25}$

$\dfrac{8}{25}$ of the states can claim at least one Ritz-Carlton hotel.

b. $50 - 16 = 34$
34 states do not have a Ritz-Carlton hotel.

c. $\dfrac{34}{50} = \dfrac{2 \cdot 17}{2 \cdot 25} = \dfrac{17}{25}$

$\dfrac{17}{25}$ of the states do not have a Ritz-Carlton hotel.

59. $\dfrac{10 \text{ inches}}{24 \text{ inches}} = \dfrac{2 \cdot 5}{2 \cdot 12} = \dfrac{5}{12}$

$\dfrac{5}{12}$ of the outer wall of the Pentagon is concrete.

61. a. $50 - 32 = 18$
18 states do not have Kroger Company Stores.

b. $\dfrac{18 \text{ states}}{50 \text{ states}} = \dfrac{2 \cdot 9}{2 \cdot 25} = \dfrac{9}{25}$

$\dfrac{9}{25}$ of the states do not have Kroger Company Stores.

63. $\dfrac{22 \text{ individuals}}{320 \text{ individuals}} = \dfrac{22}{320} = \dfrac{2 \cdot 11}{2 \cdot 160} = \dfrac{11}{160}$

$\dfrac{11}{160}$ of the U.S. astronauts who had flown in space were born in Texas.

65.
$$\begin{array}{r} 91 \\ \times\ 4 \\ \hline 364 \end{array}$$

67.
$$\begin{array}{r} 387 \\ \times\ \ 6 \\ \hline 2322 \end{array}$$

69.
$$\begin{array}{r} 72 \\ \times\ 35 \\ \hline 360 \\ 2160\ \\ \hline 2520 \end{array}$$

71. answers may vary

73. $\dfrac{3975}{6625} = \dfrac{1325 \cdot 3}{1325 \cdot 5} = \dfrac{3}{5}$

75. $\dfrac{36 \text{ donors}}{100 \text{ donors}} = \dfrac{4 \cdot 9}{4 \cdot 25} = \dfrac{9}{25}$

$\dfrac{9}{25}$ of blood donors have blood type A Rh-positive.

77. $3 + 1 = 4$

$\dfrac{4 \text{ donors}}{100 \text{ donors}} = \dfrac{4 \cdot 1}{4 \cdot 25} = \dfrac{1}{25}$

$\dfrac{1}{25}$ of blood donors have an AB blood type.

79. The piece representing education is labeled $\dfrac{1}{10}$, so $\dfrac{1}{10}$ of entering college freshmen plan to major in education.

81. answers may vary

83. The piece representing National Memorials is labeled $\dfrac{2}{25}$, so $\dfrac{2}{25}$ of National Park Service areas are National Memorials.

85. answers may vary

87. 8691, 786, 2235, 105, 222, 900, and 1470 are divisible by 3 because the sum of each number's digits is divisible by 3. 786, 22, 222, 900, and 1470 are divisible by 2 because they are even numbers. 786, 222, 900, and 1470 are divisible by both 2 and 3.

89. 6; answers may vary

Integrated Review

1. 3 of the 6 parts are shaded: $\dfrac{3}{6}$

2. Each part is $\dfrac{1}{4}$ of a whole and 7 parts are shaded, or 1 whole and 3 more parts: $\dfrac{7}{4}$ or $1\dfrac{3}{4}$

3. People getting fewer than 8 hours of sleep $\rightarrow 73$
People in survey $\rightarrow \overline{85}$

$\dfrac{73}{85}$ of the people in a survey get fewer than 8 hours of sleep.

4.

5. $\dfrac{11}{11} = 1$

6. $\dfrac{17}{1} = 17$

7. $\dfrac{0}{3} = 0$

8. $\dfrac{7}{0}$ is undefined.

9. $3\dfrac{1}{8} = \dfrac{8 \cdot 3 + 1}{8} = \dfrac{25}{8}$

10. $5\dfrac{3}{5} = \dfrac{5 \cdot 5 + 3}{5} = \dfrac{28}{5}$

11. $9\dfrac{6}{7} = \dfrac{7 \cdot 9 + 6}{7} = \dfrac{69}{7}$

12. $20\dfrac{1}{7} = \dfrac{7 \cdot 20 + 1}{7} = \dfrac{141}{7}$

13.
$$\begin{array}{r} 2 \text{ R } 6 \\ 7\overline{)\,20} \\ \underline{-14} \\ 6 \end{array}$$
$\dfrac{20}{7} = 2\dfrac{6}{7}$

14.
$$\begin{array}{r} 5 \\ 11\overline{)\,55} \\ \underline{-55} \\ 0 \end{array}$$
$\dfrac{55}{11} = 5$

15.
$$\begin{array}{r} 4 \text{ R } 7 \\ 8\overline{)\,39} \\ \underline{-32} \\ 7 \end{array}$$
$\dfrac{39}{8} = 4\dfrac{7}{8}$

16.
$$\begin{array}{r} 8 \text{ R } 10 \\ 11\overline{)\,98} \\ \underline{-88} \\ 10 \end{array}$$
$\dfrac{98}{11} = 8\dfrac{10}{11}$

17. $1 \cdot 35 = 35$
$5 \cdot 7 = 35$
The factors of 35 are 1, 5, 7, and 35.

18. $1 \cdot 40 = 40$
$2 \cdot 20 = 40$
$4 \cdot 10 = 40$
$5 \cdot 8 = 40$
The factors of 40 are 1, 2, 4, 5, 8, 10, 20, and 40.

19. Composite since its factors are 1, 2, 3, 4, 6, 8, 9, 12, 18, 24, 36, and 72.

20. Prime, since its only factors are 1 and 13.

21.
$$\begin{array}{r} 13 \\ 5\overline{)65} \end{array}$$
$65 = 5 \cdot 13$

22.
$$\begin{array}{r} 7 \\ 5\overline{)35} \\ 2\overline{)70} \end{array}$$
$70 = 2 \cdot 5 \cdot 7$

23.
$$2\overline{)6}^{\;3}$$
$$2\overline{)12}$$
$$2\overline{)24}$$
$$2\overline{)48}$$
$$2\overline{)96}$$

$96 = 2^5 \cdot 3$

24.
$$3\overline{)33}^{\;11}$$
$$2\overline{)66}$$
$$2\overline{)132}$$

$132 = 2^2 \cdot 3 \cdot 11$

25.
$$3\overline{)21}^{\;7}$$
$$3\overline{)63}$$
$$3\overline{)126}$$
$$2\overline{)252}$$

$252 = 2^2 \cdot 3^2 \cdot 7$

26. Prime, since its only factors are 1 and 31.

27.
$$5\overline{)35}^{\;7}$$
$$3\overline{)105}$$
$$3\overline{)315}$$

$315 = 3^2 \cdot 5 \cdot 7$

28.
$$7\overline{)49}^{\;7}$$
$$3\overline{)147}$$
$$3\overline{)441}$$

$441 = 3^2 \cdot 7^2$

29.
$$11\overline{)143}^{\;13}$$
$$2\overline{)286}$$

$286 = 2 \cdot 11 \cdot 13$

30. Prime, since its only factor are 1 and 41.

31. $\dfrac{2}{14} = \dfrac{2 \cdot 1}{2 \cdot 7} = \dfrac{1}{7}$

32. $\dfrac{24}{20} = \dfrac{4 \cdot 6}{4 \cdot 5} = \dfrac{6}{5}$ or $1\dfrac{1}{5}$

33. $\dfrac{18}{38} = \dfrac{2 \cdot 9}{2 \cdot 19} = \dfrac{9}{19}$

34. $\dfrac{42}{110} = \dfrac{2 \cdot 21}{2 \cdot 55} = \dfrac{21}{55}$

35. $\dfrac{56}{60} = \dfrac{4 \cdot 14}{4 \cdot 15} = \dfrac{14}{15}$

36. $\dfrac{72}{80} = \dfrac{8 \cdot 9}{8 \cdot 10} = \dfrac{9}{10}$

37. $\dfrac{54}{135} = \dfrac{27 \cdot 2}{27 \cdot 5} = \dfrac{2}{5}$

38. $\dfrac{90}{240} = \dfrac{30 \cdot 3}{30 \cdot 8} = \dfrac{3}{8}$

39. $\dfrac{165}{210} = \dfrac{15 \cdot 11}{15 \cdot 14} = \dfrac{11}{14}$

40. $\dfrac{245}{385} = \dfrac{35 \cdot 7}{35 \cdot 11} = \dfrac{7}{11}$

41. Not equivalent, since the cross products are not equal: $8 \cdot 9 = 72$ and $10 \cdot 7 = 70$

42. Equivalent, since the cross products are equal: $12 \cdot 15 = 180$ and $18 \cdot 10 = 180$

43. a. $\dfrac{2 \text{ states}}{50 \text{ states}} = \dfrac{2 \cdot 1}{2 \cdot 25} = \dfrac{1}{25}$

$\dfrac{1}{25}$ of the states are not adjacent to any other states.

b. $50 - 2 = 48$
48 states are adjacent to other states.

c. $\dfrac{48 \text{ states}}{50 \text{ states}} = \dfrac{24 \cdot 2}{25 \cdot 2} = \dfrac{24}{25}$

$\dfrac{24}{25}$ of the states are adjacent to other states.

44. a. $\dfrac{165 \text{ films}}{475 \text{ films}} = \dfrac{5 \cdot 33}{5 \cdot 95} = \dfrac{33}{95}$

$\dfrac{33}{95}$ of the films were rated PG-13.

b. $475 - 165 = 310$

310 films were rated other than PG-13.

c. $\dfrac{310 \text{ films}}{475 \text{ films}} = \dfrac{5 \cdot 62}{5 \cdot 95} = \dfrac{62}{95}$

$\dfrac{62}{95}$ of the films were rated other than PG-13.

Section 2.4 Practice

1. $\dfrac{3}{8} \cdot \dfrac{5}{7} = \dfrac{3 \cdot 5}{8 \cdot 7} = \dfrac{15}{56}$

2. $\dfrac{1}{3} \cdot \dfrac{1}{6} = \dfrac{1 \cdot 1}{3 \cdot 6} = \dfrac{1}{18}$

3. $\dfrac{6}{55} \cdot \dfrac{5}{8} = \dfrac{6 \cdot 5}{55 \cdot 8} = \dfrac{\overset{1}{\cancel{2}} \cdot 3 \cdot \overset{1}{\cancel{5}}}{\cancel{5} \cdot 11 \cdot \cancel{2} \cdot 2 \cdot 2} = \dfrac{3}{11 \cdot 2 \cdot 2} = \dfrac{3}{44}$

4. $\dfrac{4}{15} \cdot \dfrac{3}{8} = \dfrac{4 \cdot 3}{15 \cdot 8} = \dfrac{\overset{1}{\cancel{4}} \cdot \overset{1}{\cancel{3}}}{\cancel{3} \cdot 5 \cdot 2 \cdot \cancel{4}} = \dfrac{1}{5 \cdot 2} = \dfrac{1}{10}$

5. $\dfrac{2}{5} \cdot \dfrac{20}{7} = \dfrac{2 \cdot 20}{5 \cdot 7} = \dfrac{2 \cdot 4 \cdot \overset{1}{\cancel{5}}}{\cancel{5} \cdot 7} = \dfrac{8}{7}$

6. $\dfrac{4}{11} \cdot \dfrac{33}{16} = \dfrac{4 \cdot 33}{11 \cdot 16} = \dfrac{\overset{1}{\cancel{4}} \cdot 3 \cdot \overset{1}{\cancel{11}}}{\cancel{11} \cdot \cancel{4} \cdot 4} = \dfrac{3}{4}$

7. $\dfrac{1}{6} \cdot \dfrac{3}{10} \cdot \dfrac{25}{16} = \dfrac{1 \cdot 3 \cdot 25}{6 \cdot 10 \cdot 16} = \dfrac{\overset{1}{\cancel{3}} \cdot \overset{1}{\cancel{5}} \cdot 5}{2 \cdot \cancel{3} \cdot 2 \cdot \cancel{5} \cdot 16} = \dfrac{5}{64}$

8. $2\dfrac{1}{2} = \dfrac{5}{2}$

$2\dfrac{1}{2} \cdot \dfrac{8}{15} = \dfrac{5}{2} \cdot \dfrac{8}{15} = \dfrac{\overset{1}{\cancel{5}} \cdot \overset{1}{\cancel{2}} \cdot 4}{\cancel{2} \cdot 3 \cdot \cancel{5}} = \dfrac{4}{3}$ or $1\dfrac{1}{3}$

9. $\dfrac{2}{3} \cdot 18 = \dfrac{2}{3} \cdot \dfrac{18}{1} = \dfrac{2 \cdot 18}{3 \cdot 1} = \dfrac{2 \cdot \overset{6}{\cancel{3}} \cdot 6}{\cancel{3} \cdot 1} = \dfrac{12}{1} = 12$

10. $3\dfrac{1}{5} \cdot 2\dfrac{3}{4} = \dfrac{16}{5} \cdot \dfrac{11}{4} = \dfrac{16 \cdot 11}{5 \cdot 4} = \dfrac{\overset{4}{\cancel{4}} \cdot 4 \cdot 11}{5 \cdot \cancel{4}} = \dfrac{44}{5}$ or $8\dfrac{4}{5}$

11. $5 \cdot 3\dfrac{11}{15} = \dfrac{5}{1} \cdot \dfrac{56}{15} = \dfrac{5 \cdot 56}{1 \cdot 15} = \dfrac{\overset{1}{\cancel{5}} \cdot 56}{1 \cdot 3 \cdot \cancel{5}} = \dfrac{56}{3}$ or $18\dfrac{2}{3}$

12. $\dfrac{9}{11} \cdot 0 = 0$

13. $0 \cdot 4\dfrac{1}{8} = 0$

14. $\dfrac{1}{6} \cdot 60 = \dfrac{1}{6} \cdot \dfrac{60}{1} = \dfrac{1 \cdot 60}{6 \cdot 1} = \dfrac{1 \cdot 6 \cdot 10}{6 \cdot 1} = 10$

Thus, there are 10 roller coasters in Hershey Park.

Vocabulary and Readiness Check

1. To multiply two fractions, we write $\dfrac{a}{b} \cdot \dfrac{c}{d} = \underline{\dfrac{a \cdot c}{b \cdot d}}$.

2. Using the definition of an exponent, the expression $\dfrac{2^3}{7} = \underline{\dfrac{2 \cdot 2 \cdot 2}{7}}$ while $\left(\dfrac{2}{7}\right)^3 = \underline{\dfrac{2}{7} \cdot \dfrac{2}{7} \cdot \dfrac{2}{7}}$.

3. The word "of" indicates <u>multiplication</u>.

4. $\dfrac{1}{5} \cdot 0 = \underline{0}$

Exercise Set 2.4

1. $\dfrac{1}{3} \cdot \dfrac{2}{5} = \dfrac{1 \cdot 2}{3 \cdot 5} = \dfrac{2}{15}$

3. $\dfrac{6}{5} \cdot \dfrac{1}{7} = \dfrac{6 \cdot 1}{5 \cdot 7} = \dfrac{6}{35}$

5. $\dfrac{3}{10} \cdot \dfrac{3}{8} = \dfrac{3 \cdot 3}{10 \cdot 8} = \dfrac{9}{80}$

7. $\dfrac{2}{7} \cdot \dfrac{5}{8} = \dfrac{2 \cdot 5}{7 \cdot 8} = \dfrac{2 \cdot 5}{7 \cdot 2 \cdot 4} = \dfrac{5}{7 \cdot 4} = \dfrac{5}{28}$

9. $\dfrac{16}{5} \cdot \dfrac{3}{4} = \dfrac{16 \cdot 3}{5 \cdot 4} = \dfrac{4 \cdot 4 \cdot 3}{5 \cdot 4} = \dfrac{3 \cdot 4}{5} = \dfrac{12}{5}$ or $2\dfrac{2}{5}$

11. $\dfrac{5}{28} \cdot \dfrac{2}{25} = \dfrac{5 \cdot 2}{28 \cdot 25} = \dfrac{5 \cdot 2 \cdot 1}{14 \cdot 2 \cdot 5 \cdot 5} = \dfrac{1}{14 \cdot 5} = \dfrac{1}{70}$

13. $0 \cdot \dfrac{8}{9} = 0$

15. $\dfrac{1}{10} \cdot \dfrac{1}{11} = \dfrac{1 \cdot 1}{10 \cdot 11} = \dfrac{1}{110}$

17. $\dfrac{18}{20} \cdot \dfrac{36}{99} = \dfrac{18 \cdot 36}{20 \cdot 99} = \dfrac{9 \cdot 2 \cdot 4 \cdot 9}{4 \cdot 5 \cdot 9 \cdot 11} = \dfrac{2 \cdot 9}{5 \cdot 11} = \dfrac{18}{55}$

19. $\dfrac{3}{8} \cdot \dfrac{9}{10} = \dfrac{3 \cdot 9}{8 \cdot 10} = \dfrac{27}{80}$

21. $\dfrac{11}{20} \cdot \dfrac{1}{7} \cdot \dfrac{5}{22} = \dfrac{11 \cdot 1 \cdot 5}{20 \cdot 7 \cdot 22}$

$\qquad\qquad\qquad = \dfrac{11 \cdot 1 \cdot 5}{5 \cdot 4 \cdot 7 \cdot 11 \cdot 2}$

$\qquad\qquad\qquad = \dfrac{1}{4 \cdot 7 \cdot 2}$

$\qquad\qquad\qquad = \dfrac{1}{56}$

23. $\dfrac{1}{3} \cdot \dfrac{2}{7} \cdot \dfrac{1}{5} = \dfrac{1 \cdot 2 \cdot 1}{3 \cdot 7 \cdot 5} = \dfrac{2}{105}$

25. $\dfrac{9}{20} \cdot 0 \cdot \dfrac{4}{19} = 0$

27. $\dfrac{3}{14} \cdot \dfrac{6}{25} \cdot \dfrac{5}{27} \cdot \dfrac{7}{6} = \dfrac{3 \cdot 6 \cdot 5 \cdot 7}{14 \cdot 25 \cdot 27 \cdot 6}$

$\qquad\qquad\qquad = \dfrac{3 \cdot 6 \cdot 5 \cdot 7}{2 \cdot 7 \cdot 5 \cdot 5 \cdot 3 \cdot 9 \cdot 6}$

$\qquad\qquad\qquad = \dfrac{1}{2 \cdot 5 \cdot 9}$

$\qquad\qquad\qquad = \dfrac{1}{90}$

29. $7\dfrac{7}{8}$ rounds to 8.

31. $6\dfrac{1}{5}$ rounds to 6.

33. $19\dfrac{11}{20}$ rounds to 20.

35. $12 \cdot \dfrac{1}{4} = \dfrac{12}{1} \cdot \dfrac{1}{4} = \dfrac{12 \cdot 1}{1 \cdot 4} = \dfrac{4 \cdot 3 \cdot 1}{1 \cdot 4} = \dfrac{3}{1} = 3$

37. $\dfrac{5}{8} \cdot 4 = \dfrac{5}{8} \cdot \dfrac{4}{1} = \dfrac{5 \cdot 4}{8 \cdot 1} = \dfrac{5 \cdot 4}{2 \cdot 4 \cdot 1} = \dfrac{5}{2 \cdot 1} = \dfrac{5}{2}$ or $2\dfrac{1}{2}$

39. $1\dfrac{1}{4} \cdot \dfrac{4}{25} = \dfrac{5}{4} \cdot \dfrac{4}{25} = \dfrac{5 \cdot 4}{4 \cdot 25} = \dfrac{5 \cdot 4}{4 \cdot 5 \cdot 5} = \dfrac{1}{5}$

41. $\dfrac{2}{5} \cdot 4\dfrac{1}{6} = \dfrac{2}{5} \cdot \dfrac{25}{6} = \dfrac{2 \cdot 25}{5 \cdot 6} = \dfrac{2 \cdot 5 \cdot 5}{5 \cdot 2 \cdot 3} = \dfrac{5}{3}$ or $1\dfrac{2}{3}$

43. $\dfrac{2}{3} \cdot 1 = \dfrac{2}{3}$

45. Exact: $2\dfrac{1}{5} \cdot 3\dfrac{1}{2} = \dfrac{11}{5} \cdot \dfrac{7}{2} = \dfrac{11 \cdot 7}{5 \cdot 2} = \dfrac{77}{10}$ or $7\dfrac{7}{10}$

Estimate: $2\dfrac{1}{5}$ rounds to 2, $3\dfrac{1}{2}$ rounds to 4.

$2 \cdot 4 = 8$, so the answer is reasonable.

47. Exact:

$3\dfrac{4}{5} \cdot 6\dfrac{2}{7} = \dfrac{19}{5} \cdot \dfrac{44}{7} = \dfrac{19 \cdot 44}{5 \cdot 7} = \dfrac{836}{35}$ or $23\dfrac{31}{35}$

Estimates: $3\dfrac{4}{5}$ rounds to 4, $6\dfrac{2}{7}$ rounds to 6.

$4 \cdot 6 = 24$, so the answer is reasonable.

49. $5 \cdot 2\dfrac{1}{2} = \dfrac{5}{1} \cdot \dfrac{5}{2} = \dfrac{5 \cdot 5}{1 \cdot 2} = \dfrac{25}{2}$ or $12\dfrac{1}{2}$

51. $1\dfrac{1}{5} \cdot 12\dfrac{1}{2} = \dfrac{6}{5} \cdot \dfrac{25}{2}$

$\qquad\qquad = \dfrac{6 \cdot 25}{5 \cdot 2}$

$\qquad\qquad = \dfrac{2 \cdot 3 \cdot 5 \cdot 5}{5 \cdot 2}$

$\qquad\qquad = \dfrac{3 \cdot 5}{1}$

$\qquad\qquad = 15$

53. $\dfrac{3}{4} \cdot 16 \cdot \dfrac{1}{2} = \dfrac{3}{4} \cdot \dfrac{16}{1} \cdot \dfrac{1}{2}$

$= \dfrac{3 \cdot 16 \cdot 1}{4 \cdot 1 \cdot 2}$

$= \dfrac{3 \cdot 4 \cdot 2 \cdot 2 \cdot 1}{4 \cdot 1 \cdot 2}$

$= \dfrac{3 \cdot 2}{1}$

$= \dfrac{6}{1}$

$= 6$

55. $\dfrac{3}{10} \cdot 15 \cdot 2\dfrac{1}{2} = \dfrac{3}{10} \cdot \dfrac{15}{1} \cdot \dfrac{5}{2}$

$= \dfrac{3 \cdot 15 \cdot 5}{10 \cdot 1 \cdot 2}$

$= \dfrac{3 \cdot 15 \cdot 5}{5 \cdot 2 \cdot 1 \cdot 2}$

$= \dfrac{3 \cdot 15}{2 \cdot 2}$

$= \dfrac{45}{4}$ or $11\dfrac{1}{4}$

57. $3\dfrac{1}{2} \cdot 1\dfrac{3}{4} \cdot 2\dfrac{2}{3} = \dfrac{7}{2} \cdot \dfrac{7}{4} \cdot \dfrac{8}{3}$

$= \dfrac{7 \cdot 7 \cdot 8}{2 \cdot 4 \cdot 3}$

$= \dfrac{7 \cdot 7 \cdot 2 \cdot 4}{2 \cdot 4 \cdot 3}$

$= \dfrac{7 \cdot 7}{3}$

$= \dfrac{49}{3}$ or $16\dfrac{1}{3}$

59. $\dfrac{1}{4} \cdot \dfrac{2}{15} = \dfrac{1 \cdot 2}{4 \cdot 15} = \dfrac{1 \cdot 2}{2 \cdot 2 \cdot 15} = \dfrac{1}{2 \cdot 15} = \dfrac{1}{30}$

61. $\dfrac{19}{37} \cdot 0 = 0$

63. $2\dfrac{4}{5} \cdot 1\dfrac{1}{7} = \dfrac{14}{5} \cdot \dfrac{8}{7}$

$= \dfrac{14 \cdot 8}{5 \cdot 7}$

$= \dfrac{2 \cdot 7 \cdot 8}{5 \cdot 7}$

$= \dfrac{2 \cdot 8}{5}$

$= \dfrac{16}{5}$ or $3\dfrac{1}{5}$

65. $\dfrac{3}{2} \cdot \dfrac{7}{3} = \dfrac{3 \cdot 7}{2 \cdot 3} = \dfrac{7}{2}$ or $3\dfrac{1}{2}$

67. $\dfrac{6}{15} \cdot \dfrac{5}{16} = \dfrac{6 \cdot 5}{15 \cdot 16} = \dfrac{2 \cdot 3 \cdot 5}{3 \cdot 5 \cdot 2 \cdot 8} = \dfrac{1}{8}$

69. $\dfrac{7}{72} \cdot \dfrac{9}{49} = \dfrac{7 \cdot 9}{72 \cdot 49} = \dfrac{7 \cdot 9}{9 \cdot 8 \cdot 7 \cdot 7} = \dfrac{1}{8 \cdot 7} = \dfrac{1}{56}$

71. $20 \cdot \dfrac{11}{12} = \dfrac{20}{1} \cdot \dfrac{11}{12}$

$= \dfrac{20 \cdot 11}{1 \cdot 12}$

$= \dfrac{4 \cdot 5 \cdot 11}{1 \cdot 4 \cdot 3}$

$= \dfrac{5 \cdot 11}{3}$

$= \dfrac{55}{3}$ or $18\dfrac{1}{3}$

73. $9\dfrac{5}{7} \cdot 8\dfrac{1}{5} \cdot 0 = 0$

75. $12\dfrac{4}{5} \cdot 6\dfrac{7}{8} \cdot \dfrac{26}{77} = \dfrac{64}{5} \cdot \dfrac{55}{8} \cdot \dfrac{26}{77}$

$= \dfrac{64 \cdot 55 \cdot 26}{5 \cdot 8 \cdot 77}$

$= \dfrac{8 \cdot 8 \cdot 5 \cdot 11 \cdot 26}{5 \cdot 8 \cdot 11 \cdot 7}$

$= \dfrac{8 \cdot 26}{7}$

$= \dfrac{208}{7}$ or $29\dfrac{5}{7}$

77. $\dfrac{1}{4} \cdot 200 = \dfrac{1}{4} \cdot \dfrac{200}{1}$

$= \dfrac{1 \cdot 200}{4 \cdot 1}$

$= \dfrac{1 \cdot 4 \cdot 50}{4 \cdot 1}$

$= \dfrac{50}{1}$

$= 50$

$\dfrac{1}{4}$ of 200 is 50.

79. $\dfrac{5}{6} \cdot 24 = \dfrac{5}{6} \cdot \dfrac{24}{1} = \dfrac{5 \cdot 24}{6 \cdot 1} = \dfrac{5 \cdot 6 \cdot 4}{6 \cdot 1} = \dfrac{5 \cdot 4}{1} = 20$

$\dfrac{5}{6}$ of 24 is 20.

81. $\dfrac{4}{25}$ of $800 = \dfrac{4}{25} \cdot 800$

$\qquad = \dfrac{4}{25} \cdot \dfrac{800}{1}$

$\qquad = \dfrac{4 \cdot 800}{25 \cdot 1}$

$\qquad = \dfrac{4 \cdot 25 \cdot 32}{25 \cdot 1}$

$\qquad = \dfrac{4 \cdot 32}{1}$

$\qquad = 128$

128 of the students would be expected to major in business.

83. $\dfrac{7}{25}$ of 175 million $= \dfrac{7}{25} \cdot 175,000,000$

$\qquad = \dfrac{7}{25} \cdot \dfrac{175,000,000}{1}$

$\qquad = \dfrac{7 \cdot 175,000,000}{25 \cdot 1}$

$\qquad = \dfrac{7 \cdot 25 \cdot 7,000,000}{25 \cdot 1}$

$\qquad = 49,000,000$

Approximately 49 million people ages 16–24 attended the movies.

85. $\dfrac{2}{5}$ of $2170 = \dfrac{2}{5} \cdot 2170$

$\qquad = \dfrac{2}{5} \cdot \dfrac{2170}{1}$

$\qquad = \dfrac{2 \cdot 2170}{5 \cdot 1}$

$\qquad = \dfrac{2 \cdot 5 \cdot 434}{5 \cdot 1}$

$\qquad = \dfrac{2 \cdot 434}{1}$

$\qquad = 868$

He hiked 868 miles.

87. $\dfrac{1}{2}$ of $\dfrac{3}{8} = \dfrac{1}{2} \cdot \dfrac{3}{8} = \dfrac{1 \cdot 3}{2 \cdot 8} = \dfrac{3}{16}$

The radius of the circle is $\dfrac{3}{16}$ inch.

89. $\dfrac{5}{6} \cdot 36 = \dfrac{5}{6} \cdot \dfrac{36}{1} = \dfrac{5 \cdot 36}{6 \cdot 1} = \dfrac{5 \cdot 6 \cdot 6}{6 \cdot 1} = \dfrac{5 \cdot 6}{1} = \dfrac{30}{1} = 30$

There are 30 gallons of medicated flea dip solution normally in the vat.

91. $\dfrac{1}{4} \cdot 34 = \dfrac{1}{4} \cdot \dfrac{34}{1}$

$\qquad = \dfrac{1 \cdot 34}{4 \cdot 1}$

$\qquad = \dfrac{1 \cdot 2 \cdot 17}{2 \cdot 2 \cdot 1}$

$\qquad = \dfrac{1 \cdot 17}{2 \cdot 1}$

$\qquad = \dfrac{17}{2}$ or $8\dfrac{1}{2}$

Jorge's wrist measure is $\dfrac{17}{2}$ or $8\dfrac{1}{2}$ inches.

93. $6 \cdot 3\dfrac{1}{4} = \dfrac{6}{1} \cdot \dfrac{13}{4}$

$\qquad = \dfrac{6 \cdot 13}{1 \cdot 4}$

$\qquad = \dfrac{2 \cdot 3 \cdot 13}{1 \cdot 2 \cdot 2}$

$\qquad = \dfrac{3 \cdot 13}{1 \cdot 2}$

$\qquad = \dfrac{39}{2}$ or $19\dfrac{1}{2}$

The sidewalk is $\dfrac{39}{2}$ or $19\dfrac{1}{2}$ inches wide.

95. $2\dfrac{9}{25} \cdot 1\dfrac{13}{25} = \dfrac{59}{25} \cdot \dfrac{38}{25}$

$\qquad = \dfrac{59 \cdot 38}{25 \cdot 25}$

$\qquad = \dfrac{2242}{625}$ or $3\dfrac{367}{625}$

The area of the face of the camera is $\dfrac{2242}{625}$ or $3\dfrac{367}{625}$ square inches.

97. $\dfrac{5}{14} \cdot \dfrac{1}{5} = \dfrac{5 \cdot 1}{14 \cdot 5} = \dfrac{1}{14}$

The area is $\dfrac{1}{14}$ of a square foot.

99. $1\dfrac{3}{4} \cdot 2 = \dfrac{7}{4} \cdot \dfrac{2}{1} = \dfrac{7 \cdot 2}{4 \cdot 1} = \dfrac{7 \cdot 2}{2 \cdot 2 \cdot 1} = \dfrac{7}{2}$

The area is $\dfrac{7}{2}$ or $3\dfrac{1}{2}$ square yards.

101. $\frac{8}{25} \cdot 12,000 = \frac{8}{25} \cdot \frac{12,000}{1}$

$= \frac{8 \cdot 12,000}{25 \cdot 1}$

$= \frac{8 \cdot 25 \cdot 480}{25 \cdot 1}$

$= \frac{8 \cdot 480}{1}$

$= 3840$

The family drove 3840 miles to work.

103. $\frac{1}{5} \cdot 12,000 = \frac{1}{5} \cdot \frac{12,000}{1}$

$= \frac{1 \cdot 12,000}{5 \cdot 1}$

$= \frac{1 \cdot 5 \cdot 2400}{5 \cdot 1}$

$= 2400$

The family drove 2400 miles on family business.

105.
```
     206
 8) 1648
   -16
    ‾‾
    04
    -0
    ‾‾
    48
   -48
   ‾‾
     0
```

107.
```
     56 R 12
 23) 1300
     115
     ‾‾‾
     150
     138
     ‾‾‾
      12
```

109. a. answers may vary

b. answers may vary

111. $3\frac{2}{3} \cdot 1\frac{1}{7} = \frac{11}{3} \cdot \frac{8}{7} = \frac{11 \cdot 8}{3 \cdot 7} = \frac{88}{21}$ or $4\frac{4}{21}$

113. $3\frac{1}{5}$ rounds to 3

$4\frac{5}{8}$ rounds to 5

$3 \cdot 5 = 15$

The best estimate is b.

115. 9 rounds to 9

$\frac{10}{11}$ rounds to 1

$9 \cdot 1 = 9$

The best estimate is a.

117. $\frac{3}{4} \cdot 36 = \frac{3}{4} \cdot \frac{36}{1} = \frac{3 \cdot 36}{4 \cdot 1} = \frac{3 \cdot 4 \cdot 9}{4 \cdot 1} = \frac{3 \cdot 9}{1} = 27$

There are 27 girls on the first bus.

$\frac{2}{3} \cdot 30 = \frac{2}{3} \cdot \frac{30}{1} = \frac{2 \cdot 30}{3 \cdot 1} = \frac{2 \cdot 3 \cdot 10}{3 \cdot 1} = \frac{2 \cdot 10}{1} = 20$

There are 20 boys on the second bus.

$30 - 20 = 10$

There are 10 girls on the second bus.

$27 + 10 = 37$

There are 37 girls on the two buses.

119. $\frac{3}{4} \cdot 116\frac{4}{5} = \frac{3}{4} \cdot \frac{584}{5}$

$= \frac{3 \cdot 584}{4 \cdot 5}$

$= \frac{3 \cdot 4 \cdot 146}{4 \cdot 5}$

$= \frac{3 \cdot 146}{5}$

$= \frac{438}{5}$

$= 87\frac{3}{5}$

$87\frac{3}{5}$ million American households had one or more credit cards in 2009.

Section 2.5 Practice

1. The reciprocal of $\frac{4}{9}$ is $\frac{9}{4}$.

2. The reciprocal of $\frac{15}{7}$ is $\frac{7}{15}$.

3. The reciprocal of 9, or $\frac{9}{1}$, is $\frac{1}{9}$.

4. The reciprocal of $\frac{1}{8}$ is $\frac{8}{1}$ or 8.

5. $\frac{3}{2} \div \frac{14}{5} = \frac{3}{2} \cdot \frac{5}{14} = \frac{3 \cdot 5}{2 \cdot 14} = \frac{15}{28}$

6. $\dfrac{8}{7} \div \dfrac{2}{9} = \dfrac{8}{7} \cdot \dfrac{9}{2} = \dfrac{8 \cdot 9}{7 \cdot 2} = \dfrac{4 \cdot \overset{1}{\cancel{2}} \cdot 9}{7 \cdot \underset{1}{\cancel{2}}} = \dfrac{36}{7}$ or $5\dfrac{1}{7}$

7. $\dfrac{4}{9} \div \dfrac{1}{2} = \dfrac{4}{9} \cdot \dfrac{2}{1} = \dfrac{4 \cdot 2}{9 \cdot 1} = \dfrac{8}{9}$

8. $\dfrac{4}{9} \div 7 = \dfrac{4}{9} \div \dfrac{7}{1} = \dfrac{4}{9} \cdot \dfrac{1}{7} = \dfrac{4 \cdot 1}{9 \cdot 7} = \dfrac{4}{63}$

9. $\dfrac{8}{15} \div 3\dfrac{4}{5} = \dfrac{8}{15} \div \dfrac{19}{5}$

   $= \dfrac{8}{15} \cdot \dfrac{5}{19}$

   $= \dfrac{8 \cdot 5}{15 \cdot 19}$

   $= \dfrac{8 \cdot \overset{1}{\cancel{5}}}{3 \cdot \underset{1}{\cancel{5}} \cdot 19}$

   $= \dfrac{8}{57}$

10. $3\dfrac{2}{7} \div 2\dfrac{3}{14} = \dfrac{23}{7} \div \dfrac{31}{14}$

    $= \dfrac{23}{7} \cdot \dfrac{14}{31}$

    $= \dfrac{23 \cdot 14}{7 \cdot 31}$

    $= \dfrac{23 \cdot 2 \cdot \overset{1}{\cancel{7}}}{\underset{1}{\cancel{7}} \cdot 31}$

    $= \dfrac{46}{31}$ or $1\dfrac{15}{31}$

11. $\dfrac{14}{17} \div 0$ is undefined.

12. $0 \div 2\dfrac{1}{8} = 0 \div \dfrac{17}{8} = 0$

13. $\boxed{\text{Number of outfits}}$ $\boxed{\text{is}}$ $\boxed{30}$ $\boxed{\text{divided by}}$ $\boxed{2\dfrac{1}{7}}$

    $\qquad\quad \downarrow \qquad\quad \downarrow \quad \downarrow \qquad\quad \downarrow \qquad\quad \downarrow$

    Number of outfits $=$ 30 $\div$ $2\dfrac{1}{7}$

    $30 \div 2\dfrac{1}{7} = 30 \div \dfrac{15}{7} = \dfrac{30}{1} \cdot \dfrac{7}{15} = \dfrac{30 \cdot 7}{1 \cdot 15} = \dfrac{2 \cdot \overset{1}{\cancel{15}} \cdot 7}{1 \cdot \underset{1}{\cancel{15}}} = \dfrac{14}{1} = 14$

    14 outfits can be made from a 30-yard bolt of material.

Vocabulary and Readiness Check

1. Two numbers are <u>reciprocals</u> of each other if their product is 1.

2. Every number has a reciprocal expect <u>0</u>.

3. To divide two fractions, we write $\dfrac{a}{b} \div \dfrac{c}{d} = \dfrac{a \cdot d}{b \cdot c}$.

4. The word "per" usually indicates <u>division</u>.

Exercise Set 2.5

1. The reciprocal of $\dfrac{4}{7}$ is $\dfrac{7}{4}$.

3. The reciprocal of $\dfrac{1}{11}$ is $\dfrac{11}{1}$ or 11.

5. The reciprocal of $15 = \dfrac{15}{1}$ is $\dfrac{1}{15}$.

7. The reciprocal of $\dfrac{12}{7}$ is $\dfrac{7}{12}$.

9. $\dfrac{2}{3} \div \dfrac{5}{6} = \dfrac{2}{3} \cdot \dfrac{6}{5} = \dfrac{2 \cdot 6}{3 \cdot 5} = \dfrac{2 \cdot 3 \cdot 2}{3 \cdot 5} = \dfrac{2 \cdot 2}{5} = \dfrac{4}{5}$

11. $\dfrac{8}{9} \div \dfrac{1}{2} = \dfrac{8}{9} \cdot \dfrac{2}{1} = \dfrac{8 \cdot 2}{9 \cdot 1} = \dfrac{16}{9}$ or $1\dfrac{7}{9}$

13. $\dfrac{3}{7} \div \dfrac{5}{6} = \dfrac{3}{7} \cdot \dfrac{6}{5} = \dfrac{3 \cdot 6}{7 \cdot 5} = \dfrac{18}{35}$

15. $\dfrac{3}{5} \div \dfrac{4}{5} = \dfrac{3}{5} \cdot \dfrac{5}{4} = \dfrac{3 \cdot 5}{5 \cdot 4} = \dfrac{3}{4}$

17. $\dfrac{1}{10} \div \dfrac{10}{1} = \dfrac{1}{10} \cdot \dfrac{1}{10} = \dfrac{1 \cdot 1}{10 \cdot 10} = \dfrac{1}{100}$

19. $\dfrac{7}{9} \div \dfrac{7}{3} = \dfrac{7}{9} \cdot \dfrac{3}{7} = \dfrac{7 \cdot 3}{9 \cdot 7} = \dfrac{7 \cdot 3}{3 \cdot 3 \cdot 7} = \dfrac{1}{3}$

21. $\dfrac{5}{8} \div \dfrac{3}{8} = \dfrac{5}{8} \cdot \dfrac{8}{3} = \dfrac{5 \cdot 8}{8 \cdot 3} = \dfrac{5}{3}$ or $1\dfrac{2}{3}$

23. $\dfrac{7}{45} \div \dfrac{4}{25} = \dfrac{7}{45} \cdot \dfrac{25}{4}$

$= \dfrac{7 \cdot 25}{45 \cdot 4}$

$= \dfrac{7 \cdot 5 \cdot 5}{9 \cdot 5 \cdot 4}$

$= \dfrac{7 \cdot 5}{9 \cdot 4}$

$= \dfrac{35}{36}$

25. $\dfrac{2}{37} \div \dfrac{1}{7} = \dfrac{2}{37} \cdot \dfrac{7}{1} = \dfrac{2 \cdot 7}{37 \cdot 1} = \dfrac{14}{37}$

27. $\dfrac{3}{25} \div \dfrac{27}{40} = \dfrac{3}{25} \cdot \dfrac{40}{27}$

$= \dfrac{3 \cdot 40}{25 \cdot 27}$

$= \dfrac{3 \cdot 5 \cdot 8}{5 \cdot 5 \cdot 3 \cdot 9}$

$= \dfrac{8}{5 \cdot 9}$

$= \dfrac{8}{45}$

29. $\dfrac{11}{12} \div \dfrac{11}{12} = \dfrac{11}{12} \cdot \dfrac{12}{11} = \dfrac{11 \cdot 12}{12 \cdot 11} = 1$

31. $\dfrac{8}{13} \div 0$ is undefined.

33. $0 \div \dfrac{7}{8} = 0 \cdot \dfrac{8}{7} = 0$

35. $\dfrac{25}{126} \div \dfrac{125}{441} = \dfrac{25}{126} \cdot \dfrac{441}{125}$

$= \dfrac{25 \cdot 441}{126 \cdot 125}$

$= \dfrac{5 \cdot 5 \cdot 7 \cdot 7 \cdot 9}{9 \cdot 2 \cdot 7 \cdot 5 \cdot 5 \cdot 5}$

$= \dfrac{7}{2 \cdot 5}$

$= \dfrac{7}{10}$

37.
$$\frac{2}{3} \div 4 = \frac{2}{3} \div \frac{4}{1}$$
$$= \frac{2}{3} \cdot \frac{1}{4}$$
$$= \frac{2 \cdot 1}{3 \cdot 4}$$
$$= \frac{2 \cdot 1}{3 \cdot 2 \cdot 2}$$
$$= \frac{1}{3 \cdot 2}$$
$$= \frac{1}{6}$$

39.
$$8 \div \frac{3}{5} = \frac{8}{1} \cdot \frac{5}{3} = \frac{8 \cdot 5}{1 \cdot 3} = \frac{40}{3} \text{ or } 13\frac{1}{3}$$

41.
$$2\frac{1}{2} \div \frac{1}{2} = \frac{5}{2} \div \frac{1}{2} = \frac{5}{2} \cdot \frac{2}{1} = \frac{5 \cdot 2}{2 \cdot 1} = \frac{5}{1} = 5$$

43.
$$\frac{5}{12} \div 2\frac{1}{3} = \frac{5}{12} \div \frac{7}{3}$$
$$= \frac{5}{12} \cdot \frac{3}{7}$$
$$= \frac{5 \cdot 3}{12 \cdot 7}$$
$$= \frac{5 \cdot 3}{3 \cdot 4 \cdot 7}$$
$$= \frac{5}{4 \cdot 7}$$
$$= \frac{5}{28}$$

45.
$$3\frac{3}{7} \div 3\frac{1}{3} = \frac{24}{7} \div \frac{10}{3}$$
$$= \frac{24}{7} \cdot \frac{3}{10}$$
$$= \frac{24 \cdot 3}{7 \cdot 10}$$
$$= \frac{2 \cdot 12 \cdot 3}{7 \cdot 2 \cdot 5}$$
$$= \frac{12 \cdot 3}{7 \cdot 5}$$
$$= \frac{36}{35} \text{ or } 1\frac{1}{35}$$

47.
$$1\frac{4}{9} \div 2\frac{5}{6} = \frac{13}{9} \div \frac{17}{6}$$
$$= \frac{13}{9} \cdot \frac{6}{17}$$
$$= \frac{13 \cdot 6}{9 \cdot 17}$$
$$= \frac{13 \cdot 2 \cdot 3}{3 \cdot 3 \cdot 17}$$
$$= \frac{13 \cdot 2}{3 \cdot 17}$$
$$= \frac{26}{51}$$

49.
$$0 \div 15\frac{4}{7} = 0 \div \frac{109}{7} = 0 \cdot \frac{7}{109} = 0$$

51.
$$1 \div \frac{13}{17} = \frac{1}{1} \div \frac{13}{17} = \frac{1}{1} \cdot \frac{17}{13} = \frac{1 \cdot 17}{1 \cdot 13} = \frac{17}{13} \text{ or } 1\frac{4}{13}$$

53.
$$1 \div \frac{18}{35} = 1 \cdot \frac{35}{18} = \frac{35}{18} \text{ or } 1\frac{17}{18}$$

55.
$$10\frac{5}{9} \div 16\frac{2}{3} = \frac{95}{9} \div \frac{50}{3}$$
$$= \frac{95}{9} \cdot \frac{3}{50}$$
$$= \frac{95 \cdot 3}{9 \cdot 50}$$
$$= \frac{5 \cdot 19 \cdot 3}{3 \cdot 3 \cdot 5 \cdot 10}$$
$$= \frac{19}{3 \cdot 10}$$
$$= \frac{19}{30}$$

57.
$$\frac{6}{15} \div \frac{12}{5} = \frac{6}{15} \cdot \frac{5}{12}$$
$$= \frac{6 \cdot 5}{15 \cdot 12}$$
$$= \frac{6 \cdot 5}{5 \cdot 3 \cdot 6 \cdot 2}$$
$$= \frac{1}{3 \cdot 2}$$
$$= \frac{1}{6}$$

59.
$$\frac{11}{20} \div \frac{3}{11} = \frac{11}{20} \cdot \frac{11}{3} = \frac{11 \cdot 11}{20 \cdot 3} = \frac{121}{60} \text{ or } 2\frac{1}{60}$$

61. $12 \div \dfrac{1}{8} = 12 \cdot \dfrac{8}{1} = 12 \cdot 8 = 96$

63. $\dfrac{3}{7} \div \dfrac{4}{7} = \dfrac{3}{7} \cdot \dfrac{7}{4} = \dfrac{3 \cdot 7}{7 \cdot 4} = \dfrac{3}{4}$

65. $2\dfrac{3}{8} \div 0$ is undefined.

67. $\dfrac{11}{85} \div \dfrac{7}{5} = \dfrac{11}{85} \cdot \dfrac{5}{7}$

$\qquad = \dfrac{11 \cdot 5}{85 \cdot 7}$

$\qquad = \dfrac{11 \cdot 5}{5 \cdot 17 \cdot 7}$

$\qquad = \dfrac{11}{17 \cdot 7}$

$\qquad = \dfrac{11}{119}$

69. $4\dfrac{5}{11} \div 1\dfrac{2}{5} = \dfrac{49}{11} \div \dfrac{7}{5}$

$\qquad = \dfrac{49}{11} \cdot \dfrac{5}{7}$

$\qquad = \dfrac{49 \cdot 5}{11 \cdot 7}$

$\qquad = \dfrac{7 \cdot 7 \cdot 5}{11 \cdot 7}$

$\qquad = \dfrac{7 \cdot 5}{11}$

$\qquad = \dfrac{35}{11}$ or $3\dfrac{2}{11}$

71. $\dfrac{27}{100} \div \dfrac{3}{20} = \dfrac{27}{100} \cdot \dfrac{20}{3}$

$\qquad = \dfrac{27 \cdot 20}{100 \cdot 3}$

$\qquad = \dfrac{3 \cdot 9 \cdot 20}{20 \cdot 5 \cdot 3}$

$\qquad = \dfrac{9}{5}$ or $1\dfrac{4}{5}$

73. $12\dfrac{3}{4} \div 4 = \dfrac{51}{4} \div \dfrac{4}{1} = \dfrac{51}{4} \cdot \dfrac{1}{4} = \dfrac{51}{16}$ or $3\dfrac{3}{16}$

The patient walked $3\dfrac{3}{16}$ miles per day.

75. $3\dfrac{1}{3} \div 4 = \dfrac{10}{3} \div \dfrac{4}{1}$

$\qquad = \dfrac{10}{3} \cdot \dfrac{1}{4}$

$\qquad = \dfrac{10 \cdot 1}{3 \cdot 4}$

$\qquad = \dfrac{2 \cdot 5 \cdot 1}{3 \cdot 2 \cdot 2}$

$\qquad = \dfrac{5 \cdot 1}{3 \cdot 2}$

$\qquad = \dfrac{5}{6}$

Each dose should be $\dfrac{5}{6}$ tablespoon.

77. $15\dfrac{1}{5} \div 24 = \dfrac{76}{5} \div \dfrac{24}{1}$

$\qquad = \dfrac{76}{5} \cdot \dfrac{1}{24}$

$\qquad = \dfrac{76 \cdot 1}{5 \cdot 24}$

$\qquad = \dfrac{4 \cdot 19 \cdot 1}{5 \cdot 4 \cdot 6}$

$\qquad = \dfrac{19 \cdot 1}{5 \cdot 6}$

$\qquad = \dfrac{19}{30}$

On average, $\dfrac{19}{30}$ inch of rain fell per hour.

79. $1379 \div 98\dfrac{1}{2} = \dfrac{1379}{1} \div \dfrac{197}{2}$

$\qquad = \dfrac{1379}{1} \cdot \dfrac{2}{197}$

$\qquad = \dfrac{1379 \cdot 2}{1 \cdot 197}$

$\qquad = \dfrac{197 \cdot 7 \cdot 2}{1 \cdot 197}$

$\qquad = \dfrac{7 \cdot 2}{1}$

$\qquad = 14$

The family recycled 14 pounds of aluminum cans.

81. $12 \div 2\dfrac{4}{7} = \dfrac{12}{1} \div \dfrac{18}{7}$

$= \dfrac{12}{1} \cdot \dfrac{7}{18}$

$= \dfrac{12 \cdot 7}{1 \cdot 18}$

$= \dfrac{6 \cdot 2 \cdot 7}{1 \cdot 6 \cdot 3}$

$= \dfrac{2 \cdot 7}{1 \cdot 3}$

$= \dfrac{14}{3}$ or $4\dfrac{2}{3}$

The length of the rectangle is $4\dfrac{2}{3}$ meters.

83. $\dfrac{2}{5} \cdot \dfrac{4}{7} = \dfrac{2 \cdot 4}{5 \cdot 7} = \dfrac{8}{35}$

85. $2\dfrac{2}{3} \div 1\dfrac{1}{16} = \dfrac{8}{3} \div \dfrac{17}{16}$

$= \dfrac{8}{3} \cdot \dfrac{16}{17}$

$= \dfrac{8 \cdot 16}{3 \cdot 17}$

$= \dfrac{128}{51}$ or $2\dfrac{26}{51}$

87. $5\dfrac{1}{7} \cdot \dfrac{2}{9} \cdot \dfrac{14}{15} = \dfrac{36}{7} \cdot \dfrac{2}{9} \cdot \dfrac{14}{15}$

$= \dfrac{36 \cdot 2 \cdot 14}{7 \cdot 9 \cdot 15}$

$= \dfrac{9 \cdot 4 \cdot 2 \cdot 2 \cdot 7}{7 \cdot 9 \cdot 15}$

$= \dfrac{4 \cdot 2 \cdot 2}{15}$

$= \dfrac{16}{15}$ or $1\dfrac{1}{15}$

89. $\dfrac{11}{20} \div \dfrac{20}{11} = \dfrac{11}{20} \cdot \dfrac{11}{20} = \dfrac{11 \cdot 11}{20 \cdot 20} = \dfrac{121}{400}$

91.
$$\begin{array}{r} \overset{22}{27} \\ 76 \\ +\,98 \\ \hline 201 \end{array}$$

93.
$$\begin{array}{r} 968 \\ -\,772 \\ \hline 196 \end{array}$$

95.
$$\begin{array}{r} 2000 \\ -\,431 \\ \hline 1569 \end{array}$$

97. $20\dfrac{2}{3} \div 10\dfrac{1}{2} = \dfrac{62}{3} \div \dfrac{21}{2}$

$= \dfrac{62}{3} \cdot \dfrac{2}{21}$

$= \dfrac{62 \cdot 2}{3 \cdot 21}$

$= \dfrac{124}{63}$ or $1\dfrac{61}{63}$

99. $20\dfrac{1}{4}$ rounds to 20

$\dfrac{5}{6}$ rounds to 1

$20 \div 1 = 20$
The best estimate is c.

101. $12\dfrac{2}{13}$ rounds to 12.

$3\dfrac{7}{8}$ rounds to 4.

$12 \div 4 = 3$
The best estimate is d.

103. $\dfrac{42}{25} \cdot \dfrac{125}{36} \div \dfrac{7}{6} = \dfrac{42}{25} \cdot \dfrac{125}{36} \cdot \dfrac{6}{7}$

$= \dfrac{42 \cdot 125 \cdot 6}{25 \cdot 36 \cdot 7}$

$= \dfrac{6 \cdot 7 \cdot 5 \cdot 25 \cdot 6}{25 \cdot 6 \cdot 6 \cdot 7}$

$= \dfrac{5}{1}$

$= 5$

105. $252 \div \dfrac{42}{109} = \dfrac{252}{1} \cdot \dfrac{109}{42}$

$= \dfrac{252 \cdot 109}{1 \cdot 42}$

$= \dfrac{6 \cdot 42 \cdot 109}{1 \cdot 42}$

$= \dfrac{6 \cdot 109}{1}$

$= 654$

654 aircraft make up the entire FedEx Express air fleet.

107. answers may vary

Chapter 2 Vocabulary Check

1. Two numbers are <u>reciprocals</u> of each other if their product is 1.

2. A <u>composite number</u> is a natural number greater than 1 that is not prime.

3. Fractions that represent the same portion of a whole are called <u>equivalent</u> fractions.

4. An <u>improper fraction</u> is a fraction whose numerator is greater than or equal to its denominator.

5. A <u>prime number</u> is a natural number greater than 1 whose only factors are 1 and itself.

6. A fraction is in <u>simplest form</u> when the numerator and the denominator have no factors in common other than 1.

7. A <u>proper fraction</u> is one whose numerator is less than its denominator.

8. A <u>mixed number</u> contains a whole number part and a fraction part.

9. In the fraction $\dfrac{7}{9}$, the 7 is called the <u>numerator</u> and the 9 is called the <u>denominator</u>.

10. The <u>prime factorization</u> of a number is the factorization in which all the factors are prime numbers.

11. The fraction $\dfrac{3}{0}$ is <u>undefined</u>.

12. The fraction $\dfrac{0}{5} = \underline{0}$.

13. In $\dfrac{a}{b} = \dfrac{c}{d}$, $a \cdot d$ and $b \cdot c$ are called <u>cross products</u>.

Chapter 2 Review

1. $\dfrac{11}{23}$ is a proper fraction.

2. $\dfrac{9}{8}$ is an improper fraction.

3. $\dfrac{1}{2}$ is a proper fraction.

4. $2\dfrac{1}{4}$ is a mixed number.

5. 2 of the 6 equal parts are shaded: $\dfrac{2}{6}$

6. 4 of the 7 equal parts are shaded: $\dfrac{4}{7}$

7. Each part is $\dfrac{1}{3}$ of a whole and 7 parts are shaded: $\dfrac{7}{3}$

8. Each part is $\dfrac{1}{4}$ of a whole and 13 parts are shaded: $\dfrac{13}{4}$

9. $\begin{array}{ll} \text{free throws made} & \rightarrow \quad 11 \\ \text{free throws during game} & \rightarrow \quad \overline{12} \end{array}$

    The player made $\dfrac{11}{12}$ of his free throws.

10. **a.** $131 - 23 = 108$
    108 cars on the lot are not blue.

    b. There are 131 cars, of which 108 are not blue. $\dfrac{108}{131}$ of the cars are not blue.

11. $\begin{array}{r} 3\ \text{R}\ 3 \\ 4{\overline{\smash{)}\,15}} \\ \underline{-12} \\ 3 \end{array}$

    $\dfrac{15}{4} = 3\dfrac{3}{4}$

12. $\begin{array}{r} 45\ \text{R}\ 5 \\ 6{\overline{\smash{)}\,275}} \\ \underline{-24} \\ 35 \\ \underline{-30} \\ 5 \end{array}$

    $\dfrac{275}{6} = 45\dfrac{5}{6}$

13.

$$13\overline{)\begin{array}{r}3\\39\\-39\\\hline 0\end{array}}$$

$$\frac{39}{13}=3$$

14.

$$12\overline{)\begin{array}{r}5\\60\\-60\\\hline 0\end{array}}$$

$$\frac{60}{12}=5$$

15. $1\frac{1}{5}=\frac{5\cdot1+1}{5}=\frac{6}{5}$

16. $1\frac{1}{21}=\frac{21\cdot1+1}{21}=\frac{22}{21}$

17. $2\frac{8}{9}=\frac{9\cdot2+8}{9}=\frac{26}{9}$

18. $3\frac{11}{12}=\frac{12\cdot3+11}{12}=\frac{47}{12}$

19. Composite, since the factors of 51 are 1, 3, 17, and 51.

20. Prime, since the only factors of 17 are 1 and 17.

21. $1\cdot42=42$
$2\cdot21=42$
$3\cdot14=42$
$6\cdot7=42$
The factors of 42 are 1, 2, 3, 6, 7, 14, 21, and 42.

22. $1\cdot20=20$
$2\cdot10=20$
$4\cdot5=20$
The factors of 20 are 1, 2, 4, 5, 10, and 20.

23.

$$\begin{array}{r}17\\2\overline{)34}\\2\overline{)68}\end{array}$$

$$68=2^2\cdot17$$

24.

$$\begin{array}{r}5\\3\overline{)15}\\3\overline{)45}\\2\overline{)90}\end{array}$$

$$90=2\cdot3^2\cdot5$$

25.

$$5\overline{)\begin{array}{r}157\\785\end{array}}$$

$$785=5\cdot157$$

26.

$$\begin{array}{r}17\\5\overline{)85}\\3\overline{)255}\end{array}$$

$$255=3\cdot5\cdot17$$

27. $\frac{12}{28}=\frac{3\cdot4}{7\cdot4}=\frac{3}{7}$

28. $\frac{15}{27}=\frac{3\cdot5}{3\cdot9}=\frac{5}{9}$

29. $\frac{25}{75}=\frac{25\cdot1}{25\cdot3}=\frac{1}{3}$

30. $\frac{36}{72}=\frac{36\cdot1}{36\cdot2}=\frac{1}{2}$

31. $\frac{29}{32}=\frac{29}{32}$
29 and 32 have no common factors other than 1.

32. $\frac{18}{23}=\frac{18}{23}$
18 and 23 have no common factors other than 1.

33. $\frac{48}{6}=\frac{6\cdot8}{6\cdot1}=\frac{8}{1}=8$

34. $\frac{54}{9}=\frac{6\cdot9}{1\cdot9}=\frac{6}{1}=6$

35. $\frac{8\text{ inches}}{12\text{ inches}}=\frac{8}{12}=\frac{4\cdot2}{4\cdot3}=\frac{2}{3}$

8 inches represents $\frac{2}{3}$ of a foot.

36. $15 - 6 = 9$ cars are not white.

$$\frac{9 \text{ non-white cars}}{15 \text{ total cars}} = \frac{9}{15} = \frac{3 \cdot 3}{3 \cdot 5} = \frac{3}{5}$$

$\frac{3}{5}$ of the cars are not white.

37. Not equivalent, since the cross products are not equal: $34 \cdot 4 = 136$ and $14 \cdot 10 = 140$

38. Equivalent, since the cross products are equal: $50 \cdot 9 = 450$ and $15 \cdot 30 = 450$

39. $\dfrac{3}{5} \cdot \dfrac{1}{2} = \dfrac{3 \cdot 1}{5 \cdot 2} = \dfrac{3}{10}$

40. $\dfrac{6}{7} \cdot \dfrac{5}{12} = \dfrac{6 \cdot 5}{7 \cdot 12} = \dfrac{6 \cdot 5}{7 \cdot 6 \cdot 2} = \dfrac{5}{7 \cdot 2} = \dfrac{5}{14}$

41. $\dfrac{24}{5} \cdot \dfrac{15}{8} = \dfrac{24 \cdot 15}{5 \cdot 8} = \dfrac{3 \cdot 8 \cdot 3 \cdot 5}{5 \cdot 8} = \dfrac{3 \cdot 3}{1} = 9$

42. $\dfrac{27}{21} \cdot \dfrac{7}{18} = \dfrac{27 \cdot 7}{21 \cdot 18} = \dfrac{9 \cdot 3 \cdot 7}{7 \cdot 3 \cdot 9 \cdot 2} = \dfrac{1}{2}$

43. $5 \cdot \dfrac{7}{8} = \dfrac{5}{1} \cdot \dfrac{7}{8} = \dfrac{5 \cdot 7}{1 \cdot 8} = \dfrac{35}{8}$ or $4\dfrac{3}{8}$

44. $6 \cdot \dfrac{5}{12} = \dfrac{6}{1} \cdot \dfrac{5}{12} = \dfrac{6 \cdot 5}{1 \cdot 12} = \dfrac{6 \cdot 5}{1 \cdot 6 \cdot 2} = \dfrac{5}{1 \cdot 2} = \dfrac{5}{2}$ or $2\dfrac{1}{2}$

45. $\dfrac{39}{3} \cdot \dfrac{7}{13} \cdot \dfrac{5}{21} = \dfrac{39 \cdot 7 \cdot 5}{3 \cdot 13 \cdot 21} = \dfrac{3 \cdot 13 \cdot 7 \cdot 5}{3 \cdot 13 \cdot 7 \cdot 3} = \dfrac{5}{3}$ or $1\dfrac{2}{3}$

46. $\dfrac{42}{5} \cdot \dfrac{15}{6} \cdot \dfrac{7}{9} = \dfrac{42 \cdot 15 \cdot 7}{5 \cdot 6 \cdot 9}$

$$= \dfrac{6 \cdot 7 \cdot 3 \cdot 5 \cdot 7}{5 \cdot 6 \cdot 3 \cdot 3}$$

$$= \dfrac{7 \cdot 7}{3}$$

$$= \dfrac{49}{3} \text{ or } 16\dfrac{1}{3}$$

47. Exact: $1\dfrac{5}{8} \cdot 3\dfrac{1}{5} = \dfrac{13}{8} \cdot \dfrac{16}{5}$

$$= \dfrac{13 \cdot 16}{8 \cdot 5}$$

$$= \dfrac{13 \cdot 8 \cdot 2}{8 \cdot 5}$$

$$= \dfrac{13 \cdot 2}{5}$$

$$= \dfrac{26}{5} \text{ or } 5\dfrac{1}{5}$$

Estimate: $1\dfrac{5}{8}$ rounds to 2, $3\dfrac{1}{5}$ rounds to 3.

$2 \cdot 3 = 6$

48. Exact: $3\dfrac{6}{11} \cdot 1\dfrac{7}{13} = \dfrac{39}{11} \cdot \dfrac{20}{13}$

$$= \dfrac{39 \cdot 20}{11 \cdot 13}$$

$$= \dfrac{13 \cdot 3 \cdot 20}{11 \cdot 13}$$

$$= \dfrac{3 \cdot 20}{11}$$

$$= \dfrac{60}{11} \text{ or } 5\dfrac{5}{11}$$

Estimate: $3\dfrac{6}{11}$ rounds to 4, $1\dfrac{7}{13}$ rounds to 2.

$4 \cdot 2 = 8$

49. $\dfrac{3}{4} \cdot 8 \cdot 4\dfrac{1}{8} = \dfrac{3}{4} \cdot \dfrac{8}{1} \cdot \dfrac{33}{8}$

$$= \dfrac{3 \cdot 8 \cdot 33}{4 \cdot 1 \cdot 8}$$

$$= \dfrac{3 \cdot 33}{4 \cdot 1}$$

$$= \dfrac{99}{4} \text{ or } 24\dfrac{3}{4}$$

50. $2\dfrac{1}{9} \cdot 3 \cdot \dfrac{1}{38} = \dfrac{19}{9} \cdot \dfrac{3}{1} \cdot \dfrac{1}{38}$

$$= \dfrac{19 \cdot 3 \cdot 1}{9 \cdot 1 \cdot 38}$$

$$= \dfrac{19 \cdot 3 \cdot 1}{3 \cdot 3 \cdot 1 \cdot 2 \cdot 19}$$

$$= \dfrac{1}{3 \cdot 1 \cdot 2}$$

$$= \dfrac{1}{6}$$

51. $5 \cdot 7\frac{1}{3} = \frac{5}{1} \cdot \frac{22}{3} = \frac{5 \cdot 22}{1 \cdot 3} = \frac{110}{3}$ or $36\frac{2}{3}$

A 5-ounce hamburger patty has $\frac{110}{3}$ or $36\frac{2}{3}$ grams of fat.

52. $45 \cdot \frac{3}{4} = \frac{45}{1} \cdot \frac{3}{4} = \frac{45 \cdot 3}{1 \cdot 4} = \frac{135}{4}$ or $33\frac{3}{4}$

The art teacher needs $\frac{135}{4}$ or $33\frac{3}{4}$ inches of piping.

53. $\frac{7}{10} \cdot 2\frac{1}{8} = \frac{7}{10} \cdot \frac{17}{8} = \frac{7 \cdot 17}{10 \cdot 8} = \frac{119}{80}$ or $1\frac{39}{80}$

The area is $\frac{119}{80}$ or $1\frac{39}{80}$ square inches.

54. $6\frac{7}{8} \cdot 5 = \frac{55}{8} \cdot \frac{5}{1} = \frac{55 \cdot 5}{8 \cdot 1} = \frac{275}{8}$ or $34\frac{3}{8}$

The area is $\frac{275}{8}$ or $34\frac{3}{8}$ square meters.

55. The reciprocal of 7, or $\frac{7}{1}$, is $\frac{1}{7}$.

56. The reciprocal of $\frac{1}{8}$ is $\frac{8}{1}$ or 8.

57. The reciprocal of $\frac{14}{23}$ is $\frac{23}{14}$.

58. The reciprocal of $\frac{17}{5}$ is $\frac{5}{17}$.

59. $\frac{3}{4} \div \frac{3}{8} = \frac{3}{4} \cdot \frac{8}{3} = \frac{3 \cdot 8}{4 \cdot 3} = \frac{3 \cdot 4 \cdot 2}{4 \cdot 3} = \frac{2}{1} = 2$

60. $\frac{21}{4} \div \frac{7}{5} = \frac{21}{4} \cdot \frac{5}{7}$

$= \frac{21 \cdot 5}{4 \cdot 7}$

$= \frac{3 \cdot 7 \cdot 5}{4 \cdot 7}$

$= \frac{3 \cdot 5}{4}$

$= \frac{15}{4}$ or $3\frac{3}{4}$

61. $\frac{5}{3} \div 2 = \frac{5}{3} \div \frac{2}{1} = \frac{5}{3} \cdot \frac{1}{2} = \frac{5 \cdot 1}{3 \cdot 2} = \frac{5}{6}$

62. $5 \div \frac{15}{8} = \frac{5}{1} \cdot \frac{8}{15} = \frac{5 \cdot 8}{1 \cdot 15} = \frac{5 \cdot 8}{1 \cdot 5 \cdot 3} = \frac{8}{1 \cdot 3} = \frac{8}{3}$ or $2\frac{2}{3}$

63. $6\frac{3}{4} \div 1\frac{2}{7} = \frac{27}{4} \div \frac{9}{7}$

$= \frac{27}{4} \cdot \frac{7}{9}$

$= \frac{27 \cdot 7}{4 \cdot 9}$

$= \frac{3 \cdot 9 \cdot 7}{4 \cdot 9}$

$= \frac{3 \cdot 7}{4}$

$= \frac{21}{4}$ or $5\frac{1}{4}$

64. $5\frac{1}{2} \div 2\frac{1}{11} = \frac{11}{2} \div \frac{23}{11}$

$= \frac{11}{2} \cdot \frac{11}{23}$

$= \frac{11 \cdot 11}{2 \cdot 23}$

$= \frac{121}{46}$ or $2\frac{29}{46}$

65. $341 \div 15\frac{1}{2} = \frac{341}{1} \div \frac{31}{2}$

$= \frac{341}{1} \cdot \frac{2}{31}$

$= \frac{341 \cdot 2}{1 \cdot 31}$

$= \frac{11 \cdot 31 \cdot 2}{1 \cdot 31}$

$= \frac{11 \cdot 2}{1}$

$= 22$

We might expect the truck to travel 22 miles on 1 gallon of gas.

66. $5\frac{1}{4} \div 5 = \frac{21}{4} \div \frac{5}{1} = \frac{21}{4} \cdot \frac{1}{5} = \frac{21 \cdot 1}{4 \cdot 5} = \frac{21}{20}$ or $1\frac{1}{20}$

He walks $\frac{21}{20}$ or $1\frac{1}{20}$ miles each day.

67. $\frac{0}{3}$ is a proper fraction.

68. $\frac{12}{12}$ is an improper fraction.

69. $5\frac{6}{7}$ is a mixed number.

70. $\frac{13}{9}$ is an improper fraction.

71.

$$4\overline{)125} \quad 31\text{ R }1$$
$$\underline{-12}$$
$$05$$
$$\underline{-4}$$
$$1$$

$$\frac{125}{4} = 31\frac{1}{4}$$

72.

$$9\overline{)54} \quad 6$$
$$\underline{-54}$$
$$0$$

$$\frac{54}{9} = 6$$

73. $5\frac{10}{17} = \frac{17 \cdot 5 + 10}{17} = \frac{95}{17}$

74. $7\frac{5}{6} = \frac{6 \cdot 7 + 5}{6} = \frac{47}{6}$

75. Composite, since the factors of 27 are 1, 3, 9, and 27.

76. Prime, since the only factors of 23 are 1 and 23.

77.

$$3\overline{)15} \quad 5$$
$$3\overline{)45}$$
$$2\overline{)90}$$
$$2\overline{)180}$$

$$180 = 2^2 \cdot 3^2 \cdot 5$$

78.

$$7\overline{)49} \quad 7$$
$$2\overline{)98}$$

$$90 = 2 \cdot 7^2$$

79. $\frac{45}{50} = \frac{9 \cdot 5}{10 \cdot 5} = \frac{9}{10}$

80. $\frac{30}{42} = \frac{6 \cdot 5}{6 \cdot 7} = \frac{5}{7}$

81. $\frac{140}{150} = \frac{14 \cdot 10}{15 \cdot 10} = \frac{14}{15}$

82. $\frac{84}{140} = \frac{28 \cdot 3}{28 \cdot 5} = \frac{3}{5}$

83. $\frac{7}{8} \cdot \frac{2}{3} = \frac{7 \cdot 2}{8 \cdot 3} = \frac{7 \cdot 2}{4 \cdot 2 \cdot 3} = \frac{7}{4 \cdot 3} = \frac{7}{12}$

84. $\frac{6}{15} \cdot \frac{5}{8} = \frac{6 \cdot 5}{15 \cdot 8} = \frac{2 \cdot 3 \cdot 5}{3 \cdot 5 \cdot 2 \cdot 4} = \frac{1}{4}$

85. $\frac{18}{5} \div \frac{2}{5} = \frac{18}{5} \cdot \frac{5}{2} = \frac{18 \cdot 5}{5 \cdot 2} = \frac{2 \cdot 9 \cdot 5}{5 \cdot 2} = \frac{9}{1} = 9$

86. $\frac{9}{2} \div \frac{1}{3} = \frac{9}{2} \cdot \frac{3}{1} = \frac{9 \cdot 3}{2 \cdot 1} = \frac{27}{2}$ or $13\frac{1}{2}$

87. Exact: $4\frac{1}{6} \cdot 2\frac{2}{5} = \frac{25}{6} \cdot \frac{12}{5}$

$$= \frac{25 \cdot 12}{6 \cdot 5}$$

$$= \frac{5 \cdot 5 \cdot 6 \cdot 2}{6 \cdot 5}$$

$$= \frac{5 \cdot 2}{1}$$

$$= 10$$

Estimate: $4\frac{1}{6}$ rounds to 4

$2\frac{2}{5}$ rounds to 2

$4 \cdot 2 = 8$

88. Exact: $5\dfrac{2}{3} \cdot 2\dfrac{1}{4} = \dfrac{17}{3} \cdot \dfrac{9}{4}$

$$= \dfrac{17 \cdot 9}{3 \cdot 4}$$

$$= \dfrac{17 \cdot 3 \cdot 3}{3 \cdot 4}$$

$$= \dfrac{17 \cdot 3}{4}$$

$$= \dfrac{51}{4} \text{ or } 12\dfrac{3}{4}$$

Estimate: $5\dfrac{2}{3}$ rounds to 6

$\quad\quad\quad\quad 2\dfrac{1}{4}$ rounds to 2

$\quad\quad\quad\quad 6 \cdot 2 = 12$

89. $\dfrac{7}{2} \div 1\dfrac{1}{2} = \dfrac{7}{2} \div \dfrac{3}{2} = \dfrac{7}{2} \cdot \dfrac{2}{3} = \dfrac{7 \cdot 2}{2 \cdot 3} = \dfrac{7}{3} \text{ or } 2\dfrac{1}{3}$

90. $1\dfrac{3}{5} \div \dfrac{1}{4} = \dfrac{8}{5} \cdot \dfrac{4}{1} = \dfrac{8 \cdot 4}{5 \cdot 1} = \dfrac{32}{5} \text{ or } 6\dfrac{2}{5}$

91. $5\dfrac{1}{2} \cdot 7\dfrac{4}{11} = \dfrac{11}{2} \cdot \dfrac{81}{11} = \dfrac{11 \cdot 81}{2 \cdot 11} = \dfrac{81}{2} \text{ or } 40\dfrac{1}{2}$

The area is $\dfrac{81}{2}$ or $40\dfrac{1}{2}$ square feet.

92. $23\dfrac{1}{2} \div 30\dfrac{1}{2} = \dfrac{47}{2} \div \dfrac{61}{2} = \dfrac{47}{2} \cdot \dfrac{2}{61} = \dfrac{47 \cdot 2}{2 \cdot 61} = \dfrac{47}{61}$

This is $\dfrac{47}{61}$ inch of rain per 1 hour.

Chapter 2 Test

1. 7 of the 16 equal parts are shaded: $\dfrac{7}{16}$

2. Each part is $\dfrac{1}{5}$ of a whole and 13 parts are

shaded: $\dfrac{13}{5}$

3. $7\dfrac{2}{3} = \dfrac{3 \cdot 7 + 2}{3} = \dfrac{23}{3}$

4. $3\dfrac{6}{11} = \dfrac{11 \cdot 3 + 6}{11} = \dfrac{39}{11}$

5. $\quad\begin{array}{r} 4 \text{ R } 3 \\ 5\overline{)23} \\ \underline{20} \\ 3 \end{array}$

$\dfrac{23}{5} = 4\dfrac{3}{5}$

6. $\quad\begin{array}{r} 18 \text{ R } 3 \\ 4\overline{)75} \\ \underline{-4} \\ 35 \\ \underline{-32} \\ 3 \end{array}$

$\dfrac{75}{4} = 18\dfrac{3}{4}$

7. $\dfrac{24}{210} = \dfrac{6 \cdot 4}{6 \cdot 35} = \dfrac{4}{35}$

8. $\dfrac{42}{70} = \dfrac{14 \cdot 3}{14 \cdot 5} = \dfrac{3}{5}$

9. Not equivalent, since the cross products are not equal: $7 \cdot 8 = 56$ and $11 \cdot 5 = 55$.

10. Equivalent, since the cross products are equal: $27 \cdot 14 = 378$ and $63 \cdot 6 = 378$.

11. $\begin{array}{r} 7 \\ 3\overline{)21} \\ 2\overline{)42} \\ 2\overline{)84} \end{array}$

$84 = 2^2 \cdot 3 \cdot 7$

12. $\begin{array}{r} 11 \\ 5\overline{)55} \\ 3\overline{)165} \\ 3\overline{)495} \end{array}$

$495 = 3^2 \cdot 5 \cdot 11$

13. $\dfrac{4}{4} \div \dfrac{3}{4} = \dfrac{4}{4} \cdot \dfrac{4}{3} = \dfrac{4 \cdot 4}{4 \cdot 3} = \dfrac{4}{3} \text{ or } 1\dfrac{1}{3}$

14. $\dfrac{4}{3} \cdot \dfrac{4}{4} = \dfrac{4 \cdot 4}{3 \cdot 4} = \dfrac{4}{3} \text{ or } 1\dfrac{1}{3}$

15. $2 \cdot \dfrac{1}{8} = \dfrac{2}{1} \cdot \dfrac{1}{8} = \dfrac{2 \cdot 1}{1 \cdot 8} = \dfrac{2 \cdot 1}{1 \cdot 2 \cdot 4} = \dfrac{1}{4}$

16. $\dfrac{2}{3} \cdot \dfrac{8}{15} = \dfrac{2 \cdot 8}{3 \cdot 15} = \dfrac{16}{45}$

17. $8 \div \dfrac{1}{2} = \dfrac{8}{1} \cdot \dfrac{2}{1} = \dfrac{8 \cdot 2}{1 \cdot 1} = 16$

18. $13\dfrac{1}{2} \div 3 = \dfrac{27}{2} \div \dfrac{3}{1}$

$= \dfrac{27}{2} \cdot \dfrac{1}{3}$

$= \dfrac{27 \cdot 1}{2 \cdot 3}$

$= \dfrac{3 \cdot 9 \cdot 1}{2 \cdot 3}$

$= \dfrac{9}{2}$ or $4\dfrac{1}{2}$

19. $\dfrac{3}{8} \cdot \dfrac{16}{6} \cdot \dfrac{4}{11} = \dfrac{3 \cdot 16 \cdot 4}{8 \cdot 6 \cdot 11} = \dfrac{3 \cdot 2 \cdot 8 \cdot 4}{8 \cdot 2 \cdot 3 \cdot 11} = \dfrac{4}{11}$

20. $5\dfrac{1}{4} \div \dfrac{7}{12} = \dfrac{21}{4} \cdot \dfrac{12}{7} = \dfrac{21 \cdot 12}{4 \cdot 7} = \dfrac{3 \cdot 7 \cdot 3 \cdot 4}{4 \cdot 7} = \dfrac{3 \cdot 3}{1} = 9$

21. $\dfrac{16}{3} \div \dfrac{3}{12} = \dfrac{16}{3} \cdot \dfrac{12}{3}$

$= \dfrac{16 \cdot 12}{3 \cdot 3}$

$= \dfrac{16 \cdot 3 \cdot 4}{3 \cdot 3}$

$= \dfrac{16 \cdot 4}{3}$

$= \dfrac{64}{3}$ or $21\dfrac{1}{3}$

22. $3\dfrac{1}{3} \cdot 6\dfrac{3}{4} = \dfrac{10}{3} \cdot \dfrac{27}{4}$

$= \dfrac{10 \cdot 27}{3 \cdot 4}$

$= \dfrac{2 \cdot 5 \cdot 3 \cdot 9}{3 \cdot 2 \cdot 2}$

$= \dfrac{5 \cdot 9}{2}$

$= \dfrac{45}{2}$ or $22\dfrac{1}{2}$

23. $12 \div 3\dfrac{1}{3} = \dfrac{12}{1} \div \dfrac{10}{3}$

$= \dfrac{12}{1} \cdot \dfrac{3}{10}$

$= \dfrac{12 \cdot 3}{1 \cdot 10}$

$= \dfrac{2 \cdot 6 \cdot 3}{1 \cdot 2 \cdot 5}$

$= \dfrac{6 \cdot 3}{1 \cdot 5}$

$= \dfrac{18}{5}$ or $3\dfrac{3}{5}$

24. $\dfrac{14}{5} \cdot \dfrac{25}{21} \cdot 2 = \dfrac{14}{5} \cdot \dfrac{25}{21} \cdot \dfrac{2}{1}$

$= \dfrac{14 \cdot 25 \cdot 2}{5 \cdot 21 \cdot 1}$

$= \dfrac{2 \cdot 7 \cdot 5 \cdot 5 \cdot 2}{5 \cdot 3 \cdot 7 \cdot 1}$

$= \dfrac{2 \cdot 5 \cdot 2}{3 \cdot 1}$

$= \dfrac{20}{3}$ or $6\dfrac{2}{3}$

25. $\dfrac{2}{3} \cdot 1\dfrac{8}{9} = \dfrac{2}{3} \cdot \dfrac{17}{9} = \dfrac{2 \cdot 17}{3 \cdot 9} = \dfrac{34}{27}$ or $1\dfrac{7}{27}$

The area is $\dfrac{34}{27}$ or $1\dfrac{7}{27}$ square miles.

26. $258 \div 10\dfrac{3}{4} = \dfrac{258}{1} \div \dfrac{43}{4}$

$= \dfrac{258}{1} \cdot \dfrac{4}{43}$

$= \dfrac{258 \cdot 4}{1 \cdot 43}$

$= \dfrac{43 \cdot 6 \cdot 4}{1 \cdot 43}$

$= \dfrac{24}{1}$

$= 24$

We expect the car to travel 24 miles on 1 gallon of gas.

27. $100 \cdot 53\frac{1}{3} = \frac{100}{1} \cdot \frac{160}{3}$

$\qquad\qquad = \frac{100 \cdot 160}{1 \cdot 3}$

$\qquad\qquad = \frac{16,000}{3}$ or $5333\frac{1}{3}$

$\frac{16,000}{3}$ or $5333\frac{1}{3}$ square yards of artificial turf

are necessary to cover the football field.

28. $120 \cdot \frac{3}{4} = \frac{120}{1} \cdot \frac{3}{4} = \frac{120 \cdot 3}{1 \cdot 4} = \frac{4 \cdot 30 \cdot 3}{1 \cdot 4} = \frac{30 \cdot 3}{1} = 90$

The stock sold for $90 per share after the oil spill.

Cumulative Review Chapters 1–2

1. The place value of the 3 in 396,418 is hundred-thousands.

2. 2036 is written as two thousand, thirty-six.

3. Eight hundred five in standard form is 805.

4.
$$\begin{array}{r} \overset{2}{7} \\ 6 \\ 10 \\ 3 \\ + 5 \\ \hline 31 \end{array}$$

5.
$$\begin{array}{r} \overset{1\,1}{34,285} \!\!\!\!\overset{1}{} \\ + 149,761 \\ \hline 184,046 \end{array}$$

6.
$$\begin{array}{r} \overset{1}{56} \\ 18 \\ + 43 \\ \hline 117 \end{array}$$

$$\begin{array}{r} 39 \\ 3\overline{)117} \\ \underline{-9} \\ 27 \\ \underline{-27} \\ 0 \end{array}$$

The average is 39.

7.
$$\begin{array}{r} \overset{1}{2} \\ 3 \\ 1 \\ 3 \\ + 4 \\ \hline 13 \end{array}$$

The perimeter is 13 inches.

8.
$$\begin{array}{r} 25 \\ - 8 \\ \hline 17 \end{array}$$

9.
$$\begin{array}{r} 94,113 \\ + 4\,525 \\ \hline 98,638 \end{array}$$

The number of seats in the stadium for the 2009 season is 98,638.

10. $\sqrt{25} = 5$, since $5 \cdot 5 = 25$.

11.
$$\begin{array}{r} 7826 \\ - 505 \\ \hline 7321 \end{array}$$

Check:
$$\begin{array}{r} 7321 \\ + 505 \\ \hline 7826 \end{array}$$

12. $8^2 = 8 \cdot 8 = 64$

13. a. The country with the greatest number of endangered species is Indonesia.

   b. The number of endangered species for Australia is 64, the number of endangered species for China is 83, and the number of endangered species for India is 89.

$$\begin{array}{r} 64 \\ 83 \\ + 89 \\ \hline 236 \end{array}$$

The total number of endangered species for these three countries is 236.

14.
$$\begin{array}{r} 25 \text{ R } 5 \\ 8\overline{)205} \\ \underline{-16} \\ 45 \\ \underline{-40} \\ 5 \end{array}$$

$205 \div 8 = 25$ R 5

15. To round 568 to the nearest ten, observe that the digit in the ones place is 8. Since this digit is at least 5, we add 1 to the tens place. The number 568 rounded to the tens place is 570.

16. To round 2366 to the nearest hundred, observe that the digit in the tens place is 6. Since this digit is at least 5, we add 1 to the hundreds place. The number 2366 rounded to the nearest hundred is 2400.

17.

4725	rounds to	4700
− 2879	rounds to	− 2900
		1800

The estimated difference is 1800.

18.

		2
38	rounds to	40
43	rounds to	40
126	rounds to	130
+ 92	rounds to	+ 90
		300

The estimated sum is 300.

19. a. $6 \times 1 = 6$

b. $0(8) = 0$

c. $1 \cdot 45 = 45$

d. $(75)(0) = 0$

20. $30 \div 3 \cdot 2 = 10 \cdot 2 = 20$

21. a. $3(4 + 5) = 3 \cdot 4 + 3 \cdot 5$

b. $10(6 + 8) = 10 \cdot 6 + 10 \cdot 8$

c. $2(7 + 3) = 2 \cdot 7 + 2 \cdot 3$

22.

$$\begin{array}{r} 12 \\ \times 15 \\ \hline 60 \\ 120 \\ \hline 180 \end{array}$$

23. a.

$$\begin{array}{r} 0 \\ 9\overline{)\,0} \\ -0 \\ \hline 0 \end{array}$$

Check: $0 \cdot 9 = 0$

b. $0 \div 12 = 0$
Check: $0 \cdot 12 = 0$

c. $\dfrac{0}{5} = 0$
Check: $0 \cdot 5 = 0$

d. $\dfrac{3}{0}$ is undefined.

24. Area = length · width
$= 7 \cdot 22$
$= 154$ square miles
The area is 154 square miles..

25.

$$\begin{array}{r} 208 \\ 9\overline{)\,1872} \\ -18 \\ \hline 07 \\ -0 \\ \hline 72 \\ -72 \\ \hline 0 \end{array}$$

Check:

$$\begin{array}{r} 208 \\ \times \ 9 \\ \hline 1872 \end{array}$$

26.

$$\begin{array}{r} 5000 \\ - 986 \\ \hline 4014 \end{array}$$

27.

$$\begin{array}{r} 12 \\ 19\overline{)\,238} \\ -19 \\ \hline 48 \\ -38 \\ \hline 10 \end{array}$$

Each friend will receive 12 download cards. There will be 10 download cards left over.

28.

$$\begin{array}{r} 9 \\ \times 7 \\ \hline 63 \end{array}$$

The product of 9 and 7 is 63.

29.

$$\begin{array}{r} 30 \\ \times 20 \\ \hline 0 \\ 600 \\ \hline 600 \end{array}$$

$$\begin{array}{r} 40 \\ 15\overline{)\,600} \\ -60 \\ \hline 00 \\ -0 \\ \hline 0 \end{array}$$

The new length of the garden is 40 ft.

30.

$$\begin{array}{r} \overset{1}{9} \\ + 7 \\ \hline 16 \end{array}$$

The sum of 9 and 7 is 16.

31. $7 \cdot 7 \cdot 7 = 7^3$

32. $7 \cdot 7 \cdot 7 \cdot 7 = 7^4$

33. $3 \cdot 3 \cdot 3 \cdot 3 \cdot 17 \cdot 17 \cdot 17 = 3^4 \cdot 17^3$

34. $2 \cdot 2 \cdot 3 \cdot 3 \cdot 3 \cdot 3 = 2^2 \cdot 3^4$

35. $2 \cdot 4 - 3 \div 3 = 8 - 3 \div 3 = 8 - 1 = 7$

36. $8 \cdot \sqrt{100} - 4^2 \cdot 5 = 8 \cdot 10 - 4^2 \cdot 5$
$$\begin{aligned} &= 80 - 16 \cdot 5 \\ &= 80 - 80 \\ &= 0 \end{aligned}$$

37. 2 of the 5 equal parts are shaded: $\dfrac{2}{5}$

38.

$$\begin{array}{r} 13 \\ 3\overline{)39} \\ 2\overline{)78} \\ 2\overline{)156} \end{array}$$

$156 = 2^2 \cdot 3 \cdot 13$

39. a. $4\dfrac{2}{9} = \dfrac{9 \cdot 4 + 2}{9} = \dfrac{38}{9}$

    b. $1\dfrac{8}{11} = \dfrac{11 \cdot 1 + 8}{11} = \dfrac{19}{11}$

40. $7\dfrac{4}{5} = \dfrac{5 \cdot 7 + 4}{5} = \dfrac{39}{5}$

41. $1 \cdot 20 = 20$
$2 \cdot 10 = 20$
$4 \cdot 5 = 20$
The factors of 20 are 1, 2, 4, 5, 10, and 20.

42. Equivalent, since the cross products are equal:
$20 \cdot 14 = 280$ and $35 \cdot 8 = 280$.

43. $\dfrac{42}{66} = \dfrac{6 \cdot 7}{6 \cdot 11} = \dfrac{7}{11}$

44. $\dfrac{70}{105} = \dfrac{35 \cdot 2}{35 \cdot 3} = \dfrac{2}{3}$

45. $3\dfrac{1}{3} \cdot \dfrac{7}{8} = \dfrac{10}{3} \cdot \dfrac{7}{8}$
$$\begin{aligned} &= \dfrac{10 \cdot 7}{3 \cdot 8} \\ &= \dfrac{2 \cdot 5 \cdot 7}{3 \cdot 2 \cdot 4} \\ &= \dfrac{5 \cdot 7}{3 \cdot 4} \\ &= \dfrac{35}{12} \text{ or } 2\dfrac{11}{12} \end{aligned}$$

46. $\dfrac{2}{3} \cdot 4 = \dfrac{2}{3} \cdot \dfrac{4}{1} = \dfrac{2 \cdot 4}{3 \cdot 1} = \dfrac{8}{3} \text{ or } 2\dfrac{2}{3}$

47. The reciprocal of $\dfrac{1}{3}$ is $\dfrac{3}{1}$ or 3.

48. The reciprocal of 9, or $\dfrac{9}{1}$, is $\dfrac{1}{9}$.

49. $\dfrac{5}{16} \div \dfrac{3}{4} = \dfrac{5}{16} \cdot \dfrac{4}{3} = \dfrac{5 \cdot 4}{16 \cdot 3} = \dfrac{5 \cdot 4}{4 \cdot 4 \cdot 3} = \dfrac{5}{4 \cdot 3} = \dfrac{5}{12}$

50. $1\dfrac{1}{10} \div 5\dfrac{3}{5} = \dfrac{11}{10} \div \dfrac{28}{5}$
$$\begin{aligned} &= \dfrac{11}{10} \cdot \dfrac{5}{28} \\ &= \dfrac{11 \cdot 5}{10 \cdot 28} \\ &= \dfrac{11 \cdot 5}{2 \cdot 5 \cdot 28} \\ &= \dfrac{11}{2 \cdot 28} \\ &= \dfrac{11}{56} \end{aligned}$$

Chapter 3

Section 3.1 Practice

1. $\dfrac{5}{9} + \dfrac{2}{9} = \dfrac{5+2}{9} = \dfrac{7}{9}$

2. $\dfrac{5}{8} + \dfrac{1}{8} = \dfrac{5+1}{8} = \dfrac{6}{8} = \dfrac{2 \cdot 3}{2 \cdot 4} = \dfrac{3}{4}$

3. $\dfrac{10}{11} + \dfrac{1}{11} + \dfrac{7}{11} = \dfrac{10+1+7}{11} = \dfrac{18}{11}$ or $1\dfrac{7}{11}$

4. $\dfrac{7}{12} - \dfrac{2}{12} = \dfrac{7-2}{12} = \dfrac{5}{12}$

5. $\dfrac{9}{10} - \dfrac{1}{10} = \dfrac{9-1}{10} = \dfrac{8}{10} = \dfrac{2 \cdot 4}{2 \cdot 5} = \dfrac{4}{5}$

6. $\text{perimeter} = \dfrac{3}{20} + \dfrac{3}{20} + \dfrac{3}{20} + \dfrac{3}{20}$
   $= \dfrac{3+3+3+3}{20}$
   $= \dfrac{12}{20}$
   $= \dfrac{3 \cdot 4}{5 \cdot 4}$
   $= \dfrac{3}{5}$

   The perimeter is $\dfrac{3}{5}$ mile.

7. The amount of time she practiced is the time she practiced in the morning plus the time she practiced in the evening.
   $\dfrac{3}{8} + \dfrac{1}{8} = \dfrac{3+1}{8} = \dfrac{4}{8} = \dfrac{4 \cdot 1}{4 \cdot 2} = \dfrac{1}{2}$

   She practiced $\dfrac{1}{2}$ hour that day.

8. The phrase "How much farther" tells us to subtract distances. Subtract the distance run on Wednesday from the distance run on Monday.
   $\dfrac{13}{4} - \dfrac{7}{4} = \dfrac{13-7}{4} = \dfrac{6}{4} = \dfrac{2 \cdot 3}{2 \cdot 2} = \dfrac{3}{2}$ or $1\dfrac{1}{2}$

   He ran $\dfrac{3}{2}$ or $1\dfrac{1}{2}$ miles farther on Monday.

Vocabulary and Readiness Check

1. The fractions $\dfrac{9}{11}$ and $\dfrac{13}{11}$ are called <u>like</u> fractions while $\dfrac{3}{4}$ and $\dfrac{1}{3}$ are called <u>unlike</u> fractions.

2. $\dfrac{a}{b} + \dfrac{c}{b} = \dfrac{a+c}{\underline{b}}$

3. $\dfrac{a}{b} - \dfrac{c}{b} = \dfrac{a-c}{\underline{b}}$.

4. The distance around a figure is called its <u>perimeter</u>.

5. $\dfrac{7}{8}$ and $\dfrac{7}{10}$ are unlike fractions.

6. $\dfrac{2}{3}$ and $\dfrac{4}{9}$ are unlike fractions.

7. $\dfrac{9}{10}$ and $\dfrac{1}{10}$ are like fractions.

8. $\dfrac{8}{11}$ and $\dfrac{2}{11}$ are like fractions

9. $\dfrac{2}{31}, \dfrac{30}{31}$, and $\dfrac{19}{31}$ are like fractions.

10. $\dfrac{3}{10}, \dfrac{3}{11}$, and $\dfrac{3}{13}$ are unlike fractions.

11. $\dfrac{5}{12}, \dfrac{7}{12}$, and $\dfrac{12}{11}$ are unlike fractions.

12. $\dfrac{1}{5}, \dfrac{2}{5}$, and $\dfrac{4}{5}$ are like fractions.

Exercise Set 3.1

1. $\dfrac{1}{7} + \dfrac{2}{7} = \dfrac{1+2}{7} = \dfrac{3}{7}$

3. $\dfrac{1}{10} + \dfrac{1}{10} = \dfrac{1+1}{10} = \dfrac{2}{10} = \dfrac{2 \cdot 1}{2 \cdot 5} = \dfrac{1}{5}$

5. $\dfrac{2}{9} + \dfrac{4}{9} = \dfrac{2+4}{9} = \dfrac{6}{9} = \dfrac{3 \cdot 2}{3 \cdot 3} = \dfrac{2}{3}$

7. $\dfrac{6}{20} + \dfrac{1}{20} = \dfrac{6+1}{20} = \dfrac{7}{20}$

9. $\dfrac{3}{14} + \dfrac{4}{14} = \dfrac{3+4}{14} = \dfrac{7}{14} = \dfrac{7 \cdot 1}{7 \cdot 2} = \dfrac{1}{2}$

11. $\dfrac{10}{11} + \dfrac{3}{11} = \dfrac{10+3}{11} = \dfrac{13}{11}$ or $1\dfrac{2}{11}$

13. $\dfrac{4}{13} + \dfrac{2}{13} + \dfrac{1}{13} = \dfrac{4+2+1}{13} = \dfrac{7}{13}$

15. $\dfrac{7}{18} + \dfrac{3}{18} + \dfrac{2}{18} = \dfrac{7+3+2}{18} = \dfrac{12}{18} = \dfrac{6 \cdot 2}{6 \cdot 3} = \dfrac{2}{3}$

17. $\dfrac{10}{11} - \dfrac{4}{11} = \dfrac{10-4}{11} = \dfrac{6}{11}$

19. $\dfrac{4}{5} - \dfrac{1}{5} = \dfrac{4-1}{5} = \dfrac{3}{5}$

21. $\dfrac{7}{4} - \dfrac{3}{4} = \dfrac{7-3}{4} = \dfrac{4}{4} = 1$

23. $\dfrac{7}{8} - \dfrac{1}{8} = \dfrac{7-1}{8} = \dfrac{6}{8} = \dfrac{2 \cdot 3}{2 \cdot 4} = \dfrac{3}{4}$

25. $\dfrac{25}{12} - \dfrac{15}{12} = \dfrac{25-15}{12} = \dfrac{10}{12} = \dfrac{2 \cdot 5}{2 \cdot 6} = \dfrac{5}{6}$

27. $\dfrac{11}{10} - \dfrac{3}{10} = \dfrac{11-3}{10} = \dfrac{8}{10} = \dfrac{2 \cdot 4}{2 \cdot 5} = \dfrac{4}{5}$

29. $\dfrac{86}{90} - \dfrac{85}{90} = \dfrac{86-85}{90} = \dfrac{1}{90}$

31. $\dfrac{27}{33} - \dfrac{8}{33} = \dfrac{27-8}{33} = \dfrac{19}{33}$

33. $\dfrac{8}{21} + \dfrac{5}{21} = \dfrac{8+5}{21} = \dfrac{13}{21}$

35. $\dfrac{99}{100} - \dfrac{9}{100} = \dfrac{99-9}{100} = \dfrac{90}{100} = \dfrac{10 \cdot 9}{10 \cdot 10} = \dfrac{9}{10}$

37. $\dfrac{13}{28} - \dfrac{13}{28} = \dfrac{13-13}{28} = \dfrac{0}{28} = 0$

39. $\dfrac{3}{16} + \dfrac{7}{16} + \dfrac{2}{16} = \dfrac{3+7+2}{16} = \dfrac{12}{16} = \dfrac{4 \cdot 3}{4 \cdot 4} = \dfrac{3}{4}$

41. $\dfrac{4}{20} + \dfrac{7}{20} + \dfrac{9}{20} = \dfrac{4+7+9}{20} = \dfrac{20}{20} = 1$

The perimeter is 1 inch.

43. Opposite sides of a rectangle have equal length.

$$\dfrac{7}{12} + \dfrac{5}{12} + \dfrac{7}{12} + \dfrac{5}{12} = \dfrac{7+5+7+5}{12}$$
$$= \dfrac{24}{12}$$
$$= \dfrac{12 \cdot 2}{12 \cdot 1}$$
$$= \dfrac{2}{1}$$
$$= 2$$

The perimeter is 2 meters.

45. To find the remaining amount of track to be inspected, subtract the $\dfrac{5}{20}$ mile that has already been inspected from the $\dfrac{19}{20}$ mile total that must be inspected.

$$\dfrac{19}{20} - \dfrac{5}{20} = \dfrac{19-5}{20} = \dfrac{14}{20} = \dfrac{2 \cdot 7}{2 \cdot 10} = \dfrac{7}{10}$$

$\dfrac{7}{10}$ of a mile of track remains to be inspected.

47. Add the lengths of his workouts.

$$\dfrac{7}{8} + \dfrac{5}{8} = \dfrac{7+5}{8} = \dfrac{12}{8} = \dfrac{4 \cdot 3}{4 \cdot 2} = \dfrac{3}{2} \text{ or } 1\dfrac{1}{2}$$

Emil worked out $\dfrac{3}{2}$ or $1\dfrac{1}{2}$ hours that day.

49. To find the fractional part that should come from vegetables and fruits, add the fractional part from vegetables, $\dfrac{5}{24}$, to the fractional part from fruit, $\dfrac{4}{24}$.

$$\dfrac{5}{24} + \dfrac{4}{24} = \dfrac{5+4}{24} = \dfrac{9}{24} = \dfrac{3 \cdot 3}{3 \cdot 8} = \dfrac{3}{8}$$

$\dfrac{3}{8}$ of a college student's daily servings should come from vegetables and fruit.

51. To find how much greater, subtract the fractional part from fruit, $\frac{4}{24}$, from the fractional part from grains, $\frac{6}{24}$.

$$\frac{6}{24} - \frac{4}{24} = \frac{6-4}{24} = \frac{2}{24} = \frac{2 \cdot 1}{2 \cdot 12} = \frac{1}{12}$$

The fractional part of college student's daily servings that come from grains is $\frac{1}{12}$ more than from fruit.

53. To find the fraction of teenagers with cell phones who do social networking tasks at least once a week, add the fraction of teenagers with cell phones who do social networking tasks daily, $\frac{6}{100}$, to the fraction of teenagers with cell phones who do social networking tasks once a week to several times a week, $\frac{7}{100}$.

$$\frac{6}{100} + \frac{7}{100} = \frac{6+7}{100} = \frac{13}{100}$$

The fraction of teenagers with cell phones who do social networking tasks at least once a week is $\frac{13}{100}$.

55. To find the fraction of states that had maximum speed limits less than 70 mph, subtract the fraction that had speed limits of 70 mph, $\frac{9}{25}$ from the fraction that had speed limits up to and including 70 mph, $\frac{19}{25}$.

$$\frac{19}{25} - \frac{9}{25} = \frac{19-9}{25} = \frac{10}{25} = \frac{5 \cdot 2}{5 \cdot 5} = \frac{2}{5}$$

$\frac{2}{5}$ of the states had maximum speed limits less than 70 mph.

57. North America takes up $\frac{16}{100}$ of the world's land area, while South America takes up $\frac{12}{100}$ of the land area.

$$\frac{16}{100} + \frac{12}{100} = \frac{16+12}{100} = \frac{28}{100} = \frac{4 \cdot 7}{4 \cdot 25} = \frac{7}{25}$$

$\frac{7}{25}$ of the world's land area is within North America and South America.

59. Antarctica's fractional part is $\frac{9}{100}$, while Europe's fractional part is $\frac{7}{100}$.

$$\frac{9}{100} - \frac{7}{100} = \frac{9-7}{100} = \frac{2}{100} = \frac{2 \cdot 1}{2 \cdot 50} = \frac{1}{50}$$

Antarctica contains $\frac{1}{50}$ more of the world's land area than Europe.

61. To find the fraction of U.S. theaters that are single screen or miniplexes, add the fraction of single screens, $\frac{4}{100}$, to the fraction of miniplexes, $\frac{21}{100}$.

$$\frac{4}{100} + \frac{21}{100} = \frac{4+21}{100} = \frac{25}{100} = \frac{25 \cdot 1}{25 \cdot 4} = \frac{1}{4}$$

The fraction of U.S. theaters that are single screen or miniplexes is $\frac{1}{4}$.

63. $2\overline{)10}$ quotient 5

$10 = 2 \cdot 5$

65. $2\overline{)4}$ quotient 2
$2\overline{)8}$

$8 = 2 \cdot 2 \cdot 2 = 2^3$

67. $5\overline{)55}$ quotient 11

$55 = 5 \cdot 11$

69. $\dfrac{3}{8} + \dfrac{7}{8} - \dfrac{5}{8} = \dfrac{3+7-5}{8} = \dfrac{10-5}{8} = \dfrac{5}{8}$

71. $\dfrac{4}{11} + \dfrac{5}{11} - \dfrac{3}{11} + \dfrac{2}{11} = \dfrac{4+5-3+2}{11}$

$= \dfrac{9-3+2}{11}$

$= \dfrac{6+2}{11}$

$= \dfrac{8}{11}$

73. $\dfrac{2}{7} + \dfrac{9}{7} = \dfrac{2+9}{7} = \dfrac{11}{7}$ or $1\dfrac{4}{7}$

75. answers may vary

77. The sum of all daily servings' fractions is 1; answers may vary.

79. $\dfrac{3}{8} + \dfrac{3}{8} - \dfrac{4}{8} = \dfrac{3+3-4}{8} = \dfrac{6-4}{8} = \dfrac{2}{8} = \dfrac{2 \cdot 1}{2 \cdot 4} = \dfrac{1}{4}$

Mike was $\dfrac{1}{4}$ of a mile from home.

Section 3.2 Practice

1. Multiples of 15: 15, 30, 45, 60, 75, 90, 105, 120, 135, $\boxed{150}$, 165,...
Multiples of 50: 50, 100, $\boxed{150}$, 200, ...
The LCM of 15 and 50 is 150.

2. $10 \cdot 1 = 10$ Not a multiple of 8.
$10 \cdot 2 = 20$ Not a multiple of 8.
$10 \cdot 3 = 30$ Not a multiple of 8.
$10 \cdot 4 = 40$ A multiple of 8.
The LCM of 8 and 10 is 40.

3. $16 \cdot 1 = 16$ A multiple of 8.
The LCM of 8 and 16 is 16.

4. $30 \cdot 1 = 30$ Not a multiple of 25.
$30 \cdot 2 = 60$ Not a multiple of 25.
$30 \cdot 3 = 90$ Not a multiple of 25.
$30 \cdot 4 = 120$ Not a multiple of 25.
$30 \cdot 5 = 150$ A multiple of 25.
The LCM of 25 and 30 is 150.

5. $40 = \boxed{2 \cdot 2 \cdot 2} \cdot \boxed{5}$
$108 = 2 \cdot 2 \cdot \boxed{3 \cdot 3 \cdot 3}$
The LCM of 40 and 108 is
$2 \cdot 2 \cdot 2 \cdot 3 \cdot 3 \cdot 3 \cdot 5 = 1080$.

6. $20 = 2 \cdot 2 \cdot 5$
$24 = \boxed{2 \cdot 2 \cdot 2} \cdot 3$
$45 = \boxed{3 \cdot 3} \cdot \boxed{5 \cdot 5}$
The LCM of 20, 24, and 45 is
$2 \cdot 2 \cdot 2 \cdot 3 \cdot 3 \cdot 5 \cdot 5 = 360$.

7. $7 = 7$ or $7 = \boxed{7}$
$21 = \boxed{3} \cdot \boxed{7}$ $21 = \boxed{3} \cdot 7$
The LCM of 7 and 21 is $3 \cdot 7 = 21$.

8. Since $56 = 8 \cdot 7$, multiply by $\dfrac{7}{7}$.

$\dfrac{7}{8} = \dfrac{7}{8} \cdot \dfrac{7}{7} = \dfrac{7 \cdot 7}{8 \cdot 7} = \dfrac{49}{56}$

9. Since $15 = 5 \cdot 3$, multiply by $\dfrac{3}{3}$.

$\dfrac{3}{5} = \dfrac{3}{5} \cdot \dfrac{3}{3} = \dfrac{3 \cdot 3}{5 \cdot 3} = \dfrac{9}{15}$

10. $4 = \dfrac{4}{1} \cdot \dfrac{6}{6} = \dfrac{4 \cdot 6}{1 \cdot 6} = \dfrac{24}{6}$

Vocabulary and Readiness Check

1. Fractions that represent the same portion of a whole are called <u>equivalent</u> fractions.

2. The smallest positive number that is a multiple of all numbers in the list is called the <u>least common multiple (LCM)</u>.

3. A <u>multiple</u> of a number is the product of that number and a natural number.

Exercise Set 3.2

1. Multiplies of 3: 3, 6, 9, $\boxed{12}$, 15, ...
Multiplies of 4: 4, 8, $\boxed{12}$, 16, ...
LCM: 12

3. Multiples of 9: 9, 18, 27, 36, $\boxed{45}$, 54, ...
Multiples of 15: 15, 30, $\boxed{45}$, 60, ...
LCM: 45

5. Multiples of 12: 12, 24, $\boxed{36}$, 48, ...
Multiples of 18: 18, $\boxed{36}$, 54, ...
LCM: 36

7. $24 = \boxed{2 \cdot 2 \cdot 2} \cdot 3$
$36 = 2 \cdot 2 \cdot \boxed{3 \cdot 3}$
LCM: $2 \cdot 2 \cdot 2 \cdot 3 \cdot 3 = 72$

9. $18 = \boxed{2} \cdot \boxed{3 \cdot 3}$
$21 = 3 \cdot \boxed{7}$
LCM: $2 \cdot 3 \cdot 3 \cdot 7 = 126$

11. $15 = \boxed{3} \cdot 5$
$25 = \boxed{5 \cdot 5}$
LCM: $3 \cdot 5 \cdot 5 = 75$

13. $8 = \boxed{2 \cdot 2 \cdot 2}$
$24 = 2 \cdot 2 \cdot 2 \cdot \boxed{3}$
LCM: $2 \cdot 2 \cdot 2 \cdot 3 = 24$

15. $6 = \boxed{2} \cdot \boxed{3}$
$7 = \boxed{7}$
LCM: $2 \cdot 3 \cdot 7 = 42$

17. $8 = \boxed{2 \cdot 2 \cdot 2}$
$6 = 2 \cdot 3$
$27 = \boxed{3 \cdot 3 \cdot 3}$
LCM: $2 \cdot 2 \cdot 2 \cdot 3 \cdot 3 \cdot 3 = 216$

19. $25 = \boxed{5 \cdot 5}$
$15 = \boxed{3} \cdot 5$
$6 = \boxed{2} \cdot 3$
LCM: $2 \cdot 3 \cdot 5 \cdot 5 = 150$

21. $34 = 2 \cdot \boxed{17}$
$68 = \boxed{2 \cdot 2} \cdot 17$
LCM: $2 \cdot 2 \cdot 17 = 68$

23. $84 = \boxed{2 \cdot 2} \cdot \boxed{3} \cdot 7$
$294 = 2 \cdot 3 \cdot \boxed{7 \cdot 7}$
LCM: $2 \cdot 2 \cdot 3 \cdot 7 \cdot 7 = 588$

25. $30 = 2 \cdot 3 \cdot 5$
$36 = \boxed{2 \cdot 2} \cdot \boxed{3 \cdot 3}$
$50 = 2 \cdot \boxed{5 \cdot 5}$
LCM: $2 \cdot 2 \cdot 3 \cdot 3 \cdot 5 \cdot 5 = 900$

27. $50 = 2 \cdot \boxed{5 \cdot 5}$
$72 = \boxed{2 \cdot 2 \cdot 2} \cdot \boxed{3 \cdot 3}$
$120 = 2 \cdot 2 \cdot 2 \cdot 3 \cdot 5$
LCM: $2 \cdot 2 \cdot 2 \cdot 3 \cdot 3 \cdot 5 \cdot 5 = 1800$

29. $11 = 11$
$33 = \boxed{3} \cdot 11$
$121 = \boxed{11 \cdot 11}$
LCM: $3 \cdot 11 \cdot 11 = 363$

31. $4 = \boxed{2 \cdot 2}$
$6 = 2 \cdot \boxed{3}$
$10 = 2 \cdot \boxed{5}$
$15 = 3 \cdot 5$
LCM: $2 \cdot 2 \cdot 3 \cdot 5 = 60$

33. $\dfrac{4}{7} = \dfrac{4 \cdot 5}{7 \cdot 5} = \dfrac{20}{35}$

35. $\dfrac{2}{3} = \dfrac{2 \cdot 7}{3 \cdot 7} = \dfrac{14}{21}$

37. $5 = \dfrac{5}{1} = \dfrac{5 \cdot 3}{1 \cdot 3} = \dfrac{15}{3}$

39. $\dfrac{1}{2} = \dfrac{1 \cdot 15}{2 \cdot 15} = \dfrac{15}{30}$

41. $\dfrac{10}{7} = \dfrac{10 \cdot 3}{7 \cdot 3} = \dfrac{30}{21}$

43. $\dfrac{3}{4} = \dfrac{3 \cdot 7}{4 \cdot 7} = \dfrac{21}{28}$

45. $\dfrac{2}{3} = \dfrac{2 \cdot 15}{3 \cdot 15} = \dfrac{30}{45}$

47. $\dfrac{4}{9} = \dfrac{4 \cdot 9}{9 \cdot 9} = \dfrac{36}{81}$

49. $\dfrac{15}{13} = \dfrac{15 \cdot 6}{13 \cdot 6} = \dfrac{90}{78}$

51. $\dfrac{14}{17} = \dfrac{14 \cdot 4}{17 \cdot 4} = \dfrac{56}{68}$

53. books & magazines: $\dfrac{27}{50} = \dfrac{27 \cdot 2}{50 \cdot 2} = \dfrac{54}{100}$

clothing & accessories: $\dfrac{1}{2} = \dfrac{1 \cdot 50}{2 \cdot 50} = \dfrac{50}{100}$

computer hardware: $\dfrac{23}{50} = \dfrac{23 \cdot 2}{50 \cdot 2} = \dfrac{46}{100}$

computer software: $\dfrac{1}{2} = \dfrac{1 \cdot 50}{2 \cdot 50} = \dfrac{50}{100}$

drugs, health & beauty aids: $\dfrac{3}{20} = \dfrac{3 \cdot 5}{20 \cdot 5} = \dfrac{15}{100}$

electronics and appliances: $\dfrac{13}{20} = \dfrac{13 \cdot 5}{20 \cdot 5} = \dfrac{65}{100}$

food, beer, wine: $\dfrac{9}{20} = \dfrac{9 \cdot 5}{20 \cdot 5} = \dfrac{45}{100}$

home furnishings: $\dfrac{13}{25} = \dfrac{13 \cdot 4}{25 \cdot 4} = \dfrac{52}{100}$

music and videos: $\dfrac{3}{5} = \dfrac{3 \cdot 20}{5 \cdot 20} = \dfrac{60}{100}$

office equipment & supplies: $\dfrac{61}{100}$

sporting goods: $\dfrac{12}{25} = \dfrac{12 \cdot 4}{25 \cdot 4} = \dfrac{48}{100}$

toys and hobbies and games: $\dfrac{1}{2} = \dfrac{1 \cdot 50}{2 \cdot 50} = \dfrac{50}{100}$

55. $\dfrac{15}{100}$ is the smallest fraction, so drugs, health and beauty aids has the smallest fraction sold online.

57. $\dfrac{7}{10} - \dfrac{2}{10} = \dfrac{7-2}{10} = \dfrac{5}{10} = \dfrac{5 \cdot 1}{5 \cdot 2} = \dfrac{1}{2}$

59. $\dfrac{1}{5} + \dfrac{1}{5} = \dfrac{1+1}{5} = \dfrac{2}{5}$

61. $\dfrac{23}{18} - \dfrac{15}{18} = \dfrac{23-15}{18} = \dfrac{8}{18} = \dfrac{2 \cdot 4}{2 \cdot 9} = \dfrac{4}{9}$

63. $\dfrac{2}{9} + \dfrac{1}{9} + \dfrac{6}{9} = \dfrac{2+1+6}{9} = \dfrac{9}{9} = 1$

65. $\dfrac{37}{165} = \dfrac{37 \cdot 22}{165 \cdot 22} = \dfrac{814}{3630}$

67. answers may vary

69. a. $\dfrac{10}{15} = \dfrac{5 \cdot 2}{5 \cdot 3} = \dfrac{2}{3}$

    b. $\dfrac{40}{60} = \dfrac{20 \cdot 2}{20 \cdot 3} = \dfrac{2}{3}$

    c. $\dfrac{16}{20} = \dfrac{4 \cdot 4}{4 \cdot 5} = \dfrac{4}{5}$

    d. $\dfrac{200}{300} = \dfrac{100 \cdot 2}{100 \cdot 3} = \dfrac{2}{3}$

a, b, and d are equivalent to $\dfrac{2}{3}$.

Section 3.3 Practice

1. The LCM of 6 and 18 is 18, so the LCD is 18.

$\dfrac{1}{6} = \dfrac{1}{6} \cdot \dfrac{3}{3} = \dfrac{3}{18}$

$\dfrac{5}{18} = \dfrac{5}{18}$

$\dfrac{1}{6} + \dfrac{5}{18} = \dfrac{3}{18} + \dfrac{5}{18} = \dfrac{8}{18} = \dfrac{2 \cdot 4}{2 \cdot 9} = \dfrac{4}{9}$

2. The LCD of 6 and 9 is 18.

$\dfrac{5}{6} = \dfrac{5}{6} \cdot \dfrac{3}{3} = \dfrac{15}{18}$

$\dfrac{2}{9} = \dfrac{2}{9} \cdot \dfrac{2}{2} = \dfrac{4}{18}$

$\dfrac{5}{6} + \dfrac{2}{9} = \dfrac{15}{18} + \dfrac{4}{18} = \dfrac{19}{18}$ or $1\dfrac{1}{18}$

3. The LCD of 5 and 9 is 45.

$\dfrac{2}{5} + \dfrac{4}{9} = \dfrac{2}{5} \cdot \dfrac{9}{9} + \dfrac{4}{9} \cdot \dfrac{5}{5} = \dfrac{18}{45} + \dfrac{20}{45} = \dfrac{38}{45}$

4. The LCD of 4, 5, and 10 is 20.

$\dfrac{1}{4} + \dfrac{4}{5} + \dfrac{9}{10} = \dfrac{1}{4} \cdot \dfrac{5}{5} + \dfrac{4}{5} \cdot \dfrac{4}{4} + \dfrac{9}{10} \cdot \dfrac{2}{2}$

$= \dfrac{5}{20} + \dfrac{16}{20} + \dfrac{18}{20}$

$= \dfrac{39}{20}$ or $1\dfrac{19}{20}$

5. The LCD of 12 and 24 is 24.

$\dfrac{7}{12} = \dfrac{7}{12} \cdot \dfrac{2}{2} = \dfrac{14}{24}$

$\dfrac{5}{24} = \dfrac{5}{24}$

$\dfrac{7}{12} - \dfrac{5}{24} = \dfrac{14}{24} - \dfrac{5}{24} = \dfrac{9}{24} = \dfrac{3 \cdot 3}{3 \cdot 8} = \dfrac{3}{8}$

6. The LCD of 10 and 7 is 70.

$\dfrac{9}{10} = \dfrac{9}{10} \cdot \dfrac{7}{7} = \dfrac{63}{70}$

$\dfrac{3}{7} = \dfrac{3}{7} \cdot \dfrac{10}{10} = \dfrac{30}{70}$

$\dfrac{9}{10} - \dfrac{3}{7} = \dfrac{63}{70} - \dfrac{30}{70} = \dfrac{33}{70}$

7. The LCD of 8 and 6 is 24.

$\dfrac{7}{8} - \dfrac{5}{6} = \dfrac{7}{8} \cdot \dfrac{3}{3} - \dfrac{5}{6} \cdot \dfrac{4}{4} = \dfrac{21}{24} - \dfrac{20}{24} = \dfrac{1}{24}$

8. The total amount of cement is the sum of the amounts needed in the three locations. The LCD of 5, 10, and 15 is 30.

$$\frac{3}{5}+\frac{2}{10}+\frac{2}{15}=\frac{3}{5}\cdot\frac{6}{6}+\frac{2}{10}\cdot\frac{3}{3}+\frac{2}{15}\cdot\frac{2}{2}$$
$$=\frac{18}{30}+\frac{6}{30}+\frac{4}{30}$$
$$=\frac{28}{30}$$
$$=\frac{14}{15}$$

She will need $\frac{14}{15}$ of a cubic yard of cement.

9. The phrase "Find the difference in length" tells us to subtract the lengths of the boards. The LCD of 5 and 3 is 15.

$$\frac{4}{5}-\frac{2}{3}=\frac{4}{5}\cdot\frac{3}{3}-\frac{2}{3}\cdot\frac{5}{5}=\frac{12}{15}-\frac{10}{15}=\frac{2}{15}$$

The lengths of the boards differ by $\frac{2}{15}$ of a foot.

Calculator Explorations

1. $\dfrac{1}{16}+\dfrac{2}{5}=\dfrac{37}{80}$

2. $\dfrac{3}{20}+\dfrac{2}{25}=\dfrac{23}{100}$

3. $\dfrac{4}{9}+\dfrac{7}{8}=\dfrac{95}{72}$

4. $\dfrac{9}{11}+\dfrac{5}{12}=\dfrac{163}{132}$

5. $\dfrac{10}{17}+\dfrac{12}{19}=\dfrac{394}{323}$

6. $\dfrac{14}{31}+\dfrac{15}{21}=\dfrac{253}{217}$

Vocabulary and Readiness Check

1. To add or subtract unlike fractions, we first write the fractions as equivalent fractions with a common denominator. The common denominator we use is called the least common denominator.

2. The LCD for $\dfrac{1}{6}$ and $\dfrac{5}{8}$ is underline{24}.

3. $\dfrac{1}{6}+\dfrac{5}{8}=\dfrac{1}{6}\cdot\dfrac{4}{4}+\dfrac{5}{8}\cdot\dfrac{3}{3}=\dfrac{4}{24}+\dfrac{15}{24}=\dfrac{19}{24}.$

4. $\dfrac{5}{8}-\dfrac{1}{6}=\dfrac{5}{8}\cdot\dfrac{3}{3}-\dfrac{1}{6}\cdot\dfrac{4}{4}=\dfrac{15}{24}-\dfrac{4}{24}=\dfrac{11}{24}.$

Exercise Set 3.3

1. The LCD of 3 and 6 is 6.

$$\frac{2}{3}+\frac{1}{6}=\frac{2}{3}\cdot\frac{2}{2}+\frac{1}{6}=\frac{4}{6}+\frac{1}{6}=\frac{5}{6}$$

3. The LCD of 2 and 3 is 6.

$$\frac{1}{2}+\frac{1}{3}=\frac{1}{2}\cdot\frac{3}{3}+\frac{1}{3}\cdot\frac{2}{2}=\frac{3}{6}+\frac{2}{6}=\frac{5}{6}$$

5. The LCD of 11 and 33 is 33.

$$\frac{2}{11}+\frac{2}{33}=\frac{2}{11}\cdot\frac{3}{3}+\frac{2}{33}=\frac{6}{33}+\frac{2}{33}=\frac{8}{33}$$

7. The LCD of 14 and 7 is 14.

$$\frac{3}{14}+\frac{3}{7}=\frac{3}{14}+\frac{3}{7}\cdot\frac{2}{2}=\frac{3}{14}+\frac{6}{14}=\frac{9}{14}$$

9. The LCD of 35 and 7 is 35.

$$\frac{11}{35}+\frac{2}{7}=\frac{11}{35}+\frac{2}{7}\cdot\frac{5}{5}$$
$$=\frac{11}{35}+\frac{10}{35}$$
$$=\frac{21}{35}$$
$$=\frac{7\cdot3}{7\cdot5}$$
$$=\frac{3}{5}$$

11. The LCD of 25 and 35 is 175.

$$\frac{8}{25}+\frac{7}{35}=\frac{8}{25}\cdot\frac{7}{7}+\frac{7}{35}\cdot\frac{5}{5}$$
$$=\frac{56}{175}+\frac{35}{175}$$
$$=\frac{91}{175}$$
$$=\frac{7\cdot13}{7\cdot25}$$
$$=\frac{13}{25}$$

13. The LCD of 15 and 12 is 60.

$$\frac{7}{15}+\frac{5}{12}=\frac{7}{15}\cdot\frac{4}{4}+\frac{5}{12}\cdot\frac{5}{5}$$
$$=\frac{28}{60}+\frac{25}{60}$$
$$=\frac{53}{60}$$

15. The LCD of 28 and 21 is 84.

$$\frac{2}{28}+\frac{2}{21}=\frac{2}{28}\cdot\frac{3}{3}+\frac{2}{21}\cdot\frac{4}{4}$$
$$=\frac{6}{84}+\frac{8}{84}$$
$$=\frac{14}{84}$$
$$=\frac{14\cdot 1}{14\cdot 6}$$
$$=\frac{1}{6}$$

17. The LCD of 44 and 36 is 396.

$$\frac{9}{44}+\frac{17}{36}=\frac{9}{44}\cdot\frac{9}{9}+\frac{17}{36}\cdot\frac{11}{11}$$
$$=\frac{81}{396}+\frac{187}{396}$$
$$=\frac{268}{396}$$
$$=\frac{4\cdot 67}{4\cdot 99}$$
$$=\frac{67}{99}$$

19. The LCD of 11 and 13 is 143.

$$\frac{5}{11}+\frac{3}{13}=\frac{5}{11}\cdot\frac{13}{13}+\frac{3}{13}\cdot\frac{11}{11}=\frac{65}{143}+\frac{33}{143}=\frac{98}{143}$$

21. The LCD of 3, 9, and 27 is 27.

$$\frac{1}{3}+\frac{1}{9}+\frac{1}{27}=\frac{1}{3}\cdot\frac{9}{9}+\frac{1}{9}\cdot\frac{3}{3}+\frac{1}{27}$$
$$=\frac{9}{27}+\frac{3}{27}+\frac{1}{27}$$
$$=\frac{13}{27}$$

23. The LCD of 7, 8, and 2 is 56.

$$\frac{5}{7}+\frac{1}{8}+\frac{1}{2}=\frac{5}{7}\cdot\frac{8}{8}+\frac{1}{8}\cdot\frac{7}{7}+\frac{1}{2}\cdot\frac{28}{28}$$
$$=\frac{40}{56}+\frac{7}{56}+\frac{28}{56}$$
$$=\frac{75}{56}\text{ or }1\frac{19}{56}$$

25. The LCD of 36, 4, and 6 is 36.

$$\frac{5}{36}+\frac{3}{4}+\frac{1}{6}=\frac{5}{36}+\frac{3}{4}\cdot\frac{9}{9}+\frac{1}{6}\cdot\frac{6}{6}$$
$$=\frac{5}{36}+\frac{27}{36}+\frac{6}{36}$$
$$=\frac{38}{36}$$
$$=\frac{2\cdot 19}{2\cdot 18}$$
$$=\frac{19}{18}\text{ or }1\frac{1}{18}$$

27. The LCD of 20, 5, and 3 is 60.

$$\frac{13}{20}+\frac{3}{5}+\frac{1}{3}=\frac{13}{20}\cdot\frac{3}{3}+\frac{3}{5}\cdot\frac{12}{12}+\frac{1}{3}\cdot\frac{20}{20}$$
$$=\frac{39}{60}+\frac{36}{60}+\frac{20}{60}$$
$$=\frac{95}{60}$$
$$=\frac{5\cdot 19}{5\cdot 12}$$
$$=\frac{19}{12}\text{ or }1\frac{7}{12}$$

29. The LCD of 8 and 16 is 16.

$$\frac{7}{8}-\frac{3}{16}=\frac{7}{8}\cdot\frac{2}{2}-\frac{3}{16}=\frac{14}{16}-\frac{3}{16}=\frac{11}{16}$$

31. The LCD of 6 and 7 is 42.

$$\frac{5}{6}-\frac{3}{7}=\frac{5}{6}\cdot\frac{7}{7}-\frac{3}{7}\cdot\frac{6}{6}=\frac{35}{42}-\frac{18}{42}=\frac{17}{42}$$

33. The LCD of 7 and 8 is 56.

$$\frac{5}{7}-\frac{1}{8}=\frac{5}{7}\cdot\frac{8}{8}-\frac{1}{8}\cdot\frac{7}{7}=\frac{40}{56}-\frac{7}{56}=\frac{33}{56}$$

35. The LCD of 11 and 9 is 99.

$$\frac{9}{11}-\frac{4}{9}=\frac{9}{11}\cdot\frac{9}{9}-\frac{4}{9}\cdot\frac{11}{11}$$
$$=\frac{81}{99}-\frac{44}{99}$$
$$=\frac{37}{99}$$

37. The LCD of 35 and 7 is 35.

$$\frac{11}{35}-\frac{2}{7}=\frac{11}{35}-\frac{2}{7}\cdot\frac{5}{5}=\frac{11}{35}-\frac{10}{35}=\frac{1}{35}$$

39. The LCD of 12 and 9 is 36.

$$\frac{5}{12}-\frac{1}{9}=\frac{5}{12}\cdot\frac{3}{3}-\frac{1}{9}\cdot\frac{4}{4}=\frac{15}{36}-\frac{4}{36}=\frac{11}{36}$$

41. The LCD of 15 and 12 is 60.

$$\frac{7}{15} - \frac{5}{12} = \frac{7}{15} \cdot \frac{4}{4} - \frac{5}{12} \cdot \frac{5}{5}$$
$$= \frac{28}{60} - \frac{25}{60}$$
$$= \frac{3}{60}$$
$$= \frac{3 \cdot 1}{3 \cdot 20}$$
$$= \frac{1}{20}$$

43. The LCD of 28 and 21 is 84.

$$\frac{3}{28} - \frac{2}{21} = \frac{3}{28} \cdot \frac{3}{3} - \frac{2}{21} \cdot \frac{4}{4} = \frac{9}{84} - \frac{8}{84} = \frac{1}{84}$$

45. The LCD of 100 and 1000 is 1000.

$$\frac{1}{100} - \frac{1}{1000} = \frac{1}{100} \cdot \frac{10}{10} - \frac{1}{1000}$$
$$= \frac{10}{1000} - \frac{1}{1000}$$
$$= \frac{9}{1000}$$

47. The LCD of 44 and 36 is 396.

$$\frac{21}{44} - \frac{11}{36} = \frac{21}{44} \cdot \frac{9}{9} - \frac{11}{36} \cdot \frac{11}{11}$$
$$= \frac{189}{396} - \frac{121}{396}$$
$$= \frac{68}{396}$$
$$= \frac{4 \cdot 17}{4 \cdot 99}$$
$$= \frac{17}{99}$$

49. The LCD of 12 and 9 is 36.

$$\frac{5}{12} + \frac{1}{9} = \frac{5}{12} \cdot \frac{3}{3} + \frac{1}{9} \cdot \frac{4}{4} = \frac{15}{36} + \frac{4}{36} = \frac{19}{36}$$

51. The LCD of 35 and 7 is 35.

$$\frac{17}{35} - \frac{2}{7} = \frac{17}{35} - \frac{2}{7} \cdot \frac{5}{5}$$
$$= \frac{17}{35} - \frac{10}{35}$$
$$= \frac{7}{35}$$
$$= \frac{7 \cdot 1}{7 \cdot 5}$$
$$= \frac{1}{5}$$

53. The LCD of 28 and 40 is 280.

$$\frac{9}{28} - \frac{3}{40} = \frac{9}{28} \cdot \frac{10}{10} - \frac{3}{40} \cdot \frac{7}{7}$$
$$= \frac{90}{280} - \frac{21}{280}$$
$$= \frac{69}{280}$$

55. The LCD of 3, 45, and 5 is 45.

$$\frac{2}{3} + \frac{4}{45} + \frac{4}{5} = \frac{2}{3} \cdot \frac{15}{15} + \frac{4}{45} + \frac{4}{5} \cdot \frac{9}{9}$$
$$= \frac{30}{45} + \frac{4}{45} + \frac{36}{45}$$
$$= \frac{70}{45}$$
$$= \frac{5 \cdot 14}{5 \cdot 9}$$
$$= \frac{14}{9} \text{ or } 1\frac{5}{9}$$

57. Add the lengths of the four sides. The LCD of 3 and 5 is 15.

$$\frac{4}{5} + \frac{1}{3} + \frac{4}{5} + \frac{1}{3} = \frac{4}{5} \cdot \frac{3}{3} + \frac{1}{3} \cdot \frac{5}{5} + \frac{4}{5} \cdot \frac{3}{3} + \frac{1}{3} \cdot \frac{5}{5}$$
$$= \frac{12}{15} + \frac{5}{15} + \frac{12}{15} + \frac{5}{15}$$
$$= \frac{34}{15} \text{ or } 2\frac{4}{15}$$

The perimeter is $\frac{34}{15}$ or $2\frac{4}{15}$ cm.

59. Add the lengths of the four sides. The LCD of 2, 4, and 5 is 20.

$$\frac{1}{4} + \frac{1}{5} + \frac{1}{2} + \frac{3}{4} = \frac{1}{4} \cdot \frac{5}{5} + \frac{1}{5} \cdot \frac{4}{4} + \frac{1}{2} \cdot \frac{10}{10} + \frac{3}{4} \cdot \frac{5}{5}$$
$$= \frac{5}{20} + \frac{4}{20} + \frac{10}{20} + \frac{15}{20}$$
$$= \frac{34}{20}$$
$$= \frac{2 \cdot 17}{2 \cdot 10}$$
$$= \frac{17}{10} \text{ or } 1\frac{7}{10}$$

The perimeter is $\frac{17}{10}$ or $1\frac{7}{10}$ meters.

61. The LCD is 100.

$$\frac{17}{100}-\frac{1}{10}=\frac{17}{100}-\frac{1\cdot 10}{10\cdot 10}=\frac{17}{100}-\frac{10}{100}=\frac{7}{100}$$

The sloth can travel $\dfrac{7}{100}$ mph faster in trees.

63. $1-\dfrac{3}{16}-\dfrac{3}{16}=\dfrac{16}{16}-\dfrac{3}{16}-\dfrac{3}{16}=\dfrac{10}{16}=\dfrac{5}{8}$

The inner diameter is $\dfrac{5}{8}$ inch.

65. Add the lengths of the 3 sections. The LCD of 16, 2, and 32 is 32.

$$\frac{5}{16}+\frac{1}{2}+\frac{5}{32}=\frac{5}{16}\cdot\frac{2}{2}+\frac{1}{2}\cdot\frac{16}{16}+\frac{5}{32}$$
$$=\frac{10}{32}+\frac{16}{32}+\frac{5}{32}$$
$$=\frac{31}{32}$$

The total length of the diagram is $\dfrac{31}{32}$ inch.

67. Subtract the fraction of Thin Mints from the fraction of Thin Mints and Samoas together. The LCD of 25 and 4 is 100.

$$\frac{11}{25}-\frac{1}{4}=\frac{11}{25}\cdot\frac{4}{4}-\frac{1}{4}\cdot\frac{25}{25}=\frac{44}{100}-\frac{25}{100}=\frac{19}{100}$$

The fraction of Girl Scout cookies that are Samoas is $\dfrac{19}{100}$.

69. The LCD is 50.

$$\frac{13}{50}+\frac{1}{2}=\frac{13}{50}+\frac{1}{2}\cdot\frac{25}{25}$$
$$=\frac{13}{50}+\frac{25}{50}$$
$$=\frac{38}{50}$$
$$=\frac{2\cdot 19}{2\cdot 25}$$
$$=\frac{19}{25}$$

The Pacific and Atlantic Oceans account for $\dfrac{19}{25}$ of the world's water surfaces.

71. The piece representing Lakes/Seashores is labeled $\dfrac{1}{25}$, so $\dfrac{1}{25}$ of the areas maintained by the National Park Service are National Lakes or National Seashores.

73. $1-\dfrac{21}{100}=\dfrac{100}{100}-\dfrac{21}{100}=\dfrac{79}{100}$

$\dfrac{79}{100}$ of areas maintained by the National Park Service are NOT National Monuments.

75. $1\dfrac{1}{2}\cdot 3\dfrac{1}{3}=\dfrac{3}{2}\cdot\dfrac{10}{3}=\dfrac{3\cdot 10}{2\cdot 3}=\dfrac{10}{2}=5$

77. $4\div 7\dfrac{1}{4}=\dfrac{4}{1}\div\dfrac{29}{4}=\dfrac{4}{1}\cdot\dfrac{4}{29}=\dfrac{4\cdot 4}{1\cdot 29}=\dfrac{16}{29}$

79. $3\cdot 2\dfrac{1}{9}=\dfrac{3}{1}\cdot\dfrac{19}{9}=\dfrac{3\cdot 19}{1\cdot 9}=\dfrac{3\cdot 19}{1\cdot 3\cdot 3}=\dfrac{19}{3}$ or $6\dfrac{1}{3}$

81. a. $\dfrac{3}{5}$

$\dfrac{4}{5}$

$\dfrac{7}{10}$

b. Yes, there is an error.

c. $\dfrac{3}{5}+\dfrac{4}{5}=\dfrac{3+4}{5}=\dfrac{7}{5}$ or $1\dfrac{2}{5}\ \left(\text{not }\dfrac{7}{10}\right)$

83. The LCD is 540.

$$\frac{2}{3}-\frac{1}{4}-\frac{2}{540}=\frac{2}{3}\cdot\frac{180}{180}-\frac{1}{4}\cdot\frac{135}{135}-\frac{2}{540}$$
$$=\frac{360}{540}-\frac{135}{540}-\frac{2}{540}$$
$$=\frac{225}{540}-\frac{2}{540}$$
$$=\frac{223}{540}$$

85. The LCD is 1760.

$$\frac{30}{55}+\frac{1000}{1760}=\frac{30\cdot 32}{55\cdot 32}+\frac{1000}{1760}$$
$$=\frac{960}{1760}+\frac{1000}{1760}$$
$$=\frac{1960}{1760}$$
$$=\frac{49\cdot 40}{44\cdot 40}$$
$$=\frac{49}{44}\text{ or }1\dfrac{5}{44}$$

87. answers may vary

Integrated Review

1. $5 = \boxed{5}$
   $6 = \boxed{2} \cdot \boxed{3}$
   LCM: $2 \cdot 3 \cdot 5 = 30$

2. $3 = \boxed{3}$
   $7 = \boxed{7}$
   LCM: $3 \cdot 7 = 21$

3. $2 = \boxed{2}$
   $14 = 2 \cdot \boxed{7}$
   LCM: $2 \cdot 7 = 14$

4. $5 = 5$
   $25 = \boxed{5 \cdot 5}$
   LCM: $5 \cdot 5 = 25$

5. $4 = \boxed{2 \cdot 2}$
   $20 = 2 \cdot 2 \cdot 5$
   $25 = \boxed{5 \cdot 5}$
   LCM: $2 \cdot 2 \cdot 5 \cdot 5 = 100$

6. $6 = \boxed{2} \cdot 3$
   $18 = 2 \cdot \boxed{3 \cdot 3}$
   $30 = 2 \cdot 3 \cdot \boxed{5}$
   LCM $= 2 \cdot 3 \cdot 3 \cdot 5 = 90$

7. $\dfrac{3}{8} = \dfrac{3}{8} \cdot \dfrac{3}{3} = \dfrac{9}{24}$

8. $\dfrac{7}{9} = \dfrac{7}{9} \cdot \dfrac{4}{4} = \dfrac{28}{36}$

9. $\dfrac{1}{4} = \dfrac{1}{4} \cdot \dfrac{10}{10} = \dfrac{10}{40}$

10. $\dfrac{2}{5} = \dfrac{2}{5} \cdot \dfrac{6}{6} = \dfrac{12}{30}$

11. $\dfrac{11}{15} = \dfrac{11}{15} \cdot \dfrac{5}{5} = \dfrac{55}{75}$

12. $\dfrac{5}{6} = \dfrac{5}{6} \cdot \dfrac{8}{8} = \dfrac{40}{48}$

13. $\dfrac{3}{8} + \dfrac{1}{8} = \dfrac{3+1}{8} = \dfrac{4}{8} = \dfrac{4 \cdot 1}{4 \cdot 2} = \dfrac{1}{2}$

14. $\dfrac{7}{10} - \dfrac{3}{10} = \dfrac{7-3}{10} = \dfrac{4}{10} = \dfrac{2 \cdot 2}{2 \cdot 5} = \dfrac{2}{5}$

15. $\dfrac{17}{24} - \dfrac{3}{24} = \dfrac{17-3}{24} = \dfrac{14}{24} = \dfrac{2 \cdot 7}{2 \cdot 12} = \dfrac{7}{12}$

16. $\dfrac{4}{15} + \dfrac{9}{15} = \dfrac{4+9}{15} = \dfrac{13}{15}$

17. The LCD of 4 and 2 is 4.
    $\dfrac{1}{4} + \dfrac{1}{2} = \dfrac{1}{4} + \dfrac{1}{2} \cdot \dfrac{2}{2} = \dfrac{1}{4} + \dfrac{2}{4} = \dfrac{3}{4}$

18. The LCD of 3 and 5 is 15.
    $\dfrac{1}{3} - \dfrac{1}{5} = \dfrac{1}{3} \cdot \dfrac{5}{5} - \dfrac{1}{5} \cdot \dfrac{3}{3} = \dfrac{5}{15} - \dfrac{3}{15} = \dfrac{2}{15}$

19. The LCD of 9 and 5 is 45.
    $\dfrac{7}{9} - \dfrac{2}{5} = \dfrac{7}{9} \cdot \dfrac{5}{5} - \dfrac{2}{5} \cdot \dfrac{9}{9} = \dfrac{35}{45} - \dfrac{18}{45} = \dfrac{17}{45}$

20. The LCD of 10 and 25 is 50.
    $\dfrac{3}{10} + \dfrac{2}{25} = \dfrac{3}{10} \cdot \dfrac{5}{5} + \dfrac{2}{25} \cdot \dfrac{2}{2} = \dfrac{15}{50} + \dfrac{4}{50} = \dfrac{19}{50}$

21. The LCD of 8 and 20 is 40.
    $\dfrac{7}{8} + \dfrac{1}{20} = \dfrac{7}{8} \cdot \dfrac{5}{5} + \dfrac{1}{20} \cdot \dfrac{2}{2} = \dfrac{35}{40} + \dfrac{2}{40} = \dfrac{37}{40}$

22. The LCD of 12 and 18 is 36.
    $\dfrac{5}{12} - \dfrac{2}{18} = \dfrac{5}{12} \cdot \dfrac{3}{3} - \dfrac{2}{18} \cdot \dfrac{2}{2} = \dfrac{15}{36} - \dfrac{4}{36} = \dfrac{11}{36}$

23. $\dfrac{1}{11} - \dfrac{1}{11} = \dfrac{1-1}{11} = \dfrac{0}{11}$

24. $\dfrac{3}{17} - \dfrac{2}{17} = \dfrac{3-2}{17} = \dfrac{1}{17}$

25. The LCD of 11 and 3 is 33.
    $\dfrac{9}{11} - \dfrac{2}{3} = \dfrac{9}{11} \cdot \dfrac{3}{3} - \dfrac{2}{3} \cdot \dfrac{11}{11} = \dfrac{27}{33} - \dfrac{22}{33} = \dfrac{5}{33}$

26. The LCD of 6 and 7 is 42.
    $\dfrac{1}{6} - \dfrac{1}{7} = \dfrac{1}{6} \cdot \dfrac{7}{7} - \dfrac{1}{7} \cdot \dfrac{6}{6} = \dfrac{7}{42} - \dfrac{6}{42} = \dfrac{1}{42}$

27. The LCD of 9 and 18 is 18.
    $\dfrac{2}{9} + \dfrac{1}{18} = \dfrac{2}{9} \cdot \dfrac{2}{2} + \dfrac{1}{18} = \dfrac{4}{18} + \dfrac{1}{18} = \dfrac{5}{18}$

28. The LCD of 13 and 26 is 26.

$$\frac{4}{13}+\frac{2}{26}=\frac{4}{13}\cdot\frac{2}{2}+\frac{2}{26}$$
$$=\frac{8}{26}+\frac{2}{26}$$
$$=\frac{10}{26}$$
$$=\frac{2\cdot5}{2\cdot13}$$
$$=\frac{5}{13}$$

29. The LCD of 9, 18, and 3 is 18.

$$\frac{2}{9}+\frac{1}{18}+\frac{1}{3}=\frac{2}{9}\cdot\frac{2}{2}+\frac{1}{18}+\frac{1}{3}\cdot\frac{6}{6}$$
$$=\frac{4}{18}+\frac{1}{18}+\frac{6}{18}$$
$$=\frac{11}{18}$$

30. The LCD of 10, 5, and 25 is 50.

$$\frac{3}{10}+\frac{1}{5}+\frac{6}{25}=\frac{3}{10}\cdot\frac{5}{5}+\frac{1}{5}\cdot\frac{10}{10}+\frac{6}{25}\cdot\frac{2}{2}$$
$$=\frac{15}{50}+\frac{10}{50}+\frac{12}{50}$$
$$=\frac{37}{50}$$

31. The LCD of 10 and 3 is 30.

$$\frac{9}{10}+\frac{2}{3}=\frac{9}{10}\cdot\frac{3}{3}+\frac{2}{3}\cdot\frac{10}{10}=\frac{37}{30}+\frac{20}{30}=\frac{47}{30} \text{ or } 1\frac{17}{30}$$

32. The LCD of 10 and 3 is 30.

$$\frac{9}{10}-\frac{2}{3}=\frac{9}{10}\cdot\frac{3}{3}-\frac{2}{3}\cdot\frac{10}{10}=\frac{27}{30}-\frac{20}{30}=\frac{7}{30}$$

33. $\dfrac{9}{10}\cdot\dfrac{2}{3}=\dfrac{9\cdot2}{10\cdot3}=\dfrac{3\cdot3\cdot2}{2\cdot5\cdot3}=\dfrac{3}{5}$

34. $\dfrac{9}{10}\div\dfrac{2}{3}=\dfrac{9}{10}\cdot\dfrac{3}{2}=\dfrac{9\cdot3}{10\cdot2}=\dfrac{27}{20} \text{ or } 1\dfrac{7}{20}$

35. The LCD of 25 and 70 is 350.

$$\frac{21}{25}-\frac{3}{70}=\frac{21}{25}\cdot\frac{14}{14}-\frac{3}{70}\cdot\frac{5}{5}=\frac{294}{350}-\frac{15}{350}=\frac{279}{350}$$

36. The LCD of 25 and 70 is 350.

$$\frac{21}{25}+\frac{3}{70}=\frac{21}{25}\cdot\frac{14}{14}+\frac{3}{70}\cdot\frac{5}{5}=\frac{294}{350}+\frac{15}{350}=\frac{309}{350}$$

37. $\dfrac{21}{25}\div\dfrac{3}{70}=\dfrac{21}{25}\cdot\dfrac{70}{3}$
$$=\frac{3\cdot7\cdot5\cdot14}{5\cdot5\cdot3}$$
$$=\frac{7\cdot14}{5}$$
$$=\frac{98}{5} \text{ or } 19\frac{3}{5}$$

38. $\dfrac{21}{25}\cdot\dfrac{3}{70}=\dfrac{21\cdot3}{25\cdot70}=\dfrac{3\cdot7\cdot3}{25\cdot7\cdot10}=\dfrac{3\cdot3}{25\cdot10}=\dfrac{9}{250}$

39. $3\dfrac{7}{8}\cdot2\dfrac{2}{3}=\dfrac{31}{8}\cdot\dfrac{8}{3}=\dfrac{31\cdot8}{8\cdot3}=\dfrac{31}{3} \text{ or } 10\dfrac{1}{3}$

40. $3\dfrac{7}{8}\div2\dfrac{2}{3}=\dfrac{31}{8}\div\dfrac{8}{3}=\dfrac{31}{8}\cdot\dfrac{3}{8}=\dfrac{31\cdot3}{8\cdot8}=\dfrac{93}{64} \text{ or } 1\dfrac{29}{64}$

41. The LCD of 9, 27, and 2 is 54.

$$\frac{2}{9}+\frac{5}{27}+\frac{1}{2}=\frac{2}{9}\cdot\frac{6}{6}+\frac{5}{27}\cdot\frac{2}{2}+\frac{1}{2}\cdot\frac{27}{27}$$
$$=\frac{12}{54}+\frac{10}{54}+\frac{27}{54}$$
$$=\frac{49}{54}$$

42. The LCD of 8, 16, and 3 is 48.

$$\frac{3}{8}+\frac{11}{16}+\frac{2}{3}=\frac{3}{8}\cdot\frac{6}{6}+\frac{11}{16}\cdot\frac{3}{3}+\frac{2}{3}\cdot\frac{16}{16}$$
$$=\frac{18}{48}+\frac{33}{48}+\frac{32}{48}$$
$$=\frac{83}{48} \text{ or } 1\frac{35}{48}$$

43. $11\dfrac{7}{10}\div3\dfrac{3}{100}=\dfrac{117}{10}\div\dfrac{303}{100}$
$$=\frac{117}{10}\cdot\frac{100}{303}$$
$$=\frac{3\cdot39\cdot10\cdot10}{10\cdot3\cdot101}$$
$$=\frac{39\cdot10}{101}$$
$$=\frac{390}{101} \text{ or } 3\frac{87}{101}$$

44. $7\dfrac{1}{4} \cdot 3\dfrac{1}{5} = \dfrac{29}{4} \cdot \dfrac{16}{5}$

$\qquad\qquad = \dfrac{29 \cdot 16}{4 \cdot 5}$

$\qquad\qquad = \dfrac{29 \cdot 4}{1 \cdot 5}$

$\qquad\qquad = \dfrac{116}{5}$ or $23\dfrac{1}{5}$

45. The LCD of 15 and 27 is 135.

$\dfrac{14}{15} - \dfrac{4}{27} = \dfrac{14}{15} \cdot \dfrac{9}{9} - \dfrac{4}{27} \cdot \dfrac{5}{5} = \dfrac{126}{135} - \dfrac{20}{135} = \dfrac{106}{135}$

46. The LCD of 14 and 32 is 224.

$\dfrac{9}{14} - \dfrac{11}{32} = \dfrac{9}{14} \cdot \dfrac{16}{16} - \dfrac{11}{32} \cdot \dfrac{7}{7} = \dfrac{144}{224} - \dfrac{77}{224} = \dfrac{67}{224}$

Section 3.4 Practice

1. The LCD of 5 and 6 is 30.

$$\begin{array}{rcccc}
4\dfrac{2}{5} & = & 4\dfrac{2 \cdot 6}{5 \cdot 6} & = & 4\dfrac{12}{30} \\
+5\dfrac{1}{6} & = & 5\dfrac{1 \cdot 5}{6 \cdot 5} & = & 5\dfrac{5}{30} \\
\hline
& & & & 9\dfrac{17}{30}
\end{array}$$

2. The LCD of 14 and 7 is 14.

$$\begin{array}{rcc}
2\dfrac{5}{14} & = & 2\dfrac{5}{14} \\
+5\dfrac{6}{7} & = & 5\dfrac{12}{14} \\
\hline
& & 7\dfrac{17}{14} = 7 + 1\dfrac{3}{14} = 8\dfrac{3}{14}
\end{array}$$

3. The LCD of 7 and 5 is 35.

$$\begin{array}{rcc}
10 & = & 10 \\
2\dfrac{6}{7} & = & 2\dfrac{30}{35} \\
+3\dfrac{1}{5} & = & 3\dfrac{7}{35} \\
\hline
& & 15\dfrac{37}{35} = 15 + 1\dfrac{2}{35} = 16\dfrac{2}{35}
\end{array}$$

4. The LCD of 9 and 18 is 18.

$$\begin{array}{rcc}
29\dfrac{7}{9} & = & 29\dfrac{14}{18} \\
-13\dfrac{5}{18} & = & -13\dfrac{5}{18} \\
\hline
& & 16\dfrac{9}{18} = 16\dfrac{1}{2}
\end{array}$$

5. The LCD of 15 and 5 is 15.

$$\begin{array}{rcccc}
9\dfrac{7}{15} & = & 9\dfrac{7}{15} & = & 8\dfrac{22}{15} \\
-5\dfrac{3}{5} & = & -5\dfrac{9}{15} & = & -5\dfrac{9}{15} \\
\hline
& & & & 3\dfrac{13}{15}
\end{array}$$

6.

$$\begin{array}{rcc}
25 & = & 24\dfrac{9}{9} \\
-10\dfrac{2}{9} & = & -10\dfrac{2}{9} \\
\hline
& & 14\dfrac{7}{9}
\end{array}$$

7. To find the total weight, add the two weights. The LCD of 2 and 3 is 6.

$$\begin{array}{rcc}
2\dfrac{1}{2} & = & 2\dfrac{3}{6} \\
+3\dfrac{2}{3} & = & 3\dfrac{4}{6} \\
\hline
& & 5\dfrac{7}{6} = 5 + 1\dfrac{1}{6} = 6\dfrac{1}{6}
\end{array}$$

The total weight of the two trout is $6\dfrac{1}{6}$ pounds.

8. The phrase "How much larger" tells us to subtract the largest sugar maple girth from the largest American beech girth. The LCD of 4 and 12 is 12.

$$\begin{array}{rcccc}
23\dfrac{1}{4} & = & 23\dfrac{3}{12} & = & 22\dfrac{15}{12} \\
-19\dfrac{5}{12} & = & -19\dfrac{5}{12} & = & -19\dfrac{5}{12} \\
\hline
& & & & 3\dfrac{10}{12} = 3\dfrac{5}{6}
\end{array}$$

The girth of the largest known American beech tree is $3\dfrac{5}{6}$ feet larger than the girth of the largest known sugar maple tree.

Vocabulary and Readiness Check

1. The number $5\frac{3}{4}$ is called a <u>mixed number</u>.

2. For $5\frac{3}{4}$, the 5 is called the <u>whole number</u> part and $\frac{3}{4}$ is called the <u>fraction</u> part.

3. To estimate operations on mixed numbers, we <u>round</u> mixed numbers to the nearest whole number.

4. The mixed number $2\frac{5}{8}$ written as an <u>improper</u> fraction is $\frac{21}{8}$.

5. $3\frac{7}{8}$ rounds to 4 and $2\frac{1}{5}$ rounds to 2. The best estimate of $3\frac{7}{8} + 2\frac{1}{5}$ is $4 + 2 = 6$, choice a.

6. $3\frac{7}{8}$ rounds to 4 and $2\frac{1}{5}$ rounds to 2. The best estimate of $3\frac{7}{8} - 2\frac{1}{5}$ is $4 - 2 = 2$, choice d.

7. $8\frac{1}{3}$ rounds to 8 and $1\frac{1}{2}$ rounds to 2. The best estimate of $8\frac{1}{3} - 1\frac{1}{2}$ is $8 - 2 = 6$, choice c.

8. $8\frac{1}{3}$ rounds to 8 and $1\frac{1}{2}$ rounds to 2. The best estimate of $8\frac{1}{3} + 1\frac{1}{2}$ is $8 + 2 = 10$, choice b.

Exercise Set 3.4

1. Exact:
$$\begin{array}{r} 4\frac{7}{10} \\ +\ 2\frac{1}{10} \\ \hline 6\frac{8}{10} = 6\frac{4}{5} \end{array}$$
Estimate: $4\frac{7}{10}$ rounds to 5. $2\frac{1}{10}$ rounds to 2. $5 + 2 = 7$

3. The LCD of 14 and 7 is 14.
Exact: $\quad 10\frac{3}{14} \qquad 10\frac{3}{14}$
$$\begin{array}{r} +\ 3\frac{4}{7} \qquad +\ 3\frac{8}{14} \\ \hline 13\frac{11}{14} \end{array}$$
Estimate: $10\frac{3}{14}$ rounds to 10. $3\frac{4}{7}$ rounds to 4. $10 + 4 = 14$

5. The LCD of 5 and 25 is 25.
$$\begin{array}{r} 9\frac{1}{5} \qquad 9\frac{5}{25} \\ +\ 8\frac{2}{25} \quad +\ 8\frac{2}{25} \\ \hline 17\frac{7}{25} \end{array}$$

7. The LCD of 2 and 8 is 8.
$$\begin{array}{r} 3\frac{1}{2} \qquad 3\frac{4}{8} \\ +\ 4\frac{1}{8} \quad +\ 4\frac{1}{8} \\ \hline 7\frac{5}{8} \end{array}$$

9. The LCD of 6 and 8 is 24.
$$\begin{array}{r} 1\frac{5}{6} \qquad 1\frac{20}{24} \\ +\ 5\frac{3}{8} \quad +\ 5\frac{9}{24} \\ \hline 6\frac{29}{24} = 6 + 1\frac{5}{24} = 7\frac{5}{24} \end{array}$$

11. The LCD of 5 and 3 is 15.

$$8\frac{2}{5} \qquad 8\frac{6}{15}$$

$$+11\frac{2}{3} \qquad +11\frac{10}{15}$$

$$\overline{ } \qquad \overline{19\frac{16}{15} = 19+1\frac{1}{15} = 20\frac{1}{15}}$$

13.

$$11\frac{3}{5}$$

$$+7\frac{2}{5}$$

$$\overline{18\frac{5}{5} = 18+1 = 19}$$

15. The LCD of 10 and 27 is 270.

$$40\frac{9}{10} \qquad 40\frac{243}{270}$$

$$+15\frac{8}{27} \qquad +15\frac{80}{270}$$

$$\overline{ } \qquad \overline{55\frac{323}{270} = 55+1\frac{53}{270} = 56\frac{53}{270}}$$

17. The LCD of 8, 6, and 4 is 24.

$$3\frac{5}{8} \qquad 3\frac{15}{24}$$

$$2\frac{1}{6} \qquad 2\frac{4}{24}$$

$$+7\frac{3}{4} \qquad +7\frac{18}{24}$$

$$\overline{ } \qquad \overline{12\frac{37}{24} = 12+1\frac{13}{24} = 13\frac{13}{24}}$$

19. The LCD of 14 and 12 is 84.

$$12\frac{3}{14} \qquad 12\frac{18}{84}$$

$$10 \qquad\quad 10$$

$$+25\frac{5}{12} \qquad +25\frac{35}{84}$$

$$\overline{ } \qquad \overline{47\frac{53}{84}}$$

21. Exact: $4\frac{7}{10}$

$$-2\frac{1}{10}$$

$$\overline{2\frac{6}{10} = 2\frac{3}{5}}$$

Estimate: $4\frac{7}{10}$ rounds to 5. $2\frac{1}{10}$ rounds to 2.

$$5-2 = 3$$

23. The LCD of 14 and 7 is 14.

Exact: $10\frac{13}{14} \qquad 10\frac{13}{14}$

$$-3\frac{4}{7} \qquad -3\frac{8}{14}$$

$$\overline{ } \qquad \overline{7\frac{5}{14}}$$

Estimate: $10\frac{13}{14}$ rounds to 11. $3\frac{8}{14}$ rounds to 4.

$$11-4 = 7$$

25. The LCD of 5 and 25 is 25.

$$9\frac{1}{5} \qquad 9\frac{5}{25} \qquad 8\frac{30}{25}$$

$$-8\frac{6}{25} \qquad -8\frac{6}{25} \qquad -8\frac{6}{25}$$

$$\overline{} \qquad \overline{} \qquad \overline{\frac{24}{25}}$$

27. The LCD of 3 and 5 is 15.

$$5\frac{2}{3} \qquad 5\frac{10}{15}$$

$$-3\frac{1}{5} \qquad -3\frac{3}{15}$$

$$\overline{} \qquad \overline{2\frac{7}{15}}$$

29. The LCD of 7 and 14 is 14.

$$15\frac{4}{7} \qquad 15\frac{8}{14} \qquad 14\frac{22}{14}$$

$$-9\frac{11}{14} \qquad -9\frac{11}{14} \qquad -9\frac{11}{14}$$

$$\overline{} \qquad \overline{} \qquad \overline{5\frac{11}{14}}$$

31. The LCD of 18 and 24 is 72.

$$47\frac{4}{18} \qquad 47\frac{16}{72} \qquad 46\frac{88}{72}$$
$$-23\frac{19}{24} \qquad -23\frac{57}{72} \qquad -23\frac{57}{72}$$
$$\overline{\phantom{-23\frac{19}{24}}} \qquad \overline{\phantom{-23\frac{57}{72}}} \qquad \overline{23\frac{31}{72}}$$

33.

$$10 \qquad\qquad 9\frac{5}{5}$$
$$-8\frac{1}{5} \qquad -8\frac{1}{5}$$
$$\overline{\phantom{-8\frac{1}{5}}} \qquad \overline{1\frac{4}{5}}$$

35. The LCD of 5 and 15 is 15.

$$11\frac{3}{5} \qquad 11\frac{9}{15} \qquad 10\frac{24}{15}$$
$$-9\frac{11}{15} \qquad -9\frac{11}{15} \qquad -9\frac{11}{15}$$
$$\overline{\phantom{-9\frac{11}{15}}} \qquad \overline{\phantom{-9\frac{11}{15}}} \qquad \overline{1\frac{13}{15}}$$

37.

$$6 \qquad\qquad 5\frac{9}{9}$$
$$-2\frac{4}{9} \qquad -2\frac{4}{9}$$
$$\overline{\phantom{-2\frac{4}{9}}} \qquad \overline{3\frac{5}{9}}$$

39. The LCD of 6 and 12 is 12.

$$63\frac{1}{6} \qquad 63\frac{2}{12} \qquad 62\frac{14}{12}$$
$$-47\frac{5}{12} \qquad -47\frac{5}{12} \qquad -47\frac{5}{12}$$
$$\overline{\phantom{-47\frac{5}{12}}} \qquad \overline{\phantom{-47\frac{5}{12}}} \qquad \overline{15\frac{9}{12}} = 15\frac{3}{4}$$

41. The LCD of 6 and 12 is 12.

$$15\frac{1}{6} \qquad 15\frac{2}{12}$$
$$+13\frac{5}{12} \qquad +13\frac{5}{12}$$
$$\overline{\phantom{+13\frac{5}{12}}} \qquad \overline{28\frac{7}{12}}$$

43. $22\dfrac{7}{8}$

$\dfrac{-7\phantom{\dfrac{7}{8}}}{15\dfrac{7}{8}}$

45. $5\dfrac{8}{9}$

$+\,2\dfrac{1}{9}$

$\overline{7\dfrac{9}{9}=7+1=8}$

47. The LCD of 20 and 30 is 60.

$33\dfrac{11}{20}$ $\qquad$ $33\dfrac{33}{60}$ $\qquad$ $32\dfrac{93}{60}$

$-15\dfrac{19}{30}$ $\qquad$ $-15\dfrac{38}{60}$ $\qquad$ $-15\dfrac{38}{60}$

$\phantom{-15\dfrac{19}{30}\qquad -15\dfrac{38}{60}\qquad}$ $17\dfrac{55}{60}=17\dfrac{11}{12}$

49.

how much wider	is	larger entrance hole	minus	smaller entrance hole
↓	↓	↓	↓	↓
how much wider	=	$1\dfrac{9}{16}$	–	$1\dfrac{1}{2}$

The LCD of 2 and 16 is 16.

$1\dfrac{9}{16}$ $\qquad$ $1\dfrac{9}{16}$

$-1\dfrac{1}{2}$ $\qquad$ $-1\dfrac{8}{16}$

$\phantom{-1\dfrac{1}{2}\qquad}$ $\dfrac{1}{16}$

The entrance holes for Mountain Bluebirds should be $\dfrac{1}{16}$ inch wider.

51. Subtract the sum of the lengths of the two pieces cut off from the original length of pipe. The LCD of 2 and 4 is 4.

$2\dfrac{1}{2}$ $\qquad$ $2\dfrac{2}{4}$

$+\,3\dfrac{1}{4}$ $\qquad$ $+\,3\dfrac{1}{4}$

$\phantom{+\,3\dfrac{1}{4}\qquad}$ $5\dfrac{3}{4}$

The LCD of 4 and 3 is 12.

$$15\frac{2}{3} \qquad 15\frac{8}{12} \qquad 14\frac{20}{12}$$
$$-5\frac{3}{4} \qquad -5\frac{9}{12} \qquad -5\frac{9}{12}$$
$$\qquad\qquad\qquad\qquad\qquad 9\frac{11}{12}$$

Subtract the remaining piece length from the length she now needs.

$$10 \qquad\qquad 9\frac{12}{12}$$
$$-9\frac{11}{12} \qquad -9\frac{11}{12}$$
$$\qquad\qquad\qquad\qquad \frac{1}{12}$$

The remaining piece will not be long enough. It will be $\frac{1}{12}$ of a foot short.

53. Subtract Yuma's average annual rainfall from Tucson's average annual rainfall. The LCD of 4 and 5 is 20.

$$11\frac{1}{4} \qquad 11\frac{5}{20} \qquad 10\frac{25}{20}$$
$$-3\frac{3}{5} \qquad -3\frac{12}{20} \qquad -3\frac{12}{20}$$
$$\qquad\qquad\qquad\qquad\qquad 7\frac{13}{20}$$

Annually, Tucson gets an average of $7\frac{13}{20}$ inches more rain than Yuma.

55. Add the practice times. The LCD of 2, 3, 4, and 6 is 12.

$$2\frac{1}{2} \qquad 2\frac{6}{12}$$
$$1\frac{2}{3} \qquad 1\frac{8}{12}$$
$$2\frac{1}{4} \qquad 2\frac{3}{12}$$
$$+3\frac{5}{6} \qquad +3\frac{10}{12}$$
$$\qquad\qquad 8\frac{27}{12} = 8 + 2\frac{3}{12} = 10\frac{3}{12} = 10\frac{1}{4}$$

The total practice time is $10\frac{1}{4}$ hours.

57. Subtract the time for a personal return from the time for a small business return.
The LCD of 2 and 8 is 8.

$$5\frac{7}{8} \qquad 5\frac{7}{8}$$
$$-3\frac{1}{2} \qquad -3\frac{4}{8}$$
$$\qquad\qquad 2\frac{3}{8}$$

It takes him $2\frac{3}{8}$ hours longer to prepare a small business return.

59. Add the height of the figure to the height of the pedestal. The LCD of 20 and 50 is 100.

$$46\frac{1}{20} \qquad 46\frac{5}{100}$$
$$+46\frac{47}{50} \qquad +46\frac{94}{100}$$
$$\qquad\qquad 92\frac{99}{100}$$

The overall height of the Statue of Liberty is $92\frac{99}{100}$ meters.

61. Subtract the length of the Hood Canal Bridge from the length of the Evergreen Point Bridge.

$$2526 \qquad\qquad 2525\frac{3}{3}$$
$$-2173\frac{2}{3} \qquad -2173\frac{2}{3}$$
$$\qquad\qquad\qquad\qquad 352\frac{1}{3}$$

The Evergreen Point Bridge is $352\frac{1}{3}$ yards longer than the Hood Canal Bridge.

63. Add the duration for the three eclipses together. The LCD of 3, 15, and 10 is 30.

$$2\frac{2}{3} \qquad 2\frac{20}{30}$$
$$4\frac{7}{15} \qquad 4\frac{14}{30}$$
$$+2\frac{3}{10} \qquad +2\frac{9}{30}$$
$$\qquad\quad 8\frac{43}{30} = 8 + \frac{43}{30} = 8 + 1\frac{13}{30} = 9\frac{13}{30}$$

The total duration for the three eclipses is $9\frac{13}{30}$ min.

65. Subtract the duration of the August 21, 2017 eclipse from the duration of the April 8, 2024 eclipse. The LCD of 3 and 15 is 15.

$$
\begin{array}{ccc}
4\dfrac{7}{15} & 4\dfrac{7}{15} & 3\dfrac{22}{15} \\[2mm]
-2\dfrac{2}{3} & -2\dfrac{10}{15} & -2\dfrac{10}{15} \\[2mm]
\hline
& & 1\dfrac{12}{15}=1\dfrac{4}{5}
\end{array}
$$

The April 8, 2024 eclipse will be $1\dfrac{4}{5}$ minutes longer than the August 21, 2017 eclipse.

67. Add the lengths of the sides.

$$
\begin{array}{c}
2\dfrac{1}{3} \\[2mm]
2\dfrac{1}{3} \\[2mm]
+\,2\dfrac{1}{3} \\[2mm]
\hline
6\dfrac{3}{3}=6+1=7
\end{array}
$$

The perimeter is 7 miles.

69. Add the lengths of the sides. The LCD of 3 and 8 is 24.

$$
\begin{array}{cc}
3 & 3 \\[1mm]
5\dfrac{1}{3} & 5\dfrac{8}{24} \\[2mm]
5 & 5 \\[1mm]
+\,7\dfrac{7}{8} & +\,7\dfrac{21}{24} \\[2mm]
\hline
& 20\dfrac{29}{24}=20+1\dfrac{5}{24}=21\dfrac{5}{24}
\end{array}
$$

The perimeter is $21\dfrac{5}{24}$ meters.

71. $2^3 = 2\cdot 2\cdot 2 = 8$

73. $5^2 = 5\cdot 5 = 25$

75. $20 \div 10 \cdot 2 = 2 \cdot 2 = 4$

77. $\begin{aligned}[t]
2+3(8\cdot 7-1) &= 2+3(56-1) \\
&= 2+3(55) \\
&= 2+165 \\
&= 167
\end{aligned}$

79. $3\dfrac{5}{5}=3+1=4$

81. $9\dfrac{10}{16}=9+\dfrac{2\cdot 5}{2\cdot 8}=9\dfrac{5}{8}$

83. a. $9\dfrac{5}{5}=9+1=10$

b. $9\dfrac{100}{100}=9+1=10$

c. $6\dfrac{44}{11}=6+4=10$

d. $8\dfrac{13}{13}=8+1=9$

a, b, and c are equivalent to 10.

85. answers may vary

87. Add the weights of nuts and candy for the Supreme box. The LCD of 4 and 2 is 4.

$$
\begin{array}{cc}
2\dfrac{1}{4} & 2\dfrac{1}{4} \\[2mm]
+\,3\dfrac{1}{2} & +\,3\dfrac{2}{4} \\[2mm]
\hline
& 5\dfrac{3}{4}
\end{array}
$$

Add the weights of nuts and candy for the Deluxe box. The LCD of 8 and 4 is 8.

$$
\begin{array}{cc}
1\dfrac{3}{8} & 1\dfrac{3}{8} \\[2mm]
+\,4\dfrac{1}{4} & +\,4\dfrac{2}{8} \\[2mm]
\hline
& 5\dfrac{5}{8}
\end{array}
$$

Subtract the total weight of the Deluxe box from the total weight of the Supreme box. The LCD of 4 and 8 is 8.

$$
\begin{array}{cc}
5\dfrac{3}{4} & 5\dfrac{6}{8} \\[2mm]
-5\dfrac{5}{8} & -5\dfrac{5}{8} \\[2mm]
\hline
& \dfrac{1}{8}
\end{array}
$$

The Supreme box is heavier by $\dfrac{1}{8}$ of a pound.

Section 3.5 Practice

1. The LCD of 9 and 11 is 99.

$$\frac{8}{9} = \frac{8}{9} \cdot \frac{11}{11} = \frac{88}{99}$$

$$\frac{10}{11} = \frac{10}{11} \cdot \frac{9}{9} = \frac{90}{99}$$

Since $88 < 90$, then $\frac{88}{99} < \frac{90}{99}$ or $\frac{8}{9} < \frac{10}{11}$.

2. The LCD of 5 and 9 is 45.

$$\frac{3}{5} = \frac{3}{5} \cdot \frac{9}{9} = \frac{27}{45}$$

$$\frac{2}{9} = \frac{2}{9} \cdot \frac{5}{5} = \frac{10}{45}$$

Since $27 > 10$, then $\frac{27}{45} > \frac{10}{45}$ or $\frac{3}{5} > \frac{2}{9}$.

3. $\left(\frac{1}{5}\right)^2 = \frac{1}{5} \cdot \frac{1}{5} = \frac{1}{25}$

4. $\left(\frac{2}{3}\right)^3 = \frac{2}{3} \cdot \frac{2}{3} \cdot \frac{2}{3} = \frac{8}{27}$

5. $\left(\frac{1}{4}\right)^2 \left(\frac{2}{3}\right)^3 = \left(\frac{1}{4} \cdot \frac{1}{4}\right) \cdot \left(\frac{2}{3} \cdot \frac{2}{3} \cdot \frac{2}{3}\right)$

$$= \frac{1 \cdot 1 \cdot 2 \cdot 2 \cdot 2}{2 \cdot 2 \cdot 2 \cdot 2 \cdot 3 \cdot 3 \cdot 3}$$

$$= \frac{1}{54}$$

6. $\frac{3}{7} \div \frac{10}{11} = \frac{3}{7} \cdot \frac{11}{10} = \frac{3 \cdot 11}{7 \cdot 10} = \frac{33}{70}$

7. The LCD of 15 and 5 is 15.

$$\frac{4}{15} + \frac{2}{5} = \frac{4}{15} + \frac{2}{5} \cdot \frac{3}{3}$$

$$= \frac{4}{15} + \frac{6}{15}$$

$$= \frac{10}{15}$$

$$= \frac{5 \cdot 2}{5 \cdot 3}$$

$$= \frac{2}{3}$$

8. $\frac{2}{3} \cdot \frac{9}{10} = \frac{2 \cdot 3 \cdot 3}{3 \cdot 2 \cdot 5} = \frac{3}{5}$

9. The LCD of 12 and 5 is 60.

$$\frac{11}{12} - \frac{2}{5} = \frac{11}{12} \cdot \frac{5}{5} - \frac{2}{5} \cdot \frac{12}{12} = \frac{55}{60} - \frac{24}{60} = \frac{31}{60}$$

10. $\frac{2}{9} \div \frac{4}{7} \cdot \frac{3}{10} = \frac{2}{9} \cdot \frac{7}{4} \cdot \frac{3}{10}$

$$= \frac{2 \cdot 7}{9 \cdot 2 \cdot 2} \cdot \frac{3}{10}$$

$$= \frac{7}{18} \cdot \frac{3}{10}$$

$$= \frac{7 \cdot 3}{3 \cdot 6 \cdot 10}$$

$$= \frac{7}{60}$$

11. The LCD of 5 and 25 is 25.

$$\left(\frac{2}{5}\right)^2 \div \left(\frac{3}{5} - \frac{11}{25}\right) = \left(\frac{2}{5}\right)^2 \div \left(\frac{15}{25} - \frac{11}{25}\right)$$

$$= \left(\frac{2}{5}\right)^2 \div \frac{4}{25}$$

$$= \frac{4}{25} \div \frac{4}{25}$$

$$= \frac{4}{25} \cdot \frac{25}{4}$$

$$= \frac{4 \cdot 25}{25 \cdot 4}$$

$$= 1$$

12. The average is the sum, divided by 3. The LCD of 2, 8, and 24 is 24.

$$\left(\frac{1}{2} + \frac{3}{8} + \frac{7}{24}\right) \div 3 = \left(\frac{12}{24} + \frac{9}{24} + \frac{7}{24}\right) \div 3$$

$$= \frac{28}{24} \div 3$$

$$= \frac{28}{24} \cdot \frac{1}{3}$$

$$= \frac{4 \cdot 7 \cdot 1}{4 \cdot 6 \cdot 3}$$

$$= \frac{7}{18}$$

Vocabulary and Readiness Check

1. To simplify $\frac{1}{2} + \frac{2}{3} \cdot \frac{7}{8}$, which operation do we perform first? <u>multiplication</u>

2. To simplify $\frac{1}{2} \div \frac{2}{3} \cdot \frac{7}{8}$, which operation do we perform first? <u>division</u>

3. To simplify $\dfrac{7}{8} \cdot \left(\dfrac{1}{2} - \dfrac{2}{3}\right)$, which operation do we perform first? <u>subtraction</u>

4. To simplify $9 - \left(\dfrac{3}{4}\right)^2$, which operation do we perform first? <u>evaluate the exponential expression</u>

Exercise Set 3.5

1. Since $7 > 6$, then $\dfrac{7}{9} > \dfrac{6}{9}$.

3. Since $3 < 5$, then $\dfrac{3}{3} < \dfrac{5}{3}$.

5. The LCD of 42 and 21 is 42.

$\dfrac{9}{42}$ has a denominator of 42.

$\dfrac{5}{21} = \dfrac{5}{21} \cdot \dfrac{2}{2} = \dfrac{10}{42}$

Since $9 < 10$, then $\dfrac{9}{42} < \dfrac{10}{42}$, so $\dfrac{9}{42} < \dfrac{5}{21}$.

7. The LCD of 8 and 16 is 16.

$\dfrac{9}{8} = \dfrac{9}{8} \cdot \dfrac{2}{2} = \dfrac{18}{16}$

$\dfrac{17}{16}$ has a denominator of 16.

Since $18 > 17$, then $\dfrac{18}{16} > \dfrac{17}{16}$, so $\dfrac{9}{8} > \dfrac{17}{16}$.

9. The LCD of 4 and 3 is 12.

$\dfrac{3}{4} = \dfrac{3}{4} \cdot \dfrac{3}{3} = \dfrac{9}{12}$

$\dfrac{2}{3} = \dfrac{2}{3} \cdot \dfrac{4}{4} = \dfrac{8}{12}$

Since $9 > 8$, then $\dfrac{9}{12} > \dfrac{8}{12}$, so $\dfrac{3}{4} > \dfrac{2}{3}$.

11. The LCD of 5 and 14 is 70.

$\dfrac{3}{5} = \dfrac{3}{5} \cdot \dfrac{14}{14} = \dfrac{42}{70}$

$\dfrac{9}{14} = \dfrac{9}{14} \cdot \dfrac{5}{5} = \dfrac{45}{70}$

Since $42 < 45$, then $\dfrac{42}{70} < \dfrac{45}{70}$, so $\dfrac{3}{5} < \dfrac{9}{14}$.

13. The LCD of 10 and 11 is 110.

$\dfrac{1}{10} = \dfrac{1}{10} \cdot \dfrac{11}{11} = \dfrac{11}{110}$

$\dfrac{1}{11} = \dfrac{1}{11} \cdot \dfrac{10}{10} = \dfrac{10}{110}$

Since $11 > 10$, then $\dfrac{11}{110} > \dfrac{10}{110}$, so $\dfrac{1}{10} > \dfrac{1}{11}$.

15. The LCD of 100 and 25 is 100.

$\dfrac{27}{100}$ has a denominator of 100.

$\dfrac{7}{25} = \dfrac{7}{25} \cdot \dfrac{4}{4} = \dfrac{28}{100}$

Since $27 < 28$, then $\dfrac{27}{100} < \dfrac{28}{100}$, so $\dfrac{27}{100} < \dfrac{7}{25}$.

17. $\left(\dfrac{1}{2}\right)^4 = \dfrac{1}{2} \cdot \dfrac{1}{2} \cdot \dfrac{1}{2} \cdot \dfrac{1}{2} = \dfrac{1}{16}$

19. $\left(\dfrac{2}{5}\right)^3 = \dfrac{2}{5} \cdot \dfrac{2}{5} \cdot \dfrac{2}{5} = \dfrac{8}{125}$

21. $\left(\dfrac{4}{7}\right)^3 = \dfrac{4}{7} \cdot \dfrac{4}{7} \cdot \dfrac{4}{7} = \dfrac{64}{343}$

23. $\left(\dfrac{2}{9}\right)^2 = \dfrac{2}{9} \cdot \dfrac{2}{9} = \dfrac{4}{81}$

25. $\left(\dfrac{3}{4}\right)^2 \cdot \left(\dfrac{2}{3}\right)^3 = \left(\dfrac{3}{4} \cdot \dfrac{3}{4}\right)\left(\dfrac{2}{3} \cdot \dfrac{2}{3} \cdot \dfrac{2}{3}\right)$

$= \dfrac{3 \cdot 3 \cdot 2 \cdot 2 \cdot 2}{4 \cdot 4 \cdot 3 \cdot 3 \cdot 3}$

$= \dfrac{3 \cdot 3 \cdot 2 \cdot 2 \cdot 2}{2 \cdot 2 \cdot 2 \cdot 2 \cdot 3 \cdot 3 \cdot 3}$

$= \dfrac{1}{2 \cdot 3}$

$= \dfrac{1}{6}$

27. $\dfrac{9}{10}\left(\dfrac{2}{5}\right)^2 = \dfrac{9}{10} \cdot \left(\dfrac{2}{5} \cdot \dfrac{2}{5}\right)$

$= \dfrac{9 \cdot 2 \cdot 2}{10 \cdot 5 \cdot 5}$

$= \dfrac{9 \cdot 2 \cdot 2}{2 \cdot 5 \cdot 5 \cdot 5}$

$= \dfrac{18}{125}$

29. The LCD of 15 and 5 is 15.

$$\frac{2}{15}+\frac{3}{5}=\frac{2}{15}+\frac{3}{5}\cdot\frac{3}{3}=\frac{2}{15}+\frac{9}{15}=\frac{11}{15}$$

31. $\dfrac{3}{7}\cdot\dfrac{1}{5}=\dfrac{3\cdot1}{7\cdot5}=\dfrac{3}{35}$

33. $1-\dfrac{4}{9}=\dfrac{9}{9}-\dfrac{4}{9}=\dfrac{5}{9}$

35. The LCD of 9 and 11 is 99.

$$4\frac{2}{9}+5\frac{9}{11}=4\frac{22}{99}+5\frac{81}{99}$$
$$=9\frac{103}{99}$$
$$=9+1\frac{4}{99}$$
$$=10\frac{4}{99}$$

37. The LCD of 6 and 4 is 24.

$$\frac{5}{6}-\frac{3}{4}=\frac{5}{6}\cdot\frac{4}{4}-\frac{3}{4}\cdot\frac{6}{6}$$
$$=\frac{20}{24}-\frac{18}{24}$$
$$=\frac{2}{24}$$
$$=\frac{2\cdot1}{2\cdot12}$$
$$=\frac{1}{12}$$

39. $\dfrac{6}{11}\div\dfrac{2}{3}=\dfrac{6}{11}\cdot\dfrac{3}{2}=\dfrac{6\cdot3}{11\cdot2}=\dfrac{2\cdot3\cdot3}{11\cdot2}=\dfrac{3\cdot3}{11}=\dfrac{9}{11}$

41. $0\cdot\dfrac{9}{10}=0$

43. $0\div\dfrac{9}{10}=0\cdot\dfrac{10}{9}=0$

45. $\dfrac{20}{35}\cdot\dfrac{7}{10}=\dfrac{20\cdot7}{35\cdot10}=\dfrac{10\cdot2\cdot7}{7\cdot5\cdot10}=\dfrac{2}{5}$

47. The LCD of 7 and 11 is 77.

$$\frac{4}{7}-\frac{6}{11}=\frac{4}{7}\cdot\frac{11}{11}-\frac{6}{11}\cdot\frac{7}{7}=\frac{44}{77}-\frac{42}{77}=\frac{2}{77}$$

49. $\dfrac{1}{5}+\dfrac{1}{3}\cdot\dfrac{1}{4}=\dfrac{1}{5}+\dfrac{1\cdot1}{3\cdot4}=\dfrac{1}{5}+\dfrac{1}{12}$

The LCD of 5 and 12 is 60.

$$\frac{1}{5}+\frac{1}{12}=\frac{1}{5}\cdot\frac{12}{12}+\frac{1}{12}\cdot\frac{5}{5}=\frac{12}{60}+\frac{5}{60}=\frac{17}{60}$$

51. $\dfrac{5}{6}\div\dfrac{1}{3}\cdot\dfrac{1}{4}=\dfrac{5}{6}\cdot\dfrac{3}{1}\cdot\dfrac{1}{4}=\dfrac{15}{6}\cdot\dfrac{1}{4}=\dfrac{15}{24}=\dfrac{3\cdot5}{3\cdot8}=\dfrac{5}{8}$

53. $\dfrac{1}{5}\left(2\dfrac{5}{6}-\dfrac{1}{3}\right)=\dfrac{1}{5}\left(\dfrac{17}{6}-\dfrac{1}{3}\cdot\dfrac{2}{2}\right)$

$$=\frac{1}{5}\left(\frac{17}{6}-\frac{2}{6}\right)$$
$$=\frac{1}{5}\left(\frac{15}{6}\right)$$
$$=\frac{1\cdot15}{5\cdot6}$$
$$=\frac{1\cdot5\cdot3}{5\cdot3\cdot2}$$
$$=\frac{1}{2}$$

55. $2\cdot\left(\dfrac{1}{4}+\dfrac{1}{5}\right)+2=2\cdot\left(\dfrac{1}{4}\cdot\dfrac{5}{5}+\dfrac{1}{5}\cdot\dfrac{4}{4}\right)+2$

$$=2\cdot\left(\frac{5}{20}+\frac{4}{20}\right)+2$$
$$=2\cdot\left(\frac{9}{20}\right)+2$$
$$=\frac{2}{1}\cdot\frac{9}{20}+2$$
$$=\frac{2\cdot9}{1\cdot20}+2$$
$$=\frac{2\cdot9}{1\cdot2\cdot10}+2$$
$$=\frac{9}{10}+2$$
$$=2\frac{9}{10}$$

57. $\left(\dfrac{3}{4}\right)^2 \div \left(\dfrac{3}{4} - \dfrac{1}{12}\right) = \left(\dfrac{3}{4}\right)^2 \div \left(\dfrac{3}{4} \cdot \dfrac{3}{3} - \dfrac{1}{12}\right)$

$= \left(\dfrac{3}{4}\right)^2 \div \left(\dfrac{9}{12} - \dfrac{1}{12}\right)$

$= \left(\dfrac{3}{4}\right)^2 \div \left(\dfrac{8}{12}\right)$

$= \left(\dfrac{3}{4}\right)^2 \div \left(\dfrac{4 \cdot 2}{4 \cdot 3}\right)$

$= \left(\dfrac{3}{4}\right)^2 \div \left(\dfrac{2}{3}\right)$

$= \left(\dfrac{3}{4} \cdot \dfrac{3}{4}\right) \div \left(\dfrac{2}{3}\right)$

$= \dfrac{9}{16} \div \dfrac{2}{3}$

$= \dfrac{9}{16} \cdot \dfrac{3}{2}$

$= \dfrac{9 \cdot 3}{16 \cdot 2}$

$= \dfrac{27}{32}$

59. $\left(\dfrac{2}{3} - \dfrac{5}{9}\right)^2 = \left(\dfrac{2}{3} \cdot \dfrac{3}{3} - \dfrac{5}{9}\right)^2$

$= \left(\dfrac{6}{9} - \dfrac{5}{9}\right)^2$

$= \left(\dfrac{1}{9}\right)^2$

$= \dfrac{1}{9} \cdot \dfrac{1}{9}$

$= \dfrac{1}{81}$

61. $\dfrac{5}{9} \cdot \dfrac{1}{2} + \dfrac{2}{3} \cdot \dfrac{5}{6} = \dfrac{5 \cdot 1}{9 \cdot 2} + \dfrac{2 \cdot 5}{3 \cdot 6}$

$= \dfrac{5}{18} + \dfrac{10}{18}$

$= \dfrac{15}{18}$

$= \dfrac{3 \cdot 5}{3 \cdot 6}$

$= \dfrac{5}{6}$

63. $\dfrac{27}{16} \cdot \left(\dfrac{2}{3}\right)^2 - \dfrac{3}{20} = \dfrac{27}{16} \cdot \left(\dfrac{2}{3} \cdot \dfrac{2}{3}\right) - \dfrac{3}{20}$

$= \dfrac{3 \cdot 3 \cdot 3 \cdot 2 \cdot 2}{2 \cdot 2 \cdot 2 \cdot 2 \cdot 3 \cdot 3} - \dfrac{3}{20}$

$= \dfrac{3}{2 \cdot 2} - \dfrac{3}{20}$

$= \dfrac{3}{4} - \dfrac{3}{20}$

$= \dfrac{3}{4} \cdot \dfrac{5}{5} - \dfrac{3}{20}$

$= \dfrac{15}{20} - \dfrac{3}{20}$

$= \dfrac{12}{20}$

$= \dfrac{4 \cdot 3}{4 \cdot 5}$

$= \dfrac{3}{5}$

65. $\dfrac{3}{13} \div \dfrac{9}{26} - \dfrac{7}{24} \cdot \dfrac{8}{14} = \dfrac{3}{13} \cdot \dfrac{26}{9} - \dfrac{7}{24} \cdot \dfrac{8}{14}$

$= \dfrac{3 \cdot 26}{13 \cdot 9} - \dfrac{7 \cdot 8}{24 \cdot 14}$

$= \dfrac{3 \cdot 13 \cdot 2}{13 \cdot 3 \cdot 3} - \dfrac{7 \cdot 8}{8 \cdot 3 \cdot 7 \cdot 2}$

$= \dfrac{2}{3} - \dfrac{1}{6}$

$= \dfrac{2}{3} \cdot \dfrac{2}{2} - \dfrac{1}{6}$

$= \dfrac{4}{6} - \dfrac{1}{6}$

$= \dfrac{3}{6}$

$= \dfrac{1}{2}$

67. $\dfrac{3}{14}+\dfrac{10}{21}\div\left(\dfrac{3}{7}\right)\left(\dfrac{9}{4}\right)=\dfrac{3}{14}+\dfrac{10}{21}\cdot\dfrac{7}{3}\left(\dfrac{9}{4}\right)$

$\qquad\qquad = \dfrac{3}{14}+\dfrac{10\cdot7}{21\cdot3}\cdot\dfrac{9}{4}$

$\qquad\qquad = \dfrac{3}{14}+\dfrac{10\cdot7}{7\cdot3\cdot3}\cdot\dfrac{9}{4}$

$\qquad\qquad = \dfrac{3}{14}+\dfrac{10}{3\cdot3}\cdot\dfrac{9}{4}$

$\qquad\qquad = \dfrac{3}{14}+\dfrac{2\cdot5\cdot3\cdot3}{3\cdot3\cdot2\cdot2}$

$\qquad\qquad = \dfrac{3}{14}+\dfrac{5}{2}$

$\qquad\qquad = \dfrac{3}{14}+\dfrac{5}{2}\cdot\dfrac{7}{7}$

$\qquad\qquad = \dfrac{3}{14}+\dfrac{35}{14}$

$\qquad\qquad = \dfrac{38}{14}$

$\qquad\qquad = \dfrac{2\cdot19}{2\cdot7}$

$\qquad\qquad = \dfrac{19}{7}$ or $2\dfrac{5}{7}$

69. $\left(\dfrac{3}{4}+\dfrac{1}{8}\right)^2-\left(\dfrac{1}{2}+\dfrac{1}{8}\right)$

$=\left(\dfrac{3}{4}\cdot\dfrac{2}{2}+\dfrac{1}{8}\right)^2-\left(\dfrac{1}{2}\cdot\dfrac{4}{4}+\dfrac{1}{8}\right)$

$=\left(\dfrac{6}{8}+\dfrac{1}{8}\right)^2-\left(\dfrac{4}{8}+\dfrac{1}{8}\right)$

$=\left(\dfrac{7}{8}\right)^2-\left(\dfrac{5}{8}\right)$

$=\left(\dfrac{7}{8}\cdot\dfrac{7}{8}\right)-\dfrac{5}{8}$

$=\dfrac{49}{64}-\dfrac{5}{8}$

$=\dfrac{49}{64}-\dfrac{5}{8}\cdot\dfrac{8}{8}$

$=\dfrac{49}{64}-\dfrac{40}{64}$

$=\dfrac{9}{64}$

71. The average is the sum, divided by 2.

$\left(\dfrac{5}{6}+\dfrac{2}{3}\right)\div2=\left(\dfrac{5}{6}+\dfrac{2}{3}\cdot\dfrac{2}{2}\right)\div2$

$\qquad\qquad = \left(\dfrac{5}{6}+\dfrac{4}{6}\right)\div2$

$\qquad\qquad = \dfrac{9}{6}\div\dfrac{2}{1}$

$\qquad\qquad = \dfrac{9}{6}\cdot\dfrac{1}{2}$

$\qquad\qquad = \dfrac{9\cdot1}{6\cdot2}$

$\qquad\qquad = \dfrac{3\cdot3\cdot1}{2\cdot3\cdot2}$

$\qquad\qquad = \dfrac{3\cdot1}{2\cdot2}$

$\qquad\qquad = \dfrac{3}{4}$

The average is $\dfrac{3}{4}$.

73. The average is the sum, divided by 3.

$\left(\dfrac{1}{5}+\dfrac{3}{10}+\dfrac{3}{20}\right)\div3=\left(\dfrac{1}{5}\cdot\dfrac{4}{4}+\dfrac{3}{10}\cdot\dfrac{2}{2}+\dfrac{3}{20}\right)\div3$

$\qquad\qquad = \left(\dfrac{4}{20}+\dfrac{6}{20}+\dfrac{3}{20}\right)\div3$

$\qquad\qquad = \dfrac{13}{20}\div3$

$\qquad\qquad = \dfrac{13}{20}\cdot\dfrac{1}{3}$

$\qquad\qquad = \dfrac{13}{60}$

75. Find the average by adding the three fractions and dividing the sum by 3.

$\left(\dfrac{23}{50}+\dfrac{1}{2}+\dfrac{3}{5}\right)\div3=\left(\dfrac{23}{50}+\dfrac{1}{2}\cdot\dfrac{25}{25}+\dfrac{3}{5}\cdot\dfrac{10}{10}\right)\div3$

$\qquad\qquad = \left(\dfrac{23}{50}+\dfrac{25}{50}+\dfrac{30}{50}\right)\div3$

$\qquad\qquad = \dfrac{78}{50}\div3$

$\qquad\qquad = \dfrac{78}{50}\cdot\dfrac{1}{3}$

$\qquad\qquad = \dfrac{26}{50}$

$\qquad\qquad = \dfrac{13}{25}$

The average fraction of online sales is $\dfrac{13}{25}$.

77. "Increased by" is most likely to translate to addition (A).

79. "Triple" is most likely to translate to multiplication (M).

81. "Subtracted from" is most likely to translate to subtraction (S).

83. "Quotient" is most likely to translate to division (D).

85. "Times" is most likely to translate to multiplication (M).

87. "Total" is most likely to translate to addition (A).

89. $\dfrac{2^3}{3} = \dfrac{2 \cdot 2 \cdot 2}{3} = \dfrac{8}{3}$

$\left(\dfrac{2}{3}\right)^3 = \dfrac{2}{3} \cdot \dfrac{2}{3} \cdot \dfrac{2}{3} = \dfrac{8}{27}$

$\dfrac{2^3}{3}$ and $\left(\dfrac{2}{3}\right)^3$ do not simplify to the same value.

answers may vary

91. The operations should be done in the order: subtraction, multiplication, addition, division

93. The operations should be done in the order: division, multiplication, subtraction, addition

95. Compare the fractions by writing as like fractions.
The LCD of 250 and 100 is 500.

$\dfrac{114}{250} = \dfrac{114}{250} \cdot \dfrac{2}{2} = \dfrac{228}{500}$

$\dfrac{49}{100} = \dfrac{49}{100} \cdot \dfrac{5}{5} = \dfrac{245}{500}$

Since $228 < 245$, then $\dfrac{228}{500} < \dfrac{245}{500}$, so $\dfrac{114}{250} < \dfrac{49}{100}$. Standard mail accounts for a greater portion of the mail handled by weight.

97. Compare the fractions by writing as like fractions.
The LCD of 160 and 64 is 320.

$\dfrac{11}{160} = \dfrac{11}{160} \cdot \dfrac{2}{2} = \dfrac{22}{320}$

$\dfrac{5}{64} = \dfrac{5}{64} \cdot \dfrac{5}{5} = \dfrac{25}{320}$

Since $22 < 25$, then $\dfrac{22}{320} < \dfrac{25}{320}$, so $\dfrac{11}{160} < \dfrac{5}{64}$. New York is the birthplace of the greater number of astronauts.

Section 3.6 Practice

1. The volume of a box is the product of its length, width, and height.

   volume of a box $\;$ is $\;$ length $\cdot$ width $\cdot$ height
   $\downarrow\qquad\qquad\downarrow\qquad\downarrow\qquad\downarrow$

   volume of a box $= \;4\dfrac{1}{3}$ ft $\cdot\; 1\dfrac{1}{2}$ ft $\cdot\; 3\dfrac{1}{3}$ ft

   $= \dfrac{13}{3}\cdot\dfrac{3}{2}\cdot\dfrac{10}{3}$ cubic feet

   $= \dfrac{13\cdot3\cdot2\cdot5}{3\cdot2\cdot3}$ cubic feet

   $= \dfrac{13\cdot5}{3}$ cubic feet

   $= \dfrac{65}{3}$ or $21\dfrac{2}{3}$ cubic feet

   The volume of the box is $21\dfrac{2}{3}$ cubic feet.

2. The phrase "total width" tells us to add.

   total width $\;$ is $\;$ first width $+$ second width $+$ third width
   $\downarrow\qquad\quad\downarrow\qquad\downarrow\qquad\qquad\downarrow\qquad\qquad\downarrow$

   total width $= \;\dfrac{11}{16}$ in. $+\;\dfrac{5}{8}$ in. $\;+\;\dfrac{11}{16}$ in.

   $\dfrac{11}{16}+\dfrac{5}{8}+\dfrac{11}{6}=\dfrac{11}{16}+\dfrac{5\cdot2}{8\cdot2}+\dfrac{11}{16}=\dfrac{11}{16}+\dfrac{10}{16}+\dfrac{11}{16}=\dfrac{32}{16}=2$

   The total width is 2 inches.

3. The phrase "the rest of the land" tells us that initially we are to subtract.

   acreage for lots $\;$ is $\;$ total acreage $\;$ minus $\;$ acreage for roads and wetlands
   $\downarrow\qquad\qquad\downarrow\qquad\downarrow\qquad\qquad\downarrow\qquad\qquad\downarrow$

   acreage for lots $=\qquad 25\qquad -\qquad\qquad 6\dfrac{2}{3}$

   $\begin{array}{rcl}25 &=& 24\dfrac{3}{3}\\[2mm] -\,6\dfrac{2}{3} &=& -\,6\dfrac{2}{3}\\[2mm]\hline && 18\dfrac{1}{3}\end{array}$

   Now calculate how many $\dfrac{5}{6}$-acre lots the available land can be divided into.

   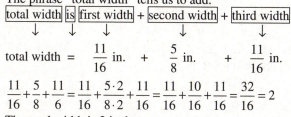 number of $\dfrac{5}{6}$-acre lots $\;$ is $\;$ acreage for lots $\;$ divided by $\;$ size of each lot
   $\downarrow\qquad\qquad\downarrow\qquad\qquad\downarrow\qquad\qquad\downarrow$

   number of $\dfrac{5}{6}$-acre lots $=\qquad 18\dfrac{1}{3}\qquad\div\qquad \dfrac{5}{6}$

   $18\dfrac{1}{3}\div\dfrac{5}{6}=\dfrac{55}{3}\cdot\dfrac{6}{5}=\dfrac{5\cdot11\cdot2\cdot3}{3\cdot5}=\dfrac{22}{1}=22$

   The land can be divided into $22\;\dfrac{5}{6}$-acre lots.

Exercise Set 3.6

1. "Sum" translates as addition.

$$\frac{1}{2}+\frac{1}{3}$$

3. "Quotient" translates as division.

$$20 \div 6\frac{2}{5}$$

5. "Subtract" indicates subtraction.

$$\frac{15}{16}-\frac{5}{8}$$

7. "Increased by" translates as addition.

$$\frac{21}{68}+\frac{7}{34}$$

9. "Product" translates as multiplication.

$$8\frac{1}{3}\cdot\frac{7}{9}$$

11.

sugar needed	is	double	recipe amount of sugar
↓	↓	↓	↓
sugar needed	=	$2\cdot$	$1\frac{1}{3}$

$$=\frac{2}{1}\cdot\frac{5}{3}=\frac{10}{3} \text{ or } 3\frac{1}{3}$$

You need $\frac{10}{3}$ or $3\frac{1}{3}$ cups of sugar.

13.

Wall height	is	number of bricks	times	brick width	plus	layers of mortar	times	mortar width
↓	↓	↓	↓	↓	↓	↓	↓	↓
Height	=	4	·	$2\frac{3}{4}$	+	3	·	$\frac{1}{2}$

$$=4\cdot\frac{11}{4}+3\cdot\frac{1}{2}=11+\frac{3}{2}=11+1+\frac{1}{2}=12\frac{1}{2}$$

The wall is $12\frac{1}{2}$ inches high.

15.

Miles per gallon	is	miles driven	divided by	gallons of gas used
↓	↓	↓	↓	↓
Miles per gallon	=	$290\frac{1}{4}$	÷	$13\frac{1}{2}$

$$= \frac{1161}{4} \div \frac{27}{2} = \frac{1161}{4} \cdot \frac{2}{27} = \frac{43 \cdot 27 \cdot 2}{2 \cdot 2 \cdot 27} = \frac{43}{2} \text{ or } 21\frac{1}{2}$$

Doug and Claudia Scaggs got $\frac{43}{2}$ or $21\frac{1}{2}$ miles per gallon in their vehicle.

17.

Bill life	is	$\frac{1}{20}$	times	coin life
↓	↓	↓	↓	↓
Life	=	$\frac{1}{20}$	·	30

$$= \frac{30}{20} = \frac{10 \cdot 3}{10 \cdot 2} = \frac{3}{2} \text{ or } 1\frac{1}{2}$$

The life expectancy of circulating paper money is $1\frac{1}{2}$ years.

19.

Extra width of Spain's gauge	is	width of Spain's gauge	minus	width of U.S. gauge
↓	↓	↓	↓	↓
Extra width	=	$65\frac{9}{10}$	–	$56\frac{1}{2}$

$$= 65\frac{9}{10} - 56\frac{5}{10} = 9\frac{4}{10} = 9\frac{2}{5}$$

Spain's standard gauge is $9\frac{2}{5}$ inches wider than the U.S. standard gauge.

21.

Total needed	is	amount for large shirt	plus	number of small shirts	times	amount for small shirt
↓	↓	↓	↓	↓	↓	↓
Total	=	$1\frac{1}{2}$	+	5	·	$\frac{3}{4}$

$$= 1\frac{1}{2} + 5 \cdot \frac{3}{4} = \frac{3}{2} + \frac{15}{4} = \frac{3}{2} \cdot \frac{2}{2} + \frac{15}{4} = \frac{6}{4} + \frac{15}{4} = \frac{21}{4} \text{ or } 5\frac{1}{4}$$

The amount of cloth needed is $5\frac{1}{4}$ yards, so the 5-yard remnant is not enough. Another $\frac{1}{4}$ yard of material is required.

23.

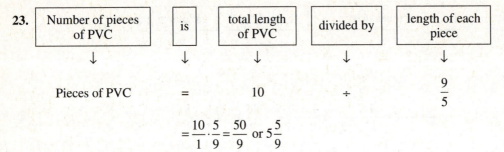

Number of pieces of PVC	is	total length of PVC	divided by	length of each piece
↓	↓	↓	↓	↓
Pieces of PVC	=	10	÷	$\dfrac{9}{5}$

$$= \frac{10}{1} \cdot \frac{5}{9} = \frac{50}{9} \text{ or } 5\frac{5}{9}$$

The plumber can cut $5\dfrac{9}{5}$-foot pieces from a 10-foot piece of PVC.

25.

Outer diameter	is	left thickness	plus	inner diameter	plus	right thickness
↓	↓	↓	↓	↓	↓	↓
Outer diameter	=	$\dfrac{3}{16}$	+	$\dfrac{3}{4}$	+	$\dfrac{3}{16}$

$$= \frac{3}{16} + \frac{3}{4} + \frac{3}{16} = \frac{3}{16} + \frac{3}{4} \cdot \frac{4}{4} + \frac{3}{16} = \frac{3}{16} + \frac{12}{16} + \frac{3}{16} = \frac{18}{16} = \frac{2 \cdot 9}{2 \cdot 8} = \frac{9}{8} \text{ or } 1\frac{1}{8}$$

The outer diameter is $\dfrac{9}{8}$ or $1\dfrac{1}{8}$ inches.

27.

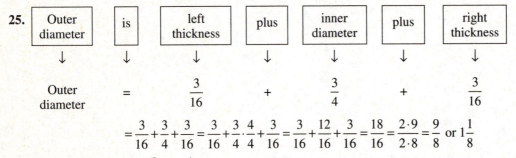

Total amount of flour	is	number of recipes	times	flour needed for each recipe
↓	↓	↓	↓	↓
Total	=	$1\dfrac{1}{2}$	·	$2\dfrac{1}{2}$

$$= \frac{3}{2} \cdot \frac{5}{2} = \frac{15}{4} \text{ or } 3\frac{3}{4}$$

The amount of flour needed is $\dfrac{15}{4}$ or $3\dfrac{3}{4}$ cups.

29.

Area of photograph	is	length	times	width
↓	↓	↓	↓	↓
Area	=	$4\dfrac{1}{2}$	·	$2\dfrac{1}{2}$

$$= 4\frac{1}{2} \cdot 2\frac{1}{2} = \frac{9}{2} \cdot \frac{5}{2} = \frac{9 \cdot 5}{2 \cdot 2} = \frac{45}{4} \text{ or } 11\frac{1}{4}$$

The area of the photograph is $11\dfrac{1}{4}$ square inches.

31.

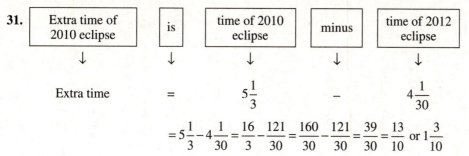

$$\text{Extra time} = 5\frac{1}{3} - 4\frac{1}{30}$$

$$=5\frac{1}{3}-4\frac{1}{30}=\frac{16}{3}-\frac{121}{30}=\frac{160}{30}-\frac{121}{30}=\frac{39}{30}=\frac{13}{10}\text{ or }1\frac{3}{10}$$

The 2010 eclipse is $\frac{13}{10}$ or $1\frac{3}{10}$ minutes longer than the 2012 eclipse.

33.

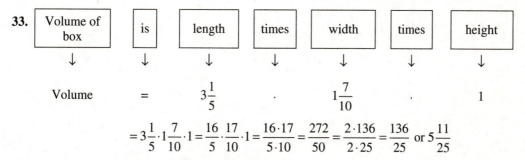

$$\text{Volume} = 3\frac{1}{5} \cdot 1\frac{7}{10} \cdot 1$$

$$=3\frac{1}{5}\cdot1\frac{7}{10}\cdot1=\frac{16}{5}\cdot\frac{17}{10}\cdot1=\frac{16\cdot17}{5\cdot10}=\frac{272}{50}=\frac{2\cdot136}{2\cdot25}=\frac{136}{25}\text{ or }5\frac{11}{25}$$

The volume of the box is $5\frac{11}{25}$ cubic inches.

35.

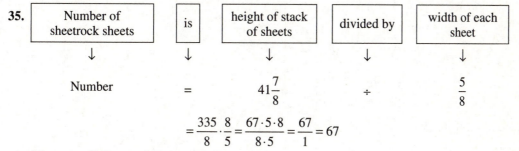

$$\text{Number} = 41\frac{7}{8} \div \frac{5}{8}$$

$$=\frac{335}{8}\cdot\frac{8}{5}=\frac{67\cdot5\cdot8}{8\cdot5}=\frac{67}{1}=67$$

There are 67 sheets of sheetrock in the stack.

37. a.

$$\begin{array}{|c|c|c|c|c|}
\hline
\text{Length}\\\text{needed} & \text{is} & \text{number}\\\text{of}\\\text{pieces} & \text{times} & \text{length of}\\\text{each}\\\text{piece}\\
\hline
\end{array}$$

$$\downarrow \qquad \downarrow \qquad \downarrow \qquad \downarrow \qquad \downarrow$$

$$\text{Length} = 12 \cdot \frac{3}{4}$$

$$=12\cdot\frac{3}{4}=\frac{12}{1}\cdot\frac{3}{4}=\frac{4\cdot3\cdot3}{1\cdot4}=\frac{3\cdot3}{1}=9$$

Since he needs a total of 9 feet of tubing, the 10-foot piece is enough.

b. $10 - 9 = 1$

He will have 1 foot of tubing left over.

39.

Average	is	sum of values	divided by	number of values
↓	↓	↓	↓	↓

$$\text{Average} \quad = \quad \left(2\frac{1}{8}+2\frac{7}{8}+3\frac{1}{4}+3\frac{1}{2}\right) \quad \div \quad 4$$

$$=\left(2\frac{1}{8}+2\frac{7}{8}+3\frac{2}{8}+3\frac{4}{8}\right)\div 4=\left(10\frac{14}{8}\right)\div 4=\left(11\frac{6}{8}\right)\div 4=\left(11\frac{3}{4}\right)\div\frac{4}{1}=\frac{47}{4}\cdot\frac{1}{4}=\frac{47}{16}=2\frac{15}{16}$$

The average cub weight is $2\frac{15}{16}$ pounds.

41.

Area	is	length	times	width
↓	↓	↓	↓	↓

$$\text{Area} \quad = \quad \frac{3}{16} \quad \cdot \quad \frac{3}{8}$$

$$=\frac{3\cdot 3}{16\cdot 8}=\frac{9}{128}$$

The area is $\frac{9}{128}$ square inch.

Perimeter	is	two	times	length	plus	two	times	width
↓	↓	↓	↓	↓	↓	↓	↓	↓

$$\text{Perimeter} \quad = \quad 2 \quad \cdot \quad \frac{3}{16} \quad + \quad 2 \quad \cdot \quad \frac{3}{8}$$

$$=\frac{2\cdot 3}{16}+\frac{2\cdot 3}{8}=\frac{6}{16}+\frac{6}{8}=\frac{6}{16}+\frac{12}{16}=\frac{18}{16}=\frac{9}{8}=1\frac{1}{8}$$

The perimeter is $1\frac{1}{8}$ inches.

43.

Area	is	length of side	times	length of side
↓	↓	↓	↓	↓

$$\text{Area} \quad = \quad \frac{5}{9} \quad \cdot \quad \frac{5}{9}$$

$$=\frac{5\cdot 5}{9\cdot 9}=\frac{25}{81}$$

The area is $\frac{25}{81}$ square meter.

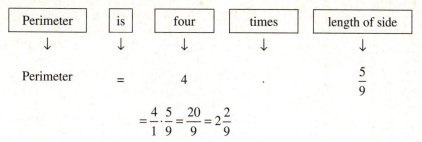

$$= \frac{4}{1} \cdot \frac{5}{9} = \frac{20}{9} = 2\frac{2}{9}$$

The perimeter is $2\frac{2}{9}$ meters.

45.

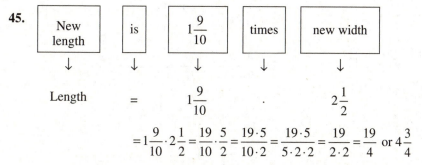

$$= 1\frac{9}{10} \cdot 2\frac{1}{2} = \frac{19}{10} \cdot \frac{5}{2} = \frac{19 \cdot 5}{10 \cdot 2} = \frac{19 \cdot 5}{5 \cdot 2 \cdot 2} = \frac{19}{2 \cdot 2} = \frac{19}{4} \text{ or } 4\frac{3}{4}$$

The length of the new flag is $4\frac{3}{4}$ feet.

47.

Width of each stripe	is	width of flag	divided by	number of stripes
↓	↓	↓	↓	↓
Width	=	$2\frac{1}{2}$	÷	13

$$= \frac{5}{2} \div \frac{13}{1} = \frac{5}{2} \cdot \frac{1}{13} = \frac{5}{26}$$

The width of each stripe is $\frac{5}{26}$ of a foot.

49. $\sqrt{9} = 3$, because $3 \cdot 3 = 9$.

51. $9^2 = 9 \cdot 9 = 81$

53. $8 \div 4 \cdot 2 = 2 \cdot 2 = 4$

55. $3^2 - 2^2 + 5^2 = 3 \cdot 3 - 2 \cdot 2 + 5 \cdot 5 = 9 - 4 + 25 = 5 + 25 = 30$

57. $5 + 3[14 - (12 \div 3)] = 5 + 3[14 - 4] = 5 + 3[10] = 5 + 30 = 35$

59. no; no; answers may vary

61. Length of small rectangle:

$$6\frac{7}{9}-3\frac{8}{9}=5\frac{16}{9}-3\frac{8}{9}=2\frac{8}{9}$$

Width of small rectangle:

$$6\frac{2}{3}-4\frac{4}{9}=6\frac{6}{9}-4\frac{4}{9}=2\frac{2}{9}$$

Area of small rectangle:

$$2\frac{8}{9}\cdot2\frac{2}{9}=\frac{26}{9}\cdot\frac{20}{9}=\frac{520}{81}=6\frac{34}{81}$$

Area of large rectangle:

$$6\frac{7}{9}\cdot4\frac{4}{9}=\frac{61}{9}\cdot\frac{40}{9}=\frac{2440}{81}=30\frac{10}{81}$$

The total area of the figure: $6\frac{34}{81}+30\frac{10}{81}=36\frac{44}{81}$

The total area of the figure is $36\frac{44}{81}$ square feet.

63. Customers ordering a $7 lunch:

$$\frac{3}{10}\cdot\frac{240}{1}=\frac{3\cdot24\cdot10}{10\cdot1}=72$$

Customers ordering a $5 lunch:

$$\frac{5}{12}\cdot\frac{240}{1}=\frac{5\cdot12\cdot20}{12\cdot1}=100$$

Customers ordering a $9 lunch:
$240-72-100=168-100=68$
68 customers ordered a $9 lunch.

65. 3 years = 3 · 365 · 24 hours = 26,280 hours
Time now to produce one million coins:

$$\frac{11}{13,140}\cdot\frac{26,280}{1}=\frac{11\cdot2\cdot13,140}{13,140\cdot1}=22$$

Today, it only takes 22 hours to produce one million coins.

Chapter 3 Vocabulary Check

1. Fractions that have the same denominator are called <u>like</u> fractions.

2. The <u>least common multiple</u> is the smallest number that is a multiple of all numbers in a list of numbers.

3. <u>Equivalent</u> fractions represent the same portion of a whole.

4. A <u>mixed number</u> has a whole number part and a fraction part.

5. The symbol $\geq$ means is greater than.

6. The symbol $\leq$ means is less than.

7. The LCM of the denominators in a list of fractions is called the <u>least common denominator</u>.

8. Fractions that have different denominators are called <u>unlike</u> fractions.

9. A shorthand notation for repeated multiplication of the same factor is a(n) <u>exponent</u>.

Chapter 3 Review

1. $\dfrac{7}{11}+\dfrac{3}{11}=\dfrac{7+3}{11}=\dfrac{10}{11}$

2. $\dfrac{4}{50}+\dfrac{2}{50}=\dfrac{4+2}{50}=\dfrac{6}{50}=\dfrac{2\cdot3}{2\cdot25}=\dfrac{3}{25}$

3. $\dfrac{11}{15}-\dfrac{1}{15}=\dfrac{11-1}{15}=\dfrac{10}{15}=\dfrac{5\cdot2}{5\cdot3}=\dfrac{2}{3}$

4. $\dfrac{4}{21}-\dfrac{1}{21}=\dfrac{4-1}{21}=\dfrac{3}{21}=\dfrac{3\cdot1}{3\cdot7}=\dfrac{1}{7}$

5. $\dfrac{4}{15}+\dfrac{3}{15}+\dfrac{2}{15}=\dfrac{4+3+2}{15}=\dfrac{9}{15}=\dfrac{3\cdot3}{3\cdot5}=\dfrac{3}{5}$

6. $\dfrac{3}{20}+\dfrac{7}{20}+\dfrac{2}{20}=\dfrac{3+7+2}{20}=\dfrac{12}{20}=\dfrac{4\cdot3}{4\cdot5}=\dfrac{3}{5}$

7. $\dfrac{1}{12}+\dfrac{11}{12}=\dfrac{1+11}{12}=\dfrac{12}{12}=1$

8. $\dfrac{3}{4}+\dfrac{1}{4}=\dfrac{3+1}{4}=\dfrac{4}{4}=1$

9. $\dfrac{11}{25}+\dfrac{6}{25}+\dfrac{2}{25}=\dfrac{11+6+2}{25}=\dfrac{19}{25}$

10. $\dfrac{4}{21}+\dfrac{1}{21}+\dfrac{11}{21}=\dfrac{4+1+11}{21}=\dfrac{16}{21}$

11. Add the fractional amounts of homework.

$$\frac{3}{8}+\frac{2}{8}+\frac{1}{8}=\frac{3+2+1}{8}=\frac{6}{8}=\frac{2\cdot3}{2\cdot4}=\frac{3}{4}$$

Mark did $\dfrac{3}{4}$ of his homework that evening.

12. Perimeter $= 2 \cdot \text{length} + 2 \cdot \text{width}$

$$= 2 \cdot \frac{9}{16} + 2 \cdot \frac{3}{16}$$

$$= \frac{2}{1} \cdot \frac{9}{16} + \frac{2}{1} \cdot \frac{3}{16}$$

$$= \frac{2 \cdot 9}{1 \cdot 16} + \frac{2 \cdot 3}{1 \cdot 16}$$

$$= \frac{18}{16} + \frac{6}{16}$$

$$= \frac{24}{16}$$

$$= \frac{8 \cdot 3}{8 \cdot 2}$$

$$= \frac{3}{2} \text{ or } 1\frac{1}{2}$$

The perimeter of the Simpson's land is $\frac{3}{2}$ or $1\frac{1}{2}$ miles.

13. $5 = \boxed{5}$
$11 = \boxed{11}$
$\text{LCM} = 5 \cdot 11 = 55$

14. $20 = \boxed{2 \cdot 2} \cdot \boxed{5}$
$30 = 2 \cdot \boxed{3} \cdot 5$
$\text{LCM} = 2 \cdot 2 \cdot 3 \cdot 5 = 60$

15. $20 = 2 \cdot 2 \cdot \boxed{5}$
$24 = \boxed{2 \cdot 2 \cdot 2} \cdot \boxed{3}$
$\text{LCM} = 2 \cdot 2 \cdot 2 \cdot 3 \cdot 5 = 120$

16. $16 = \boxed{2 \cdot 2 \cdot 2 \cdot 2}$
$5 = \boxed{5}$
$\text{LCM} = 2 \cdot 2 \cdot 2 \cdot 2 \cdot 5 = 80$

17. $12 = \boxed{2 \cdot 2} \cdot 3$
$21 = 3 \cdot \boxed{7}$
$63 = \boxed{3 \cdot 3} \cdot 7$
$\text{LCM} = 2 \cdot 2 \cdot 3 \cdot 3 \cdot 7 = 252$

18. $6 = 2 \cdot 3$
$8 = \boxed{2 \cdot 2 \cdot 2}$
$18 = 2 \cdot \boxed{3 \cdot 3}$
$\text{LCM} = 2 \cdot 2 \cdot 2 \cdot 3 \cdot 3 = 72$

19. $\dfrac{7}{8} = \dfrac{7}{8} \cdot \dfrac{8}{8} = \dfrac{56}{64}$

20. $\dfrac{2}{3} = \dfrac{2}{3} \cdot \dfrac{10}{10} = \dfrac{20}{30}$

21. $\dfrac{7}{11} = \dfrac{7}{11} \cdot \dfrac{3}{3} = \dfrac{21}{33}$

22. $\dfrac{10}{13} = \dfrac{10}{13} \cdot \dfrac{2}{2} = \dfrac{20}{26}$

23. $\dfrac{4}{15} = \dfrac{4}{15} \cdot \dfrac{4}{4} = \dfrac{16}{60}$

24. $\dfrac{5}{12} = \dfrac{5}{12} \cdot \dfrac{5}{5} = \dfrac{25}{60}$

25. The LCD of 18 and 9 is 18.
$$\frac{7}{18} + \frac{2}{9} = \frac{7}{18} + \frac{2}{9} \cdot \frac{2}{2} = \frac{7}{18} + \frac{4}{18} = \frac{11}{18}$$

26. The LCD of 15 and 5 is 15.
$$\frac{4}{15} + \frac{1}{5} = \frac{4}{15} + \frac{1}{5} \cdot \frac{3}{3} = \frac{4}{15} + \frac{3}{15} = \frac{7}{15}$$

27. The LCD of 13 and 26 is 26.
$$\frac{4}{13} - \frac{1}{26} = \frac{4}{13} \cdot \frac{2}{2} - \frac{1}{26} = \frac{8}{26} - \frac{1}{26} = \frac{7}{26}$$

28. The LCD of 12 and 9 is 36.
$$\frac{7}{12} - \frac{1}{9} = \frac{7}{12} \cdot \frac{3}{3} - \frac{1}{9} \cdot \frac{4}{4} = \frac{21}{36} - \frac{4}{36} = \frac{17}{36}$$

29. The LCD of 3 and 14 is 42.
$$\frac{1}{3} + \frac{9}{14} = \frac{1}{3} \cdot \frac{14}{14} + \frac{9}{14} \cdot \frac{3}{3} = \frac{14}{42} + \frac{27}{42} = \frac{41}{42}$$

30. The LCD of 18 and 24 is 72.
$$\frac{7}{18} + \frac{5}{24} = \frac{7}{18} \cdot \frac{4}{4} + \frac{5}{24} \cdot \frac{3}{3} = \frac{28}{72} + \frac{15}{72} = \frac{43}{72}$$

31. The LCD of 15 and 9 is 45.
$$\frac{11}{15} - \frac{4}{9} = \frac{11}{15} \cdot \frac{3}{3} - \frac{4}{9} \cdot \frac{5}{5} = \frac{33}{45} - \frac{20}{45} = \frac{13}{45}$$

32. The LCD of 14 and 35 is 70.
$$\frac{9}{14} - \frac{3}{35} = \frac{9}{14} \cdot \frac{5}{5} - \frac{3}{35} \cdot \frac{2}{2} = \frac{45}{70} - \frac{6}{70} = \frac{39}{70}$$

33. Perimeter $= 2 \cdot \text{length} + 2 \cdot \text{width}$

$$= \frac{2}{1} \cdot \frac{5}{6} + \frac{2}{1} \cdot \frac{2}{9}$$

$$= \frac{2 \cdot 5}{1 \cdot 6} + \frac{2 \cdot 2}{1 \cdot 9}$$

$$= \frac{2 \cdot 5}{1 \cdot 2 \cdot 3} + \frac{4}{9}$$

$$= \frac{5}{3} + \frac{4}{9}$$

$$= \frac{5}{3} \cdot \frac{3}{3} + \frac{4}{9}$$

$$= \frac{15}{9} + \frac{4}{9}$$

$$= \frac{19}{9} \text{ or } 2\frac{1}{9}$$

The perimeter of the rectangle is $\frac{19}{9}$ or

$2\frac{1}{9}$ meters.

34. Perimeter is the distance around. The LCD of 5 and 10 is 10.

$$\text{Perimeter} = \frac{1}{5} + \frac{3}{5} + \frac{7}{10}$$

$$= \frac{1}{5} \cdot \frac{2}{2} + \frac{3}{5} \cdot \frac{2}{2} + \frac{7}{10}$$

$$= \frac{2}{10} + \frac{6}{10} + \frac{7}{10}$$

$$= \frac{15}{10}$$

$$= \frac{5 \cdot 3}{5 \cdot 2}$$

$$= \frac{3}{2} \text{ or } 1\frac{1}{2}$$

The perimeter of the triangle is $\frac{3}{2}$ or $1\frac{1}{2}$ feet.

35. Subtract the length of the shorter scarf from the length of the longer scarf. The LCD of 3 and 12 is 12.

$$\frac{2}{3} - \frac{5}{12} = \frac{2}{3} \cdot \frac{4}{4} - \frac{5}{12} = \frac{8}{12} - \frac{5}{122} = \frac{3}{12} = \frac{3 \cdot 1}{3 \cdot 4} = \frac{1}{4}$$

The difference in lengths of the two scarves is $\frac{1}{4}$ of a yard.

36. Add the fractional amounts of cleaning. The LCD of 5 and 10 is 10.

$$\frac{3}{5} + \frac{1}{10} = \frac{3}{5} \cdot \frac{2}{2} + \frac{1}{10} = \frac{6}{10} + \frac{1}{10} = \frac{7}{10}$$

Truman cleaned $\frac{7}{10}$ of the house.

37. The LCD of 7 and 21 is 21.

$$31\frac{2}{7} + 14\frac{10}{21} = 31\frac{2 \cdot 3}{7 \cdot 3} + 14\frac{10}{21}$$

$$= 31\frac{6}{21} + 14\frac{10}{21}$$

$$= 45\frac{16}{21}$$

38. $24\frac{4}{5} + 35\frac{1}{5} = 59\frac{5}{5} = 59 + 1 = 60$

39. The LCD of 22 and 11 is 22.

$$69\frac{5}{22} - 36\frac{7}{11} = 69\frac{5}{22} - 36\frac{7 \cdot 2}{11 \cdot 2}$$

$$= 69\frac{5}{22} - 36\frac{14}{22}$$

$$= 68\frac{27}{22} - 36\frac{14}{22}$$

$$= 32\frac{13}{22}$$

40. The LCD of 20 and 6 is 60.

$$36\frac{3}{20} - 32\frac{5}{6} = 36\frac{3 \cdot 3}{20 \cdot 3} - 32\frac{5 \cdot 10}{6 \cdot 10}$$

$$= 36\frac{9}{60} - 32\frac{50}{60}$$

$$= 35\frac{69}{60} - 32\frac{50}{60}$$

$$= 3\frac{19}{60}$$

41. The LCD of 9, 18, and 3 is 18.

$$
\begin{array}{cc}
29\frac{2}{9} & 29\frac{4}{18} \\
27\frac{7}{18} & 27\frac{7}{18} \\
+ 54\frac{2}{3} & + 54\frac{12}{18} \\
\hline
& 110\frac{23}{18} = 110 + 1\frac{5}{18} = 111\frac{5}{18}
\end{array}
$$

42. The LCD of 8, 6, and 12 is 24.

$$7\frac{3}{8} \qquad 7\frac{9}{24}$$
$$9\frac{5}{6} \qquad 9\frac{20}{24}$$
$$+\,3\frac{1}{12} \qquad +\,3\frac{2}{24}$$
$$\rule{2cm}{0.4pt} \qquad \rule{2cm}{0.4pt}$$
$$19\frac{31}{24} = 19 + 1\frac{7}{24} = 20\frac{7}{24}$$

43. The LCD of 5 and 7 is 35.

$$9\frac{3}{5} \qquad 9\frac{21}{35}$$
$$-\,4\frac{1}{7} \qquad -\,4\frac{5}{35}$$
$$\rule{2cm}{0.4pt} \qquad \rule{2cm}{0.4pt}$$
$$5\frac{16}{35}$$

44. The LCD of 11 and 5 is 55.

$$8\frac{3}{11} \qquad 8\frac{15}{55}$$
$$-\,5\frac{1}{5} \qquad -\,5\frac{11}{55}$$
$$\rule{2cm}{0.4pt} \qquad \rule{2cm}{0.4pt}$$
$$3\frac{4}{55}$$

45. Subtract last year's snowfall from the average annual snowfall. The LCD of 10 and 2 is 10.

$$62\frac{3}{10} \qquad 62\frac{3}{10} \qquad 61\frac{13}{10}$$
$$-\,54\frac{1}{2} \qquad -\,54\frac{5}{10} \qquad -\,54\frac{5}{10}$$
$$\rule{1.5cm}{0.4pt} \qquad \rule{1.5cm}{0.4pt} \qquad \rule{1.5cm}{0.4pt}$$
$$7\frac{8}{10} = 7\frac{4}{5}$$

Last year's snowfall was $7\frac{4}{5}$ inches below the average annual snowfall.

46. Subtract Dinah's ounces per can from Amy's ounces per can. The LCD of 5 and 8 is 40.

$$15\frac{5}{8} \qquad 15\frac{25}{40}$$
$$-\,15\frac{3}{5} \qquad -\,15\frac{24}{40}$$
$$\rule{2cm}{0.4pt} \qquad \rule{2cm}{0.4pt}$$
$$\frac{1}{40}$$

Amy's brand weighs $\frac{1}{40}$ of an ounce more than Dinah's.

47. Perimeter is distance around.

$$\text{Perimeter} = 1\frac{1}{4} + 1\frac{1}{4} + 1\frac{1}{4} + 1\frac{1}{4} = 4\frac{4}{4} = 4 + 1 = 5$$

The perimeter of the shelf paper is 5 feet.

48. Perimeter is distance around. The LCD of 3 and 2 is 6.

$$\text{Perimeter} = 3\frac{1}{3} + 3\frac{1}{3} + 2\frac{1}{4} + 2\frac{1}{4}$$
$$= 6\frac{2}{3} + 4\frac{2}{4}$$
$$= 6\frac{2}{3} + 4\frac{1}{2}$$
$$= 6\frac{2\cdot 2}{3\cdot 2} + 4\frac{1\cdot 3}{2\cdot 3}$$
$$= 6\frac{4}{6} + 4\frac{3}{6}$$
$$= 10\frac{7}{6}$$
$$= 10 + 1\frac{1}{6}$$
$$= 11\frac{1}{6}$$

The perimeter of the gift wrap is $11\frac{1}{6}$ feet.

49. Since 5 < 6, then $\dfrac{5}{11} < \dfrac{6}{11}$.

50. Since 4 > 3, then $\dfrac{4}{35} > \dfrac{3}{35}$.

51. The LCD of 14 and 42 is 42.

$$\frac{5}{14} = \frac{5}{14}\cdot\frac{3}{3} = \frac{15}{42}$$

$\dfrac{16}{42}$ has the denominator 42.

Since 15 < 16, then $\dfrac{15}{42} < \dfrac{16}{42}$, so $\dfrac{5}{14} < \dfrac{16}{42}$.

52. The LCD of 35 and 105 is 105.

$$\frac{6}{35} = \frac{6}{35}\cdot\frac{3}{3} = \frac{18}{105}$$

$\dfrac{17}{105}$ has the denominator 105.

Since 18 > 17, then $\dfrac{18}{105} > \dfrac{17}{105}$, so $\dfrac{6}{35} > \dfrac{17}{105}$.

53. The LCD of 8 and 7 is 56.

$$\frac{7}{8} = \frac{7}{8} \cdot \frac{7}{7} = \frac{49}{56}$$

$$\frac{6}{7} = \frac{6}{7} \cdot \frac{8}{8} = \frac{48}{56}$$

Since 49 > 48, then $\frac{49}{56} > \frac{48}{56}$, so $\frac{7}{8} > \frac{6}{7}$.

54. The LCD of 10 and 3 is 30.

$$\frac{7}{10} = \frac{7}{10} \cdot \frac{3}{3} = \frac{21}{30}$$

$$\frac{2}{3} = \frac{2}{3} \cdot \frac{10}{10} = \frac{20}{30}$$

Since 21 > 20, then $\frac{21}{30} > \frac{20}{30}$, so $\frac{7}{10} > \frac{2}{3}$.

55. $\left(\frac{3}{7}\right)^2 = \frac{3}{7} \cdot \frac{3}{7} = \frac{9}{49}$

56. $\left(\frac{4}{5}\right)^3 = \frac{4}{5} \cdot \frac{4}{5} \cdot \frac{4}{5} = \frac{64}{125}$

57. $\left(\frac{1}{2}\right)^4 \cdot \left(\frac{3}{5}\right)^2 = \frac{1}{2} \cdot \frac{1}{2} \cdot \frac{1}{2} \cdot \frac{1}{2} \cdot \frac{3}{5} \cdot \frac{3}{5} = \frac{9}{400}$

58. $\left(\frac{1}{3}\right)^2 \cdot \left(\frac{9}{10}\right)^2 = \frac{1}{3} \cdot \frac{1}{3} \cdot \frac{9}{10} \cdot \frac{9}{10} = \frac{1 \cdot 1 \cdot 3 \cdot 3 \cdot 9}{3 \cdot 3 \cdot 10 \cdot 10} = \frac{9}{100}$

59. $\frac{5}{13} \div \frac{1}{2} \cdot \frac{4}{5} = \frac{5}{13} \cdot \frac{2}{1} \cdot \frac{4}{5} = \frac{5 \cdot 2 \cdot 4}{13 \cdot 1 \cdot 5} = \frac{8}{13}$

60. $\frac{8}{11} \div \frac{1}{3} \cdot \frac{11}{12} = \frac{8}{11} \cdot \frac{3}{1} \cdot \frac{11}{12} = \frac{2 \cdot 4 \cdot 3 \cdot 11}{11 \cdot 1 \cdot 3 \cdot 4} = \frac{2}{1} = 2$

61. $\left(\frac{6}{7} - \frac{3}{14}\right)^2 = \left(\frac{6}{7} \cdot \frac{2}{2} - \frac{3}{14}\right)^2$

$$= \left(\frac{12}{14} - \frac{3}{14}\right)^2$$

$$= \left(\frac{9}{14}\right)^2$$

$$= \frac{9}{14} \cdot \frac{9}{14}$$

$$= \frac{81}{196}$$

62. $\left(\frac{1}{3}\right)^2 - \frac{2}{27} = \frac{1}{3} \cdot \frac{1}{3} - \frac{2}{27}$

$$= \frac{1}{9} - \frac{2}{27}$$

$$= \frac{1}{9} \cdot \frac{3}{3} - \frac{2}{27}$$

$$= \frac{3}{27} - \frac{2}{27}$$

$$= \frac{1}{27}$$

63. $\frac{8}{9} - \frac{1}{8} \div \frac{3}{4} = \frac{8}{9} - \frac{1}{8} \cdot \frac{4}{3}$

$$= \frac{8}{9} - \frac{1 \cdot 4}{2 \cdot 4 \cdot 3}$$

$$= \frac{8}{9} - \frac{1}{6}$$

$$= \frac{8}{9} \cdot \frac{2}{2} - \frac{1}{6} \cdot \frac{3}{3}$$

$$= \frac{16}{18} - \frac{3}{18}$$

$$= \frac{13}{18}$$

64. $\frac{9}{10} - \frac{1}{9} \div \frac{2}{3} = \frac{9}{10} - \frac{1}{9} \cdot \frac{3}{2}$

$$= \frac{9}{10} - \frac{1 \cdot 3}{3 \cdot 3 \cdot 2}$$

$$= \frac{9}{10} - \frac{1}{6}$$

$$= \frac{9}{10} \cdot \frac{3}{3} - \frac{1}{6} \cdot \frac{5}{5}$$

$$= \frac{27}{30} - \frac{5}{30}$$

$$= \frac{22}{30}$$

$$= \frac{2 \cdot 11}{2 \cdot 15}$$

$$= \frac{11}{15}$$

65. $\frac{2}{7} \cdot \left(\frac{1}{5} + \frac{3}{10}\right) = \frac{2}{7} \cdot \left(\frac{1}{5} \cdot \frac{2}{2} + \frac{3}{10}\right)$

$\quad = \frac{2}{7}\left(\frac{2}{10} + \frac{3}{10}\right)$

$\quad = \frac{2}{7} \cdot \frac{5}{10}$

$\quad = \frac{2 \cdot 5}{7 \cdot 2 \cdot 5}$

$\quad = \frac{1}{7}$

66. $\frac{9}{10} \div \left(\frac{1}{5} + \frac{1}{20}\right) = \frac{9}{10} \div \left(\frac{1}{5} \cdot \frac{4}{4} + \frac{1}{20}\right)$

$\quad = \frac{9}{10} \div \left(\frac{4}{20} + \frac{1}{202}\right)$

$\quad = \frac{9}{10} \div \frac{5}{20}$

$\quad = \frac{9}{10} \div \frac{1}{4}$

$\quad = \frac{9}{10} \cdot \frac{4}{1}$

$\quad = \frac{9 \cdot 2 \cdot 2}{5 \cdot 2 \cdot 1}$

$\quad = \frac{18}{5}$ or $3\frac{3}{5}$

67. $\left(\frac{3}{4} + \frac{1}{2}\right) \div \left(\frac{4}{9} + \frac{1}{3}\right) = \left(\frac{3}{4} + \frac{1}{2} \cdot \frac{2}{2}\right) \div \left(\frac{4}{9} + \frac{1}{3} \cdot \frac{3}{3}\right)$

$\quad = \left(\frac{3}{4} + \frac{2}{4}\right) \div \left(\frac{4}{9} + \frac{3}{9}\right)$

$\quad = \frac{5}{4} \div \frac{7}{9}$

$\quad = \frac{5}{4} \cdot \frac{9}{7}$

$\quad = \frac{45}{28}$ or $1\frac{17}{28}$

68. $\left(\frac{3}{8} - \frac{1}{16}\right) \div \left(\frac{1}{2} - \frac{1}{8}\right) = \left(\frac{3}{8} \cdot \frac{2}{2} - \frac{1}{16}\right) \div \left(\frac{1}{2} \cdot \frac{4}{4} - \frac{1}{8}\right)$

$\quad = \left(\frac{6}{16} - \frac{1}{16}\right) \div \left(\frac{4}{8} - \frac{1}{8}\right)$

$\quad = \frac{5}{16} \div \frac{3}{8}$

$\quad = \frac{5}{16} \cdot \frac{8}{3}$

$\quad = \frac{5 \cdot 8}{2 \cdot 8 \cdot 3}$

$\quad = \frac{5}{6}$

69. $\frac{6}{7} \cdot \frac{5}{2} - \frac{3}{4} \cdot \frac{1}{2} = \frac{3 \cdot 2 \cdot 5}{7 \cdot 2} - \frac{3 \cdot 1}{4 \cdot 2}$

$\quad = \frac{15}{7} - \frac{3}{8}$

$\quad = \frac{15}{7} \cdot \frac{8}{8} - \frac{3}{8} \cdot \frac{7}{7}$

$\quad = \frac{120}{56} - \frac{21}{56}$

$\quad = \frac{99}{56}$ or $1\frac{43}{56}$

70. $\frac{9}{10} \cdot \frac{1}{3} - \frac{2}{5} \cdot \frac{1}{11} = \frac{3 \cdot 3 \cdot 1}{10 \cdot 3} - \frac{2 \cdot 1}{5 \cdot 11}$

$\quad = \frac{3}{10} - \frac{2}{55}$

$\quad = \frac{3}{10} \cdot \frac{11}{11} - \frac{2}{55} \cdot \frac{2}{2}$

$\quad = \frac{33}{110} - \frac{4}{110}$

$\quad = \frac{29}{110}$

71. The average is the sum, divided by 3.

$\left(\frac{2}{3} + \frac{5}{6} + \frac{1}{9}\right) \div 3 = \left(\frac{2}{3} \cdot \frac{6}{6} + \frac{5}{6} \cdot \frac{3}{3} + \frac{1}{9} \cdot \frac{2}{2}\right) \div 3$

$\quad = \left(\frac{12}{18} + \frac{15}{18} + \frac{2}{18}\right) \div 3$

$\quad = \frac{29}{18} \div \frac{3}{1}$

$\quad = \frac{29}{18} \cdot \frac{1}{3}$

$\quad = \frac{29}{54}$

72. The average is the sum, divided by 3.

$\left(\frac{4}{5} + \frac{9}{10} + \frac{3}{20}\right) \div 3 = \left(\frac{4}{5} \cdot \frac{4}{4} + \frac{9}{10} \cdot \frac{2}{2} + \frac{3}{20}\right) \div 3$

$\quad = \left(\frac{16}{20} + \frac{18}{20} + \frac{3}{20}\right) \div 3$

$\quad = \frac{37}{20} \div \frac{3}{1}$

$\quad = \frac{37}{20} \cdot \frac{1}{3}$

$\quad = \frac{37}{60}$

73. Multiply the ratio times the number of Saturn's moons.

$$\frac{3}{4} \cdot 28 = \frac{3}{4} \cdot \frac{28}{1} = \frac{3 \cdot 28}{4 \cdot 1} = \frac{3 \cdot 4 \cdot 7}{4 \cdot 1} = \frac{21}{1} = 21$$

Uranus has 21 moons.

74. Add the areas of two pieces of land. The LCD of 4 and 8 is 8.

$$\begin{array}{r} 9\dfrac{3}{4} \\ +\,5\dfrac{7}{8} \\ \hline \end{array} \qquad \begin{array}{r} 9\dfrac{6}{8} \\ +\,5\dfrac{7}{8} \\ \hline 14\dfrac{13}{8} = 14 + 1\dfrac{5}{8} = 15\dfrac{5}{8} \end{array}$$

James Hardaway now owns $15\dfrac{5}{8}$ acres of land.

75. Subtract the inside diameter from the outside diameter then divide by two.

$$\left(18\frac{7}{8} - 10\frac{3}{8}\right) \div 2 = 8\frac{4}{8} \div 2$$
$$= 8\frac{1}{2} \div \frac{2}{1}$$
$$= \frac{17}{2} \cdot \frac{1}{2}$$
$$= \frac{17}{4} \text{ or } 4\frac{1}{4}$$

Each measurement is $4\dfrac{1}{4}$ inches.

76. Subtract the known height from the total height. The LCD of 5 and 10 is 10.

$$\begin{array}{r} 1\dfrac{3}{10} \\ -\,\dfrac{3}{5} \\ \hline \end{array} \quad \begin{array}{r} 1\dfrac{3}{10} \\ -\,\dfrac{6}{10} \\ \hline \end{array} \quad \begin{array}{r} \dfrac{13}{10} \\ -\,\dfrac{6}{10} \\ \hline \dfrac{7}{10} \end{array}$$

The unknown length is $\dfrac{7}{10}$ of a yard.

77. Perimeter $= 2 \cdot \text{length} + 2 \cdot \text{width}$
$$= 2 \cdot \frac{1}{2} + 2 \cdot \frac{3}{11}$$
$$= \frac{2}{1} \cdot \frac{1}{2} + \frac{2}{1} \cdot \frac{3}{11}$$
$$= \frac{2}{2} + \frac{6}{11}$$
$$= 1 + \frac{6}{11}$$
$$= 1\frac{6}{11}$$

The perimeter is $1\dfrac{6}{11}$ miles.

$$\text{Area} = \text{length} \cdot \text{width} = \frac{1}{2} \cdot \frac{3}{11} = \frac{1 \cdot 3}{2 \cdot 11} = \frac{3}{22}$$

The area is $\dfrac{3}{22}$ of a square mile.

78. Perimeter $= 2 \cdot \text{length} + 2 \cdot \text{width}$
$$= 2 \cdot \frac{5}{12} + 2 \cdot \frac{3}{4}$$
$$= \frac{2}{1} \cdot \frac{5}{12} + \frac{2}{1} \cdot \frac{3}{4}$$
$$= \frac{10}{12} + \frac{6}{4}$$
$$= \frac{5}{6} + \frac{3}{2}$$
$$= \frac{5}{6} + \frac{3}{2} \cdot \frac{3}{3}$$
$$= \frac{5}{6} + \frac{9}{6}$$
$$= \frac{14}{6}$$
$$= \frac{7}{3} \text{ or } 2\frac{1}{3}$$

The perimeter is $\dfrac{7}{3}$ or $2\dfrac{1}{3}$ meters.

$$\text{Area} = \text{length} \cdot \text{width}$$
$$= \frac{5}{12} \cdot \frac{3}{4}$$
$$= \frac{5 \cdot 3}{12 \cdot 4}$$
$$= \frac{5 \cdot 3}{3 \cdot 4 \cdot 4}$$
$$= \frac{5}{16}$$

The area is $\dfrac{5}{16}$ of a square meter.

79. $15 = 3 \cdot \boxed{5}$
$30 = \boxed{2} \cdot 3 \cdot 5$
$45 = \boxed{3 \cdot 3} \cdot 5$
LCM: $2 \cdot 3 \cdot 3 \cdot 5 = 90$

80. $6 = 2 \cdot \boxed{3}$
$15 = 3 \cdot \boxed{5}$
$20 = \boxed{2 \cdot 2} \cdot 5$
LCM: $2 \cdot 2 \cdot 3 \cdot 5 = 60$

81. $\dfrac{5}{6} = \dfrac{5}{6} \cdot \dfrac{8}{8} = \dfrac{40}{48}$

82. $\dfrac{7}{8} = \dfrac{7}{8} \cdot \dfrac{9}{9} = \dfrac{63}{72}$

83. $\dfrac{5}{12} - \dfrac{3}{12} = \dfrac{5-3}{12} = \dfrac{2}{12} = \dfrac{2 \cdot 1}{2 \cdot 6} = \dfrac{1}{6}$

84. $\dfrac{3}{10} - \dfrac{1}{10} = \dfrac{3-1}{10} = \dfrac{2}{10} = \dfrac{2 \cdot 1}{2 \cdot 5} = \dfrac{1}{5}$

85. The LCD of 3 and 4 is 12.
$\dfrac{2}{3} + \dfrac{1}{4} = \dfrac{2}{3} \cdot \dfrac{4}{4} + \dfrac{1}{4} \cdot \dfrac{3}{3} = \dfrac{8}{12} + \dfrac{3}{12} = \dfrac{8+3}{12} = \dfrac{11}{12}$

86. The LCD of 11 and 55 is 55.
$\dfrac{5}{11} + \dfrac{2}{55} = \dfrac{5}{11} \cdot \dfrac{5}{5} + \dfrac{2}{55} = \dfrac{25}{55} + \dfrac{2}{55} = \dfrac{27}{55}$

87. The LCD of 4 and 3 is 12.

$$\begin{array}{cc} 7\dfrac{3}{4} & 7\dfrac{9}{12} \\ +5\dfrac{2}{3} & +5\dfrac{8}{12} \\ \hline & 12\dfrac{17}{12} = 12 + 1\dfrac{5}{12} = 13\dfrac{5}{12} \end{array}$$

88. The LCD of 8 and 2 is 8.

$$\begin{array}{cc} 2\dfrac{7}{8} & 2\dfrac{7}{8} \\ +9\dfrac{1}{2} & +9\dfrac{4}{8} \\ \hline & 11\dfrac{11}{8} = 11 + 1\dfrac{3}{8} = 12\dfrac{3}{8} \end{array}$$

89. The LCD of 5 and 7 is 35.

$$\begin{array}{cc} 12\dfrac{3}{5} & 12\dfrac{21}{35} \\ -9\dfrac{1}{7} & -9\dfrac{5}{35} \\ \hline & 3\dfrac{16}{35} \end{array}$$

90. The LCD of 21 and 7 is 21.

$$\begin{array}{cc} 32\dfrac{10}{21} & 32\dfrac{10}{21} \\ -24\dfrac{3}{7} & -24\dfrac{9}{21} \\ \hline & 8\dfrac{1}{21} \end{array}$$

91.
$$\dfrac{2}{5} + \left(\dfrac{2}{5}\right)^2 - \dfrac{3}{25} = \dfrac{2}{5} + \left(\dfrac{2}{5} \cdot \dfrac{2}{5}\right) - \dfrac{3}{25}$$
$$= \dfrac{2}{5} + \dfrac{4}{25} - \dfrac{3}{25}$$
$$= \dfrac{2}{5} \cdot \dfrac{5}{5} + \dfrac{4}{25} - \dfrac{3}{25}$$
$$= \dfrac{10}{25} + \dfrac{4}{25} - \dfrac{3}{25}$$
$$= \dfrac{14}{25} - \dfrac{3}{25}$$
$$= \dfrac{11}{25}$$

92.
$$\dfrac{1}{4} + \left(\dfrac{1}{2}\right)^2 - \dfrac{3}{8} = \dfrac{1}{4} + \left(\dfrac{1}{2} \cdot \dfrac{1}{2}\right) - \dfrac{3}{8}$$
$$= \dfrac{1}{4} + \dfrac{1}{4} - \dfrac{3}{8}$$
$$= \dfrac{2}{4} - \dfrac{3}{8}$$
$$= \dfrac{2}{4} \cdot \dfrac{2}{2} - \dfrac{3}{8}$$
$$= \dfrac{4}{8} - \dfrac{3}{8}$$
$$= \dfrac{1}{8}$$

93. $\left(\dfrac{5}{6}-\dfrac{3}{4}\right)^2 = \left(\dfrac{5}{6}\cdot\dfrac{2}{2}-\dfrac{3}{4}\cdot\dfrac{3}{3}\right)^2$

$\qquad = \left(\dfrac{10}{12}-\dfrac{9}{12}\right)^2$

$\qquad = \left(\dfrac{1}{12}\right)^2$

$\qquad = \dfrac{1}{12}\cdot\dfrac{1}{12}$

$\qquad = \dfrac{1}{144}$

94. $\left(2-\dfrac{2}{3}\right)^3 = \left(1\dfrac{3}{3}-\dfrac{2}{3}\right)^3$

$\qquad = \left(1\dfrac{1}{3}\right)^3$

$\qquad = \left(\dfrac{4}{3}\right)^3$

$\qquad = \dfrac{4}{3}\cdot\dfrac{4}{3}\cdot\dfrac{4}{3}$

$\qquad = \dfrac{64}{27}$ or $2\dfrac{10}{27}$

95. $\dfrac{2}{3}\div\left(\dfrac{3}{5}+\dfrac{5}{3}\right) = \dfrac{2}{3}\div\left(\dfrac{3}{5}\cdot\dfrac{3}{3}+\dfrac{5}{3}\cdot\dfrac{5}{5}\right)$

$\qquad = \dfrac{2}{3}\div\left(\dfrac{9}{15}+\dfrac{25}{15}\right)$

$\qquad = \dfrac{2}{3}\div\dfrac{34}{15}$

$\qquad = \dfrac{2}{3}\cdot\dfrac{15}{34}$

$\qquad = \dfrac{2\cdot15}{3\cdot34}$

$\qquad = \dfrac{2\cdot3\cdot5}{3\cdot2\cdot17}$

$\qquad = \dfrac{5}{17}$

96. $\dfrac{3}{8}\left(\dfrac{2}{3}-\dfrac{4}{9}\right) = \dfrac{3}{8}\left(\dfrac{2}{3}\cdot\dfrac{3}{3}-\dfrac{4}{9}\right)$

$\qquad = \dfrac{3}{8}\left(\dfrac{6}{9}-\dfrac{4}{9}\right)$

$\qquad = \dfrac{3}{8}\cdot\dfrac{2}{9}$

$\qquad = \dfrac{3\cdot2}{8\cdot9}$

$\qquad = \dfrac{3\cdot2}{2\cdot4\cdot3\cdot3}$

$\qquad = \dfrac{1}{12}$

97. The LCD of 14 and 3 is 42.

$\dfrac{3}{14}=\dfrac{3}{14}\cdot\dfrac{3}{3}=\dfrac{9}{42}$

$\dfrac{2}{3}=\dfrac{2}{3}\cdot\dfrac{14}{14}=\dfrac{28}{42}$

Since $9 < 28$, then $\dfrac{9}{42}<\dfrac{28}{42}$, so $\dfrac{3}{14}<\dfrac{2}{3}$.

98. the LCD of 23 and 16 is 368.

$\dfrac{7}{23}=\dfrac{7}{23}\cdot\dfrac{16}{16}=\dfrac{112}{368}$

$\dfrac{3}{16}=\dfrac{3}{16}\cdot\dfrac{23}{23}=\dfrac{69}{368}$

Since $112 > 69$, then $\dfrac{112}{368}>\dfrac{69}{368}$, so $\dfrac{7}{23}>\dfrac{3}{16}$.

99. Add the study times.

$\dfrac{3}{8}+\dfrac{1}{8}=\dfrac{3+1}{8}=\dfrac{4}{8}=\dfrac{4\cdot1}{4\cdot2}=\dfrac{1}{2}$

Gregor studied $\dfrac{1}{2}$ of an hour.

100. Add the package weights. The LCD of 4 and 5 is 20.

$$\begin{array}{ll} 3\dfrac{3}{4} & 3\dfrac{15}{20} \\[2mm] +2\dfrac{3}{5} & +2\dfrac{12}{20} \\ \hline & 5\dfrac{27}{20}=5+1\dfrac{7}{20}=6\dfrac{7}{20} \end{array}$$

The combined weight of the packages is $6\dfrac{7}{20}$ pounds.

101. Subtract the length of the piece cut off from the length of the reel.

$$50 - 5\frac{1}{2} = 49\frac{2}{2} - 5\frac{1}{2} = 44\frac{1}{2}$$

The length of the ribbon remaining on the reel is

$44\frac{1}{2}$ yards.

102. Divide the length of the board by 5.

$$10\frac{2}{3} \div 5 = \frac{32}{3} \div \frac{5}{1} = \frac{32}{3} \cdot \frac{1}{5} = \frac{32}{15} \text{ or } 2\frac{2}{15}$$

The length of each piece is $\frac{32}{15}$ or $2\frac{2}{15}$ feet.

103. Multiply the amount of cilantro by 5.

$$1\frac{1}{2} \cdot 5 = \frac{3}{2} \cdot \frac{5}{1} = \frac{15}{2} \text{ or } 7\frac{1}{2}$$

The amount of cilantro needed is $\frac{15}{2}$ or

$7\frac{1}{2}$ tablespoons.

104. Add the amounts mixed, then subtract the amount drunk.

$$\left(\frac{5}{8} + \frac{1}{8}\right) - \frac{3}{8} = \frac{6}{8} - \frac{3}{8} = \frac{3}{8}$$

There was $\frac{3}{8}$ of a gallon of the punch remaining.

Chapter 3 Test

1. $4 = \boxed{2 \cdot 2}$
$15 = \boxed{3} \cdot \boxed{5}$
LCM: $2 \cdot 2 \cdot 3 \cdot 5 = 60$

2. $8 = \boxed{2 \cdot 2 \cdot 2}$
$9 = \boxed{3 \cdot 3}$
$12 = 2 \cdot 2 \cdot 3$
LCM: $2 \cdot 2 \cdot 2 \cdot 3 \cdot 3 = 72$

3. The LCD of 6 and 30 is 30.

$$\frac{5}{6} = \frac{5}{6} \cdot \frac{5}{5} = \frac{25}{30}$$

$\frac{26}{30}$ has the denominator 30.

Since $25 < 26$, then $\frac{25}{30} < \frac{26}{30}$, so $\frac{5}{6} < \frac{26}{30}$.

4. The LCD of 8 and 9 is 72.

$$\frac{7}{8} = \frac{7}{8} \cdot \frac{9}{9} = \frac{63}{72}$$

$$\frac{8}{9} = \frac{8}{9} \cdot \frac{8}{8} = \frac{64}{72}$$

Since $63 < 64$, then $\frac{63}{72} < \frac{64}{72}$, so $\frac{7}{8} < \frac{8}{9}$.

5. $\dfrac{7}{9} + \dfrac{1}{9} = \dfrac{7+1}{9} = \dfrac{8}{9}$

6. $\dfrac{8}{15} - \dfrac{2}{15} = \dfrac{8-2}{15} = \dfrac{6}{15} = \dfrac{3 \cdot 2}{3 \cdot 5} = \dfrac{2}{5}$

7. The LCD of 10 and 5 is 10.

$$\frac{9}{10} + \frac{2}{5} = \frac{9}{10} + \frac{2}{5} \cdot \frac{2}{2} = \frac{9}{10} + \frac{4}{10} = \frac{13}{10} \text{ or } 1\frac{3}{10}$$

8. The LCD of 6 and 14 is 42.

$$\frac{1}{6} + \frac{3}{14} = \frac{1}{6} \cdot \frac{7}{7} + \frac{3}{14} \cdot \frac{3}{3}$$
$$= \frac{7}{42} + \frac{9}{42}$$
$$= \frac{16}{42}$$
$$= \frac{2 \cdot 8}{2 \cdot 21}$$
$$= \frac{8}{21}$$

9. The LCD of 8 and 3 is 24.

$$\frac{7}{8} - \frac{1}{3} = \frac{7}{8} \cdot \frac{3}{3} - \frac{1}{3} \cdot \frac{8}{8} = \frac{21}{24} - \frac{8}{24} = \frac{13}{24}$$

10. The LCD of 21 and 7 is 21.

$$\frac{17}{21} - \frac{1}{7} = \frac{17}{21} - \frac{1}{7} \cdot \frac{3}{3} = \frac{17}{21} - \frac{3}{21} = \frac{14}{21} = \frac{7 \cdot 2}{7 \cdot 3} = \frac{2}{3}$$

11. The LCD of 20 and 3 is 60.

$$\frac{9}{20} + \frac{2}{3} = \frac{9}{20} \cdot \frac{3}{3} + \frac{2}{3} \cdot \frac{20}{20}$$
$$= \frac{27}{60} + \frac{40}{60}$$
$$= \frac{67}{60} \text{ or } 1\frac{7}{60}$$

12. The LCD of 25 and 2 is 50.

$$\frac{16}{25} - \frac{1}{2} = \frac{16}{25} \cdot \frac{2}{2} - \frac{1}{2} \cdot \frac{25}{25} = \frac{32}{50} - \frac{25}{50} = \frac{7}{50}$$

13. The LCD of 12, 8, and 24 is 24.

$$\frac{11}{12}+\frac{3}{8}+\frac{5}{24}=\frac{11}{12}\cdot\frac{2}{2}+\frac{3}{8}\cdot\frac{3}{3}+\frac{5}{24}$$
$$=\frac{22}{24}+\frac{9}{24}+\frac{5}{24}$$
$$=\frac{36}{24}$$
$$=\frac{12\cdot3}{12\cdot2}$$
$$=\frac{3}{2}\text{ or }1\frac{1}{2}$$

14. The LCD of 8, 5, and 4 is 40.

$$
\begin{array}{ll}
3\frac{7}{8} & 3\frac{35}{40} \\
7\frac{2}{5} & 7\frac{16}{40} \\
+2\frac{3}{4} & +2\frac{30}{40} \\
\hline
& 12\frac{81}{40}=12+2\frac{1}{40}=14\frac{1}{40}
\end{array}
$$

15. The LCD of 9 and 15 is 45.

$$
\begin{array}{ll}
8\frac{2}{9} & 8\frac{10}{45} \\
12 & 12 \\
+10\frac{1}{15} & +10\frac{3}{45} \\
\hline
& 30\frac{13}{45}
\end{array}
$$

16. The LCD of 6 and 8 is 24.

$$
\begin{array}{lll}
5\frac{1}{6} & 5\frac{4}{24} & 4\frac{28}{24} \\
-3\frac{7}{8} & -3\frac{21}{24} & -3\frac{21}{24} \\
\hline
& & 1\frac{7}{24}
\end{array}
$$

17.

$$
\begin{array}{ll}
19 & 18\frac{11}{11} \\
-2\frac{3}{11} & -2\frac{3}{11} \\
\hline
& 16\frac{8}{11}
\end{array}
$$

18. $\frac{2}{7}\cdot\left(6-\frac{1}{6}\right)=\frac{2}{7}\left(5\frac{6}{6}-\frac{1}{6}\right)$
$$=\frac{2}{7}\cdot\left(5\frac{5}{6}\right)$$
$$=\frac{2}{7}\cdot\frac{35}{6}$$
$$=\frac{2\cdot35}{7\cdot6}$$
$$=\frac{2\cdot5\cdot7}{7\cdot2\cdot3}$$
$$=\frac{5}{3}\text{ or }1\frac{2}{3}$$

19. $\left(\frac{2}{3}\right)^4=\frac{2}{3}\cdot\frac{2}{3}\cdot\frac{2}{3}\cdot\frac{2}{3}=\frac{16}{81}$

20. $\frac{1}{2}\div\frac{2}{3}\cdot\frac{3}{4}=\frac{1}{2}\cdot\frac{3}{2}\cdot\frac{3}{4}=\frac{3}{4}\cdot\frac{3}{4}=\frac{9}{16}$

21. $\left(\frac{4}{5}\right)^2+\left(\frac{1}{2}\right)^3=\left(\frac{4}{5}\cdot\frac{4}{5}\right)+\left(\frac{1}{2}\cdot\frac{1}{2}\cdot\frac{1}{2}\right)$
$$=\frac{16}{25}+\frac{1}{8}$$
$$=\frac{16}{25}\cdot\frac{8}{8}+\frac{1}{8}\cdot\frac{25}{25}$$
$$=\frac{128}{200}+\frac{25}{200}$$
$$=\frac{153}{200}$$

22. $\left(\frac{3}{4}\right)^2\div\left(\frac{2}{3}+\frac{5}{6}\right)=\left(\frac{3}{4}\cdot\frac{3}{4}\right)\div\left(\frac{2}{3}\cdot\frac{2}{2}+\frac{5}{6}\right)$
$$=\frac{9}{16}\div\left(\frac{4}{6}+\frac{5}{6}\right)$$
$$=\frac{9}{16}\div\frac{9}{6}$$
$$=\frac{9}{16}\cdot\frac{6}{9}$$
$$=\frac{9\cdot6}{16\cdot9}$$
$$=\frac{9\cdot2\cdot3}{2\cdot8\cdot9}$$
$$=\frac{3}{8}$$

23. The average is the sum divided by 3. The LCD of 6, 3, and 12 is 12.

$$\left(\frac{5}{6}+\frac{4}{3}+\frac{7}{12}\right)\div 3=\left(\frac{5}{6}\cdot\frac{2}{2}+\frac{4}{3}\cdot\frac{4}{4}+\frac{7}{12}\right)\div 3$$
$$=\left(\frac{10}{12}+\frac{16}{12}+\frac{7}{12}\right)\div 3$$
$$=\frac{33}{12}\div\frac{3}{1}$$
$$=\frac{33}{12}\cdot\frac{1}{3}$$
$$=\frac{33\cdot 1}{12\cdot 3}$$
$$=\frac{3\cdot 11\cdot 1}{12\cdot 3}$$
$$=\frac{11}{12}$$

24. Subtract the length of the cut piece from the total length of the plank. The LCD of 4 and 2 is 4.

$$6\frac{1}{2} \qquad 6\frac{2}{4} \qquad 5\frac{6}{4}$$
$$-2\frac{3}{4} \qquad -2\frac{3}{4} \qquad -2\frac{3}{4}$$
$$\overline{} \qquad \overline{} \qquad \overline{3\frac{3}{4}}$$

The length of the remaining piece is $3\frac{3}{4}$ feet.

25. Divide the number of gallons of fuel by the number of hours.

$$58\frac{3}{4}\div 7\frac{1}{2}=\frac{235}{4}\div\frac{15}{2}$$
$$=\frac{235}{4}\cdot\frac{2}{15}$$
$$=\frac{235\cdot 2}{4\cdot 15}$$
$$=\frac{5\cdot 47\cdot 2}{2\cdot 2\cdot 5\cdot 3}$$
$$=\frac{47}{2\cdot 3}$$
$$=\frac{47}{6}\text{ or }7\frac{5}{6}$$

$7\frac{5}{6}$ gallons of fuel were used each hour.

26. Add the fractions for housing and food. The LCD of 25 and 50 is 50.

$$\frac{8}{25}+\frac{7}{50}=\frac{8}{25}\cdot\frac{2}{2}+\frac{7}{50}=\frac{16}{50}+\frac{7}{50}=\frac{23}{50}$$

The fraction spent for food and housing is $\frac{23}{50}$.

27. Add the fractions for education, transportation, and clothing. The LCD of 50, 5, and 25 is 50.

$$\frac{1}{50}+\frac{1}{5}+\frac{1}{25}=\frac{1}{50}+\frac{1}{5}\cdot\frac{10}{10}+\frac{1}{25}\cdot\frac{2}{2}$$
$$=\frac{1}{50}+\frac{10}{50}+\frac{2}{50}$$
$$=\frac{13}{50}$$

The fraction spent for education, transportation, and clothing is $\frac{13}{50}$.

28. Multiply the fraction for health care by the money the family spent on all the items.

$$\frac{3}{50}\cdot 47{,}000=\frac{3}{50}\cdot\frac{47{,}000}{1}$$
$$=\frac{3\cdot 47{,}000}{50\cdot 1}$$
$$=\frac{3\cdot 50\cdot 940}{50\cdot 1}$$
$$=\frac{3\cdot 940}{1}$$
$$=2820$$

$2820 would be expected to be spent on health care.

29. Perimeter $=2\cdot\text{length}+2\cdot\text{width}$
$$=2\cdot 1+2\cdot\frac{2}{3}$$
$$=2+\frac{2}{1}\cdot\frac{2}{3}$$
$$=2+\frac{4}{3}$$
$$=2+1+\frac{1}{3}$$
$$=3\frac{1}{3}$$

The perimeter is $3\frac{1}{3}$ feet.

$$\text{Area}=\text{length}\cdot\text{width}=1\cdot\frac{2}{3}=\frac{2}{3}$$

The area is $\frac{2}{3}$ square foot.

30. Perimeter is the distance around. The LCD of 15 and 3 is 15.

$$\text{Perimeter} = \frac{2}{15} + \frac{4}{15} + \frac{6}{15} + \frac{1}{3} + \frac{8}{15}$$

$$= \frac{2}{15} + \frac{4}{15} + \frac{6}{15} + \frac{1}{3} \cdot \frac{5}{5} + \frac{8}{15}$$

$$= \frac{2}{15} + \frac{4}{15} + \frac{6}{15} + \frac{5}{15} + \frac{8}{15}$$

$$= \frac{25}{15}$$

$$= \frac{5 \cdot 5}{5 \cdot 3}$$

$$= \frac{5}{3} \text{ or } 1\frac{2}{3}$$

The perimeter is $\frac{5}{3}$ or $1\frac{2}{3}$ inches.

Cumulative Review Chapters 1–3

1. 85 is written as eighty-five.

2. 107 is written as one hundred seven.

3. 126 is written as one hundred twenty-six.

4. 5026 is written as five thousand twenty-six.

5.
$$\begin{array}{r} 23 \\ + 136 \\ \hline 159 \end{array}$$

6. Perimeter = $3 + 7 + 9 = 19$
The perimeter is 19 inches.

7.
$$\begin{array}{r} 543 \\ - 29 \\ \hline 514 \end{array} \quad \text{Check: } \begin{array}{r} \overset{1}{514} \\ + 29 \\ \hline 543 \end{array}$$

8.
$$\begin{array}{r} 121 \text{ R } 1 \\ 27\overline{)3268} \\ -27 \\ \hline 56 \\ -54 \\ \hline 28 \\ -27 \\ \hline 1 \end{array}$$

9. To round 278,362 to the nearest thousand observe that the digit in the hundreds place is 3. Since this digit is less than 5, we do not add 1 to the digit in the thousands place. The number 278,362 rounded to the nearest thousand is 278,000.

10. $1 \cdot 30 = 30, 2 \cdot 15 = 30, 3 \cdot 10 = 30, 5 \cdot 6 = 30$
The factors of 30 are 1, 2, 3, 5, 6, 10, 15, and 30.

11.
$$\begin{array}{r} 236 \\ \times 86 \\ \hline 1\,416 \\ 18\,880 \\ \hline 20,296 \end{array}$$

12. $236 \times 86 \times 0 = 0$

13. a.
$$\begin{array}{r} 7 \\ 1\overline{)7} \\ -7 \\ \hline 0 \end{array}$$
Check: $7 \cdot 1 = 7$

b. $12 \div 1 = 12$
Check: $12 \cdot 1 = 12$

c. $\frac{6}{6} = 1$
Check: $1 \cdot 6 = 6$

d. $9 \div 9 = 1$
Check: $9 \cdot 1 = 9$

e. $\frac{20}{1} = 20$
Check: $20 \cdot 1 = 20$

f.
$$\begin{array}{r} 1 \\ 18\overline{)18} \\ -18 \\ \hline 0 \end{array}$$
Check: $1 \cdot 18 = 18$

14. There are four numbers. The average is the sum, divided by 4.

$$\begin{array}{r} \overset{1\,3}{} \\ 25 \\ 17 \\ 19 \\ + 39 \\ \hline 100 \end{array} \qquad \begin{array}{r} 25 \\ 4\overline{)100} \\ -8 \\ \hline 20 \\ -20 \\ \hline 0 \end{array}$$

The average is 25.

15.
$$\begin{array}{r} 306 \\ + 732 \\ \hline 1038 \end{array}$$
The Snake River is 1038 miles long.

16. $\sqrt{121} = 11$, because $11 \cdot 11 = 121$.

17. $9^2 = 9 \cdot 9 = 81$

18. $5^3 = 5 \cdot 5 \cdot 5 = 125$

19. $3^4 = 3 \cdot 3 \cdot 3 \cdot 3 = 81$

20. $10^3 = 10 \cdot 10 \cdot 10 = 1000$

21. Each part is $\dfrac{1}{3}$ of a whole and there are 4 parts shaded or 1 whole and 1 more part.

improper fraction: $\dfrac{4}{3}$; mixed number: $1\dfrac{1}{3}$

22. Each part is $\dfrac{1}{4}$ of a whole and there are 11 parts shaded or 2 wholes and 3 more parts.

improper fraction: $\dfrac{11}{4}$; mixed number: $2\dfrac{3}{4}$

23. Each part is $\dfrac{1}{2}$ of a whole and here are 5 parts shaded, or 2 wholes and 1 more part.

improper fraction: $\dfrac{5}{2}$; mixed number: $2\dfrac{1}{2}$

24. Each part is $\dfrac{1}{3}$ of a whole and there are 14 parts shaded, or 4 wholes and 2 more parts.

improper fraction: $\dfrac{14}{3}$; mixed number: $4\dfrac{2}{3}$

25. 3 is prime since its only factors are 1 and 3.
9 is composite since its factors are 1, 3, and 9.
11 is prime since its only factors are 1 and 11.
17 is prime since its only factors are 1 and 17.
26 is composite since its factors are 1, 2, 13, and 26.

26.
$$\dfrac{6^2 + 4 \cdot 4 + 2^3}{37 - 5^2} = \dfrac{6 \cdot 6 + 4 \cdot 4 + 2 \cdot 2 \cdot 2}{37 - 5 \cdot 5}$$
$$= \dfrac{36 + 16 + 8}{37 - 25}$$
$$= \dfrac{52 + 8}{12}$$
$$= \dfrac{60}{12}$$
$$= \dfrac{12 \cdot 5}{12 \cdot 1}$$
$$= \dfrac{5}{1}$$
$$= 5$$

27.
$$3\overline{)15}^{\;5}$$
$$3\overline{)45}$$
$$2\overline{)90}$$
$$2\overline{)180}$$
$$180 = 2 \cdot 2 \cdot 3 \cdot 3 \cdot 5 = 2^2 \cdot 3^2 \cdot 5$$

28.
$$\begin{array}{r} 87 \\ -25 \\ \hline 62 \end{array}$$
The difference of 87 and 25 is 62.

29. $\dfrac{72}{26} = \dfrac{2 \cdot 36}{2 \cdot 13} = \dfrac{36}{13}$ or $2\dfrac{10}{13}$

30. $9\dfrac{7}{8} = \dfrac{8 \cdot 9 + 7}{8} = \dfrac{72 + 7}{8} = \dfrac{79}{8}$

31. Equivalent, since the cross products are equal; $16 \cdot 25 = 400$ and $10 \cdot 40 = 400$.

32. The LCD of 7 and 9 is 63.
$$\dfrac{4}{7} = \dfrac{4}{7} \cdot \dfrac{9}{9} = \dfrac{36}{63}$$
$$\dfrac{5}{9} = \dfrac{5}{9} \cdot \dfrac{7}{7} = \dfrac{35}{63}$$
Since $36 > 35$, then $\dfrac{36}{63} > \dfrac{35}{63}$, so $\dfrac{4}{7} > \dfrac{5}{9}$.

33. $\dfrac{2}{3} \cdot \dfrac{5}{11} = \dfrac{2 \cdot 5}{3 \cdot 11} = \dfrac{10}{33}$

34. $2\dfrac{5}{8} \cdot \dfrac{4}{7} = \dfrac{21}{8} \cdot \dfrac{4}{7} = \dfrac{21 \cdot 4}{8 \cdot 7} = \dfrac{3 \cdot 7 \cdot 4}{2 \cdot 4 \cdot 7} = \dfrac{3}{2}$ or $1\dfrac{1}{2}$

35. $\dfrac{1}{4} \cdot \dfrac{1}{2} = \dfrac{1 \cdot 1}{4 \cdot 2} = \dfrac{1}{8}$

36. $7 \cdot 5\dfrac{2}{7} = \dfrac{7}{1} \cdot \dfrac{37}{7} = \dfrac{7 \cdot 37}{1 \cdot 7} = \dfrac{37}{1} = 37$

37. $\dfrac{11}{18} \div 2\dfrac{5}{6} = \dfrac{11}{18} \div \dfrac{17}{6}$
$\quad = \dfrac{11}{18} \cdot \dfrac{6}{17}$
$\quad = \dfrac{11 \cdot 6}{18 \cdot 17}$
$\quad = \dfrac{11 \cdot 6}{3 \cdot 6 \cdot 17}$
$\quad = \dfrac{11}{51}$

38. $\dfrac{15}{19} \div \dfrac{3}{5} = \dfrac{15}{19} \cdot \dfrac{5}{3} = \dfrac{15 \cdot 5}{19 \cdot 3} = \dfrac{3 \cdot 5 \cdot 5}{19 \cdot 3} = \dfrac{25}{19}$ or $1\dfrac{6}{19}$

39. $5\dfrac{2}{3} \div 2\dfrac{5}{9} = \dfrac{17}{3} \div \dfrac{23}{9}$
$\quad = \dfrac{17}{3} \cdot \dfrac{9}{23}$
$\quad = \dfrac{17 \cdot 9}{3 \cdot 23}$
$\quad = \dfrac{17 \cdot 3 \cdot 3}{3 \cdot 23}$
$\quad = \dfrac{51}{23}$ or $2\dfrac{5}{23}$

40. $\dfrac{8}{11} \div \dfrac{1}{22} = \dfrac{8}{11} \cdot \dfrac{22}{1} = \dfrac{8 \cdot 22}{11 \cdot 1} = \dfrac{8 \cdot 2 \cdot 11}{11 \cdot 1} = \dfrac{16}{1} = 16$

41. $\dfrac{3}{16} + \dfrac{7}{16} = \dfrac{3+7}{16} = \dfrac{10}{16} = \dfrac{2 \cdot 5}{2 \cdot 8} = \dfrac{5}{8}$

42. $\dfrac{11}{20} - \dfrac{7}{20} = \dfrac{11-7}{20} = \dfrac{4}{20} = \dfrac{4 \cdot 1}{4 \cdot 5} = \dfrac{1}{5}$

43. $6 = 2 \cdot \boxed{3}$
$8 = \boxed{2 \cdot 2 \cdot 2}$
LCM: $2 \cdot 2 \cdot 2 \cdot 3 = 24$

44. $7 = \boxed{7}$
$5 = \boxed{5}$
LCM: $5 \cdot 7 = 35$

45. The LCD of 2, 3, and 6 is 6.
$\dfrac{1}{2} + \dfrac{2}{3} + \dfrac{5}{6} = \dfrac{1}{2} \cdot \dfrac{3}{3} + \dfrac{2}{3} \cdot \dfrac{2}{2} + \dfrac{5}{6}$
$\quad = \dfrac{3}{6} + \dfrac{4}{6} + \dfrac{5}{6}$
$\quad = \dfrac{12}{6}$
$\quad = \dfrac{6 \cdot 2}{6 \cdot 1}$
$\quad = \dfrac{2}{1}$
$\quad = 2$

46. $\left(\dfrac{5}{9}\right)^2 = \dfrac{5}{9} \cdot \dfrac{5}{9} = \dfrac{25}{81}$

47. The LCD of 7 and 21 is 21.
$$\begin{array}{cc} 9\dfrac{3}{7} & 9\dfrac{9}{21} \\[2mm] -5\dfrac{2}{21} & -5\dfrac{2}{21} \\ \hline & 4\dfrac{7}{21} = 4\dfrac{1}{3} \end{array}$$

48. The LCD of 100 and 25 is 100.
$\dfrac{31}{100} - \dfrac{5}{25} = \dfrac{31}{100} - \dfrac{5}{25} \cdot \dfrac{4}{4} = \dfrac{31}{100} - \dfrac{20}{100} = \dfrac{11}{100}$

49. $\left(\dfrac{2}{3}\right)^2 \div \left(\dfrac{8}{27} + \dfrac{2}{3}\right) = \left(\dfrac{2}{3} \cdot \dfrac{2}{3}\right) \div \left(\dfrac{8}{27} + \dfrac{2}{3} \cdot \dfrac{9}{9}\right)$
$\quad = \dfrac{4}{9} \div \left(\dfrac{8}{27} + \dfrac{18}{27}\right)$
$\quad = \dfrac{4}{9} \div \dfrac{26}{27}$
$\quad = \dfrac{4}{9} \cdot \dfrac{27}{26}$
$\quad = \dfrac{4 \cdot 27}{9 \cdot 26}$
$\quad = \dfrac{2 \cdot 2 \cdot 3 \cdot 9}{9 \cdot 2 \cdot 13}$
$\quad = \dfrac{6}{13}$

50.

$$\frac{1}{10} \div \frac{7}{8} \cdot \frac{2}{5} = \frac{1}{10} \cdot \frac{8}{7} \cdot \frac{2}{5}$$

$$= \frac{8}{70} \cdot \frac{2}{5}$$

$$= \frac{8 \cdot 2}{70 \cdot 5}$$

$$= \frac{8 \cdot 2}{2 \cdot 35 \cdot 5}$$

$$= \frac{8}{175}$$

Chapter 4

Section 4.1 Practice

1. 8.7 in words is eight and seven tenths.

2. 97.28 in words is ninety-seven and twenty-eight hundredths.

3. 302.105 in words is three hundred two and one hundred five thousandths.

4. 72.1085 in words is seventy-two and one thousand eighty-five ten-thousandths.

5. The check should be paid to "CLECO' for the amount "207.40," which is written in words as "Two hundred seven and 40/100."

6. Three hundred and ninety-six hundredths is 300.96.

7. Thirty-nine and forty-two thousandths is 39.042.

8. $0.037 = \dfrac{37}{1000}$

9. $14.97 = 14\dfrac{97}{100}$

10. $0.12 = \dfrac{12}{100} = \dfrac{3}{25}$

11. $57.8 = 57\dfrac{8}{10} = 57\dfrac{4}{5}$

12. $209.986 = 209\dfrac{986}{1000} = 209\dfrac{493}{500}$

13. $\dfrac{58}{100} = 0.58$

14. $\dfrac{59}{100} = 0.59$

15. $\dfrac{6}{1000} = 0.006$

16. $\dfrac{172}{10} = 17.2$

Vocabulary and Readiness Check

1. The number "twenty and eight hundredths" is written in <u>words</u> and "20.08" is written in <u>standard form</u>.

2. Like fractions, <u>decimals</u> are used to denote part of a whole.

3. When writing a decimal number in words, the decimal point is written as <u>and</u>.

4. The place value <u>tenths</u> is to the right of the decimal point while <u>tens</u> is to the left of the decimal point.

5. The place value of the digit 7 in the number 70 is tens.

6. The place value of the digit 7 in the number 700 is hundreds.

7. The place value of the digit 7 in the number 0.7 is tenths.

8. The place value of the digit 7 in the number 0.07 is hundredths.

Exercise Set 4.1

1. 6.52 in words is six and fifty-two hundredths.

3. 16.23 in words is sixteen and twenty-three hundredths.

5. 0.205 in words is two hundred five thousandths.

7. 167.009 in words is one hundred sixty-seven and nine thousandths.

9. 200.005 in words is two hundred and five thousandths.

11. 105.6 in words is one hundred five and six tenths.

13. 2.43 in words is two and forty-three hundredths.

15. 87.97 in words is eighty-seven and ninety-seven hundredths.

17. 114.5 in words is one hundred fourteen and five tenths.

19. The check should be paid to "R.W. Financial" for the amount of "321.42," which is written in words as "Three hundred twenty-one and 42/100."

21. The check should be paid to "Bell South," for the amount of "59.68," which is written in words as "Fifty-nine and 68/100."

23. Six and five tenths is 6.5.

25. Nine and eight hundredths is 9.08.

27. Seven hundred five and six hundred twenty-five thousandths is 705.625.

29. Forty-six ten-thousandths is 0.0046.

31. Thirty-two and fifty-two hundredths is 32.52.

33. One and three-tenths is 1.3.

35. $0.3 = \dfrac{3}{10}$

37. $0.27 = \dfrac{27}{100}$

39. $0.8 = \dfrac{8}{10} = \dfrac{4}{5}$

41. $0.15 = \dfrac{15}{100} = \dfrac{3}{20}$

43. $5.47 = 5\dfrac{47}{100}$

45. $0.048 = \dfrac{48}{1000} = \dfrac{6}{125}$

47. $7.008 = 7\dfrac{8}{1000} = 7\dfrac{1}{125}$

49. $15.802 = 15\dfrac{802}{1000} = 15\dfrac{401}{500}$

51. $0.3005 = \dfrac{3005}{10,000} = \dfrac{601}{2000}$

53. $487.32 = 487\dfrac{32}{100} = 487\dfrac{8}{25}$

55. $\dfrac{6}{10} = 0.6$

57. $\dfrac{45}{100} = 0.45$

59. $\dfrac{37}{10} = 3.7$

61. $\dfrac{268}{1000} = 0.268$

63. $\dfrac{9}{100} = 0.09$

65. $\dfrac{4026}{1000} = 4.026$

67. $\dfrac{28}{1000} = 0.028$

69. $\dfrac{563}{10} = 56.3$

71. In standard form, $\dfrac{43}{100}$ is 0.43, while in words it is forty-three hundredths.

73. In standard form, eight tenths is 0.8, while as a fraction it is $\dfrac{8}{10}$ or $\dfrac{4}{5}$.

75. In words, 0.077 is seventy-seven thousandths, while as a fraction it is $\dfrac{77}{1000}$.

77. To round 47,261 to the nearest ten, observe that the digit in the ones place is 1. Since this digit is less than 5, we do not add 1 to the digit in the tens place. The number 47,261 rounded to the nearest ten is 47,260.

79. To round 47,261 to the nearest thousand, observe that the digit in the hundreds place is 2. Since this digit is less than 5, we do not add 1 to the digit in the thousands place. The number 47,261 rounded to the nearest thousand is 47,000.

81. answers may vary

83. 0.00026849576 in words is twenty-six million, eight hundred forty-nine thousand, five hundred seventy-six hundred-billionths.

85. $17\dfrac{268}{1000} = 17.268$

Section 4.2 Practice

1. Compare 13.208 and 13.281:
   The tens places are the same.
   The ones places are the same.
   The tenths places are the same.
   The hundredths places are different.
   Since 0 < 8, then 13.208 < 13.281.

2. Compare 0.124 and 0.086:
   1 is greater than 0 in the tenths place, so
   0.124 > 0.086.

3. 0.61   0.076
    ↑   ↑
    6 > 0 so
   0.61 > 0.076

4. By comparing the ones digits, 13.9 is the smallest number. The other numbers are:
   14.605   14.65 = 14.650   14.006
   By comparing digits to the right of the decimal place, arrange the numbers from smallest to largest: 13.9, 14.006, 14.605, 14.65.

5. In 123.7814, the digit to the right of the thousandths place is 4. Since 4 is less than 5, delete all digits to the right of the thousandths place. 123.7814 rounded to the nearest thousandth is 123.781.

6. In 123.7817, the digit to the right of the tenths place is 8. Since 8 is at least 5, add one to the digit in the tenths place and delete all digits to the right. 123.7817 rounded to the nearest tenth is 123.8.

7. In $3.1589, the digit to the right of the cents (hundredths) place is at least 5, so we add one to the hundredths digit and delete all digits to the right. $3.1589 rounded to the nearest cent is $3.16.

8. In $1.095, the digit to the right of the cents (hundredths) digit is 5, so add 1 to the hundredths digit and delete all digits to the right. Since 9 + 1 = 10, replace the digit 9 by 0 and carry the 1 to the place value to the left. $1.095 rounded to the nearest cent is $1.10.

9. Rounding to the nearest dollar means rounding to the ones place. In $24.62, the digit to the right of the ones place is 6, so add 1 to the digit in the ones place and delete the digits to the right. $24.62 rounded to the nearest dollar is $25.

10. To round 3.14159265 to the nearest ten-thousandth, observe that the digit in the hundred-thousandths place is 9. Since this digit is at least 5, we add 1 to the digit in the ten-thousandths place. The number 3.14159265 rounded to the nearest the ten-thousandth is 3.1416, or $\pi \approx 3.1416$.

Vocabulary and Readiness Check

1. Another name for the distance around a circle is its <u>circumference</u>.

2. $\pi = \dfrac{\text{circumference of a circle}}{\text{diameter of a circle}}$

3. The decimal point in a whole number is <u>after</u> the last digit.

4. The whole number 7 = <u>7.0</u>.

Exercise Set 4.2

1. 0.15   0.16
    ↑   ↑
    5 < 6 so
   0.15 < 0.16

3. 0.57   0.54
    ↑   ↑
    7 > 4 so
   0.57 > 0.54

5. 0.098   0.1
    ↑   ↑
    0 <   1 so
   0.098 < 0.1

7. 0.54900 = 0.549

9. 167.908   167.980
      ↑   ↑
      0 <   8 so
   167.908 < 167.980

11. 420,000 > 0.000042

13. Smallest to largest: 0.006, 0.0061, 0.06

15. Smallest to largest: 0.03, 0.042, 0.36

17. Smallest to largest: 1.01, 1.09, 1.1, 1.16

19. Smallest to largest: 20.905, 21.001, 21.03, 21.12

21. To round 0.57 to the nearest tenth, observe that the digit in the hundredths place is 7. Since this digit is at least 5, we need to add 1 to the digit in the tenths place. The number 0.57 rounded to the nearest tenth is 0.6.

23. To round 0.234 to the nearest hundredth, observe that the digit in the thousandths place is 4. Since this digit is less than 5, we do not add 1 to the digit in the hundredths place. The number 0.234 rounded to the nearest hundredth is 0.23.

25. To round 0.5942 to the nearest thousandth, observe that the digit in the ten-thousandths place is 2. Since this digit is less than 5, we do not add 1 to the digit in the thousandths place. The number 0.5942 rounded to the nearest thousandth is 0.594.

27. To round 98,207.23 to the nearest ten, observe that the digit in the ones place is 7. Since this digit is at least 5, we add 1 to the digit in the tens place. The number 98,207.23 rounded to the nearest ten is 98,210.

29. To round 12.342 to the nearest tenth, observe that the digit in the hundredths place is 4. Since this digit is less than 5, we do not add 1 to the digit in the tenths place. The number 12.342 rounded to the nearest tenth is 12.3.

31. To round 17.667 to the nearest hundredth, observe that the digit in the thousandths place is 7. Since this digit is at least 5, we add 1 to the digit in the hundredths place. The number 17.667 rounded to the nearest hundredth is 17.67.

33. To round 0.501 to the nearest tenth, observe that the digit in the hundredths place is 0. Since this digit is less than 5, we do not add 1 to the digit in the tenths place. The number 0.501 rounded to the nearest tenth is 0.5.

35. To round 0.1295 to the nearest thousandth, observe that the digit in the ten-thousandths place is 5. Since this digit is at least 5, we add 1 to the digit in the thousandths place. Since 9 + 1 = 10, we replace the 9 with a 0 and then add 1 to the digit in the hundreds place. The number 0.1295 rounded to the nearest thousandth is 0.130.

37. To round 3829.34 to the nearest ten, observe that the digit in the ones place is 9. Since this digit is at least 5, we add 1 to the digit in the tens place. The number 3829.34 rounded to the nearest ten is 3830.

39. To round 0.067 to the nearest hundredth, observe that the digit in the thousandths place is 7. Since this digit is at least 5, we add 1 to the digit in the hundredths place. The number 0.067 rounded to the nearest hundredth is 0.07. The amount is $0.07.

41. To round 42,650.14 to the nearest one, observe that the digit in the tenths place is 1. Since this digit is less than 5, we do not add 1 to the digit in the ones place. The number 42,650.14 rounded to the nearest one is 42,650. The amount is $42,650.

43. To round 26.95 to the nearest one, observe that the digit in the tenths place is 9. Since this digit is at least 5, we add 1 to the digit in the ones place. The number 26.95 rounded to the nearest one is 27. The amount is $27.

45. To round 0.1992 to the nearest hundredth, observe that the digit in the thousandths place is 9. Since this digit is at least 5, we need to add 1 to the digit in the hundredths place. The number 0.1992 rounded to the nearest hundredth is 0.20. The amount is $0.20.

47. To round 0.4064 to the nearest tenth, observe that the digit in the hundredths place is 0. Since this digit is less than 5, we do not add 1 to the digit in the tenths place. The number 0.4064 rounded to the nearest tenth is 0.4. The thickness is 0.4 centimeter.

49. To round 1.5581 to the nearest hundredth, observe that the digit in the thousandths place is 8. Since this is at least 5, we add 1 to the digit in the hundredths place. The number 1.5581 rounded to the nearest hundredth is 1.56. The time was 1.56 hours.

51. To round 47.89 to the nearest one, observe that the digit in the tenths place is 8. Since this digit is at least 5, we add 1 to the digit in the ones place. The number 47.89 rounded to the nearest one is 48. The price is $48.

53. To round 1.736 to the nearest hundredth, observe that the digit in the thousandths place is 6. Since this is at least 5, we add 1 to the digit in the hundredths place. The number 1.736 rounded to the nearest hundredth is 1.74. The time was 1.74 minutes.

55. To round 24.6229 to the nearest thousandth, observe that the digit in the ten-thousandths place is 9. Since this digit is at least 5, we add 1 to the digit in the thousandths place. 24.6229 rounded to the nearest thousandth is 24.623. The day length is 24.623 hours.

57. To round 2.75 to the nearest tenth, observe that the digit in the hundredths place is 5. Since this digit is at least 5, we add 1 to the digit in the tenths place. The number 2.75 rounded to the nearest tenth is 2.8. The ride lasts 2.8 minutes.

59. $\begin{array}{r} 3452 \\ + 2314 \\ \hline 5766 \end{array}$

61. $\begin{array}{r} 94 \\ - 23 \\ \hline 71 \end{array}$

63. $\begin{array}{r} 482 \\ - 239 \\ \hline 243 \end{array}$

65. To round 2849.1738 to the nearest hundred, observe that the digit in the tens place is 4. Since this digit is less than 5, we do not add 1 to the digit in the hundreds place. The number 2849.1738 rounded to the nearest hundred is 2800, which is choice b.

67. To round 2849.1738 to the nearest hundredth, observe that the digit in the thousandths place is 3. Since this digit is less than 5, we do not add 1 to the digit in the hundredths place. 2849.1738 rounded to the nearest hundredth is 2849.17, which is choice a.

69. The fastest average speed was 41.654 kilometers per hour.

$$41.654 = 41\frac{654}{1000} = 41\frac{327}{500}$$

Lance Armstrong achieved this speed.

71. Comparing the place value of each speed from 2006 through 2009, and listing from fastest to slowest yields: 40.789, 40.788, 40.413, 39.233.

73. answers may vary

75. answers may vary

77. 0.26499 and 0.25786 rounded to the nearest hundredth are 0.26. 0.26559 rounds to 0.27 and 0.25186 rounds to 0.25.

79. From smallest to largest, 0.10299, 0.1037, 0.1038, 0.9

81. Round to the nearest hundred million and then add.
$$\begin{array}{r} 700 \\ 600 \\ 500 \\ 500 \\ 400 \\ + 400 \\ \hline 3100 \end{array}$$
Estimate the total amount of money earned to be $3100 million.

Section 4.3 Practice

1. a. $\begin{array}{r} 15.520 \\ + 2.371 \\ \hline 17.891 \end{array}$

   b. $\begin{array}{r} 20.060 \\ + 17.612 \\ \hline 37.672 \end{array}$

   c. $\begin{array}{r} 0.125 \\ + 122.800 \\ \hline 122.925 \end{array}$

2. a. $\begin{array}{r} \scriptstyle 11\ 1 \\ 34.5670 \\ 129.4300 \\ + 2.8903 \\ \hline 166.8873 \end{array}$

   b. $\begin{array}{r} \scriptstyle 1 \\ 11.210 \\ 46.013 \\ + 362.526 \\ \hline 419.749 \end{array}$

3. $\begin{array}{r} \scriptstyle 1 \\ 26.072 \\ + 119.000 \\ \hline 145.072 \end{array}$

4. a.
$$\begin{array}{r} \scriptstyle 7\ 11\ 17 \\ 8\,2\,.\,7\,5 \\ -\,1\,5\,.\,9\,0 \\ \hline 6\,6\,.\,8\,5 \end{array}$$

Check:
$$\begin{array}{r} \scriptstyle 11 \\ 66.85 \\ +\,15.90 \\ \hline 82.75 \end{array}$$

b.
$$\begin{array}{r} \scriptstyle 5\ 10 \\ 12\,6\,.\,0\,3\,2 \\ -\,95.710 \\ \hline 30.322 \end{array}$$

Check:
$$\begin{array}{r} 30.322 \\ +\,95.710 \\ \hline 126.032 \end{array}$$

5. a.
$$\begin{array}{r} \scriptstyle 4\ 17\ 10 \\ 5\,.\,8\,0 \\ -\,3.9\,2 \\ \hline 1.8\,8 \end{array}$$

Check:
$$\begin{array}{r} \scriptstyle 1\ 1 \\ 1.88 \\ +\,3.92 \\ \hline 5.80 \end{array}$$

b.
$$\begin{array}{r} \scriptstyle 6\ 11\ 10 \\ 9\,.\,7\,2\,0 \\ -\,4.0\,6\,8 \\ \hline 5.6\,5\,2 \end{array}$$

Check:
$$\begin{array}{r} \scriptstyle 11 \\ 5.652 \\ +\,4.068 \\ \hline 9.720 \end{array}$$

6. a.
$$\begin{array}{r} \scriptstyle 4\ 12\ 9\ 10 \\ 5\,3\,.\,0\,0 \\ -\,2\,9.3\,1 \\ \hline 2\,3.6\,9 \end{array}$$

Check:
$$\begin{array}{r} \scriptstyle 1\ 1 \\ 23.69 \\ +\,29.31 \\ \hline 53.00 \end{array}$$

b.
$$\begin{array}{r} \scriptstyle 11\ 9\ 9\ 10 \\ 1\,2\,0\,.\,0\,0 \\ -\,6\,8.2\,2 \\ \hline 5\,1.7\,8 \end{array}$$

Check:
$$\begin{array}{r} \scriptstyle 11\ 1 \\ 51.78 \\ +\,68.22 \\ \hline 120.00 \end{array}$$

7. a.

Exact	Estimate 1	Estimate 2
$\begin{array}{r}\scriptstyle 11\\48.10\\+\,326.97\\\hline 375.07\end{array}$	$\begin{array}{r}50\\+\,300\\\hline 350\end{array}$	$\begin{array}{r}50\\+\,330\\\hline 380\end{array}$

b.

Exact	Estimate 1	Estimate 2
$\begin{array}{r}\scriptstyle 7\ 10\ 8\ 10\\18.0\,9\,0\\-\,0.7\,4\,6\\\hline 17.3\,4\,4\end{array}$	$\begin{array}{r}20\\-\,1\\\hline 19\end{array}$	$\begin{array}{r}18\\-\,1\\\hline 17\end{array}$

8. Add the payment, insurance, and the average gasoline bill amounts.
$$\begin{array}{r} \scriptstyle 111\ 1 \\ 536.52 \\ 52.68 \\ +\,87.50 \\ \hline 676.70 \end{array}$$
The total monthly cost is $676.70.

9. Subtract the average height in Israel from the average height in the Netherlands.
$$\begin{array}{r} \scriptstyle 1\ 12 \\ 7\,2\,.8 \\ -\,6\,8.2 \\ \hline 3.6 \end{array}$$
The average height in the Netherlands is 3.6 inches greater than the average height in Israel.

Calculator Explorations

1. $315.782 + 12.96 = 328.742$

2. $29.68 + 85.902 = 115.582$

3. $6.249 - 1.0076 = 5.2414$

4. $5.238 - 0.682 = 4.556$

5.
$$\begin{array}{r} 12.555 \\ 224.987 \\ 5.2 \\ +\,622.65 \\ \hline 865.392 \end{array}$$

6.
$$\begin{array}{r} 47.006 \\ 0.17 \\ 313.259 \\ +\,139.088 \\ \hline 499.523 \end{array}$$

Vocabulary and Readiness Check

1. The number 37 equals <u>37.0</u>.

2. The decimal point in a whole number is positioned after the <u>last</u> digit.

3. In 89.2 − 14.9 = 74.3, the number 74.3 is called the <u>difference</u>, 89.2 is the <u>minuend</u>, and 14.9 is the <u>subtrahend</u>.

4. To add or subtract decimals, we line up the decimal points <u>vertically</u>.

5. True or false: The number 5.6 is closer to 5 than 6 on the number line. <u>false</u>

6. True or false: The number 10.48 is closer to 10 than 11 on the number line. <u>true</u>

Exercise Set 4.3

1.
$$\begin{array}{r} 1.3 \\ + 2.2 \\ \hline 3.5 \end{array}$$

3.
$$\begin{array}{r} 5.7 \\ + 1.13 \\ \hline 6.83 \end{array}$$

5.
$$\begin{array}{r} 0.003 \\ + 0.091 \\ \hline 0.094 \end{array}$$

7.
$$\begin{array}{r} {}^{1\,1\,1} \\ 19.23 \\ + 602.782 \\ \hline 622.012 \end{array}$$

9.
$$\begin{array}{r} {}^{1} \\ 490.00 \\ + 93.09 \\ \hline 583.09 \end{array}$$

11. Exact:
$$\begin{array}{r} {}^{1\,1} \\ 234.89 \\ + 230.67 \\ \hline 465.56 \end{array}$$
Estimate:
$$\begin{array}{r} 230 \\ + 230 \\ \hline 460 \end{array}$$

13. Exact:
$$\begin{array}{r} {}^{1\quad 1\,1} \\ 100.009 \\ 6.080 \\ + 9.034 \\ \hline 115.123 \end{array}$$
Estimate:
$$\begin{array}{r} {}^{1} \\ 100 \\ 6 \\ + 9 \\ \hline 115 \end{array}$$

15.
$$\begin{array}{r} {}^{1\,1} \\ 24.6 \\ 2.39 \\ + 0.0678 \\ \hline 27.0578 \end{array}$$

17.
$$\begin{array}{r} {}^{1\quad 1} \\ 45.023 \\ 3.006 \\ + 8.403 \\ \hline 56.432 \end{array}$$

19.
$$\begin{array}{r} 8.8 \\ - 2.3 \\ \hline 6.5 \end{array}$$
Check:
$$\begin{array}{r} 6.5 \\ + 2.3 \\ \hline 8.8 \end{array}$$

21.
$$\begin{array}{r} 18.0 \\ - 2.7 \\ \hline 15.3 \end{array}$$
Check:
$$\begin{array}{r} {}^{1} \\ 15.3 \\ + 2.7 \\ \hline 18.0 \end{array}$$

23.
$$\begin{array}{r} 654.9 \\ - 56.67 \\ \hline 598.23 \end{array}$$
Check:
$$\begin{array}{r} {}^{11\quad 1} \\ 598.23 \\ + 56.67 \\ \hline 654.90 = 654.9 \end{array}$$

25. Exact:
$$\begin{array}{r} 5.9 \\ - 4.07 \\ \hline 1.83 \end{array}$$
Estimate:
$$\begin{array}{r} 6 \\ - 4 \\ \hline 2 \end{array}$$
Check:
$$\begin{array}{r} {}^{1} \\ 1.83 \\ + 4.07 \\ \hline 5.90 = 5.90 \end{array}$$

27.
$$\begin{array}{r} 923.5 \\ - 61.9 \\ \hline 861.6 \end{array}$$
Check:
$$\begin{array}{r} {}^{1\quad 1} \\ 861.6 \\ + 61.9 \\ \hline 923.5 \end{array}$$

29.
$$\begin{array}{r} 500.34 \\ - 123.45 \\ \hline 376.89 \end{array}$$
Check:
$$\begin{array}{r} {}^{1\,1\,1\,1} \\ 376.89 \\ + 123.45 \\ \hline 500.34 \end{array}$$

31. Exact:
$$\begin{array}{r} 1000.0 \\ - 123.4 \\ \hline 876.6 \end{array}$$
Estimate:
$$\begin{array}{r} 1000 \\ - 100 \\ \hline 900 \end{array}$$
Check:
$$\begin{array}{r} {}^{1\,1\,1\,1} \\ 876.6 \\ + 123.4 \\ \hline 1000.0 \end{array}$$

33.
$$\begin{array}{r} 200.0 \\ - 5.6 \\ \hline 194.4 \end{array}$$
Check:
$$\begin{array}{r} {}^{1\,1\,1} \\ 194.4 \\ + 5.6 \\ \hline 200.0 \end{array}$$

35.
$$\begin{array}{r} 3.0000 \\ - 0.0012 \\ \hline 2.9988 \end{array}$$
Check:
$$\begin{array}{r} {}^{1\,1\,1\,1} \\ 2.9988 \\ + 0.0012 \\ \hline 3.0000 \end{array}$$

37.
$$\begin{array}{r} 23.0 \\ -\ 6.7 \\ \hline 16.3 \end{array}$$
Check:
$$\begin{array}{r} \overset{11}{16.3} \\ +\ 6.7 \\ \hline 23.0 \end{array}$$

39.
$$\begin{array}{r} \overset{1\ 1}{86.05} \\ +\ 1.978 \\ \hline 88.028 \end{array}$$

41.
$$\begin{array}{r} 86.050 \\ -\ 1.978 \\ \hline 84.072 \end{array}$$

43.
$$\begin{array}{r} \overset{1}{150} \\ +\ 93.17 \\ \hline 243.17 \end{array}$$

45.
$$\begin{array}{r} 150.00 \\ -\ 93.17 \\ \hline 56.83 \end{array}$$

47.
$$\begin{array}{r} 12.10 \\ -\ 8.94 \\ \hline 3.16 \end{array}$$

49. Change = 40 − 32.48
$$\begin{array}{r} 40.00 \\ -\ 32.48 \\ \hline 7.52 \end{array}$$
If Ann-Margaret paid with two $20 bills, her change was $7.52.

51. The phrase "Find the total" indicates that we should add the amounts.
$$\begin{array}{r} \overset{2\,1\,1\,1}{275.36} \\ 83.00 \\ 81.60 \\ +\ 14.75 \\ \hline 454.71 \end{array}$$
The total monthly cost is $454.71.

53. The word "change" indicates that we should subtract the amounts.
$$\begin{array}{r} 2.979 \\ -\ 2.839 \\ \hline 0.140 \end{array}$$
The change in price was $0.14.

55. The perimeter is the distance around. The sides of a square have the same length.
$$\begin{array}{r} \overset{2\ \ 1}{7.14} \\ 7.14 \\ 7.14 \\ +\ 7.14 \\ \hline 28.56 \end{array}$$
The perimeter is 28.56 meters.

57. the perimeter is the distance around.
$$\begin{array}{r} \overset{1\,2}{3.6} \\ 3.6 \\ 1.5 \\ 1.5 \\ \hline 10.2 \end{array}$$
The perimeter is 10.2 inches.

59. The phrase "How much warmer" indicates we should subtract the average of all U.S. average temperatures from the average U.S. temperature in 1998.
$$\begin{array}{r} 55.08 \\ -\ 52.85 \\ \hline 2.23 \end{array}$$
The average U.S. temperature in 1998 was 2.23 degrees Fahrenheit warmer than the average of all U.S. average temperatures.

61. The phrase "How much faster" indicates that we should subtract the average wind speed from the record speed.
$$\begin{array}{r} 321.0 \\ -\ 35.2 \\ \hline 285.8 \end{array}$$
The highest wind speed is 285.8 miles per hour faster than the average wind speed.

63. The phrase "faster than" indicates we should add.
$$\begin{array}{r} \overset{11\ 11}{633.468} \\ +\ 129.567 \\ \hline 763.035 \end{array}$$
Green's record setting speed was 763.035 miles per hour.

65. The perimeter is the distance around.
$$\begin{array}{r} 4.5 \\ 2.4 \\ 4.5 \\ +\ 2.4 \\ \hline 13.8 \end{array}$$
The perimeter is 13.8 inches.

67. The phrase "Find the increase" indicates we should subtract the 2008 ticket price from the 2009 ticket price.

$$\begin{array}{r} 7.50 \\ -\ 7.18 \\ \hline 0.32 \end{array}$$

The increase in average U.S. movie-theater ticket price from 2008 to 2009 was $0.32.

69. To find the increase, subtract.

$$\begin{array}{r} 75.3 \\ -\ 62.9 \\ \hline 12.4 \end{array}$$

The increase was 12.4 million or 12,400,000 users.

71. To find the amount of snow in Blue Canyon, add 111.6 to the amount in Marquette.

$$\begin{array}{r} 129.2 \\ +\ 111.6 \\ \hline 240.8 \end{array}$$

Blue Canyon receives on average 240.8 inches each year.

73. Add the lengths of the sides to get the perimeter.

$$\begin{array}{r} 12.40 \\ 29.34 \\ +\ 25.70 \\ \hline 67.44 \end{array}$$

67.44 feet of border material is needed.

75. The phrase "How much more" indicates we should subtract the average cost in 2005 from the average cost in 2008.

$$\begin{array}{r} 3.813 \\ -\ 2.338 \\ \hline 1.475 \end{array}$$

The average cost of a gallon of gasoline in 2008 was $1.475 more than the average cost in 2005.

77. The phrase "Find the total" indicates that we should add durations of the flights.

$$\begin{array}{r} {\scriptstyle 2\ 1\ 2\ 1\ 1} \\ 330.583 \\ 94.567 \\ 147.000 \\ +\ 142.900 \\ \hline 715.050 \end{array}$$

James A. Lovell spent 715.05 hours in spaceflight.

79. The tallest bar indicates the greatest chocolate consumption per person, so Switzerland has the greatest chocolate consumption per person.

81.
$$\begin{array}{r} 22.36 \\ -\ 17.93 \\ \hline 4.43 \end{array}$$

The difference in consumption is 4.43 pounds per year.

83.

Country	Pounds of Chocolate per Person
Switzerland	22.36
Austria	20.13
Ireland	19.47
Germany	18.04
Norway	17.93

85. $23 \cdot 2 = 46$

87. $43 \cdot 90 = 3870$

89. $\left(\dfrac{2}{3}\right)^2 = \dfrac{2}{3} \cdot \dfrac{2}{3} = \dfrac{2 \cdot 2}{3 \cdot 3} = \dfrac{4}{9}$

91. It is incorrect. Align the decimals.

$$\begin{array}{r} 9.200 \\ 8.630 \\ +\ 4.005 \\ \hline 21.835 \end{array}$$

93. $10.68 - (2.3 + 2.3) = 10.68 - 4.60 = 6.08$
The unknown length is 6.08 inches.

95. 3 nickels, 3 dimes, and 3 quarters:
$0.05 + 0.05 + 0.05 + 0.10 + 0.10 + 0.10 + 0.25 + 0.25 + 0.25 = 1.2$
The value of the coins shown is $1.20.

97. 1 nickel, 1 dime, and 2 pennies:
$0.05 + 0.10 + 0.01 + 0.01 = 0.17$
3 nickels and 2 pennies:
$0.05 + 0.05 + 0.05 + 0.01 + 0.01 = 0.17$
1 dime and 7 pennies:
$0.10 + 0.01 + 0.01 + 0.01 + 0.01 + 0.01 + 0.01 + 0.01 = 0.17$
2 nickels and 7 pennies:
$0.05 + 0.05 + 0.01 + 0.01 + 0.01 + 0.01 + 0.01 + 0.01 + 0.01 = 0.17$

99. answers may vary

101. answers may vary

Section 4.4 Practice

1.
$$
\begin{array}{r}
45.9 \\
\times\, 0.42 \\
\hline
918 \\
18\,360 \\
\hline
19.278
\end{array}
$$
1 decimal place
2 decimal places

3 decimal places

2.
$$
\begin{array}{r}
0.112 \\
\times\,\ 0.6 \\
\hline
0.0672
\end{array}
$$
3 decimal places
1 decimal place
4 decimal places

3.
$$
\begin{array}{r}
0.0721 \\
\times\ \ \ 48 \\
\hline
5768 \\
2\,8840 \\
\hline
3.4608
\end{array}
$$
4 decimal places
0 decimal places

4 decimal places

4. Exact
$$
\begin{array}{r}
30.26 \\
\times\ \ 2.98 \\
\hline
2\,4208 \\
27\,2340 \\
60\,5200 \\
\hline
90.1748
\end{array}
$$
Estimate
$$
\begin{array}{r}
30 \\
\times\,3 \\
\hline
90
\end{array}
$$

5. $23.7 \times 10 = 237$

6. $203.004 \times 100 = 20,300.4$

7. $1.15 \times 1000 = 1150$

8. $7.62 \times 0.1 = 0.762$

9. $1.9 \times 0.01 = 0.019$

10. $7682 \times 0.001 = 7.682$

11. 321.2 million $= 321.2 \times 1$ million
$= 321.2 \times 1,000,000$
$= 321,200,000$

12. Circumference $= 2 \cdot \pi \cdot$ radius
$= 2 \cdot \pi \cdot 11$ meters
$= 22\pi$ meters
$\approx 22(3.14)$ meters
≈ 69.08 meters
The circumference is 22π meters which is approximately 69.08 meters.

13. Total amount is amount per square yard times the number of square yards.
Total amount $= 5.6 \times 60.5 = 338.8$
She needs 338.8 ounces of fertilizer.

Vocabulary and Readiness Check

1. When multiplying decimals, the number of decimal places in the product is equal to the <u>sum</u> of the number of decimal place in the factors.

2. In $8.6 \times 5 = 43$, the number 43 is called the <u>product</u> while 8.6 and 5 are each called a <u>factor</u>.

3. When multiplying a decimal number by powers of 10 such as 10, 100, 1000, and so on, we move the decimal point in the number to the <u>right</u> the same number of places as there are <u>zeros</u> in the power of 10.

4. When multiplying a decimal number by powers of 10 such as 0.1, 0.01, and so on, we move the decimal point in the number to the <u>left</u> the same number of places as there are <u>decimal places</u> in the power of 10.

5. The distance around a circle is called its <u>circumference</u>.

6.
$$
\begin{array}{r}
0.46 \\
\times\, 0.81
\end{array}
$$
There will be $2 + 2 = 4$ decimal places in the product.

7.
$$
\begin{array}{r}
57.9 \\
\times\, 0.36
\end{array}
$$
There will be $1 + 2 = 3$ decimal places in the product.

8.
$$
\begin{array}{r}
0.428 \\
\times\ \ 0.2
\end{array}
$$
There will be $3 + 1 = 4$ decimal places in the product.

9.
$$
\begin{array}{r}
0.0073 \\
\times\ \ \ \ 21
\end{array}
$$
There will be $4 + 0 = 4$ decimal places in the product.

10.
$$
\begin{array}{r}
0.028 \\
\times\, 1.36
\end{array}
$$
There will be $3 + 2 = 5$ decimal places in the product.

11.
$$
\begin{array}{r}
5.1296 \\
\times\, 7.3987
\end{array}
$$
There will be $4 + 4 = 8$ decimal places in the product.

Exercise Set 4.4

1. 0.2
$\times 0.6$
―――
0.12

3. 1.2
$\times 0.5$
―――
$0.60 = 0.6$

5. 0.26
$\times \ \ 5$
―――
$1.30 = 1.3$

7. Exact: 5.3   Estimate: 5
    $\times 4.2$ $\times 4$
    ―――   ―――
    106   20
    2120
    ―――
    22.26

9. 0.576
$\times \ \ 0.7$
―――
0.4032

11. Exact: 1.0047   Estimate: 1
     $\times \ \ \ 8.2$ $\times 8$
     ―――   ―――
     20094   8
     8 03760
     ―――
     8.23854

13. 490.2
$\times \ 0.023$
―――
1 4706
9 8040
―――
11.2746

15. 16.003
$\times \ \ \ 5.31$
―――
16003
4 80090
80 01500
―――
84.97593

17. $6.5 \times 10 = 65$

19. $6.5 \times 0.1 = 0.65$

21. $7.2 \times 0.01 = 0.072$

23. $7.093 \times 100 = 709.3$

25. $6.046 \times 1000 = 6046$

27. $37.62 \times 0.001 = 0.03762$

29. 0.123
$\times \ \ 0.4$
―――
0.0492

31. $0.123 \times 100 = 12.3$

33. 8.6
$\times 0.15$
―――
430
860
―――
$1.290 = 1.29$

35. $9.6 \times 0.01 = 0.096$

37. $562.3 \times 0.001 = 0.5623$

39. 5.62
$\times \ \ 7.7$
―――
3 934
39 340
―――
43.274

41. 5.5 billion $= 5.5 \times 1$ billion
      $= 5.5 \times 1,000,000,000$
      $= 5,500,000,000$
The silos hold enough to make
5,500,000,000 bars.

43. 49.8 million $= 49.8 \times 1$ million
      $= 49.8 \times 1,000,000$
      $= 49,800,000$
The Blue Streak has given more than
49,800,000 rides.

45. 314 thousand $= 314 \times 1$ thousand
      $= 314 \times 1000$
      $= 314,000$
An estimated 314,000 people visited the park
each week.

47. Circumference $= 2 \cdot \pi \cdot$ radius
      $C = 2 \cdot \pi \cdot 4 = 8\pi$
      $C \approx 8(3.14) = 25.12$
The circumference is 8π meters, which is
approximately 25.12 meters.

49. Circumference $= \pi \cdot$ diameter
      $C = \pi \cdot 10 = 10\pi$
      $C \approx 10(3.14) = 31.4$
The circumference is 10π centimeters, which is
approximately 31.4 centimeters.

51. Circumference = $2 \cdot \pi \cdot$ radius
$$C = 2 \cdot \pi \cdot 9.1 = 18.2\pi$$
$$C \approx 18.2(3.14) = 57.148$$

The circumference is 18.2π yards, which is approximately 57.148 yards.

53. pay before taxes = $\underline{17.88} \times \underline{40}$

$$
\begin{array}{r}
17.88 \\
\times\ \ \ 40 \\
\hline
715.20
\end{array}
$$

His pay before taxes last week was \$715.20.

55. Multiply the number of miles driven by the cost per mile.

$$
\begin{array}{r}
8750 \\
\times\ 0.54 \\
\hline
35000 \\
437500 \\
\hline
4725.00
\end{array}
$$

The cost to drive a medium sedan in 2009 was \$4725.

57. Multiply the number of ounces by the number of grams of saturated fat in 1 ounce.

$$
\begin{array}{r}
6.2 \\
\times\ 4 \\
\hline
24.8
\end{array}
$$

There are 24.8 grams of saturated fat in a 4-ounce serving of cream cheese.

59. Area = length $\cdot$ width

$$
\begin{array}{r}
4.5 \\
\times\ 2.4 \\
\hline
1\ 80 \\
9\ 00 \\
\hline
10.80
\end{array}
$$

The area is 10.8 square inches.

61. Circumference = $\pi \cdot$ diameter
$$C = \pi \cdot 250 = 250\pi$$

$$
\begin{array}{r}
250 \\
\times\ \ 3.14 \\
\hline
10\ 00 \\
25\ 00 \\
750\ 00 \\
\hline
785.00
\end{array}
$$

The circumference is 250π feet, which is approximately 785 feet.

63. $C = \pi \cdot d$
$$C = \pi \cdot 135 = 135\pi$$

$$
\begin{array}{r}
135 \\
\times\ 3.14 \\
\hline
5\ 40 \\
13\ 50 \\
405\ 00 \\
\hline
423.90
\end{array}
$$

He travels 135π meters or approximately 423.9 meters.

65. Multiply her height in meters by the number of inches in 1 meter.

$$
\begin{array}{r}
39.37 \\
\times\ \ 1.65 \\
\hline
1\ 9685 \\
23\ 6220 \\
39\ 3700 \\
\hline
64.9605
\end{array}
$$

She is approximately 64.9605 inches tall.

67. a. Circumference = $2 \cdot \pi \cdot$ radius
Smaller circle:
$$C = 2 \cdot \pi \cdot 10 = 20\pi$$
$$C \approx 20(3.14) = 62.8$$
The circumference of the smaller circle is approximately 62.8 meters.
Larger circle:
$$C = 2 \cdot \pi \cdot 20 = 40\pi$$
$$C \approx 40(3.14) = 125.6$$
The circumference of the larger circle is approximately 125.6 meters.

b. Yes, the circumference gets doubled when the radius is doubled.

69. Area = length $\times$ width

$$
\begin{array}{r}
3.6 \\
\times\ 1.5 \\
\hline
1\ 80 \\
3\ 60 \\
\hline
5.40
\end{array}
$$

The area is 5.4 square inches.

71.
$$
\begin{array}{r}
26 \\
5\overline{)\ 130} \\
\underline{-10} \\
30 \\
\underline{-30} \\
6
\end{array}
$$

73.
$$56\overline{)2016}$$
36
−168
336
−336
0

75.
$$365\overline{)2920}$$
8
−2920
0

77. $\dfrac{24}{7} \div \dfrac{8}{21} = \dfrac{24}{7} \cdot \dfrac{21}{8}$

$= \dfrac{24 \cdot 21}{7 \cdot 8}$

$= \dfrac{8 \cdot 3 \cdot 7 \cdot 3}{7 \cdot 8}$

$= \dfrac{3 \cdot 3}{1}$

$= 9$

79.
3.60
+ 0.04
3.64

81.
3.60
− 0.04
3.56

83.
0.221
×   0.5
0.1105

85. $20.6(1.86 \times 100,000) = 20.6(186,000)$
$= 3,831,600$
Radio waves travel 3,831,600 miles in 20.6 seconds.

87. answers may vary

89. answers may vary

Integrated Review

1.
1
1.60
+ 0.97
2.57

2.
1
3.20
+ 0.85
4.05

3.
9.8
− 0.9
8.9

4.
10.2
− 6.7
3.5

5.
0.8
× 0.2
0.16

6.
0.6
× 0.4
0.24

7.
11
8.00
2.16
+ 0.90
11.06

8.
6.00
3.12
+ 0.60
9.72

9.
9.6
× 0.5
4.80 = 4.8

10.
8.7
× 0.7
6.09

11.
123.60
− 48.04
75.56

12.
325.20
− 36.08
289.12

13.
25.000
+ 0.026
25.026

14.
0.125
44.000
44.125

15.
100.0
− 17.3
82.7

16.
$$\begin{array}{r} 300.0 \\ -\ 26.1 \\ \hline 273.9 \end{array}$$

17. $2.8 \times 100 = 280$

18. $1.6 \times 1000 = 1600$

19.
$$\begin{array}{r} \overset{1\ 1}{} \\ 96.210 \\ 7.028 \\ +\ 121.700 \\ \hline 224.938 \end{array}$$

20.
$$\begin{array}{r} \overset{2\ 1}{} \\ 0.268 \\ 1.930 \\ +\ 142.881 \\ \hline 145.079 \end{array}$$

21.
$$\begin{array}{r} 1.2 \\ \times\ 5 \\ \hline 6.0 = 6 \end{array}$$
The product is 6.

22.
$$\begin{array}{r} 1.2 \\ +\ 5.0 \\ \hline 6.2 \end{array}$$
The sum is 6.2.

23.
$$\begin{array}{r} 12.004 \\ \times\ 2.3 \\ \hline 3\ 6012 \\ 24\ 0080 \\ \hline 27.6092 \end{array}$$

24.
$$\begin{array}{r} 28.006 \\ \times\ 5.2 \\ \hline 5\ 6012 \\ 140\ 0300 \\ \hline 145.6312 \end{array}$$

25.
$$\begin{array}{r} 10.0 \\ -\ 4.6 \\ \hline 5.4 \end{array}$$

26.
$$\begin{array}{r} 18.00 \\ -\ 0.26 \\ \hline 17.74 \end{array}$$

27.
$$\begin{array}{r} \overset{1\ 1\ \ \ 1}{} \\ 268.19 \\ +\ 146.25 \\ \hline 414.44 \end{array}$$

28.
$$\begin{array}{r} \overset{1\ 1}{} \\ 860.18 \\ +\ 434.85 \\ \hline 1295.03 \end{array}$$

29.
$$\begin{array}{r} 160.00 \\ -\ 43.19 \\ \hline 116.81 \end{array}$$

30.
$$\begin{array}{r} 120.00 \\ -\ 101.21 \\ \hline 18.79 \end{array}$$

31. $15.62 \times 10 = 156.2$

32.
$$\begin{array}{r} 15.62 \\ +\ 10.00 \\ \hline 25.62 \end{array}$$

33.
$$\begin{array}{r} 15.62 \\ -\ 10.00 \\ \hline 5.62 \end{array}$$

34.
$$\begin{array}{r} 117.26 \\ \times\ 2.6 \\ \hline 70\ 356 \\ 234\ 520 \\ \hline 304.876 \end{array}$$

35.
$$\begin{array}{r} 117.26 \\ -\ 2.60 \\ \hline 114.66 \end{array}$$

36.
$$\begin{array}{r} 117.26 \\ +\ 2.60 \\ \hline 119.86 \end{array}$$

37.
$$\begin{array}{r} 0.0072 \\ \times\ 0.06 \\ \hline 0.000432 \end{array}$$

38.
$$\begin{array}{r} 0.0025 \\ \times\ 0.03 \\ \hline 0.000075 \end{array}$$

39.
$$\begin{array}{r} 0.0072 \\ +\ 0.0600 \\ \hline 0.0672 \end{array}$$

40.
$$\begin{array}{r} 0.0300 \\ -\ 0.0025 \\ \hline 0.0275 \end{array}$$

41. $0.862 \times 1000 = 862$

42. $2.93 \times 0.01 = 0.0293$

43.
53.7	rounds to	50
79.2	rounds to	80
+ 71.2	rounds to	70
		200

The distance between Garden City and Wichita is approximately 200 miles.

Section 4.5 Practice

1.
$$
\begin{array}{r}
46.3 \\
8\overline{)370.4} \\
\underline{-32} \\
50 \\
\underline{-48} \\
2\,4 \\
\underline{-2\,4} \\
0
\end{array}
$$

Check:
$$
\begin{array}{r}
46.3 \\
\times\ 8 \\
\hline
370.4
\end{array}
$$

2.
$$
\begin{array}{r}
0.71 \\
48\overline{)34.08} \\
\underline{-33\,6} \\
48 \\
\underline{-48} \\
0
\end{array}
$$

Check:
$$
\begin{array}{r}
0.71 \\
\times\ 48 \\
\hline
5\,68 \\
28\,40 \\
\hline
34.08
\end{array}
$$

3. a.
$$
\begin{array}{r}
0.05 \\
8\overline{)0.40} \\
\underline{-40} \\
0
\end{array}
$$

Check:
$$
\begin{array}{r}
0.05 \\
\times\ 8 \\
\hline
0.40
\end{array}
$$

b.
$$
\begin{array}{r}
1.135 \\
12\overline{)13.620} \\
\underline{-12} \\
1\,6 \\
\underline{-1\,2} \\
42 \\
\underline{-36} \\
60 \\
\underline{-60} \\
0
\end{array}
$$

Check:
$$
\begin{array}{r}
1.135 \\
\times\ 12 \\
\hline
2\,270 \\
11\,350 \\
\hline
13.620
\end{array}
$$

4. $5.6\overline{)166.88}$ becomes
$$
\begin{array}{r}
29.8 \\
56\overline{)1668.8} \\
\underline{-112} \\
548 \\
\underline{-504} \\
48\,8 \\
\underline{-48\,8} \\
0
\end{array}
$$

5. $0.16\overline{)1.976}$ becomes
$$
\begin{array}{r}
12.35 \\
16\overline{)197.60} \\
\underline{-16} \\
37 \\
\underline{-32} \\
56 \\
\underline{-48} \\
80 \\
\underline{-80} \\
0
\end{array}
$$

6. $0.57\overline{)23.4}$ becomes
$$
\begin{array}{r}
41.052 \approx 41.05 \\
57\overline{)2340.000} \\
\underline{-228} \\
60 \\
\underline{-57} \\
300 \\
\underline{-285} \\
150 \\
\underline{-114} \\
36
\end{array}
$$

7. $91.5\overline{)713.7}$ becomes
$$
\begin{array}{r}
7.8 \\
915\overline{)7137.0} \\
\underline{-6450} \\
732\,0 \\
\underline{-732\,0} \\
0
\end{array}
$$

Estimate:
$$
\begin{array}{r}
8 \\
90\overline{)720}
\end{array}
$$

8. $\dfrac{128.3}{1000} = 0.1283$

9. $\dfrac{0.56}{10} = 0.056$

10. Divide the number of square feet of lawn by the number of square feet one bag covers.
$$
\begin{array}{r}
11.84 \\
1250\overline{)14800.00} \\
\underline{-1250} \\
2300 \\
\underline{-1250} \\
10500 \\
\underline{-10000} \\
5000 \\
\underline{-5000} \\
0
\end{array}
$$

He needs 11.84 bags of fertilizer. If he can only buy whole bags, he needs 12 whole bags.

11. $897.8 \div 100 \times 10 = 8.978 \times 10 = 89.78$

12. $8.69(3.2 - 1.8) = 8.69(1.4) = 12.166$

13. $\dfrac{20.06 - (1.2)^2 \div 10}{0.02} = \dfrac{20.06 - 1.44 \div 10}{0.02}$
$$= \dfrac{20.06 - 0.144}{0.02}$$
$$= \dfrac{19.916}{0.02}$$
$$= 995.8$$

Calculator Explorations

1. 102.62×41.8
Estimate: $100 \times 42 = 4200$
The given result is not reasonable.

2. $174.835 \div 47.9$
Estimate: $200 \div 50 = 4$
The given result is reasonable.

3. $1025.68 - 125.42$
Estimate: $1000 - 100 = 900$
The given result is reasonable.

4. $562.781 + 2.96$
Estimate: $563 + 3 = 566$
The given result is not reasonable.

Vocabulary and Readiness Check

1. In $6.5 \div 5 = 1.3$, the number 1.3 is called the
quotient, 5 is the divisor, and 6.5 is the dividend.

2. To check a division exercise, we can perform the
following multiplication:
quotient · divisor = dividend.

3. To divide a decimal number by a power of 10
such as 10, 100, 1000, or so on, we move the
decimal point in the number to the left the same
number of places as there are zeros in the power
of 10.

4. True or false: If $1.058 \div 0.46 = 2.3$, then
$2.3 \times 0.46 = 1.058$ true

5. $\dfrac{5.9}{1} = 5.9$

6. $\dfrac{0.7}{0.7} = 1$

7. $\dfrac{0}{9.86} = 0$

8. $\dfrac{2.36}{0}$ is undefined.

9. $\dfrac{7.261}{7.261} = 1$

10. $\dfrac{8.25}{1} = 8.25$

11. $\dfrac{11.1}{0}$ is undefined.

12. $\dfrac{0}{89.96} = 0$

Exercise Set 4.5

1.
$$
\begin{array}{r}
4.6 \\
3{\overline{\smash{\big)}\,13.8}} \\
\underline{-12} \\
1\,8
\end{array}
$$

3.
$$
\begin{array}{r}
0.094 \\
5{\overline{\smash{\big)}\,0.470}} \\
\underline{-45} \\
20 \\
\underline{-20} \\
0
\end{array}
$$

5. $0.06{\overline{\smash{\big)}\,18}}$ becomes
$$
\begin{array}{r}
300 \\
6{\overline{\smash{\big)}\,1800}} \\
\underline{-18} \\
000
\end{array}
$$

7. $0.82{\overline{\smash{\big)}\,4.756}}$ becomes
$$
\begin{array}{r}
5.8 \\
82{\overline{\smash{\big)}\,475.6}} \\
\underline{-410} \\
65\,6 \\
\underline{-65\,6} \\
0
\end{array}
$$

9. Exact: $5.5\overline{)36.3}$ becomes

$$\begin{array}{r} 6.6 \\ 55\overline{)\ 363.0} \\ -330 \\ \hline 33\ 0 \\ -33\ 0 \\ \hline 0 \end{array}$$

Estimate: $\begin{array}{r} 6 \\ 6\overline{)36} \end{array}$

11. $$\begin{array}{r} 0.413 \\ 15\overline{)\ 6.195} \\ -6\ 0 \\ \hline 19 \\ -15 \\ \hline 45 \\ -45 \\ \hline 0 \end{array}$$

13. $$\begin{array}{r} 0.045 \\ 12\overline{)0.540} \\ -48 \\ \hline 60 \\ -60 \\ \hline 0 \end{array}$$

15. $0.6\overline{)4.2}$ becomes $$\begin{array}{r} 7 \\ 6\overline{)\ 42} \\ -42 \\ \hline 0 \end{array}$$

17. $0.27\overline{)1.296}$ becomes $$\begin{array}{r} 4.8 \\ 27\overline{)\ 129.6} \\ -108 \\ \hline 21\ 6 \\ -21\ 6 \\ \hline 0 \end{array}$$

19. $0.02\overline{)42}$ becomes $$\begin{array}{r} 2100 \\ 2\overline{)\ 4200} \\ -4 \\ \hline 02 \\ -2 \\ \hline 000 \end{array}$$

21. $0.6\overline{)18}$ becomes $$\begin{array}{r} 30 \\ 6\overline{)\ 180} \\ -18 \\ \hline 00 \end{array}$$

23. $0.005\overline{)35}$ becomes $$\begin{array}{r} 7000 \\ 5\overline{)\ 35000} \\ -35 \\ \hline 0000 \end{array}$$

25. Exact: $7.2\overline{)70.56}$ becomes $$\begin{array}{r} 9.8 \\ 72\overline{)\ 705.6} \\ -648 \\ \hline 57\ 6 \\ -57\ 6 \\ \hline 0 \end{array}$$

Estimate: $\begin{array}{r} 10 \\ 7\overline{)70} \end{array}$

27. $5.4\overline{)51.84}$ becomes $$\begin{array}{r} 9.6 \\ 54\overline{)\ 518.4} \\ -486 \\ \hline 32\ 4 \\ -32\ 4 \\ \hline 0 \end{array}$$

29. $0.027\overline{)1.215}$ becomes $$\begin{array}{r} 45 \\ 27\overline{)\ 1215} \\ -108 \\ \hline 135 \\ -135 \\ \hline 0 \end{array}$$

31. $0.25\overline{)13.648}$ becomes $$\begin{array}{r} 54.592 \\ 25\overline{)\ 1364.800} \\ -125 \\ \hline 114 \\ -100 \\ \hline 14\ 8 \\ -12\ 5 \\ \hline 2\ 30 \\ -2\ 25 \\ \hline 50 \\ -50 \\ \hline 0 \end{array}$$

33. $3.78\overline{)0.02079}$ becomes $$\begin{array}{r} 0.0055 \\ 378\overline{)\ 2.0790} \\ -1\ 890 \\ \hline 1890 \\ -1890 \\ \hline 0 \end{array}$$

35. $2.4\overline{)429.34}$ becomes

$$
\begin{array}{r}
178.8 \approx 179 \\
24\overline{)4293.4} \\
\underline{-24} \\
189 \\
\underline{-168} \\
213 \\
\underline{-192} \\
21\ 4 \\
\underline{-19\ 2} \\
2\ 2
\end{array}
$$

37. $0.023\overline{)0.549}$ becomes

$$
\begin{array}{r}
23.869 \approx 23.87 \\
23\overline{)549.000} \\
\underline{-46} \\
89 \\
\underline{-69} \\
20\ 0 \\
\underline{-18\ 4} \\
1\ 60 \\
\underline{-1\ 38} \\
220 \\
\underline{-207} \\
13
\end{array}
$$

39. $0.4\overline{)45.23}$ becomes

$$
\begin{array}{r}
113.07 \approx 113.1 \\
4\overline{)452.30} \\
\underline{-4} \\
05 \\
\underline{-4} \\
12 \\
\underline{-12} \\
0\ 3 \\
\underline{-0} \\
30 \\
\underline{-28} \\
2
\end{array}
$$

41. $\dfrac{54.982}{100} = 0.54982$

43. $\dfrac{26.87}{10} = 2.687$

45. $\dfrac{12.9}{1000} = 0.0129$

47.

$$
\begin{array}{r}
12.6 \\
7\overline{)88.2} \\
\underline{-7} \\
18 \\
\underline{-14} \\
4\ 2 \\
\underline{-4\ 2} \\
0
\end{array}
$$

49. $\dfrac{13.1}{10} = 1.31$

51. $6.8\overline{)83.13}$ becomes

$$
\begin{array}{r}
12.225 \\
68\overline{)831.300} \\
\underline{-68} \\
151 \\
\underline{-136} \\
15\ 3 \\
\underline{-13\ 6} \\
1\ 70 \\
\underline{-1\ 36} \\
340 \\
\underline{-340} \\
0
\end{array}
$$

53. $\dfrac{456.25}{10,000} = 0.045625$

55. Number of quarts = $\underline{546} \div \underline{52}$

$$
\begin{array}{r}
10.5 \approx 11 \\
52\overline{)546.0} \\
\underline{-52} \\
26 \\
\underline{-0} \\
26\ 0 \\
\underline{-26\ 0} \\
0
\end{array}
$$

Since he must buy whole quarts, 11 quarts are needed.

57.

$$
\begin{array}{r}
202.14 \approx 202.1 \\
39\overline{)\ 7883.50} \\
-78 \\
\hline
08 \\
-0 \\
\hline
83 \\
-78 \\
\hline
5\,5 \\
-3\,9 \\
\hline
1\,60 \\
-1\,56 \\
\hline
4
\end{array}
$$

She needs to buy 202.1 pounds of fertilizer.

59.

$$39.37\overline{)200}\ \text{becomes}\quad
\begin{array}{r}
5.08 \approx 5.1 \\
3937\overline{)\ 20000.00} \\
-19685 \\
\hline
315\,0 \\
-\ \ \ 0 \\
\hline
31500 \\
-31496 \\
\hline
4
\end{array}
$$

There are approximately 5.1 meters in 200 inches.

61. Divide the number of crayons by 64.

$$
\begin{array}{r}
11.40 \approx 11.4 \\
64\overline{)\ 730.00} \\
-64 \\
\hline
90 \\
-64 \\
\hline
26\,0 \\
-25\,6 \\
\hline
40
\end{array}
$$

740 crayons is approximately 11.4 boxes.

63. $6 \times 4 = 24$
There are 24 teaspoons in 4 fluid ounces.

65. From Exercise 63, we know that there are 24 teaspoons in 4 fluid ounces. Thus, there are 48 half teaspoons (0.5 tsp) or doses in 4 fluid ounces. To see how long the medicine will last, if a dose is taken every 4 hours, there are $24 \div 4 = 6$ doses taken per day. 48 (doses) $\div$ 6 (per day) = 8 days. The medicine will last 8 days.

67. There are 52 weeks in 1 year.

$$
\begin{array}{r}
248.07 \approx 248.1 \\
52\overline{)\ 12{,}900.00} \\
-10\,4 \\
\hline
2\,50 \\
-2\,08 \\
\hline
420 \\
-416 \\
\hline
4\,0 \\
-0 \\
\hline
400 \\
-364 \\
\hline
36
\end{array}
$$

Americans aged 18–22 drive, on average, 248.1 miles per week.

69. Divide the number of miles by the number of hours.

$$
\begin{array}{r}
134.6 \\
24\overline{)\ 3230.4} \\
-24 \\
\hline
83 \\
-72 \\
\hline
110 \\
-96 \\
\hline
14\,4 \\
-14\,4 \\
\hline
0
\end{array}
$$

Their average speed was 134.6 miles per hour.

71. Divide the total number of points by the number of games.

$$
\begin{array}{r}
18.484 \approx 18.48 \\
33\overline{)\ 610.000} \\
-33 \\
\hline
280 \\
-264 \\
\hline
160 \\
-132 \\
\hline
280 \\
-264 \\
\hline
160 \\
-132 \\
\hline
28
\end{array}
$$

She scored an average of 18.48 points per game.

73. $0.7(6 - 2.5) = 0.7(3.5) = 2.45$

75. $\dfrac{0.29+1.69}{3} = \dfrac{1.98}{3} = 0.66$

77. $30.03 + 5.1 \times 9.9 = 30.03 + 50.49 = 80.52$

79. $7.8 - 4.83 \div 2.1 + 9.2 = 7.8 - 2.3 + 9.2$
$$= 5.5 + 9.2$$
$$= 14.7$$

81. $93.07 \div 10 \times 100 = 9.307 \times 100 = 930.7$

83. $\dfrac{7.8+1.1\times100-3.6}{0.2} = \dfrac{7.8+110-3.6}{0.2}$
$$= \dfrac{117.8-3.6}{0.2}$$
$$= \dfrac{114.2}{0.2}$$
$$= 571$$

85. $5(20.6 - 2.06) - (0.8)^2 = 5(18.54) - (0.8)^2$
$$= 5(18.54) - 0.64$$
$$= 92.70 - 0.64$$
$$= 92.06$$

87. $6 \div 0.1 + 8.9 \times 10 - 4.6 = 60 + 89 - 4.6$
$$= 149 - 4.6$$
$$= 144.4$$

89. $0.9 = \dfrac{9}{10}$

91. $0.05 = \dfrac{5}{100} = \dfrac{1}{20}$

93. $0.3\overline{)1.278}$ becomes

$$
\begin{array}{r}
4.26 \\
3\overline{)12.78} \\
\underline{-12} \\
0\,7 \\
\underline{-6} \\
18 \\
\underline{-18} \\
0
\end{array}
$$

95.
$$
\begin{array}{r}
1.278 \\
+\ 0.300 \\
\hline
1.578
\end{array}
$$

97.
$$
\begin{array}{r}
8.6 \\
\times\ 3.1 \\
\hline
86 \\
25\,80 \\
\hline
26.66
\end{array}
$$

99.
$$
\begin{array}{r}
1000.00 \\
-\ 95.71 \\
\hline
904.29
\end{array}
$$

101. 8.62×41.7 is approximately $9 \times 40 = 360$, which is choice c.

103. $78.6 \div 97$ is approximately $78.6 \div 100 = 0.786$, which is choice b.

105. Add the numbers, then divide by 4.
$$
\begin{array}{r}
\overset{2}{} \\
86 \\
78 \\
91 \\
+\ 87 \\
\hline
342
\end{array}
$$

$$
\begin{array}{r}
85.5 \\
4\overline{)342.0} \\
\underline{-32} \\
22 \\
\underline{-20} \\
2\,0 \\
2\,0 \\
\underline{} \\
0
\end{array}
$$

The average is 85.5.

107. Area = (length)(width)

$4.5\overline{)38.7}$ becomes
$$
\begin{array}{r}
8.6 \\
45\overline{)387.0} \\
\underline{-360} \\
27\,0 \\
\underline{-27\,0} \\
0
\end{array}
$$

The length is 8.6 feet.

109. answers may vary

111. $1.15\overline{)75}$ becomes
$$
\begin{array}{r}
65.21 \approx 65.2 \\
115\overline{)7500.00} \\
\underline{-690} \\
600 \\
\underline{-575} \\
250 \\
\underline{-230} \\
200 \\
\underline{-115} \\
85
\end{array}
$$

$$\begin{array}{r} 82.60 \approx 82.6 \\ 1.15\overline{)95} \text{ becomes } 115\overline{)\ 9500.00} \\ \underline{-920} \\ 300 \\ \underline{-230} \\ 700 \\ \underline{-690} \\ 100 \\ \underline{-0} \\ 100 \end{array}$$

The range of wind speeds is 65.2–82.6 knots.

113. First find the length for one round of wire. Then multiply by 4.

$$\begin{array}{r} 24.280 \\ 15.675 \\ 24.280 \\ +\ 15.675 \\ \hline 79.910 \end{array} \qquad \begin{array}{r} 79.91 \\ \times\quad 4 \\ \hline 319.64 \end{array}$$

He will need 319.64 meters of wire.

Section 4.6 Practice

1. a.
$$\begin{array}{r} 0.4 \\ 5\overline{)\ 2.0} \\ \underline{-2\ 0} \\ 0 \end{array}$$

$\dfrac{2}{5}$ as a decimal is 0.4.

b.
$$\begin{array}{r} 0.225 \\ 40\overline{)\ 9.000} \\ \underline{-8\ 0} \\ 1\ 00 \\ \underline{-80} \\ 200 \\ \underline{-200} \\ 0 \end{array}$$

$\dfrac{9}{40}$ as a decimal is 0.225.

2. a.
$$\begin{array}{r} 0.8333... \\ 6\overline{)\ 5.0000} \\ \underline{-4\ 8} \\ 20 \\ \underline{-18} \\ 20 \\ \underline{-18} \\ 20 \\ \underline{-18} \\ 2 \end{array}$$

$\dfrac{5}{6}$ as a decimal is $0.8\overline{3}$.

b.
$$\begin{array}{r} 0.222... \\ 9\overline{)\ 2.000} \\ \underline{-1\ 8} \\ 20 \\ \underline{-18} \\ 20 \\ \underline{-18} \\ 2 \end{array}$$

$\dfrac{2}{9}$ as a decimal is $0.\overline{2}$.

3.
$$\begin{array}{r} 2.1538 \approx 2.154 \\ 13\overline{)\ 28.0000} \\ \underline{-26} \\ 20 \\ \underline{-13} \\ 70 \\ \underline{-65} \\ 50 \\ \underline{-39} \\ 110 \\ \underline{-104} \\ 6 \end{array}$$

$\dfrac{28}{13}$ as a decimal is approximately 2.154.

4.
$$\begin{array}{r} 0.3125 \\ 16\overline{)\ 5.0000} \\ \underline{-4\ 8} \\ 20 \\ \underline{-16} \\ 40 \\ \underline{-32} \\ 80 \\ \underline{-80} \\ 0 \end{array}$$

or

$3\dfrac{5}{16} = \dfrac{53}{16}$

$$\begin{array}{r} 3.3125 \\ 16\overline{)\,53.0000} \\ \underline{-48} \\ 5\,0 \\ \underline{-4\,8} \\ 20 \\ \underline{-16} \\ 40 \\ \underline{-32} \\ 80 \\ \underline{-80} \\ 0 \end{array}$$

$3\dfrac{5}{16}$ as a decimal is 3.3125.

5. $\dfrac{3}{5} = \dfrac{3 \cdot 2}{5 \cdot 2} = \dfrac{6}{10} = 0.6$

6. $\dfrac{3}{50} = \dfrac{3 \cdot 2}{50 \cdot 2} = \dfrac{6}{100} = 0.06$

7. $\dfrac{1}{5} = \dfrac{1 \cdot 2}{5 \cdot 2} = \dfrac{2}{10} = 0.2$

Since $0.2 < 0.25$, then $\dfrac{1}{5} < 0.25$.

8. a. $\dfrac{1}{2} = \dfrac{1 \cdot 5}{2 \cdot 5} = \dfrac{5}{10} = 0.5$

Since $0.5 < 0.54$, then $\dfrac{1}{2} < 0.54$.

b.
$$\begin{array}{r} 0.444... \\ 9\overline{)\,4.000} \\ \underline{-3\,6} \\ 40 \\ \underline{-36} \\ 40 \\ \underline{-36} \\ 4 \end{array}$$

$0.\overline{4} = \dfrac{4}{9}$

c.
$$\begin{array}{r} 0.7142 \approx 0.714 \\ 7\overline{)\,5.0000} \\ \underline{-4\,9} \\ 10 \\ \underline{-7} \\ 30 \\ \underline{-28} \\ 20 \\ \underline{-14} \\ 6 \end{array}$$

Since $0.714 < 0.72$, then $\dfrac{5}{7} < 0.72$.

9. a.

Original Numbers	$\dfrac{1}{3}$	0.302	$\dfrac{3}{8}$
Decimals	$0.\overline{3}$	0.302	0.375
Compare in order	2nd	1st	3rd

Written in order, the numbers are 0.302, $\dfrac{1}{3}, \dfrac{3}{8}$.

b.

Original Numbers	1.26	$1\dfrac{1}{4}$	$1\dfrac{2}{5}$
Decimals	1.26	1.25	1.4
Compare in order	2nd	1st	3rd

Written in order, the numbers are $1\dfrac{1}{4}$, 1.26, $1\dfrac{2}{5}$.

c.

Original Numbers	0.4	0.41	$\dfrac{5}{7}$
Decimals	0.4	0.41	≈ 0.71
Compare in order	1st	2nd	3rd

Written in order, the numbers are 0.4, 0.41, $\dfrac{5}{7}$.

10. Area $= \dfrac{1}{2} \cdot$ base $\cdot$ height

$= \dfrac{1}{2} \cdot 7 \cdot 2.1$

$= 0.5 \cdot 7 \cdot 2.1$

$= 7.35$

The area of the triangle is 7.35 square meters.

Vocabulary and Readiness Check

1. The number $0.\overline{5}$ means 0.555. <u>false</u>

2. To write $\dfrac{9}{19}$ as a decimal, perform the division $9\overline{)19}$. <u>false</u>

3. $(1.2)^2$ means $(1.2)(1.2)$ or 1.44. <u>true</u>

4. To simplify $8.6(9.6 - 4.8)$, we first subtract. <u>true</u>

Exercise Set 4.6

1. $\dfrac{1}{5} = \dfrac{1 \cdot 2}{5 \cdot 2} = \dfrac{2}{10} = 0.2$

3. $\dfrac{17}{25} = \dfrac{17 \cdot 4}{25 \cdot 4} = \dfrac{68}{100} = 0.68$

5. $\dfrac{3}{4} = \dfrac{3 \cdot 25}{4 \cdot 25} = \dfrac{75}{100} = 0.75$

7. $\dfrac{2}{25} = \dfrac{2 \cdot 4}{25 \cdot 4} = \dfrac{8}{100} = 0.08$

9. $\dfrac{6}{5} = \dfrac{6 \cdot 20}{5 \cdot 20} = \dfrac{120}{100} = 1.20 = 1.2$

11.
```
    0.91666...
12) 11.00000      11/12 = 0.916̄
   -10 8
      20
     -12
      80
     -72
      80
     -72
      80
     -72
       8
```
$\dfrac{11}{12} = 0.91\overline{6}$

13.
```
    0.425
40) 17.000
   -16 0
     1 00
     -80
      200
     -200
        0
```
$\dfrac{17}{40} = 0.425$

15. $\dfrac{9}{20} = \dfrac{9 \cdot 5}{20 \cdot 5} = \dfrac{45}{100} = 0.45$

17.
```
   0.333...
3) 1.000
   -9
    10
    -9
    10
    -9
     1
```
$\dfrac{1}{3} = 0.\overline{3}$

19.
```
    0.4375
16) 7.0000
   -6 4
     60
    -48
     120
    -112
       80
      -80
        0
```
$\dfrac{7}{16} = 0.4375$

21.
```
    0.636363...
11) 7.000000
   -6 6
     40
    -33
     70
    -66
     40
    -33
     70
    -66
     40
    -33
      7
```
$\dfrac{7}{11} = 0.\overline{63}$

23. $5\dfrac{17}{20} = \dfrac{117}{20}$

$$20)\overline{\begin{array}{l}5.85\\117.00\end{array}}$$
$$\underline{-100}$$
$$170$$
$$\underline{-160}$$
$$1\,00$$
$$\underline{-1\,00}$$
$$0$$

$5\dfrac{17}{20} = 5.85$

25.
$$125)\overline{\begin{array}{l}0.624\\78.000\end{array}}$$
$$\underline{-75\,0}$$
$$3\,00$$
$$\underline{-2\,50}$$
$$500$$
$$\underline{-500}$$
$$0$$

$\dfrac{78}{125} = 0.624$

27. $\dfrac{1}{3} = 0.333... \approx 0.33$

29. $\dfrac{7}{16} = 0.4375 \approx 0.44$

31. $\dfrac{7}{11} = 0.636363... \approx 0.6$

33.
$$91)\overline{\begin{array}{l}0.615 \approx 0.62\\56.000\end{array}}$$
$$\underline{-54\,6}$$
$$1\,40$$
$$\underline{-91}$$
$$490$$
$$\underline{-455}$$
$$35$$

$\dfrac{56}{91} \approx 0.62$

35.
$$97)\overline{\begin{array}{l}0.731 \approx 0.73\\71.000\end{array}}$$
$$\underline{-67\,9}$$
$$3\,10$$
$$\underline{-2\,91}$$
$$190$$
$$\underline{-97}$$
$$93$$

37.
$$50)\overline{\begin{array}{l}0.02\\1.00\end{array}}$$
$$\underline{-1\,00}$$
$$0$$

39. $\underset{\uparrow}{0.562} \qquad \underset{\uparrow}{0.569}$

$2 \quad < \quad 9$ so,

$0.562 < 0.569$

41.
$$200)\overline{\begin{array}{l}0.215\\43.000\end{array}}$$
$$\underline{-40\,0}$$
$$3\,00$$
$$\underline{-2\,00}$$
$$1\,000$$
$$\underline{-1\,000}$$
$$0$$

$0.215 = \dfrac{43}{200}$

43. $\dfrac{9}{100} = 0.09$ and $0.09 < 0.0932$ so $\dfrac{9}{100} < 0.0932$.

45.
$$6)\overline{\begin{array}{l}0.833...\\5.000\end{array}}$$
$$\underline{-4\,8}$$
$$20$$
$$\underline{-18}$$
$$20$$
$$\underline{-18}$$
$$2$$

$\dfrac{5}{6} = 0.8\overline{3}$ and $0.\overline{6} < 0.8\overline{3}$, so $0.\overline{6} < \dfrac{5}{6}$.

47.
$$91)\overline{\begin{array}{l}0.5604 \approx 0.560\\51.0000\end{array}}$$
$$\underline{-45\,5}$$
$$5\,50$$
$$\underline{-5\,46}$$
$$40$$
$$\underline{-0}$$
$$400$$
$$\underline{-364}$$
$$36$$

$\dfrac{51}{91} \approx 0.560$ and $0.560 < 0.56444...$, so

$\dfrac{51}{91} < 0.56\overline{4}$.

49.
$$\begin{array}{r} 0.111 \approx 0.11 \\ 9\overline{)1.000} \\ \underline{-9} \\ 10 \\ \underline{-9} \\ 10 \\ \underline{-9} \\ 1 \end{array}$$

$\dfrac{1}{9} \approx 0.11$ and $0.11 > 0.1$, so $\dfrac{1}{9} > 0.1$.

51.
$$\begin{array}{r} 1.3846 \approx 1.385 \\ 13\overline{)18.0000} \\ \underline{-13} \\ 5\,0 \\ \underline{-3\,9} \\ 1\,10 \\ \underline{-1\,04} \\ 60 \\ \underline{-52} \\ 80 \\ \underline{-78} \\ 2 \end{array}$$

$\dfrac{18}{13} \approx 1.385$ and $1.38 < 1.385$, so $1.38 < \dfrac{18}{13}$.

53.
$$\begin{array}{r} 7.125 \\ 64\overline{)456.000} \\ \underline{-448} \\ 8\,0 \\ \underline{-6\,4} \\ 1\,60 \\ \underline{-1\,28} \\ 320 \\ \underline{-320} \\ 0 \end{array}$$

$\dfrac{456}{64} = 7.125$ and $7.123 < 7.125$, so

$7.123 < \dfrac{456}{64}$.

55. 0.32, 0.34, 0.35

57. $0.49 = 0.490$
0.49, 0.491, 0.498

59. $\dfrac{3}{4} = 0.75$

0.73, $\dfrac{3}{4}$, 0.78

61. $\dfrac{4}{7} \approx 0.571$

0.412, 0.453, $\dfrac{4}{7}$

63. $\dfrac{42}{8} = 5.25$

5.23, $\dfrac{42}{8}$, 5.34

65. $\dfrac{12}{5} = 2.400$

$2.37 = 2.370$

$\dfrac{17}{8} = 2.125$

$\dfrac{17}{8}$, 2.37, $\dfrac{12}{5}$

67. Area $= \dfrac{1}{2} \times \text{base} \times \text{height}$

$= \dfrac{1}{2} \times 5.7 \times 9$

$= 0.5 \times 5.7 \times 9$

$= 25.65$

The area is 25.65 square inches.

69. Area $= \dfrac{1}{2} \times \text{base} \times \text{height}$

$= \dfrac{1}{2} \times 5.2 \times 3.6$

$= 0.5 \times 5.2 \times 3.6$

$= 9.36$

The area is 9.36 square centimeters.

71. Area $= \text{length} \times \text{width}$

$= 0.62 \times \dfrac{2}{5}$

$= 0.62 \times 0.4$

$= 0.248$

The area is 0.248 square yard.

73. $2^3 = 2 \cdot 2 \cdot 2 = 8$

75. $6^2 \cdot 2 = 6 \cdot 6 \cdot 2 = 72$

77. $\left(\dfrac{1}{3}\right)^4 = \dfrac{1}{3} \cdot \dfrac{1}{3} \cdot \dfrac{1}{3} \cdot \dfrac{1}{3} = \dfrac{1}{81}$

79. $\left(\dfrac{3}{5}\right)^2 = \dfrac{3}{5} \cdot \dfrac{3}{5} = \dfrac{9}{25}$

81. $\left(\dfrac{2}{5}\right)\left(\dfrac{5}{2}\right)^2 = \dfrac{2}{5} \cdot \dfrac{5}{2} \cdot \dfrac{5}{2} = \dfrac{5}{2}$

83. $1.0 = 1$

85. $1.00001 > 1$

87. $\dfrac{99}{100} < 1$

89. The fraction of radio stations with a country music format was $\dfrac{2112}{13,750}$.

$$
\begin{array}{r}
0.1536 \approx 0.154 \\
13750\overline{)\ 2112.0000} \\
-1375\ 0 \\
\hline
737\ 00 \\
-687\ 50 \\
\hline
49\ 500 \\
-41\ 250 \\
\hline
8\ 2500 \\
-8\ 2500 \\
\hline
0
\end{array}
$$

$\dfrac{2112}{13,750} \approx 0.154$

91.

		3
2112	rounds to	2100
1335	rounds to	1300
780	rounds to	800
838	rounds to	800
745	rounds to	700
+ 449	rounds to	+ 400
		6100

The total number of stations with the top six formats in 2009 was approximately 6100 stations.

93. answers may vary

95. answers may vary

97. $(9.6)(5) - \dfrac{3}{4} = (9.6)(5) - 0.75$
$\phantom{(9.6)(5) - \dfrac{3}{4}} = 48 - 0.75$
$\phantom{(9.6)(5) - \dfrac{3}{4}} = 47.25$

99. $\left(\dfrac{1}{10}\right)^2 + (1.6)(2.1) = \dfrac{1}{10} \cdot \dfrac{1}{10} + (1.6)(2.1)$
$\phantom{\left(\dfrac{1}{10}\right)^2 + (1.6)(2.1)} = \dfrac{1}{100} + (1.6)(2.1)$
$\phantom{\left(\dfrac{1}{10}\right)^2 + (1.6)(2.1)} = 0.01 + 3.36$
$\phantom{\left(\dfrac{1}{10}\right)^2 + (1.6)(2.1)} = 3.37$

101. $\dfrac{3}{8}(5.9 - 4.7) = 0.375(5.9 - 4.7)$
$\phantom{\dfrac{3}{8}(5.9 - 4.7)} = 0.375(1.2)$
$\phantom{\dfrac{3}{8}(5.9 - 4.7)} = 0.45$

Chapter 4 Vocabulary Check

1. Like fractional notation, <u>decimal</u> notation is used to denote a part of a whole.

2. To write fractions as decimals, divide the <u>numerator</u> by the <u>denominator</u>.

3. To add or subtract decimals, write the decimals so that the decimal points line up <u>vertically</u>.

4. When writing decimals in words, write "<u>and</u>" for the decimal point.

5. When multiplying decimals, the decimal point in the product is placed so that the number of decimal places in the product is equal to the <u>sum</u> of the number of decimal places in the factors.

6. The distance around a circle is called the <u>circumference</u>.

7. When 2 million is written as 2,000,000, we say it is written in <u>standard form</u>.

8. $\pi = \dfrac{\underline{\text{circumference}} \text{ of a circle}}{\underline{\text{diameter}} \text{ of the same circle}}$

9. In $3.4 - 2 = 1.4$, the number 1.4 is called the <u>difference</u>.

10. In $3.4 \div 2 = 1.7$, the number 1.7 is called the <u>quotient</u>.

11. In $3.4 \times 2 = 6.8$, the number 6.8 is called the <u>product</u>.

12. In $3.4 + 2 = 5.4$, the number 5.4 is called the <u>sum</u>.

Chapter 4 Review

1. The digit 4 in 23.45 is in the tenths place.

2. The digit 4 in 0.000345 is in the hundred-thousandths place.

3. 0.45 in words is forty-five hundredths.

4. 0.00345 in words is three hundred forty-five hundred-thousandths.

5. 109.23 in words is one hundred nine and twenty-three hundredths.

6. 46.007 in words is forty-six and seven thousandths.

7. Two and fifteen hundredths is 2.15.

8. Five hundred three and one hundred two thousandths is 503.102.

9. $0.16 = \dfrac{16}{100} = \dfrac{4}{25}$

10. $12.023 = 12\dfrac{23}{1000}$

11. $1.0045 = 1\dfrac{45}{10,000} = 1\dfrac{9}{2000}$

12. $25.25 = 25\dfrac{25}{100} = 25\dfrac{1}{4}$

13. $\dfrac{9}{10} = 0.9$

14. $\dfrac{25}{100} = 0.25$

15. $\dfrac{45}{1000} = 0.045$

16. $\dfrac{261}{10} = 26.1$

17.
$$0.49 \qquad 0.43$$
$$\uparrow \qquad\quad \uparrow$$
$$9 \;\; > \;\; 3$$
$$0.49 > 0.43$$

18. $0.973 = 0.9730$

19. 0.92, 8.09, 8.6

20. 0.09, 0.091, 0.1

21. To round 0.623 to the nearest tenth, observe that the digit in the hundredths place is 2. Since this digit is less than 5, we do not add 1 to the digit in the tenths place. The number 0.623 rounded to the nearest tenth is 0.6.

22. To round 0.9384 to the nearest hundredth, observe that the digit in the thousandths place is 8. Since this digit is at least 5, we add 1 to the digit in the hundredths place. The number 0.9384 rounded to the nearest hundredth is 0.94.

23. To round 0.259 to the nearest hundredth, observe that the digit in the thousandths place is 9. Since this digit is at least 5, we add 1 to the digit in the hundredths place. The number 0.259 rounded to the nearest hundredth is 0.26. The amount is $0.26.

24. To round 12.461 to the nearest hundredth, observe that the digit in the thousandths place is 1. Since this digit is less than 5, we do not add 1 to the digit in the hundredths place. The number 12.461 rounded to the nearest hundredth is 12.46. The amount is $12.46.

25. To round 31,304.35 to the nearest one, observe that the digit in the tenths place is 3. Since this is less than 5, we do not add 1 to the digit in the ones place. The number 31.304.35 rounded to the nearest one is 31,304. The amount spent was $31,304.

26. $10.75 = 10\dfrac{75}{100} = 10\dfrac{3}{4}$

27.
$$\begin{array}{r} 2.40 \\ \underline{7.12} \\ 9.52 \end{array}$$

28.
$$\begin{array}{r} 3.9 \\ -\,1.2 \\ \hline 2.7 \end{array}$$

29.
$$\begin{array}{r} {\scriptstyle 1} \\ 6.40 \\ +\,0.88 \\ \hline 7.28 \end{array}$$

30.
$$\begin{array}{r} {\scriptstyle 1\;1\;1} \\ 19.020 \\ 6.980 \\ +\,0.007 \\ \hline 26.007 \end{array}$$

31.
$$\begin{array}{r} 892.1 \\ -\ 432.4 \\ \hline 459.7 \end{array}$$

32.
$$\begin{array}{r} 100.342 \\ -\ \ \ 0.064 \\ \hline 100.278 \end{array}$$

33.
$$\begin{array}{r} 100.00 \\ -\ 34.98 \\ \hline 65.02 \end{array}$$

34.
$$\begin{array}{r} 200.00 \\ -\ 10.02 \\ \hline 189.98 \end{array}$$

35.
$$\begin{array}{r} \overset{2\ 2}{19.9} \\ 15.1 \\ 10.9 \\ +\ 6.7 \\ \hline 52.6 \end{array}$$
The distance is 52.6 miles.

36.
$$\begin{array}{r} 51.46 \\ -\ 49.02 \\ \hline 2.44 \end{array}$$
The price of oil increased $2.44.

37.
$$\begin{array}{r} \overset{2\ 2}{6.2} \\ 6.2 \\ 4.9 \\ +\ 4.9 \\ \hline 22.2 \end{array}$$
The perimeter is 22.2 inches.

38.
$$\begin{array}{r} \overset{1}{11.8} \\ 12.9 \\ +\ 14.2 \\ \hline 38.9 \end{array}$$
The perimeter is 38.9 feet.

39.
$$\begin{array}{r} 3.7 \\ \times\ \ 5 \\ \hline 18.5 \end{array}$$

40.
$$\begin{array}{r} 9.1 \\ \times\ \ 6 \\ \hline 54.6 \end{array}$$

41. $7.2 \times 10 = 72$

42. $9.345 \times 1000 = 9345$

43.
$$\begin{array}{r} 4.02 \\ \times\ 2.3 \\ \hline 1\ 206 \\ 8\ 04 \\ \hline 9.246 \end{array}$$

44.
$$\begin{array}{r} 39.02 \\ \times\ \ \ 87.3 \\ \hline 11\ 706 \\ 273\ 140 \\ 3121\ 600 \\ \hline 3406.446 \end{array}$$

45. Circumference $= 2 \cdot \pi \cdot$ radius
$$C = 2 \cdot \pi \cdot 7 = 14\pi$$
$$C \approx 14(3.14) = 43.96$$
The circumference is 14π meters, which is approximately 43.96 meters.

46.
$$\begin{array}{r} 102 \\ \times\ 0.625 \\ \hline 510 \\ 2040 \\ 61\ 200 \\ \hline 63.750 \approx 63.8 \end{array}$$
102 kilometers is approximately 63.8 miles.

47. 887 million $= 887 \times 1$ million
$$= 887 \times 1,000,000$$
$$= 887,000,000$$
Saturn is about 887,000,000 miles from the sun.

48. 600 thousand $= 600 \times 1$ thousand
$$= 600 \times 1000$$
$$= 600,000$$
A comet's tail can be over 600,000 miles long.

49.
$$\begin{array}{r} 0.0877 \\ 3\overline{)0.2631} \\ \underline{-24} \\ 23 \\ \underline{-21} \\ 21 \\ \underline{-21} \\ 0 \end{array}$$

50.

$$
\begin{array}{r}
15.825 \\
20\overline{)316.500} \\
\underline{-20} \\
116 \\
\underline{-100} \\
16\,5 \\
\underline{-16\,0} \\
50 \\
\underline{-40} \\
100 \\
\underline{-100} \\
0
\end{array}
$$

51. $0.3\overline{)21}$ becomes

$$
\begin{array}{r}
70 \\
3\overline{)210} \\
\underline{-21} \\
00
\end{array}
$$

52. $0.03\overline{)0.0063}$ becomes

$$
\begin{array}{r}
0.21 \\
3\overline{)0.63} \\
\underline{-6} \\
03 \\
\underline{-3} \\
0
\end{array}
$$

53. $0.34\overline{)2.74}$ becomes

$$
\begin{array}{r}
8.0588 \approx 8.059 \\
34\overline{)274.0000} \\
\underline{-272} \\
2\,0 \\
\underline{-0} \\
2\,00 \\
\underline{-1\,70} \\
300 \\
\underline{-272} \\
280 \\
\underline{-272} \\
8
\end{array}
$$

54. $19.8\overline{)601.92}$ becomes

$$
\begin{array}{r}
30.4 \\
198\overline{)6019.2} \\
\underline{-594} \\
79 \\
\underline{-0} \\
79\,2 \\
\underline{-79\,2} \\
0
\end{array}
$$

55. $\dfrac{2.67}{100} = 0.0267$

56. $\dfrac{93}{10} = 9.3$

57. $3.28\overline{)24}$ becomes

$$
\begin{array}{r}
7.31 \approx 7.3 \\
328\overline{)2400.00} \\
\underline{-2296} \\
104\,0 \\
\underline{-98\,4} \\
5\,60 \\
\underline{-3\,28} \\
2\,32
\end{array}
$$

There are approximately 7.3 meters in 24 feet.

58. $69.71\overline{)3136.95}$ becomes

$$
\begin{array}{r}
45 \\
6971\overline{)313695} \\
\underline{-27884} \\
34855 \\
\underline{-34855} \\
0
\end{array}
$$

The loan will be paid off in 45 months.

59. $7.6 \times 1.9 + 2.5 = 14.44 + 2.5 = 16.94$

60. $(2.3)^2 - 1.4 = 5.29 - 1.4 = 3.89$

61. $\dfrac{7 + 0.74}{0.06} = \dfrac{7.74}{0.06} = \dfrac{774}{6} = 129$

62. $\dfrac{(1.5)^2 + 0.5}{0.05} = \dfrac{2.25 + 0.5}{0.05} = \dfrac{2.75}{0.05} = \dfrac{275}{5} = 55$

63. $0.9(6.5 - 5.6) = 0.9(0.9) = 0.81$

64. $0.0726 \div 10 \times 1000 = 0.00726 \times 1000 = 7.26$

65. $\dfrac{4}{5} = \dfrac{4 \cdot 2}{5 \cdot 2} = \dfrac{8}{10} = 0.8$

66.

$$
\begin{array}{r}
0.9230 \approx 0.923 \\
13\overline{)12.0000} \\
\underline{-11\,7} \\
30 \\
\underline{-26} \\
40 \\
\underline{-39} \\
10 \\
\underline{-0} \\
10
\end{array}
$$

$\dfrac{12}{13} \approx 0.923$

67. $2\dfrac{1}{3} = \dfrac{7}{3}$

$$\begin{array}{r} 2.333... \\ 3\overline{)\,7.000} \\ \underline{-6} \\ 1\,0 \\ \underline{-9} \\ 10 \\ \underline{-9} \\ 10 \\ \underline{-9} \\ 1 \end{array}$$

$2\dfrac{1}{3} = 2.\overline{3}$ or 2.333

68.
$$\begin{array}{r} 0.21666... \\ 60\overline{)\,13.00000} \\ \underline{-12\,0} \\ 1\,00 \\ \underline{-60} \\ 400 \\ \underline{-360} \\ 400 \\ \underline{-360} \\ 400 \\ \underline{-360} \\ 40 \end{array}$$

$\dfrac{13}{60} = 0.21\overline{6}$ or 0.217

69. $0.392 = 0.3920$

70.
$$\begin{array}{r} 0.444... \\ 9\overline{)\,4.000} \\ \underline{-3\,6} \\ 40 \\ \underline{-36} \\ 40 \\ \underline{-36} \\ 4 \end{array}$$

$0.\overline{4} = \dfrac{4}{9}$

71.
$$\begin{array}{r} 0.2941 \approx 0.294 \\ 17\overline{)\,5.0000} \\ \underline{-3\,4} \\ 1\,60 \\ \underline{-1\,53} \\ 70 \\ \underline{-68} \\ 20 \\ \underline{-17} \\ 3 \end{array}$$

$\dfrac{5}{17} \approx 0.294$ and $0.293 < 0.294$ so $0.293 < \dfrac{5}{17}$.

72.
$$\begin{array}{r} 0.5714 \approx 0.571 \\ 7\overline{)\,4.0000} \\ \underline{-3\,5} \\ 50 \\ \underline{-49} \\ 10 \\ \underline{-7} \\ 30 \\ \underline{-28} \\ 2 \end{array}$$

$\dfrac{4}{7} \approx 0.571$ and $0.571 < 0.625$ so $\dfrac{4}{7} < 0.625$.

73. $\dfrac{17}{20} = \dfrac{17 \cdot 5}{20 \cdot 5} = \dfrac{85}{100} = 0.85 = 0.850$

$0.837,\ 0.839,\ \dfrac{17}{20}$

74.
$$\begin{array}{r} 1.636 \approx 1.64 \\ 11\overline{)\,18.000} \\ \underline{-11} \\ 7\,0 \\ \underline{-6\,6} \\ 40 \\ \underline{-33} \\ 70 \\ \underline{-66} \\ 4 \end{array} \qquad \begin{array}{r} 1.583 \approx 1.58 \\ 12\overline{)\,19.000} \\ \underline{-12} \\ 7\,0 \\ \underline{-6\,0} \\ 1\,00 \\ \underline{-96} \\ 40 \\ \underline{-36} \\ 4 \end{array}$$

$\dfrac{19}{12},\ 1.63,\ \dfrac{18}{11}$

75. $\text{Area} = \frac{1}{2} \times \text{base} \times \text{height}$

$= \frac{1}{2} \times 4.6 \times 3$

$= 0.5 \times 4.6 \times 3$

$= 6.9$

The area is 6.9 square feet.

76. $\text{Area} = \frac{1}{2} \times \text{base} \times \text{height}$

$= \frac{1}{2} \times 5.2 \times 2.1$

$= 0.5 \times 5.2 \times 2.1$

$= 5.46$

The area is 5.46 square inches.

77. 200.0032 in words is two hundred and thirty-two ten-thousandths.

78. Sixteen thousand twenty-five and fourteen thousandths is 16,025.014.

79. $0.00231 = \dfrac{231}{100,000}$

80. $\dfrac{6}{7} \approx 0.86$

$\dfrac{8}{9} \approx 0.89$

$0.75, \dfrac{6}{7}, \dfrac{8}{9}$

81. $\dfrac{7}{100} = 0.07$

82.
$$\begin{array}{r} 0.1125 \\ 80\overline{)\,9.0000} \\ -80 \\ \hline 1\,00 \\ -80 \\ \hline 200 \\ -160 \\ \hline 400 \\ -400 \\ \hline 0 \end{array}$$

$\dfrac{9}{80} = 0.1125$

83.
$$\begin{array}{r} 51.0571 \approx 51.057 \\ 175\overline{)\,8935.0000} \\ -875 \\ \hline 185 \\ -175 \\ \hline 10\,0 \\ -0 \\ \hline 10\,00 \\ -8\,75 \\ \hline 1\,250 \\ -1\,225 \\ \hline 250 \\ -175 \\ \hline 75 \end{array}$$

$\dfrac{8935}{175} \approx 51.057$

84. 402.00032   402.000032

$\uparrow \uparrow$

$3 > 0 \text{ so}$

$402.00032 > 402.000032$

85. 0.230505   0.23505

$\uparrow \uparrow$

$0 < 5 \text{ so}$

$0.230505 < 0.23505$

86.
$$\begin{array}{r} 0.5454 \approx 0.545 \\ 11\overline{)\,6.0000} \\ -5\,5 \\ \hline 50 \\ -44 \\ \hline 60 \\ -55 \\ \hline 50 \\ -44 \\ \hline 6 \end{array}$$

$\dfrac{6}{11} \approx 0.545$ and $0.545 < 0.55$ so $\dfrac{6}{11} < 0.55$.

87. To round 42.895 to the nearest hundredth, observe that the digit in the thousandths place is 5. Since this digit is at least 5, we add 1 to the digit in the hundredths place. Since $9 + 1 = 10$, we replace the 9 with a 0 and then add 1 to the digit in the tenths place. The number 42.895 rounded to the nearest hundredth is 42.90.

88. To round 16.34925 to the nearest thousandth, observe that the digit in the ten-thousandths place is 2. Since this digit is less than 5, we do not add 1 to the digit in the thousandths place. The number 16.34925 rounded to the nearest thousandth is 16.349.

89. To round 123.46 to the nearest one, observe that the digit in the tenths place is 4. Since this digit is less than 5, we do not add 1 to the digit in the ones place. The number 123.46 rounded to the nearest one is 123. The amount is $123.

90. To round 3645.52 to the nearest one, observe that the digit in the tenths place is 5. Since this digit is at least 5, we add 1 to the digit in the ones place. The number 3645.52 rounded to the nearest one is 3646. The amount is $3646.

91.
$$\begin{array}{r} 4.9 \\ -3.2 \\ \hline 1.7 \end{array}$$

92.
$$\begin{array}{r} 5.23 \\ -\ 2.74 \\ \hline 2.49 \end{array}$$

93.
$$\begin{array}{r} ^{11}^{1} \\ 200.490 \\ 16.820 \\ \underline{103.002} \\ 320.312 \end{array}$$

94.
$$\begin{array}{r} ^{1} \\ 0.00236 \\ 100.45000 \\ +\ 48.29000 \\ \hline 148.74236 \end{array}$$

95.
$$\begin{array}{r} 2.54 \\ \times\ 3.2 \\ \hline 508 \\ 7\ 620 \\ \hline 8.128 \end{array}$$

96.
$$\begin{array}{r} 3.45 \\ \times\ 2.1 \\ \hline 345 \\ 6\ 900 \\ \hline 7.245 \end{array}$$

97. $0.005\overline{)24.5}$ becomes
$$\begin{array}{r} 4900 \\ 5\overline{)24500} \\ \underline{-20} \\ 45 \\ \underline{-45} \\ 000 \end{array}$$

98. $2.3\overline{)54.98}$ becomes
$$\begin{array}{r} 23.9043 \approx 23.904 \\ 23\overline{)549.8000} \\ \underline{-\ 46} \\ 89 \\ \underline{-69} \\ 20\ 8 \\ \underline{-20\ 7} \\ 10 \\ \underline{-0} \\ 100 \\ \underline{-92} \\ 80 \\ \underline{-69} \\ 11 \end{array}$$

99. 77.3 rounds to 80, 115.9 rounds to 120.
Area = length × width
$A \approx 80 \times 120 = 9600$
The area is approximately 9600 square feet.

100. 1.89 rounds to 2.00
1.07 rounds to 1.00
0.99 rounds to 1.00
Estimate: $2.00 + 1.00 + 1.00 = 4.00$
$4.00 < 6.00$ so the groceries can be purchased with a $5 bill.

101. $\dfrac{(3.2)^2}{100} = \dfrac{10.24}{100} = 0.1024$

102. $(2.6 + 1.4)(4.5 - 3.6) = (4.0)(0.9) = 3.6$

Chapter 4 Test

1. 45.092 in words is forty-five and ninety-two thousandths.

2. Three thousand and fifty nine thousandths is 3000.059.

3. To round 34.8923 to the nearest tenth, observe that the digit in the hundredths place is 9. Since this digit is at least 5, we add 1 to the digit in the tenths place. The number 34.8923 rounded to the nearest tenth is 34.9.

4. To round 0.8623 to the nearest thousandth, observe that the digit in the ten-thousandths place is 3. Since this digit is less than 5, we do not add 1 to the digit in the thousandths place. The number 0.8623 rounded to the nearest thousandth is 0.862.

5. 25.0909 < 25.9090

　　↑　　　↑

　　0　<　9 so

25.0909 < 25.9090

6. $\dfrac{4}{9} \approx 0.444$

$\dfrac{4}{9}$, 0.445, 0.454

7. $0.345 = \dfrac{345}{1000} = \dfrac{69 \cdot 5}{200 \cdot 5} = \dfrac{69}{200}$

8. $24.73 = 24\dfrac{73}{100}$

9. $\dfrac{13}{20} = \dfrac{13 \cdot 5}{20 \cdot 5} = \dfrac{65}{100} = 0.65$

10. $5\dfrac{8}{9} = \dfrac{53}{9}$

$$
\begin{array}{r}
5.888\ldots \\
9\overline{)\,53.000} \\
\underline{-45} \\
8\,0 \\
\underline{-7\,2} \\
80 \\
\underline{-72} \\
80 \\
\underline{-72} \\
8
\end{array}
$$

$5\dfrac{8}{9} = 5.\overline{8}$ or 5.889

11.

$$
\begin{array}{r}
0.9411 \approx 0.941 \\
17\overline{)\,16.0000} \\
\underline{-15\,3} \\
70 \\
\underline{-68} \\
20 \\
\underline{-17} \\
30 \\
\underline{-17} \\
13
\end{array}
$$

$\dfrac{16}{17} \approx 0.941$

12.

$$
\begin{array}{r}
\overset{1\ 1}{} \\
2.893 \\
4.200 \\
\underline{+\ 10.490} \\
17.583
\end{array}
$$

13.

$$
\begin{array}{r}
20.0 \\
\underline{-\ 8.6} \\
11.4
\end{array}
$$

14.

$$
\begin{array}{r}
10.2 \\
\times\ 4.3 \\
\hline
3\,06 \\
40\,80 \\
\hline
43.86
\end{array}
$$

15. $0.23\overline{)12.88}$ becomes

$$
\begin{array}{r}
56 \\
23\overline{)\,1288} \\
\underline{-115} \\
138 \\
\underline{-138} \\
0
\end{array}
$$

16.

$$
\begin{array}{r}
0.165 \\
\times\ 0.47 \\
\hline
1155 \\
6600 \\
\hline
0.07755
\end{array}
$$

17.

$$
\begin{array}{r}
6.6728 \approx 6.673 \\
7\overline{)\,46.7100} \\
\underline{-42} \\
4\,7 \\
\underline{-4\,2} \\
51 \\
\underline{-49} \\
20 \\
\underline{-14} \\
60 \\
\underline{-56} \\
4
\end{array}
$$

18. $126.9 \times 100 = 12{,}690$

19. $\dfrac{47.3}{10} = 4.73$

20. $0.3[1.57 - (0.6)^2] = 0.3[1.57 - 0.36]$

$= 0.3(1.21)$

$= 0.363$

21. $\dfrac{0.23+1.63}{0.3} = \dfrac{1.86}{0.3}$

$$0.3\overline{)1.86} \text{ becomes } 3\overline{\smash{)}\begin{array}{r} 6.2 \\ 18.6 \\ -18 \\ \hline 0\,6 \\ -6 \\ \hline 0 \end{array}}$$

$\dfrac{0.23+1.63}{0.3} = 6.2$

22. $4583 \text{ million} = 4583 \times 1 \text{ million}$
$= 4583 \times 1{,}000{,}000$
$= 4{,}583{,}000{,}000$

23. Area $= \dfrac{1}{2} \times \text{base} \times \text{height}$
$= \dfrac{1}{2} \times 4.2 \times 1.1$
$= 0.5 \times 4.2 \times 1.1$
$= 2.31$
The area is 2.31 square miles.

24. Circumference $= 2 \cdot \pi \cdot \text{radius}$
$C = 2 \cdot \pi \cdot 9 = 18\pi$
$C \approx 18(3.14) = 56.52$

The circumference is 18π miles, which is approximately 56.52 miles.

25. a. $\begin{array}{r} 123.8 \\ \times\ \ 80 \\ \hline 9904.0 \end{array}$
The area of the lawn is 9904 square feet.

b. Multiply the number of square feet by the number of ounces per square foot.
$\begin{array}{r} 9904 \\ \times\ 0.02 \\ \hline 198.08 \end{array}$
198.08 ounces of insecticide are needed.

26. $\begin{array}{r} \overset{1\ 1}{} 14.2 \\ 16.1 \\ +\ 23.7 \\ \hline 54.0 \end{array}$
The distance is 54 miles.

Cumulative Review Chapters 1-4

1. 106,052,447 in words is one hundred six million, fifty-two thousand, four hundred forty-seven.

2. Two hundred seventy-six thousand, four is 276,004.

3. $\begin{array}{r} 94{,}113 \\ +\ 4\,525 \\ \hline 98{,}638 \end{array}$
The number of seats in the stadium for the 2009 season is 98,638.

4. $\begin{array}{r} 12 \\ \times\ 24 \\ \hline 48 \\ 240 \\ \hline 288 \end{array}$
A case of soda has 288 ounces of soda.

5. $\begin{array}{r} 900 \\ -\ 174 \\ \hline 726 \end{array}$ Check: $\begin{array}{r} \overset{1\ 1}{726} \\ +\ 174 \\ \hline 900 \end{array}$

6. $5^2 \cdot 2^3 = 5 \cdot 5 \cdot 2 \cdot 2 \cdot 2 = 200$

7. $\begin{array}{rll} 294 & \text{rounds to} & 300 \\ 625 & \text{rounds to} & 600 \\ 1071 & \text{rounds to} & 1100 \\ +\ 349 & \text{rounds to} & +\ 300 \\ \hline & & 2300 \end{array}$

8. $7 \cdot \sqrt{144} = 7 \cdot 12 = 84$

9. $\begin{array}{r} 4800 \\ \times\ \ 12 \\ \hline 9\,600 \\ 48\,000 \\ \hline 57{,}600 \end{array}$
12 DVDs hold 57,600 megabytes.

10. Perimeter: $7 + 7 + 7 + 7 = 28$
The perimeter is 28 feet.
Area: $7 \cdot 7 = 49$
The area is 49 square feet.

11. $17\overline{\smash{)}\begin{array}{r} 401 \text{ R } 2 \\ 6819 \\ -68 \\ \hline 01 \\ -0 \\ \hline 19 \\ -17 \\ \hline 2 \end{array}}$

12. $2\dfrac{5}{8} = \dfrac{8 \cdot 2 + 5}{8} = \dfrac{16 + 5}{8} = \dfrac{21}{8}$

13. $4^3 + [3^2 - (10 \div 2)] - 7 \cdot 3 = 4^3 + [3^2 - 5] - 7 \cdot 3$
$$= 64 + [9 - 5] - 7 \cdot 3$$
$$= 64 + 4 - 7 \cdot 3$$
$$= 64 + 4 - 21$$
$$= 68 - 21$$
$$= 47$$

14.
$$\begin{array}{r} 12 \text{ R } 4 \\ 5\overline{)\ 64} \\ \underline{-5} \\ 14 \\ \underline{-10} \\ 4 \end{array}$$

$\dfrac{64}{5} = 12\dfrac{4}{5}$

15. The numerator is 3. The denominator is 7.

16. $24 \div 8 \cdot 3 = 3 \cdot 3 = 9$

17. $\dfrac{6}{60} = \dfrac{6 \cdot 1}{6 \cdot 10} = \dfrac{1}{10}$

18. $(8 - 5)^2 + (10 - 8)^3 = 3^2 + 2^3$
$$= 3 \cdot 3 + 2 \cdot 2 \cdot 2$$
$$= 9 + 8$$
$$= 17$$

19. $\dfrac{3}{4} \cdot 20 = \dfrac{3}{4} \cdot \dfrac{20}{1} = \dfrac{3 \cdot 20}{4 \cdot 1} = \dfrac{3 \cdot 4 \cdot 5}{4 \cdot 1} = \dfrac{3 \cdot 5}{1} = \dfrac{15}{1}$ or 15

20. $1 + 2[30 \div (7 - 2)] = 1 + 2[30 \div 5]$
$$= 1 + 2(6)$$
$$= 1 + 12$$
$$= 13$$

21. $\dfrac{7}{8} \div \dfrac{2}{9} = \dfrac{7}{8} \cdot \dfrac{9}{2} = \dfrac{7 \cdot 9}{8 \cdot 2} = \dfrac{63}{16}$

22.
$$\begin{array}{r} {\scriptstyle 1} \\ 117 \\ 125 \\ +\ 142 \\ \hline 384 \end{array}$$

23.
$$\begin{array}{r} 128 \\ 3\overline{)\ 384} \\ \underline{-3} \\ 08 \\ \underline{-6} \\ 24 \\ \underline{-24} \\ 0 \end{array}$$

The average is 128.

23. $1\dfrac{2}{3} \cdot 2\dfrac{1}{4} = \dfrac{5}{3} \cdot \dfrac{9}{4} = \dfrac{5 \cdot 9}{3 \cdot 4} = \dfrac{5 \cdot 3 \cdot 3}{3 \cdot 4} = \dfrac{5 \cdot 3}{4} = \dfrac{15}{4}$ or $3\dfrac{3}{4}$

24.
$$\begin{array}{r} 9 \\ 36\overline{)\ 324} \\ \underline{-324} \\ 0 \end{array}$$

Each ticket cost \$9.

25. $\dfrac{3}{4} \div 5 = \dfrac{3}{4} \div \dfrac{5}{1} = \dfrac{3}{4} \cdot \dfrac{1}{5} = \dfrac{3 \cdot 1}{4 \cdot 5} = \dfrac{3}{20}$

26. $\left(\dfrac{3}{4} \div \dfrac{1}{2}\right) \cdot \dfrac{9}{10} = \left(\dfrac{3}{4} \cdot \dfrac{2}{1}\right) \cdot \dfrac{9}{10}$
$$= \dfrac{6}{4} \cdot \dfrac{9}{10}$$
$$= \dfrac{2 \cdot 3 \cdot 9}{4 \cdot 5 \cdot 2}$$
$$= \dfrac{3 \cdot 9}{4 \cdot 5}$$
$$= \dfrac{27}{20} \text{ or } 1\dfrac{7}{20}$$

27. $\dfrac{8}{9} - \dfrac{1}{9} = \dfrac{8 - 1}{9} = \dfrac{7}{9}$

28. $\dfrac{4}{15} + \dfrac{2}{15} = \dfrac{4 + 2}{15} = \dfrac{6}{15} = \dfrac{2 \cdot 3}{5 \cdot 3} = \dfrac{2}{5}$

29. $\dfrac{7}{8} - \dfrac{5}{8} = \dfrac{7 - 5}{8} = \dfrac{2}{8} = \dfrac{2 \cdot 1}{2 \cdot 4} = \dfrac{1}{4}$

30. $\dfrac{1}{20} + \dfrac{3}{20} + \dfrac{4}{20} = \dfrac{1 + 3 + 4}{20} = \dfrac{8}{20} = \dfrac{4 \cdot 2}{4 \cdot 5} = \dfrac{2}{5}$

31. $\dfrac{3}{4} = \dfrac{3}{4} \cdot \dfrac{5}{5} = \dfrac{3 \cdot 5}{4 \cdot 5} = \dfrac{15}{20}$

32. $\dfrac{7}{9} = \dfrac{7}{9} \cdot \dfrac{5}{5} = \dfrac{7 \cdot 5}{9 \cdot 5} = \dfrac{35}{45}$

33. The LCD of 15 and 10 is 30.

$$\frac{2}{15}+\frac{3}{10}=\frac{2}{15}\cdot\frac{2}{2}+\frac{3}{10}\cdot\frac{3}{3}=\frac{4}{30}+\frac{9}{30}=\frac{13}{30}$$

34. The LCD of 30 and 9 is 90.

$$\frac{7}{30}-\frac{2}{9}=\frac{7}{30}\cdot\frac{3}{3}-\frac{2}{9}\cdot\frac{10}{10}=\frac{21}{90}-\frac{20}{90}=\frac{1}{90}$$

35. The LCD of 8 and 5 is 40.

$$
\begin{array}{ll}
2\frac{3}{8} & 2\frac{15}{40} \\
+1\frac{4}{5} & +1\frac{32}{40} \\
\hline
& 3\frac{47}{40}=3+1\frac{7}{40}=4\frac{7}{40}
\end{array}
$$

The combined weight is $4\frac{7}{40}$ pounds.

36.
$$
\begin{aligned}
12\cdot2\frac{5}{16} &=\frac{12}{1}\cdot\frac{37}{16} \\
&=\frac{12\cdot37}{1\cdot16} \\
&=\frac{3\cdot4\cdot37}{4\cdot4} \\
&=\frac{3\cdot37}{4} \\
&=\frac{111}{4}\text{ or }27\frac{3}{4}
\end{aligned}
$$

The cartridges weigh $27\frac{3}{4}$ pounds.

37. $\left(\frac{1}{4}\right)^2=\frac{1}{4}\cdot\frac{1}{4}=\frac{1}{16}$

38. $\left(\frac{7}{11}\right)=\frac{7}{11}\cdot\frac{7}{11}=\frac{49}{121}$

39. $\left(\frac{1}{6}\right)^2\cdot\left(\frac{3}{4}\right)^3=\frac{1}{6}\cdot\frac{1}{6}\cdot\frac{3}{4}\cdot\frac{3}{4}\cdot\frac{3}{4}=\frac{3}{256}$

40. $\left(\frac{1}{2}\right)^3\cdot\left(\frac{4}{9}\right)^2=\frac{1}{2}\cdot\frac{1}{2}\cdot\frac{1}{2}\cdot\frac{4}{9}\cdot\frac{4}{9}=\frac{2}{81}$

41. $0.43=\frac{43}{100}$

42. $\frac{3}{4}=\frac{3\cdot25}{4\cdot25}=\frac{75}{100}=0.75$

43.
$$
\begin{array}{cc}
0.378 & 0.368 \\
\uparrow & \uparrow \\
7 & > \quad 6\text{ so}
\end{array}
$$
$$0.378 > 0.368$$

44. Five and six hundredths is 5.06.

45.
$$
\begin{array}{ll}
\begin{array}{r}
35.218 \\
-23.650 \\
\hline
11.568
\end{array}
&
\text{Check:}
\begin{array}{r}
^{1\ 1}\ \ \\
11.568 \\
+23.650 \\
\hline
35.218
\end{array}
\end{array}
$$

46.
$$
\begin{array}{r}
75.100 \\
+\ 0.229 \\
\hline
75.329
\end{array}
$$

47. $23.702\times100=2370.2$

48.
$$
\begin{array}{r}
1.7 \\
\times\ 0.07 \\
\hline
0.119
\end{array}
$$

49. $76{,}805\times0.01=768.05$

50. $0.013\overline{)0.1157}$ becomes
$$
\begin{array}{r}
8.9 \\
13\overline{)115.7} \\
-104\ \ \\
\hline
11\ 7 \\
-11\ 7 \\
\hline
0
\end{array}
$$

Chapter 5

Section 5.1 Practice

1. The ratio of 20 to 23 is $\dfrac{20}{23}$.

2. The ratio of 10.3 to 15.1 is $\dfrac{10.3}{15.1}$.

3. The ratio of $3\dfrac{1}{3}$ to $12\dfrac{1}{5}$ is $\dfrac{3\frac{1}{3}}{12\frac{1}{5}}$.

4. The ratio of $8 to $6 is $\dfrac{\$8}{\$6} = \dfrac{8}{6} = \dfrac{4 \cdot 2}{3 \cdot 2} = \dfrac{4}{3}$.

5. The ratio of 3.9 to 8.8 is
$$\dfrac{3.9}{8.8} = \dfrac{3.9}{8.8} \cdot \dfrac{10}{10} = \dfrac{3.9 \cdot 10}{8.8 \cdot 10} = \dfrac{39}{88}.$$

6. The ratio of $2\dfrac{2}{3}$ to $1\dfrac{13}{15}$ is
$$\dfrac{2\frac{2}{3}}{1\frac{13}{15}} = 2\dfrac{2}{3} \div 1\dfrac{13}{15}$$
$$= \dfrac{8}{3} \div \dfrac{28}{15}$$
$$= \dfrac{8}{3} \cdot \dfrac{15}{28}$$
$$= \dfrac{8 \cdot 15}{3 \cdot 28}$$
$$= \dfrac{4 \cdot 2 \cdot 3 \cdot 5}{3 \cdot 4 \cdot 7}$$
$$= \dfrac{2 \cdot 5}{7}$$
$$= \dfrac{10}{7}.$$

7. The ratio of work miles to total miles is
$$\dfrac{4800 \text{ miles}}{15,000 \text{ miles}} = \dfrac{8 \cdot 600}{25 \cdot 600} = \dfrac{8}{25}.$$

8. a. The ratio of the length of the shortest side to the length of the longest side is
$$\dfrac{6 \text{ meters}}{10 \text{ meters}} = \dfrac{6}{10} = \dfrac{3 \cdot 2}{5 \cdot 2} = \dfrac{3}{5}.$$

b. The perimeter is $6 + 8 + 10 = 24$ meters. The ratio of the length of the longest side to the perimeter of the triangle is
$$\dfrac{10 \text{ meters}}{24 \text{ meters}} = \dfrac{10}{24} = \dfrac{5 \cdot 2}{12 \cdot 2} = \dfrac{5}{12}.$$

Vocabulary and Readiness Check

1. The quotient of two quantities is called a ratio. <u>true</u>

2. The ratio $\dfrac{7}{5}$ means the same as the ratio $\dfrac{5}{7}$. <u>false</u>

3. The ratio $\dfrac{7.2}{8.1}$ is in simplest form. <u>false</u>

4. The ratio $\dfrac{10 \text{ feet}}{30 \text{ feet}}$ is in simplest form. <u>false</u>

5. The ratio $\dfrac{9}{10}$ is in simplest form. <u>true</u>

6. The ratio 2 to 5 equals $\dfrac{5}{2}$ in fractional notation. <u>false</u>

7. The ratio 30:41 equals $\dfrac{30}{41}$ in fractional notation. <u>true</u>

8. The ratio 15 to 45 equals $\dfrac{3}{1}$ in fractional notation. <u>false</u>

Exercise Set 5.1

1. The ratio of 11 to 14 is $\dfrac{11}{14}$.

3. The ratio of 23 to 10 is $\dfrac{23}{10}$.

5. The ratio of 151 to 201 is $\dfrac{151}{201}$.

7. The ratio of 2.8 to 7.6 is $\dfrac{2.8}{7.6}$.

9. The ratio of 5 to $7\frac{1}{2}$ is $\dfrac{5}{7\frac{1}{2}}$.

11. The ratio of $3\frac{3}{4}$ to $1\frac{2}{3}$ is $\dfrac{3\frac{3}{4}}{1\frac{2}{3}}$.

13. The ratio of 16 to 24 is $\dfrac{16}{24} = \dfrac{8 \cdot 2}{8 \cdot 3} = \dfrac{2}{3}$.

15. The ratio of 7.7 to 10 is $\dfrac{7.7}{10} = \dfrac{7.7 \cdot 10}{10 \cdot 10} = \dfrac{77}{100}$.

17. The ratio of 4.63 to 8.21 is
$\dfrac{4.63}{8.21} = \dfrac{4.63 \cdot 100}{8.21 \cdot 100} = \dfrac{463}{821}$.

19. The ratio of 9 inches to 12 inches is
$\dfrac{9 \text{ inches}}{12 \text{ inches}} = \dfrac{9}{12} = \dfrac{3 \cdot 3}{3 \cdot 4} = \dfrac{3}{4}$.

21. The ratio of 10 hours to 24 hours is
$\dfrac{10 \text{ hours}}{24 \text{ hours}} = \dfrac{10}{24} = \dfrac{2 \cdot 5}{2 \cdot 12} = \dfrac{5}{12}$.

23. The ratio of \$32 to \$100 is
$\dfrac{\$32}{\$100} = \dfrac{32}{100} = \dfrac{4 \cdot 8}{4 \cdot 25} = \dfrac{8}{25}$.

25. The ratio of 24 days to 14 days is
$\dfrac{24 \text{ days}}{14 \text{ days}} = \dfrac{24}{14} = \dfrac{2 \cdot 12}{2 \cdot 7} = \dfrac{12}{7}$.

27. The ratio of 32,000 bytes to 46,000 bytes is
$\dfrac{32,000 \text{ bytes}}{46,000 \text{ bytes}} = \dfrac{32,000}{46,000} = \dfrac{2000 \cdot 16}{2000 \cdot 23} = \dfrac{16}{23}$.

29. The ratio of 8 inches to 20 inches is
$\dfrac{8 \text{ inches}}{20 \text{ inches}} = \dfrac{8}{20} = \dfrac{4 \cdot 2}{4 \cdot 5} = \dfrac{2}{5}$.

31. The ratio of $3\frac{1}{2}$ to $12\frac{1}{4}$ is $\dfrac{3\frac{1}{2}}{12\frac{1}{4}} = 3\frac{1}{2} \div 12\frac{1}{4}$

$= \dfrac{7}{2} \div \dfrac{49}{4}$

$= \dfrac{7}{2} \cdot \dfrac{4}{49}$

$= \dfrac{7 \cdot 2 \cdot 2}{2 \cdot 7 \cdot 7}$

$= \dfrac{2}{7}$.

33. The ratio of $7\frac{3}{5}$ hours to $1\frac{9}{10}$ hours is

$\dfrac{7\frac{3}{5} \text{ hours}}{1\frac{9}{10} \text{ hours}} = \dfrac{7\frac{3}{5}}{1\frac{9}{10}}$

$= 7\frac{3}{5} \div 1\frac{9}{10}$

$= \dfrac{38}{5} \div \dfrac{19}{10}$

$= \dfrac{38}{5} \cdot \dfrac{10}{19}$

$= \dfrac{19 \cdot 2 \cdot 5 \cdot 2}{5 \cdot 19}$

$= \dfrac{2 \cdot 2}{1}$

$= \dfrac{4}{1}$.

35. The ratio of the weight of an average mature Fin Whale to the weight of an average mature Blue Whale is $\dfrac{50 \text{ tons}}{145 \text{ tons}} = \dfrac{5 \cdot 10}{5 \cdot 29} = \dfrac{10}{29}$.

37. Perimeter = 32 + 19 + 32 + 19 = 102 feet
The ratio of width to perimeter is
$\dfrac{32 \text{ feet}}{102 \text{ feet}} = \dfrac{2 \cdot 16}{2 \cdot 51} = \dfrac{16}{51}$.

39. Perimeter = 94 + 50 + 94 + 50 = 288 feet
The ratio of width to perimeter is
$\dfrac{50 \text{ feet}}{288 \text{ feet}} = \dfrac{2 \cdot 25}{2 \cdot 144} = \dfrac{25}{144}$.

41. The ratio of women to men is
$\dfrac{125 \text{ women}}{100 \text{ men}} = \dfrac{5 \cdot 25}{4 \cdot 25} = \dfrac{5}{4}$.

43. The ratio of red blood cells to platelet cells is
$$\frac{600 \text{ cells}}{40 \text{ cells}} = \frac{40 \cdot 15}{40 \cdot 1} = \frac{15}{1}.$$

45. The ratio of the longest side to the perimeter is
$$\frac{17 \text{ feet}}{(8+15+17) \text{ feet}} = \frac{17}{40}.$$

47. The ratio of digital films to total films for 2009 is
$$\frac{20 \text{ films}}{558 \text{ films}} = \frac{20}{558} = \frac{2 \cdot 10}{2 \cdot 279} = \frac{10}{279}.$$

49. The ratio of mountains over 14,000 feet tall found in Alaska to those found in Colorado is
$$\frac{19 \text{ mountains}}{57 \text{ mountains}} = \frac{19}{57} = \frac{19 \cdot 1}{19 \cdot 3} = \frac{1}{3}.$$

51. The ratio of gold medals won by the Australian athletes to the total gold medals awarded is
$$\frac{14 \text{ gold medals}}{302 \text{ gold medals}} = \frac{14}{302} = \frac{2 \cdot 7}{2 \cdot 151} = \frac{7}{151}.$$

53. There is $50 - 49 = 1$ state without Target stores. The ratio of states without Target stores to states with Target stores is $\frac{1 \text{ state}}{49 \text{ states}} = \frac{1}{49}.$

55.
$$\begin{array}{r}
2.3 \\
9\overline{)\ 20.7} \\
\underline{-18} \\
2\ 7 \\
\underline{-2\ 7} \\
0
\end{array}$$

57. $3.7\overline{)0.555}$ becomes
$$\begin{array}{r}
0.15 \\
37\overline{)\ 5.55} \\
\underline{-3\ 7} \\
1\ 85 \\
\underline{-1\ 85} \\
0
\end{array}$$

59. $\frac{7}{9}$ should be read as "the ratio of seven to nine."

61. $30:1$ should be read as "the ratio of thirty to one."

63. no; answers may vary

65. $\frac{\$3}{\$2}$ is not in simplest form: $\frac{\$3}{\$2} = \frac{3}{2}$.

67. $\frac{1 \text{ foot}}{30 \text{ inches}}$ is not in simplest form:
$$\frac{1 \text{ foot}}{30 \text{ inches}} = \frac{12 \text{ inches}}{30 \text{ inches}} = \frac{12}{30} = \frac{6 \cdot 2}{6 \cdot 5} = \frac{2}{5}$$

69. $4\frac{1}{2}$ is not in simplest form: $4\frac{1}{2} = \frac{9}{2}$.

71. The ratio of bruised tomatoes to the total batch is
$$\frac{3}{3+33} = \frac{3}{36} = \frac{3 \cdot 1}{3 \cdot 12} = \frac{1}{12}.$$
This is less than the ratio 1 to 10, since
$$\frac{1}{12} = \frac{5 \cdot 1}{5 \cdot 12} = \frac{5}{60} < \frac{6}{60} = \frac{6 \cdot 1}{6 \cdot 10} = \frac{1}{10}.$$
The shipment should not be refused.

73. a. The ratio of states with primary seat belt laws to total U.S. states is
$$\frac{30 \text{ states}}{50 \text{ states}} = \frac{30}{50} = \frac{10 \cdot 3}{10 \cdot 5} = \frac{3}{5}.$$

    b. There are $50 - 30 = 20$ states with secondary seat belt laws. The ratio of states with primary seat belt laws to states with secondary seat belt laws is
$$\frac{30 \text{ states}}{20 \text{ states}} = \frac{30}{20} = \frac{10 \cdot 3}{10 \cdot 2} = \frac{3}{2}.$$

    c. no; answers may vary

Section 5.2 Practice

1. 12 commercials every 45 minutes is
$$\frac{12 \text{ commercials}}{45 \text{ minutes}} = \frac{4 \text{ commercials}}{15 \text{ minutes}}.$$

2. $1680 for 8 weeks is $\frac{\$1680}{8 \text{ weeks}} = \frac{\$210}{1 \text{ week}}.$

3. 236 miles on 12 gallons of gas is
$$\frac{236 \text{ miles}}{12 \text{ gallons}} = \frac{59 \text{ miles}}{3 \text{ gallons}}.$$

4. 3200 feet every 8 seconds:
$$\begin{array}{r}
400 \\
8\overline{)\ 3200} \\
\underline{-3200} \\
0
\end{array}$$
The unit rate is $\frac{400 \text{ feet}}{1 \text{ second}}$ or 400 feet/second.

5. 78 bushels from 12 trees:

$$
\begin{array}{r}
6.5 \\
12\overline{)78.0} \\
\underline{72} \\
6\,0 \\
\underline{-6\,0} \\
0
\end{array}
$$

The unit rate is $\dfrac{6.5 \text{ bushels}}{1 \text{ tree}}$ or 6.5 bushels/tree.

6. \$170 for 5 days is $\dfrac{\$170}{5 \text{ days}} = \dfrac{\$34}{1 \text{ day}}$. The unit price is \$34 per day.

7.
$$
\begin{array}{r}
0.2109 \approx 0.211 \\
11\overline{)2.3200}
\end{array}
$$
$$
\begin{array}{r}
0.2243 \approx 0.224 \\
16\overline{)3.5900}
\end{array}
$$

Since $0.211 < 0.224$, the 11-ounce bag of nacho chips is a better buy than the 16-ounce bag.

Vocabulary and Readiness Check

1. A rate with a denominator of 1 is called a <u>unit</u> rate.

2. When a rate is written as money per item, a unit rate is called a <u>unit price</u>.

3. The word *per* translates to "<u>division</u>."

4. Rates are used to compare <u>different</u> types of quantities.

5. To write a rate as a unit rate, divide the <u>numerator</u> of the rate by the <u>denominator</u>.

Exercise Set 5.2

1. The rate of 5 shrubs every 15 feet is
$$
\frac{5 \text{ shrubs}}{15 \text{ feet}} = \frac{1 \text{ shrub}}{3 \text{ feet}}.
$$

3. The rate of 15 returns for 100 sales is
$$
\frac{15 \text{ returns}}{100 \text{ sales}} = \frac{3 \text{ returns}}{20 \text{ sales}}.
$$

5. The rate of 6 laser printers for 28 computers is
$$
\frac{6 \text{ laser printers}}{28 \text{ computers}} = \frac{3 \text{ laser printers}}{14 \text{ computers}}.
$$

7. The rate of 18 gallons of pesticide for 4 acres of crops is $\dfrac{18 \text{ gallons}}{4 \text{ acres}} = \dfrac{9 \text{ gallons}}{2 \text{ acres}}$.

9. The rate of 6 flight attendants for 200 passengers is $\dfrac{6 \text{ flight attendants}}{200 \text{ passengers}} = \dfrac{3 \text{ flight attendants}}{100 \text{ passengers}}$.

11. The rate of 355 calories in a 10-fluid-ounce chocolate milkshake is
$$
\frac{355 \text{ calories}}{10 \text{ fluid ounces}} = \frac{71 \text{ calories}}{2 \text{ fluid ounces}}.
$$

13.
$$
\begin{array}{r}
110 \\
3\overline{)330} \\
\underline{-3} \\
03 \\
\underline{-3} \\
00
\end{array}
$$

330 calories in a 3-ounce serving is $\dfrac{110 \text{ calories}}{1 \text{ ounce}}$ or 110 calories/ounce.

15.
$$
\begin{array}{r}
75 \\
5\overline{)375} \\
\underline{-35} \\
25 \\
\underline{-25} \\
0
\end{array}
$$

375 riders in 5 subway cars is $\dfrac{75 \text{ riders}}{1 \text{ car}}$ or 75 riders/car.

17.
$$
\begin{array}{r}
90 \\
60\overline{)5400} \\
\underline{-540} \\
00
\end{array}
$$

5400 wingbeats per 60 seconds is $\dfrac{90 \text{ wingbeats}}{1 \text{ second}}$ or 90 wingbeats/second.

19.
$$
\begin{array}{r}
50{,}000 \\
20\overline{)1{,}000{,}000} \\
\underline{-1\,00\phantom{0{,}000}} \\
0
\end{array}
$$

\$1,000,000 paid over 20 years is $\dfrac{\$50{,}000}{1 \text{ year}}$ or \$50,000/year.

21.

$$\begin{array}{r} 225,250 \\ 2\overline{)\,450,500} \\ \underline{-4} \\ 05 \\ \underline{-4} \\ 10 \\ \underline{-10} \\ 0\,5 \\ \underline{-4} \\ 10 \\ \underline{-10} \\ 0 \end{array}$$

450,500 voters for 2 senators is $\dfrac{225,250 \text{ voters}}{1 \text{ senator}}$

or 225,250 voters/senator.

23.

$$\begin{array}{r} 300 \\ 40\overline{)\,12,000} \\ \underline{-12\,0} \\ 0 \end{array}$$

12,000 good products to 40 defective is

$\dfrac{300 \text{ good}}{1 \text{ defective}}$ or 300 good/defective.

25.

$$\begin{array}{r} 4,390,000 \\ 20\overline{)\,87,800,000} \\ \underline{-80} \\ 7\,8 \\ \underline{-6\,0} \\ 1\,80 \\ \underline{-1\,80} \\ 0 \end{array}$$

$87,800,000 for 20 players is $\dfrac{\$4,390,000}{1 \text{ player}}$ or

$4,390,000/player.

27.

$$\begin{array}{r} 65.7 \approx 66 \\ 365\overline{)\,24000.0} \\ \underline{-2190} \\ 2100 \\ \underline{-1825} \\ 2750 \\ \underline{-2555} \\ 195 \end{array}$$

$$\frac{24,000 \text{ crayons}}{1 \text{ year}} = \frac{24,000 \text{ crayons}}{365 \text{ days}}$$
$$\approx \frac{66 \text{ crayons}}{1 \text{ day}}$$
$$= 66 \text{ crayons/day}$$

29.

$$\begin{array}{r} 0.14 \\ 5000\overline{)\,700.00} \\ \underline{-500\,0} \\ 200\,00 \\ \underline{-200\,00} \\ 0 \end{array}$$

$$\frac{\$700}{5000 \text{ miles}} = \frac{\$0.14}{1 \text{ mile}} = \$0.14 \,/\, \text{mile}$$

31.

$$\begin{array}{r} 59800 \\ 25\overline{)\,1495000} \\ \underline{-125} \\ 240 \\ \underline{-225} \\ 150 \\ \underline{-150} \\ 000 \end{array}$$

$$\frac{\$1,495,00}{25 \text{ houses}} = \frac{\$59,800}{1 \text{ house}} = \$59,800 \,/\, \text{house}$$

33.

$$\begin{array}{r} 6.16 \approx 6.2 \\ 24\overline{)\,148.00} \\ \underline{-144} \\ 4\,0 \\ \underline{-2\,4} \\ 1\,60 \\ \underline{-1\,44} \\ 16 \end{array}$$

$$\frac{148 \text{ tornadoes}}{24 \text{ hours}} \approx \frac{6.2 \text{ tornadoes}}{1 \text{ hour}}$$
$$= 6.2 \text{ tornadoes/hour}$$

35. a.
$$\begin{array}{r} 31.25 \\ 8\overline{)\,250.00} \\ \underline{-24} \\ 10 \\ \underline{-8} \\ 2\,0 \\ \underline{-1\,6} \\ 40 \\ \underline{-40} \\ 0 \end{array}$$

The unit rate for Charlie is 31.25 boards/hour.

b.
$$\begin{array}{r} 33.5 \\ 12\overline{)\,402.0} \\ \underline{-36} \\ 42 \\ \underline{-36} \\ 6\,0 \\ \underline{-6\,0} \\ 0 \end{array}$$

The unit rate for Suellen is 33.5 boards/hour.

c. Suellen can assemble boards faster, since 33.5 > 31.25.

37. a. $14.5\overline{)400}$ becomes
$$\begin{array}{r} 27.58 \approx 27.6 \\ 145\overline{)\,4000.00} \\ \underline{-290} \\ 1100 \\ \underline{-1015} \\ 850 \\ \underline{-725} \\ 1250 \\ \underline{-1160} \\ 90 \end{array}$$

The unit rate for the car is ≈27.6 miles/gallon.

b. $9.25\overline{)270}$ becomes
$$\begin{array}{r} 29.18 \approx 29.2 \\ 925\overline{)\,27000.00} \\ \underline{-1850} \\ 8500 \\ \underline{-8325} \\ 175\,0 \\ \underline{-92\,5} \\ 82\,50 \\ \underline{-74\,00} \\ 8\,50 \end{array}$$

The unit rate for the truck is ≈29.2 miles/gallon.

c. The truck has the better gas mileage, since 29.2 > 27.6.

39.
$$\begin{array}{r} 11.50 \\ 5\overline{)\,57.50} \\ \underline{-5} \\ 07 \\ \underline{-5} \\ 2\,5 \\ \underline{-2\,5} \\ 0 \end{array}$$

The unit price is $11.50 per DVD.

41.
$$\begin{array}{r} 0.17 \\ 7\overline{)\,1.19} \\ \underline{-7} \\ 49 \\ \underline{-49} \\ 0 \end{array}$$

The unit price is $0.17 per banana.

43.
$$\begin{array}{r} 0.1487 \approx 0.149 \\ 8\overline{)\,1.1900} \\ \underline{-8} \\ 39 \\ \underline{-32} \\ 70 \\ \underline{-64} \\ 60 \\ \underline{-56} \\ 4 \end{array}$$

The 8-ounce size costs $0.149 per ounce.

$$
\begin{array}{r}
0.1325 \approx 0.133 \\
12\overline{)1.5900} \\
\underline{-1\,2} \\
39 \\
\underline{-36} \\
30 \\
\underline{-24} \\
60 \\
\underline{-60} \\
0
\end{array}
$$

The 12-ounce size costs $0.133 per ounce.
The 12-ounce size is the better buy.

45.
$$
\begin{array}{r}
0.1056 \approx 0.106 \\
16\overline{)1.6900} \\
\underline{-1\,6} \\
090 \\
\underline{-80} \\
100 \\
\underline{-96} \\
4
\end{array}
$$

The 16-ounce size costs $0.106 per ounce.

$$
\begin{array}{r}
0.115 \\
6\overline{)0.690} \\
\underline{-6} \\
09 \\
\underline{-6} \\
30 \\
\underline{-30} \\
0
\end{array}
$$

The 6-ounce size costs $0.115 per ounce.
The 16-ounce size is the better buy.

47.
$$
\begin{array}{r}
0.1908 \approx 0.191 \\
12\overline{)2.2900} \\
\underline{-1\,2} \\
1\,09 \\
\underline{-1\,08} \\
100 \\
\underline{-96} \\
4
\end{array}
$$

The 12-ounce size costs $0.191 per ounce.

$$
\begin{array}{r}
0.1862 \approx 0.186 \\
8\overline{)1.4900} \\
\underline{-8} \\
69 \\
\underline{-64} \\
50 \\
\underline{-48} \\
20 \\
\underline{-16} \\
4
\end{array}
$$

The 8-ounce size costs $0.186 per ounce.
The 8-ounce size is the better buy.

49.
$$
\begin{array}{r}
0.0059 \approx 0.006 \\
100\overline{)0.5900} \\
\underline{-500} \\
900 \\
\underline{-900} \\
0
\end{array}
$$

The 100-count size costs $0.006 per napkin.

$$
\begin{array}{r}
0.0051 \approx 0.005 \\
180\overline{)0.9300} \\
\underline{-900} \\
300 \\
\underline{-180} \\
120
\end{array}
$$

The 180-count size costs $0.005 per napkin.
The 180-count size is the better buy.

51.
$$
\begin{array}{r}
1.7 \\
\times\ 6 \\
\hline
10.2
\end{array}
$$

53.
$$
\begin{array}{r}
3.7 \\
\times 1.2 \\
\hline
74 \\
3\,70 \\
\hline
4.44
\end{array}
$$

55. $2.3\overline{)4.37}$ becomes
$$
\begin{array}{r}
1.9 \\
23\overline{)43.7} \\
\underline{-23} \\
20\,7 \\
\underline{-20\,7} \\
0
\end{array}
$$

57. $29,543 - 29,286 = 257$

$$13.4\overline{)257} \text{ becomes } 134\overline{)2570.00}\quad \frac{19.17}{} \approx 19.2$$

$$
\begin{array}{r}
-134 \\
\hline
1230 \\
-1206 \\
\hline
24\,0 \\
-13\,4 \\
\hline
10\,60 \\
-9\,38 \\
\hline
1\,22
\end{array}
$$

257 miles were driven, averaging approximately 19.2 miles per gallon.

59. $80,242 - 79,895 = 347$

$$16.1\overline{)347} \text{ becomes } 161\overline{)3470.00}\quad \frac{21.55}{} \approx 21.6$$

$$
\begin{array}{r}
-322 \\
\hline
250 \\
-161 \\
\hline
89\,0 \\
-80\,5 \\
\hline
8\,50 \\
-8\,05 \\
\hline
45
\end{array}
$$

347 miles were driven, averaging approximately 21.6 miles per gallon.

61. $7759\overline{)11,674.00}\quad \frac{1.50}{} \approx 1.5$

$$
\begin{array}{r}
7\,759 \\
\hline
3\,915\,0 \\
-3\,879\,5 \\
\hline
35\,50
\end{array}
$$

The unit rate is 1.5 steps/foot.

63. answers may vary

65. no; answers may vary

Integrated Review

1. The ratio of 18 to 20 is $\dfrac{18}{20} = \dfrac{2 \cdot 9}{2 \cdot 10} = \dfrac{9}{10}$.

2. The ratio of 36 to 100 is $\dfrac{36}{100} = \dfrac{4 \cdot 9}{4 \cdot 25} = \dfrac{9}{25}$.

3. The ratio of 8.6 to 10 is

$$\frac{8.6}{10} = \frac{8.6 \cdot 10}{10 \cdot 10} = \frac{86}{100} = \frac{2 \cdot 43}{2 \cdot 50} = \frac{43}{50}.$$

4. The ratio of 1.6 to 4.6 is

$$\frac{1.6}{4.6} = \frac{1.6 \cdot 10}{4.6 \cdot 10} = \frac{16}{46} = \frac{2 \cdot 8}{2 \cdot 23} = \frac{8}{23}.$$

5. The ratio of \$8.65 to \$6.95 is

$$\frac{\$8.65}{\$6.95} = \frac{8.65 \cdot 100}{6.95 \cdot 100} = \frac{865}{695} = \frac{5 \cdot 173}{5 \cdot 139} = \frac{173}{139}.$$

6. The ratio of 7.2 ounces to 8.4 ounces is

$$\frac{7.2 \text{ ounces}}{8.4 \text{ ounces}} = \frac{7.2 \cdot 10}{8.4 \cdot 10} = \frac{72}{84} = \frac{12 \cdot 6}{12 \cdot 7} = \frac{6}{7}.$$

7. The ratio of $3\frac{1}{2}$ to 13 is

$$\frac{3\frac{1}{2}}{13} = 3\frac{1}{2} \div 13 = \frac{7}{2} \div \frac{13}{1} = \frac{7}{2} \cdot \frac{1}{13} = \frac{7}{26}.$$

8. The ratio of $1\frac{2}{3}$ to $2\frac{3}{4}$ is

$$\frac{1\frac{2}{3}}{2\frac{3}{4}} = 1\frac{2}{3} \div 2\frac{3}{4} = \frac{5}{3} \div \frac{11}{4} = \frac{5}{3} \cdot \frac{4}{11} = \frac{20}{33}.$$

9. The ratio of 8 inches to 12 inches is

$$\frac{8 \text{ inches}}{12 \text{ inches}} = \frac{8}{12} = \frac{4 \cdot 2}{4 \cdot 3} = \frac{2}{3}.$$

10. The ratio of 3 hours to 24 hours is

$$\frac{3 \text{ hours}}{24 \text{ hours}} = \frac{3}{24} = \frac{3 \cdot 1}{3 \cdot 8} = \frac{1}{8}.$$

11. $\dfrac{\$108.7 \text{ thousand}}{\$76.1 \text{ thousand}} = \dfrac{108.7 \cdot 10}{76.1 \cdot 10} = \dfrac{1087}{761}$

12. $\dfrac{26 \text{ World Series}}{104 \text{ World Series}} = \dfrac{26}{104} = \dfrac{26 \cdot 1}{26 \cdot 4} = \dfrac{1}{4}$

13. a. 13 of the top 25 movies were rated PG-13.

    b. 3 of the top 25 movies were rated R. 9 of the top 25 movies were rated PG.

$$\frac{3 \text{ movies}}{9 \text{ movies}} = \frac{3}{9} = \frac{3 \cdot 1}{3 \cdot 3} = \frac{1}{3}$$

14. $\dfrac{12 \text{ inches}}{18 \text{ inches}} = \dfrac{12}{18} = \dfrac{6 \cdot 2}{6 \cdot 3} = \dfrac{2}{3}$

15. 5 offices for every 20 graduate assistants is

$$\frac{5 \text{ offices}}{20 \text{ graduate assistants}} = \frac{5 \cdot 1 \text{ offices}}{5 \cdot 4 \text{ graduate assistants}}$$
$$= \frac{1 \text{ office}}{4 \text{ graduate assistants}}.$$

16. 6 lights every 15 feet is

$$\frac{6 \text{ lights}}{15 \text{ feet}} = \frac{3 \cdot 2 \text{ lights}}{3 \cdot 5 \text{ feet}} = \frac{2 \text{ lights}}{5 \text{ feet}}.$$

17. 100 U.S. senators for 50 states is

$$\frac{100 \text{ senators}}{50 \text{ states}} = \frac{50 \cdot 2 \text{ senators}}{50 \cdot 1 \text{ states}} = \frac{2 \text{ senators}}{1 \text{ state}}.$$

18. 5 teachers for every 140 students is

$$\frac{5 \text{ teachers}}{140 \text{ students}} = \frac{5 \cdot 1 \text{ teachers}}{5 \cdot 28 \text{ students}} = \frac{1 \text{ teacher}}{28 \text{ students}}.$$

19. 64 computers for every 100 households is

$$\frac{64 \text{ computers}}{100 \text{ households}} = \frac{4 \cdot 16 \text{ computers}}{4 \cdot 25 \text{ households}}$$
$$= \frac{16 \text{ computers}}{25 \text{ households}}.$$

20. 45 students for every 10 computers is

$$\frac{45 \text{ students}}{10 \text{ computers}} = \frac{5 \cdot 9 \text{ students}}{5 \cdot 2 \text{ computers}} = \frac{9 \text{ students}}{2 \text{ computers}}.$$

21. 165 miles in 3 hours is

$$\frac{165 \text{ miles}}{3 \text{ hours}} = \frac{3 \cdot 55 \text{ miles}}{3 \cdot 1 \text{ hours}} = 55 \text{ miles/hour.}$$

22. 560 feet in 4 seconds is

$$\frac{560 \text{ feet}}{4 \text{ seconds}} = \frac{4 \cdot 140 \text{ feet}}{4 \cdot 1 \text{ seconds}} = 140 \text{ feet/second.}$$

23. 63 employees for 3 fax lines is

$$\frac{63 \text{ employees}}{3 \text{ fax lines}} = \frac{3 \cdot 21 \text{ employees}}{3 \cdot 1 \text{ fax lines}}$$
$$= 21 \text{ employees/fax line.}$$

24. 85 phone calls for 5 teenagers is

$$\frac{85 \text{ phone calls}}{5 \text{ teenagers}} = \frac{5 \cdot 17 \text{ phone calls}}{5 \cdot 1 \text{ teenagers}}$$
$$= 17 \text{ phone calls/teenager.}$$

25. 115 miles every 5 gallons is

$$\frac{115 \text{ miles}}{5 \text{ gallons}} = \frac{5 \cdot 23 \text{ miles}}{5 \cdot 1 \text{ gallons}} = 23 \text{ miles/gallon.}$$

26. 112 teachers for 7 computers is

$$\frac{112 \text{ teachers}}{7 \text{ computers}} = \frac{7 \cdot 16 \text{ teachers}}{7 \cdot 1 \text{ computers}}$$
$$= 16 \text{ teachers/computer.}$$

27. 7524 books for 1254 college students is

$$\frac{7524 \text{ books}}{1254 \text{ students}} = \frac{1254 \cdot 6 \text{ books}}{1254 \cdot 1 \text{ students}}$$
$$= 6 \text{ books/student.}$$

28. 2002 pounds for 13 adults is

$$\frac{2002 \text{ pounds}}{13 \text{ adults}} = \frac{13 \cdot 154 \text{ pounds}}{13 \cdot 1 \text{ adults}}$$
$$= 154 \text{ pounds/adult.}$$

29.
$$\begin{array}{r} 0.27 \\ 8\overline{)\ 2.16} \\ \underline{-1\ 6} \\ 56 \\ \underline{-56} \\ 0 \end{array}$$

The unit rate for the 8-pound size is exactly $0.27/pound.

$$\begin{array}{r} 0.277 \approx 0.28 \\ 18\overline{)\ 4.990} \\ \underline{-3\ 6} \\ 1\ 39 \\ \underline{-1\ 26} \\ 130 \\ \underline{-126} \\ 4 \end{array}$$

The unit rate for the 18-pound size is about $0.28/pound. The 8-pound size is the better buy.

30.
$$\begin{array}{r} 0.0198 \approx 0.020 \\ 100\overline{)\ 1.9800} \\ \underline{-1\ 00} \\ 980 \\ \underline{-900} \\ 800 \\ \underline{-800} \\ 0 \end{array}$$

The unit rate for the 100-count size is about $0.020/plate.

$$\begin{array}{r} 0.0179 \approx 0.018 \\ 500\overline{)\ 8.9900} \\ \underline{-5\ 00} \\ 3\ 990 \\ \underline{-3\ 500} \\ 4900 \\ \underline{-4500} \\ 400 \end{array}$$

The unit rate for the 500-count size is about $0.018/plate. The 500-count size is the better buy.

31.
$$\begin{array}{r} 0.796 \approx 0.80 \\ 3{\overline{\smash{\big)}\,2.390}} \\ \underline{-2\,1} \\ 29 \\ \underline{-27} \\ 20 \\ \underline{-18} \\ 2 \end{array}$$

The unit rate for the 3-pack size is about $0.80/pack.
$$\begin{array}{r} 0.748 \approx 0.75 \\ 8{\overline{\smash{\big)}\,5.990}} \\ \underline{-5\,6} \\ 39 \\ \underline{-32} \\ 70 \\ \underline{-64} \\ 6 \end{array}$$

The unit rate for the 8-pack size is about $0.75/pack. The 8-pack size is the better buy.

32.
$$\begin{array}{r} 0.922 \approx 0.92 \\ 4{\overline{\smash{\big)}\,3.690}} \\ \underline{-3\,6} \\ 09 \\ \underline{-8} \\ 10 \\ \underline{-8} \\ 2 \end{array}$$

The unit rate for the 4-battery size is about $0.92/battery.
$$\begin{array}{r} 0.989 \approx 0.99 \\ 10{\overline{\smash{\big)}\,9.890}} \\ \underline{-9\,0} \\ 89 \\ \underline{-80} \\ 90 \\ \underline{-90} \\ 0 \end{array}$$

The unit rate for the 10-battery size is about $0.99/battery. The 4-battery size is the better buy.

Section 5.3 Practice

1. a.
$$\begin{array}{lllll} \text{right} & \rightarrow & \dfrac{24}{6} = \dfrac{4}{1} & \leftarrow & \text{right} \\ \text{wrong} & \rightarrow & & \leftarrow & \text{wrong} \end{array}$$

b.
$$\begin{array}{lllll} \text{Cubs fans} & \rightarrow & \dfrac{32}{18} = \dfrac{16}{9} & \leftarrow & \text{Cubs fans} \\ \text{Mets fans} & \rightarrow & & \leftarrow & \text{Mets fans} \end{array}$$

2.
$$\frac{3}{6} = \frac{4}{8}$$
$$3 \cdot 8 \overset{?}{=} 6 \cdot 4$$
$$24 = 24$$
Since the cross products are equal, the proportion is true.

3.
$$\frac{3.6}{6} = \frac{5.4}{8}$$
$$3.6 \cdot 8 \overset{?}{=} 6 \cdot 5.4$$
$$28.8 \neq 32.4$$
Since the cross products are not equal, $\dfrac{3.6}{6} \neq \dfrac{5.4}{8}$, and the proportion is false.

4.
$$\frac{4\frac{1}{5}}{2\frac{1}{3}} = \frac{3\frac{3}{10}}{1\frac{5}{6}}$$
$$4\frac{1}{5} \cdot 1\frac{5}{6} \overset{?}{=} 2\frac{1}{3} \cdot 3\frac{3}{10}$$
$$\frac{21}{5} \cdot \frac{11}{6} \overset{?}{=} \frac{7}{3} \cdot \frac{33}{10}$$
$$\frac{231}{30} = \frac{231}{30}$$
Since the cross products are equal, the proportion is true.

5.
$$\frac{15}{2} = \frac{60}{n}$$
$$15 \cdot n = 2 \cdot 60$$
$$15n = 120$$
$$n = \frac{120}{15}$$
$$n = 8$$

6.
$$\frac{8}{n} = \frac{5}{9}$$
$$8 \cdot 9 = n \cdot 5$$
$$72 = 5n$$
$$\frac{72}{5} = n$$
$$14\frac{2}{5} = n$$

7. $\dfrac{n}{6} = \dfrac{0.7}{1.2}$

$n \cdot 1.2 = 6 \cdot 0.7$

$1.2n = 4.2$

$n = \dfrac{4.2}{1.2}$

$n = 3.5$

8. $\dfrac{n}{4\frac{1}{3}} = \dfrac{4\frac{1}{2}}{1\frac{3}{4}}$

$n \cdot 1\dfrac{3}{4} = 4\dfrac{1}{3} \cdot 4\dfrac{1}{2}$

$\dfrac{7}{4}n = \dfrac{117}{6}$

$n = \dfrac{4}{7} \cdot \dfrac{117}{6}$

$n = 11\dfrac{1}{7}$

$n = 11\dfrac{1}{7}$

Vocabulary and Readiness Check

1. $\dfrac{4.2}{8.4} = \dfrac{1}{2}$ is called a <u>proportion</u> while $\dfrac{7}{8}$ is called a <u>ratio</u>.

2. In $\dfrac{a}{b} = \dfrac{c}{d}$, $a \cdot d$ and $b \cdot c$ are called <u>cross products</u>.

3. In a proportion, if cross products are equal, the proportion is <u>true</u>.

4. In a proportion, if cross products are not equal, the proportion is <u>false</u>.

5. The cross products are $2 \cdot 3 = 6$ and $6 \cdot 1 = 6$. These are equal, so the proportion is true.

6. The cross products are $3 \cdot 5 = 15$ and $15 \cdot 1 = 15$. These are equal, so the proportion is true.

7. The cross products are $1 \cdot 5 = 5$ and $3 \cdot 2 = 6$. These are not equal, so the proportion is false.

8. The cross products are $2 \cdot 5 = 10$ and $1 \cdot 11 = 11$. These are not equal, so the proportion is false.

9. The cross products are $2 \cdot 60 = 120$ and $40 \cdot 3 = 120$. These are equal, so the proportion is true.

10. The cross products are $3 \cdot 8 = 24$ and $6 \cdot 4 = 24$. These are equal, so the proportion is true.

Exercise Set 5.3

1. $\dfrac{10 \text{ diamonds}}{6 \text{ opals}} = \dfrac{5 \text{ diamonds}}{3 \text{ opals}}$

3. $\dfrac{3 \text{ printers}}{12 \text{ computers}} = \dfrac{1 \text{ printer}}{4 \text{ computers}}$

5. $\dfrac{6 \text{ eagles}}{58 \text{ sparrows}} = \dfrac{3 \text{ eagles}}{29 \text{ sparrows}}$

7. $\dfrac{2\frac{1}{4} \text{ cups flour}}{24 \text{ cookies}} = \dfrac{6\frac{3}{4} \text{ cups flour}}{72 \text{ cookies}}$

9. $\dfrac{22 \text{ vanilla wafers}}{1 \text{ cup cookie crumbs}}$
$= \dfrac{55 \text{ vanilla wafers}}{2.5 \text{ cups cookie crumbs}}$

11. $\dfrac{15}{9} \overset{?}{=} \dfrac{5}{3}$

$15 \cdot 3 \overset{?}{=} 9 \cdot 5$

$45 = 45$

true

13. $\dfrac{8}{6} \overset{?}{=} \dfrac{9}{7}$

$8 \cdot 7 \overset{?}{=} 9 \cdot 6$

$56 \neq 54$

false

15. $\dfrac{9}{36} \overset{?}{=} \dfrac{2}{8}$

$9 \cdot 8 \overset{?}{=} 36 \cdot 2$

$72 = 72$

true

17. $\dfrac{5}{8} \overset{?}{=} \dfrac{625}{1000}$

$5 \cdot 1000 \overset{?}{=} 8 \cdot 625$

$5000 = 5000$

true

19. $\dfrac{0.8}{0.3} \stackrel{?}{=} \dfrac{0.2}{0.6}$

$0.8 \cdot 0.6 \stackrel{?}{=} 0.3 \cdot 0.2$

$0.48 \neq 0.06$

false

21. $\dfrac{8}{10} \stackrel{?}{=} \dfrac{5.6}{0.7}$

$8(0.7) \stackrel{?}{=} 10(5.6)$

$5.6 \neq 56$

false

23. $\dfrac{\frac{3}{4}}{\frac{4}{3}} \stackrel{?}{=} \dfrac{\frac{1}{2}}{\frac{8}{9}}$

$\dfrac{3}{4} \cdot \dfrac{8}{9} \stackrel{?}{=} \dfrac{4}{3} \cdot \dfrac{1}{2}$

$\dfrac{24}{36} = \dfrac{4}{6}$

true

25. $\dfrac{2\frac{2}{5}}{\frac{2}{3}} \stackrel{?}{=} \dfrac{1\frac{1}{9}}{\frac{1}{4}}$

$2\dfrac{2}{5} \cdot \dfrac{1}{4} \stackrel{?}{=} 1\dfrac{1}{9} \cdot \dfrac{2}{3}$

$\dfrac{12}{5} \cdot \dfrac{1}{4} \stackrel{?}{=} \dfrac{10}{9} \cdot \dfrac{2}{3}$

$\dfrac{3}{5} \neq \dfrac{20}{27}$

false

27. $\dfrac{\frac{4}{5}}{6} \stackrel{?}{=} \dfrac{\frac{6}{5}}{9}$

$\dfrac{4}{5} \cdot 9 \stackrel{?}{=} 6 \cdot \dfrac{6}{5}$

$\dfrac{36}{5} = \dfrac{36}{5}$

true

29. $\dfrac{8}{12} = \dfrac{4}{6}$

$8 \cdot 6 \stackrel{?}{=} 12 \cdot 4$

$48 = 48$

true

31. $\dfrac{5}{2} = \dfrac{13}{5}$

$5 \cdot 5 \stackrel{?}{=} 2 \cdot 13$

$25 \neq 26$

false

33. $\dfrac{1.8}{2} = \dfrac{4.5}{5}$

$1.8(5) \stackrel{?}{=} 4.5(2)$

$9 = 9$

true

35. $\dfrac{\frac{2}{3}}{\frac{1}{5}} = \dfrac{\frac{2}{5}}{\frac{1}{9}}$

$\dfrac{2}{3} \cdot \dfrac{1}{9} \stackrel{?}{=} \dfrac{1}{5} \cdot \dfrac{2}{5}$

$\dfrac{2}{27} \neq \dfrac{2}{25}$

false

37. $\dfrac{n}{5} = \dfrac{6}{10}$

$10n = 6 \cdot 5$

$10n = 30$

$n = 3$

39. $\dfrac{18}{54} = \dfrac{3}{n}$

$18n = 54 \cdot 3$

$18n = 162$

$n = 9$

41. $\dfrac{n}{8} = \dfrac{50}{100}$

$100n = 8 \cdot 50$

$100n = 400$

$n = 4$

43. $\dfrac{8}{15} = \dfrac{n}{6}$

$8 \cdot 6 = 15n$

$48 = 15n$

$3.2 = n$

45. $\dfrac{24}{n} = \dfrac{60}{96}$

$24 \cdot 96 = 60n$

$2304 = 60n$

$38.4 = n$

47. $\dfrac{3.5}{12.5} = \dfrac{7}{n}$

$3.5n = 12.5 \cdot 7$

$3.5n = 87.5$

$n = 25$

49.

$$\frac{0.05}{12} = \frac{n}{0.6}$$

$$0.05(0.6) = 12n$$

$$0.03 = 12n$$

$$0.0025 = n$$

51.

$$\frac{8}{\frac{1}{3}} = \frac{24}{n}$$

$$8n = \frac{1}{3} \cdot 24$$

$$8n = 8$$

$$n = 1$$

53.

$$\frac{\frac{1}{3}}{\frac{3}{8}} = \frac{\frac{2}{5}}{n}$$

$$\frac{1}{3}n = \frac{3}{8} \cdot \frac{2}{5}$$

$$\frac{1}{3}n = \frac{3}{20}$$

$$n = \frac{3}{20} \div \frac{1}{3}$$

$$n = \frac{3}{20} \cdot \frac{3}{1}$$

$$n = \frac{9}{20}$$

55.

$$\frac{12}{n} = \frac{\frac{2}{3}}{\frac{6}{9}}$$

$$12 \cdot \frac{6}{9} = \frac{2}{3}n$$

$$8 = \frac{2}{3}n$$

$$n = 8 \div \frac{2}{3}$$

$$n = 8 \cdot \frac{3}{2}$$

$$n = 12$$

57.

$$\frac{n}{1\frac{1}{5}} = \frac{4\frac{1}{6}}{6\frac{2}{3}}$$

$$\left(6\frac{2}{3}\right)n = \left(1\frac{1}{5}\right)\left(4\frac{1}{6}\right)$$

$$\frac{20}{3}n = \frac{6}{5} \cdot \frac{25}{6}$$

$$\frac{20}{3}n = \frac{5}{1}$$

$$n = \frac{5}{1} \div \frac{20}{3}$$

$$n = \frac{5}{1} \cdot \frac{3}{20}$$

$$n = \frac{3}{4}$$

59.

$$\frac{25}{n} = \frac{3}{\frac{7}{30}}$$

$$25 \cdot \frac{7}{30} = 3n$$

$$\frac{35}{6} = 3n$$

$$\frac{35}{18} = n$$

61. 8.01   8.1
  ↑   ↑
  0   <   1 so
  8.01 < 8.1

63. $2\frac{1}{2} = \frac{5}{2} = \frac{15}{6}$

$2\frac{1}{3} = \frac{7}{3} = \frac{14}{6}$

$15 > 14$, so $2\frac{1}{2} > 2\frac{1}{3}$.

65. $5\frac{1}{3} = \frac{16}{3}$

$6\frac{2}{3} = \frac{20}{3}$

$16 < 20$, so $5\frac{1}{3} < 6\frac{2}{3}$

67. $\dfrac{9}{15} = \dfrac{3}{5}$

$\dfrac{9}{3} = \dfrac{15}{5}$

$\dfrac{5}{15} = \dfrac{3}{9}$

$\dfrac{15}{9} = \dfrac{5}{3}$

69. $\dfrac{6}{18} = \dfrac{1}{3}$

$\dfrac{6}{1} = \dfrac{18}{3}$

$\dfrac{3}{18} = \dfrac{1}{6}$

$\dfrac{18}{6} = \dfrac{3}{1}$

71. $\dfrac{a}{b} = \dfrac{c}{d}$

Possible answers include:

$\dfrac{d}{b} = \dfrac{c}{a}$

$\dfrac{a}{c} = \dfrac{b}{d}$

$\dfrac{b}{a} = \dfrac{d}{c}$

73. answers may vary

75. $\dfrac{3.2}{0.3} = \dfrac{n}{1.4}$

$3.2 \cdot 1.4 = 0.3 \cdot n$

$4.48 = 0.3n$

$\dfrac{4.48}{0.3} = \dfrac{0.3n}{0.3}$

$14.9 \approx n$

77. $\dfrac{n}{5.2} = \dfrac{0.08}{6}$

$n \cdot 6 = 5.2 \cdot 0.08$

$6n = 0.416$

$\dfrac{6n}{6} = \dfrac{0.416}{6}$

$n \approx 0.07$

79. $\dfrac{43}{17} = \dfrac{8}{n}$

$43 \cdot n = 17 \cdot 8$

$43n = 136$

$\dfrac{43n}{43} = \dfrac{136}{43}$

$n \approx 3.163$

81. $\dfrac{n}{7} = \dfrac{0}{8}$

$8n = 7 \cdot 0$

$8n = 0$

$n = 0$

83. $\dfrac{n}{1150} = \dfrac{588}{483}$

$483n = 1150 \cdot 588$

$483n = 676,200$

$n = 1400$

85. $\dfrac{222}{1515} = \dfrac{37}{n}$

$222n = 1515 \cdot 37$

$222n = 56,055$

$n = 252.5$

Section 5.4 Practice

1. Let n = the wall length in feet.

inches $\rightarrow \dfrac{1}{4} = \dfrac{4\frac{1}{4}}{n} \leftarrow$ inches
feet $\rightarrow$ $\leftarrow$ feet

$1 \cdot n = 4 \cdot 4\frac{1}{4}$

$n = 4 \cdot \dfrac{17}{4}$

$n = 17$

The wall represented by a $4\frac{1}{4}$-inch line on the blueprint is 17 feet long.

2. Let n = the number of gallons of gas that can be treated by 16 ounces of alcohol.

alcohol $\rightarrow \dfrac{3}{14} = \dfrac{16}{n} \leftarrow$ alcohol
gas $\rightarrow$ $\leftarrow$ gas

$3n = 14 \cdot 16$

$3n = 224$

$n = \dfrac{224}{3}$ or $74\frac{2}{3}$

A 16-ounce bottle of alcohol can treat $74\frac{2}{3}$ or $74.\overline{6}$ gallons of gas.

3. The area of the wall is $260 \cdot 4 = 1040$ square feet. Let n = the amount of paint needed to cover the retaining wall.

   $$\text{paint} \rightarrow \frac{1}{400} = \frac{n}{1040} \leftarrow \text{paint}$$
   $$\text{wall} \rightarrow \qquad \qquad \leftarrow \text{wall}$$
   $$1040 = 400n$$
   $$\frac{1040}{400} = n$$
   $$\frac{13}{5} = n$$
   $$2\frac{3}{5} = n$$

   Rounding up to the nearest whole gallon, 3 gallons of paint are needed to cover the wall.

Exercise Set 5.4

1. Let x be the number of baskets made.

   $$\text{baskets} \rightarrow \frac{45}{100} = \frac{x}{800} \leftarrow \text{baskets}$$
   $$\text{attempts} \rightarrow \qquad \qquad \leftarrow \text{attempts}$$
   $$45 \cdot 800 = 100 \cdot x$$
   $$36,000 = 100x$$
   $$\frac{36,000}{100} = \frac{100x}{100}$$
   $$360 = x$$

   He made 360 baskets.

3. Let x be the number of minutes.

   $$\text{minutes} \rightarrow \frac{30}{4} = \frac{x}{22} \leftarrow \text{minutes}$$
   $$\text{pages} \rightarrow \qquad \qquad \leftarrow \text{pages}$$
   $$30 \cdot 22 = 4 \cdot x$$
   $$660 = 4x$$
   $$\frac{660}{4} = \frac{4x}{4}$$
   $$165 = x$$

   It will take her 165 minutes to word process and spell check 22 pages.

5. Let x be the number of applications received.

   $$\text{accepted} \rightarrow \frac{2}{7} = \frac{180}{x} \leftarrow \text{accepted}$$
   $$\text{applied} \rightarrow \qquad \qquad \leftarrow \text{applied}$$
   $$2 \cdot x = 7 \cdot 180$$
   $$2x = 1260$$
   $$\frac{2x}{2} = \frac{1260}{2}$$
   $$x = 630$$

   The school received 630 applications.

7. Let x be the length of the wall.

   $$\text{inches} \rightarrow \frac{1}{8} = \frac{2\frac{7}{8}}{x} \leftarrow \text{inches}$$
   $$\text{feet} \rightarrow \qquad \qquad \leftarrow \text{feet}$$
   $$1 \cdot x = 8 \cdot 2\frac{7}{8}$$
   $$x = \frac{8}{1} \cdot \frac{23}{8}$$
   $$x = 23$$

   The wall is 23 feet long.

9. Let x be the number of square feet required.

   $$\text{floor space} \rightarrow \frac{9}{1} = \frac{x}{30} \leftarrow \text{floor space}$$
   $$\text{students} \rightarrow \qquad \qquad \leftarrow \text{students}$$
   $$9 \cdot 30 = 1 \cdot x$$
   $$270 = x$$

   30 students require 270 square feet of floor space.

11. Let x be the number of gallons.

    $$\text{miles} \rightarrow \frac{627}{12.3} = \frac{1250}{x} \leftarrow \text{miles}$$
    $$\text{gallons} \rightarrow \qquad \qquad \leftarrow \text{gallons}$$
    $$627 \cdot x = 12.3 \cdot 1250$$
    $$627x = 15,375$$
    $$\frac{627x}{627} = \frac{15,375}{627}$$
    $$x \approx 25$$

    He can expect to burn 25 gallons.

13. Let x be the distance between Milan and Rome.

    $$\text{kilometers} \rightarrow \frac{30}{1} = \frac{x}{15} \leftarrow \text{kilometers}$$
    $$\text{cm on map} \rightarrow \qquad \qquad \leftarrow \text{cm on map}$$
    $$30 \cdot 15 = 1 \cdot x$$
    $$450 = x$$

    Milan and Rome are 450 kilometers apart.

15. Let x be the number of bags.

    $$\text{bags} \rightarrow \frac{1}{3000} = \frac{x}{260 \cdot 180} \leftarrow \text{bags}$$
    $$\text{square feet} \rightarrow \qquad \qquad \leftarrow \text{square feet}$$
    $$1 \cdot 260 \cdot 180 = 3000 \cdot x$$
    $$46,800 = 3000x$$
    $$\frac{46,800}{3000} = \frac{3000x}{3000}$$
    $$15.6 = x$$
    $$16 \approx x$$

    16 bags of fertilizer should be purchased.

17. Let x be the number of hits the player is expected to get.

$$\begin{array}{l} \text{hits} \rightarrow \\ \text{at bats} \rightarrow \end{array} \frac{3}{8} = \frac{x}{40} \begin{array}{l} \leftarrow \text{hits} \\ \leftarrow \text{at bats} \end{array}$$

$$3 \cdot 40 = 8 \cdot x$$
$$120 = 8x$$
$$\frac{120}{8} = \frac{8x}{8}$$
$$15 = x$$

The player would be expected to get 15 hits.

19. Let x be the number that prefer Coke.

$$\begin{array}{l} \text{Coke} \rightarrow \\ \text{Total} \rightarrow \end{array} \frac{2}{3} = \frac{x}{40} \begin{array}{l} \leftarrow \text{Coke} \\ \leftarrow \text{Total} \end{array}$$

$$2 \cdot 40 = 3 \cdot x$$
$$80 = 3x$$
$$\frac{80}{3} = \frac{3x}{3}$$
$$27 \approx x$$

About 27 people are likely to prefer Coke.

21. Let x be the number of applications she should expect.

$$\begin{array}{l} \text{applications} \rightarrow \\ \text{ounces} \rightarrow \end{array} \frac{4}{3} = \frac{x}{14} \begin{array}{l} \leftarrow \text{applications} \\ \leftarrow \text{ounces} \end{array}$$

$$4 \cdot 14 = 3 \cdot x$$
$$56 = 3x$$
$$\frac{56}{3} = \frac{3x}{3}$$
$$18\frac{2}{3} = x$$

She should expect 18 applications from the 14-ounce bottle.

23. Let x be the number of weeks.

$$\begin{array}{l} \text{reams} \rightarrow \\ \text{weeks} \rightarrow \end{array} \frac{5}{3} = \frac{8}{x} \begin{array}{l} \leftarrow \text{reams} \\ \leftarrow \text{weeks} \end{array}$$

$$5 \cdot x = 3 \cdot 8$$
$$5x = 24$$
$$\frac{5x}{5} = \frac{24}{5}$$
$$x = 4.8$$
$$x \approx 5$$

The case is likely to last 5 weeks.

25. Let x be the number of servings he can make.

$$\begin{array}{l} \text{milk} \rightarrow \\ \text{servings} \rightarrow \end{array} \frac{1\frac{1}{2}}{4} = \frac{4}{x} \begin{array}{l} \leftarrow \text{milk} \\ \leftarrow \text{servings} \end{array}$$

$$1\frac{1}{2} \cdot x = 4 \cdot 4$$
$$\frac{3}{2}x = 16$$
$$\frac{2}{3} \cdot \frac{3}{2}x = \frac{2}{3} \cdot 16$$
$$x = \frac{32}{3} = 10\frac{2}{3}$$

He can make $10\frac{2}{3}$ servings.

27. Let x be the time to reach the restaurant (in seconds).

$$\begin{array}{l} \text{distance} \rightarrow \\ \text{time} \rightarrow \end{array} \frac{800}{60} = \frac{500}{x} \begin{array}{l} \leftarrow \text{distance} \\ \leftarrow \text{time} \end{array}$$

$$800 \cdot x = 60 \cdot 500$$
$$800x = 30,000$$
$$\frac{800x}{800} = \frac{30,000}{800}$$
$$x = 37.5$$

It will take 37.5 seconds to reach the restaurant.

29. a. Let x be the number of teaspoons of granules needed.

$$\begin{array}{l} \text{water} \rightarrow \\ \text{granules} \rightarrow \end{array} \frac{25}{1} = \frac{450}{x} \begin{array}{l} \leftarrow \text{water} \\ \leftarrow \text{granules} \end{array}$$

$$25 \cdot x = 1 \cdot 450$$
$$25x = 450$$
$$\frac{25x}{25} = \frac{450}{25}$$
$$x = 18$$

18 teaspoons of granules are needed.

b. Let x be the number of tablespoons of granules needed.

$$\begin{array}{l} \text{tsp} \rightarrow \\ \text{tbsp} \rightarrow \end{array} \frac{3}{1} = \frac{18}{x} \begin{array}{l} \leftarrow \text{tsp} \\ \leftarrow \text{tbsp} \end{array}$$

$$3 \cdot x = 1 \cdot 18$$
$$3x = 18$$
$$\frac{3x}{3} = \frac{18}{3}$$
$$x = 6$$

6 tablespoons of granules are needed.

31. Let x be the number of people.

square feet $\to \dfrac{625}{1} = \dfrac{3750}{x} \leftarrow$ square feet
people $\to$ people

$$625 \cdot x = 1 \cdot 3750$$
$$\dfrac{625x}{625} = \dfrac{3750}{625}$$
$$x = 6$$

It will provide enough oxygen for 6 people.

33. Let x be the estimated head-to-toe height of the Statue of Liberty.

height $\to \dfrac{x}{42} = \dfrac{5\frac{1}{3}}{2} \leftarrow$ height
arm length $\to$ $\leftarrow$ arm length

$$x \cdot 2 = 42 \cdot 5\dfrac{1}{3}$$
$$2x = 42 \cdot \dfrac{16}{3}$$
$$2x = 224$$
$$\dfrac{2x}{2} = \dfrac{224}{2}$$
$$x = 112$$

The estimated height is 112 feet.

$$112 - 111\dfrac{1}{12} = \dfrac{11}{12}$$

The difference is $\dfrac{11}{12}$ foot or 11 inches.

35. Let x be the number of milligrams.

milligrams $\to \dfrac{72}{3.5} = \dfrac{x}{5} \leftarrow$ milligrams
ounces $\to$ $\leftarrow$ ounces

$$72 \cdot 5 = 3.5 \cdot x$$
$$360 = 3.5x$$
$$\dfrac{360}{3.5} = \dfrac{3.5x}{3.5}$$
$$102.9 \approx x$$

There are about 102.9 milligrams of cholesterol in 5 ounces of lobster.

37. Let x be the estimated height of the Empire State Building.

height $\to \dfrac{x}{102} = \dfrac{881}{72} \leftarrow$ height
stories $\to$ $\leftarrow$ stories

$$x \cdot 72 = 102 \cdot 881$$
$$72x = 89,862$$
$$\dfrac{72x}{72} = \dfrac{89,862}{72}$$
$$x \approx 1248$$

The height of the Empire State Building is approximately 1248 feet.

39. Let x be the number of visits needing a prescription for medication.

medication $\to \dfrac{7}{10} = \dfrac{x}{620} \leftarrow$ medication
total $\to$ $\leftarrow$ total

$$7 \cdot 620 = 10 \cdot x$$
$$4340 = 10x$$
$$\dfrac{4340}{10} = \dfrac{10x}{10}$$
$$434 = x$$

Expect 434 emergency room visits to need a prescription for medication.

41. Let x be the number expected to have worked in the restaurant industry.

restaurant $\to \dfrac{x}{84} = \dfrac{1}{3} \leftarrow$ restaurant
workers $\to$ $\leftarrow$ workers

$$x \cdot 3 = 84 \cdot 1$$
$$3x = 84$$
$$\dfrac{3x}{3} = \dfrac{84}{3}$$
$$x = 28$$

You would expect 28 of the workers to have worked in the restaurant industry.

43. Let x be the cups of salt.

ice $\to \dfrac{5}{1} = \dfrac{12}{x} \leftarrow$ ice
salt $\to$ $\leftarrow$ salt

$$5 \cdot x = 1 \cdot 12$$
$$\dfrac{5x}{5} = \dfrac{12}{5}$$
$$x = 2.4$$

Mix 2.4 cups of salt with the ice.

45. a. Let x be the number of gallons of oil needed.

oil $\to \dfrac{x}{5} = \dfrac{1}{50} \leftarrow$ oil
gas $\to$ $\leftarrow$ gas

$$x \cdot 50 = 5 \cdot 1$$
$$50x = 5$$
$$\dfrac{50x}{50} = \dfrac{5}{50}$$
$$x = \dfrac{1}{10} = 0.1$$

0.1 gallon of oil is needed.

b. Let x be the number of fluid ounces.

gallons $\to \dfrac{1}{128} = \dfrac{0.1}{x} \leftarrow$ gallons
fluid ounces $\to$ $\leftarrow$ fluid ounces

$$1 \cdot x = 128 \cdot 0.1$$
$$x = 12.8$$

0.1 gallon is approximately 13 fluid ounces.

47. a. Let x be the milligrams of medicine.

milligrams $\rightarrow \dfrac{x}{275} = \dfrac{150}{20} \leftarrow$ milligrams
pounds $\rightarrow   \leftarrow$ pounds

$$x \cdot 20 = 275 \cdot 150$$
$$20x = 41{,}250$$
$$\dfrac{20x}{20} = \dfrac{41{,}250}{20}$$
$$x = 2062.5$$

The daily dose is 2062.5 milligrams.

b. $500 \times \dfrac{24}{8} = 500 \times 3 = 1500$

No, he is not receiving the proper dosage.

49. $3\overline{)15}^{\,5}$

$15 = 3 \cdot 5$

51. $\begin{array}{r} 5 \\ 2\overline{)10} \\ 2\overline{)20} \end{array}$

$20 = 2 \cdot 2 \cdot 5 = 2^2 \cdot 5$

53. $\begin{array}{r} 5 \\ 5\overline{)25} \\ 2\overline{)50} \\ 2\overline{)100} \\ 2\overline{)200} \end{array}$

$200 = 2 \cdot 2 \cdot 2 \cdot 5 \cdot 5 = 2^3 \cdot 5^2$

55. $\begin{array}{r} 2 \\ 2\overline{)4} \\ 2\overline{)8} \\ 2\overline{)16} \\ 2\overline{)32} \end{array}$

$32 = 2 \cdot 2 \cdot 2 \cdot 2 \cdot 2 = 2^5$

57. Let x be the number of ml.

mg $\rightarrow \dfrac{15}{1} = \dfrac{12}{x} \leftarrow$ mg
ml $\rightarrow   \leftarrow$ ml

$$15 \cdot x = 1 \cdot 12$$
$$15x = 12$$
$$\dfrac{15x}{15} = \dfrac{12}{15}$$
$$x = \dfrac{4}{5} = 0.8$$

0.8 ml of the medicine should be administered.

59. Let x be the number of ml.

mg $\rightarrow \dfrac{8}{1} = \dfrac{10}{x} \leftarrow$ mg
ml $\rightarrow   \leftarrow$ ml

$$8 \cdot x = 1 \cdot 10$$
$$8x = 10$$
$$\dfrac{8x}{8} = \dfrac{10}{8}$$
$$x = 1.25$$

1.25 ml of the medicine should be administered.

61. 11 muffins are approximately 1 dozen (12) muffins.

$$1.5 \cdot 8 = 12$$

Approximately 12 cups of milk will be needed.

63. feet $\rightarrow \dfrac{7}{60} = \dfrac{n}{40} \leftarrow$ feet
pounds $\rightarrow   \leftarrow$ pounds

$$7 \cdot 40 = 60 \cdot n$$
$$280 = 60n$$
$$\dfrac{280}{60} = \dfrac{60n}{60}$$
$$4\dfrac{2}{3} = n$$

The distance is $4\dfrac{2}{3}$ feet.

65. answers may vary

Chapter 5 Vocabulary Check

1. A <u>ratio</u> is the quotient of two numbers. It can be written as a fraction, using a colon, or using the word *to*.

2. $\dfrac{x}{2} = \dfrac{7}{16}$ is an example of a <u>proportion</u>.

3. A <u>unit rate</u> is a rate with a denominator of 1.

4. A <u>unit price</u> is a "money per item" unit rate.

5. A <u>rate</u> is used to compare different kinds of quantities.

6. In the proportion $\dfrac{x}{2} = \dfrac{7}{16}$, $x \cdot 16$ and $2 \cdot 7$ are called <u>cross products</u>.

7. If cross products are <u>equal</u> the proportion is true.

8. If cross products are <u>not equal</u> the proportion is false.

Chapter 5 Review

1. The ratio of 23 to 37 is $\dfrac{23}{37}$.

2. The ratio of 14 to 51 is $\dfrac{14}{51}$.

3. The ratio of 6000 people to 4800 people is
$$\frac{6000 \text{ people}}{4800 \text{ people}} = \frac{6000}{4800} = \frac{1200 \cdot 5}{1200 \cdot 4} = \frac{5}{4}.$$

4. The ratio of \$121 to \$143 is
$$\frac{\$121}{\$143} = \frac{121}{143} = \frac{11 \cdot 11}{11 \cdot 13} = \frac{11}{13}.$$

5. The ratio of 3.5 centimeters to 7.5 centimeters is
$$\frac{3.5 \text{ centimeters}}{7.5 \text{ centimeters}} = \frac{3.5}{7.5}$$
$$= \frac{10 \cdot 3.5}{10 \cdot 7.5}$$
$$= \frac{35}{75}$$
$$= \frac{5 \cdot 7}{5 \cdot 15}$$
$$= \frac{7}{15}.$$

6. The ratio of 4.25 yards to 8.75 yards is
$$\frac{4.25 \text{ yards}}{8.75 \text{ yards}} = \frac{4.25}{8.75}$$
$$= \frac{100 \cdot 4.25}{100 \cdot 8.75}$$
$$= \frac{425}{875}$$
$$= \frac{25 \cdot 17}{25 \cdot 35}$$
$$= \frac{17}{35}.$$

7. The ratio of $2\dfrac{1}{4}$ to $4\dfrac{3}{8}$ is
$$\frac{2\frac{1}{4}}{4\frac{3}{8}} = 2\frac{1}{4} \div 4\frac{3}{8}$$
$$= \frac{9}{4} \div \frac{35}{8}$$
$$= \frac{9}{4} \cdot \frac{8}{35}$$
$$= \frac{72}{140}$$
$$= \frac{4 \cdot 18}{4 \cdot 35}$$
$$= \frac{18}{35}.$$

8. The ratio of $3\dfrac{1}{2}$ to $2\dfrac{7}{10}$ is
$$\frac{3\frac{1}{2}}{2\frac{7}{10}} = 3\frac{1}{2} \div 2\frac{7}{10}$$
$$= \frac{7}{2} \div \frac{27}{10}$$
$$= \frac{7}{2} \cdot \frac{10}{27}$$
$$= \frac{70}{54}$$
$$= \frac{2 \cdot 35}{2 \cdot 27}$$
$$= \frac{35}{27}.$$

9. **a.** 9 of the top 25 movies were rated PG.

   b. The ratio of top 25 PG-rated movies to total movies was $\dfrac{9}{25}$.

10. **a.** 3 of the top 25 movies were rated R.

    b. The ratio of top 25 R-rated movies to total movies was $\dfrac{3}{25}$.

11. $\dfrac{8 \text{ stillborn births}}{1000 \text{ live births}} = \dfrac{8 \text{ still born births}}{8 \cdot 125 \text{ live births}}$
$$= \dfrac{1 \text{ stillborn birth}}{125 \text{ live births}}$$

12. $\dfrac{6 \text{ profesors}}{20 \text{ assistants}} = \dfrac{2 \cdot 3 \text{ professors}}{2 \cdot 10 \text{ assistants}} = \dfrac{3 \text{ professors}}{10 \text{ assistants}}$

13. $\dfrac{15 \text{ pages}}{6 \text{ minutes}} = \dfrac{3 \cdot 5 \text{ pages}}{3 \cdot 2 \text{ minutes}} = \dfrac{5 \text{ pages}}{2 \text{ minutes}}$

14. $\dfrac{8 \text{ computers}}{6 \text{ hours}} = \dfrac{2 \cdot 4 \text{ computers}}{2 \cdot 3 \text{ hours}} = \dfrac{4 \text{ computers}}{3 \text{ hours}}$

15. $468 \div 9 = 52$, so $\dfrac{468 \text{ miles}}{9 \text{ hours}} = 52 \text{ miles/hour}$.

16. $180 \div 12 = 15$, so $\dfrac{180 \text{ feet}}{12 \text{ seconds}} = 15 \text{ feet/second}$.

17. $27.84 \div 4 = 1.74$, so $\dfrac{\$27.84}{4 \text{ CDs}} = \$6.96/\text{CD}$.

18. $8 \div 6 = \dfrac{4}{3} = 1\dfrac{1}{3}$, so $\dfrac{8 \text{ gallons}}{6 \text{ acres}} = 1\dfrac{1}{3} \text{ gallons/acre}$.

19. $234 \div 5 = 46.8$, so $\dfrac{\$234}{5 \text{ courses}} = \$46.80/\text{course}$.

20. $104 \div 8 = 13$, so $\dfrac{104 \text{ bushels}}{8 \text{ trees}} = 13 \text{ bushels/tree}$.

21. $0.99 \div 8 \approx 0.124$, so the unit price for the 8-ounce size is $\dfrac{\$0.99}{8 \text{ ounces}} \approx \$0.124/\text{ounce}$.

$1.69 \div 12 \approx 0.141$, so the unit price for the 12-ounce size is $\dfrac{\$1.69}{12 \text{ ounces}} \approx \$0.141/\text{ounce}$.

The 8-ounce size is the better buy.

22. $1.49 \div 18 \approx 0.083$, so the unit price for the 18-ounce size is $\dfrac{\$1.49}{18 \text{ ounces}} \approx \$0.083/\text{ounce}$.

$2.39 \div 28 \approx 0.085$, so the unit price for the 28-ounce size is $\dfrac{\$2.39}{28 \text{ ounces}} \approx \$0.085/\text{ounce}$.

The 18-ounce size is the better buy.

23. $0.59 \div 16 \approx 0.037$, so the unit price for the 16-ounce size is $\dfrac{\$0.59}{16 \text{ ounces}} \approx \$0.037/\text{ounce}$.

$1.69 \div 64 \approx 0.026$, so the unit price for the 64-ounce size is $\dfrac{\$1.69}{64 \text{ ounces}} \approx \$0.026/\text{ounce}$.

$2.29 \div 128 \approx 0.018$, so the unit price for the 1-gallon size is $\dfrac{\$2.29}{128 \text{ ounces}} \approx \$0.018/\text{ounce}$.

The 1-gallon size is the better buy.

24. $0.59 \div 12 \approx 0.049$, so the unit price for the 12-ounce size is $\dfrac{\$0.59}{12 \text{ ounces}} \approx \$0.049/\text{ounce}$.

$0.79 \div 16 \approx 0.049$, so the unit price for the 16-ounce size is $\dfrac{\$0.79}{16 \text{ ounces}} \approx \$0.049/\text{ounce}$.

$1.19 \div 32 \approx 0.037$, so the unit price for the 32-ounce size is $\dfrac{\$1.19}{32 \text{ ounces}} \approx \$0.037/\text{ounce}$.

The 32-ounce size is the better buy.

25. $\dfrac{20 \text{ men}}{14 \text{ women}} = \dfrac{10 \text{ men}}{7 \text{ women}}$

26. $\dfrac{50 \text{ tries}}{4 \text{ successes}} = \dfrac{25 \text{ tries}}{2 \text{ successes}}$

27. $\dfrac{16 \text{ sandwiches}}{8 \text{ players}} = \dfrac{2 \text{ sandwiches}}{1 \text{ player}}$

28. $\dfrac{12 \text{ tires}}{3 \text{ cars}} = \dfrac{4 \text{ tires}}{1 \text{ car}}$

29. $\dfrac{21}{8} \overset{?}{=} \dfrac{14}{6}$

$21 \cdot 6 \overset{?}{=} 8 \cdot 14$

$126 \neq 112$

The proportion is false.

30. $\dfrac{3}{5} \overset{?}{=} \dfrac{60}{100}$

$3 \cdot 100 \overset{?}{=} 5 \cdot 60$

$300 = 300$

The proportion is true.

31. $\dfrac{3.75}{3} \overset{?}{=} \dfrac{7.5}{6}$

$3.75 \cdot 6 \overset{?}{=} 3 \cdot 7.5$

$22.5 = 22.5$

The proportion is true.

32. $\dfrac{3.1}{6.2} \overset{?}{=} \dfrac{0.8}{0.16}$

$3.1 \cdot 0.16 \overset{?}{=} 6.2 \cdot 0.8$

$0.496 \neq 4.96$

The proportion is false.

33. $\dfrac{n}{6} = \dfrac{15}{18}$

$18n = 6 \cdot 15$

$18n = 90$

$n = 5$

34. $\dfrac{n}{9} = \dfrac{5}{3}$

$3n = 9 \cdot 5$

$3n = 45$

$n = 15$

35. $\dfrac{4}{13} = \dfrac{10}{n}$

$4n = 13 \cdot 10$

$4n = 130$

$n = 32.5$

36. $\dfrac{8}{5} = \dfrac{9}{n}$

$8n = 5 \cdot 9$

$8n = 45$

$n = 5.625$

37. $\dfrac{8}{\frac{3}{2}} = \dfrac{n}{6}$

$8 \cdot 6 = \dfrac{3}{2}n$

$48 = \dfrac{3}{2}n$

$n = 48 \div \dfrac{3}{2}$

$n = 48 \cdot \dfrac{2}{3}$

$n = 32$

38. $\dfrac{9}{2} = \dfrac{n}{\frac{3}{2}}$

$9 \cdot \dfrac{3}{2} = 2n$

$\dfrac{27}{2} = 2n$

$n = \dfrac{27}{2} \div 2$

$n = \dfrac{27}{2} \cdot \dfrac{1}{2}$

$n = \dfrac{27}{4} = 6\dfrac{3}{4}$

39. $\dfrac{27}{\frac{9}{4}} = \dfrac{n}{5}$

$27 \cdot 5 = \dfrac{9}{4}n$

$135 = \dfrac{9}{4}n$

$n = 135 \div \dfrac{9}{4}$

$n = 135 \cdot \dfrac{4}{9}$

$n = 60$

40. $\dfrac{6}{\frac{5}{2}} = \dfrac{n}{3}$

$6 \cdot 3 = \dfrac{5}{2}n$

$18 = \dfrac{5}{2}n$

$n = 18 \div \dfrac{5}{2}$

$n = 18 \cdot \dfrac{2}{5}$

$n = \dfrac{36}{5} = 7\dfrac{1}{5}$

41. $\dfrac{0.4}{n} = \dfrac{2}{4.7}$

$0.4 \cdot 4.7 = 2n$

$1.88 = 2n$

$0.94 = n$

42. $\dfrac{7.2}{n} = \dfrac{6}{0.3}$

$7.2 \cdot 0.3 = 6n$

$2.16 = 6n$

$0.36 = n$

43.
$$\frac{n}{4\frac{1}{2}} = \frac{2\frac{1}{10}}{8\frac{2}{5}}$$

$$8\frac{2}{5} \cdot n = 4\frac{1}{2} \cdot 2\frac{1}{10}$$

$$\frac{42}{5}n = \frac{9}{2} \cdot \frac{21}{10}$$

$$\frac{42}{5}n = \frac{189}{20}$$

$$n = \frac{189}{20} \div \frac{42}{5}$$

$$n = \frac{189}{20} \cdot \frac{5}{42}$$

$$n = \frac{945}{840} = \frac{105 \cdot 9}{105 \cdot 8} = \frac{9}{8} = 1\frac{1}{8}$$

44.
$$\frac{n}{4\frac{2}{7}} = \frac{3\frac{1}{9}}{9\frac{1}{3}}$$

$$9\frac{1}{3} \cdot n = 4\frac{2}{7} \cdot 3\frac{1}{9}$$

$$\frac{28}{3}n = \frac{30}{7} \cdot \frac{28}{9}$$

$$\frac{28}{3}n = \frac{840}{63}$$

$$\frac{28}{3}n = \frac{40}{3}$$

$$n = \frac{40}{3} \div \frac{28}{3}$$

$$n = \frac{40}{3} \cdot \frac{3}{28}$$

$$n = \frac{120}{84} = \frac{12 \cdot 10}{12 \cdot 7} = \frac{10}{7} = 1\frac{3}{7}$$

45. Let n = number of completed passes.

completed $\rightarrow$
attempted $\rightarrow$
$$\frac{n}{32} = \frac{3}{7}$$
$\leftarrow$ completed
$\leftarrow$ attempted

$$7n = 32 \cdot 3$$
$$7n = 96$$
$$n = \frac{96}{7} \approx 13.7$$

He completed about 14 passes.

46. Let n = number of attempted passes.

completed $\rightarrow$
attempted $\rightarrow$
$$\frac{15}{n} = \frac{3}{7}$$
$\leftarrow$ completed
$\leftarrow$ attempted

$$15 \cdot 7 = 3n$$
$$105 = 3n$$
$$35 = n$$

He attempted 35 passes.

47. The area of the garden is
$180 \cdot 175 = 31{,}500$ square feet.
Let n = number of bags.

bags $\rightarrow$
area $\rightarrow$
$$\frac{n}{31{,}500} = \frac{1}{4000}$$
$\leftarrow$ bags
$\leftarrow$ area

$$4000n = 31{,}500$$
$$n = 7.875$$
8 bags should be purchased.

48. The area of the garden is
$250 \cdot 250 = 62{,}500$ square feet.
Let n = number of bags.

bags $\rightarrow$
area $\rightarrow$
$$\frac{n}{62{,}500} = \frac{1}{4000}$$
$\leftarrow$ bags
$\leftarrow$ area

$$4000n = 62{,}500$$
$$n = 15.625$$
16 bags should be purchased.

49. Let n = number of miles Tom can drive.

miles $\rightarrow$
gallons $\rightarrow$
$$\frac{n}{1.5} = \frac{420}{11}$$
$\leftarrow$ miles
$\leftarrow$ gallons

$$11n = 1.5 \cdot 420$$
$$11n = 630$$
$$n = \frac{630}{11} \approx 57.27$$

Since $57.27 < 65$, Tom cannot drive to the gas station.

50. Let n = number of gallons needed.

miles $\rightarrow$
gallons $\rightarrow$
$$\frac{3000}{n} = \frac{420}{11}$$
$\leftarrow$ miles
$\leftarrow$ gallons

$$3000 \cdot 11 = 420n$$
$$33{,}000 = 420n$$
$$\frac{33{,}000}{420} = n$$
$$\frac{550}{7} = n$$
$$78.57 \approx n$$

Tom needs about 79 gallons of gas.

51. Let n = value of home, in dollars.

tax $\rightarrow$
value $\rightarrow$
$$\frac{627.90}{n} = \frac{1.15}{100}$$
$\leftarrow$ tax
$\leftarrow$ value

$$627.90 \cdot 100 = 1.15n$$
$$62{,}790 = 1.15n$$
$$54{,}600 = n$$

The value of the home is $54,600.

52. Let n = tax on the town house, in dollars.

$$\begin{array}{r} \text{tax} \rightarrow \\ \text{value} \rightarrow \end{array} \frac{n}{89,000} = \frac{1.15}{100} \begin{array}{l} \leftarrow \text{tax} \\ \leftarrow \text{value} \end{array}$$

$$100n = 89,000 \cdot 1.15$$
$$100n = 102,350$$
$$n = 1023.50$$

The tax is $1023.50.

53. Let n = length of wall, in feet.

$$\begin{array}{r} \text{blueprint} \rightarrow \\ \text{wall} \rightarrow \end{array} \frac{3\frac{3}{8}}{n} = \frac{1}{12} \begin{array}{l} \leftarrow \text{blueprint} \\ \leftarrow \text{wall} \end{array}$$

$$3\frac{3}{8} \cdot 12 = n$$
$$\frac{27}{8} \cdot 12 = n$$
$$\frac{81}{2} = n$$

The wall is $\frac{81}{2}$ or $40\frac{1}{2}$ feet long.

54. Let n = blueprint measurement, in inches.

$$\begin{array}{r} \text{blueprint} \rightarrow \\ \text{wall} \rightarrow \end{array} \frac{n}{99} = \frac{1}{12} \begin{array}{l} \leftarrow \text{blueprint} \\ \leftarrow \text{wall} \end{array}$$

$$12n = 99$$
$$n = \frac{99}{12}$$
$$n = \frac{33}{4} = 8\frac{1}{4}$$

The blueprint measurement is $8\frac{1}{4}$ inches.

55. $\frac{15}{25} = \frac{5 \cdot 3}{5 \cdot 5} = \frac{3}{5}$

56. $\frac{16}{36} = \frac{4 \cdot 4}{4 \cdot 9} = \frac{4}{9}$

57. $\frac{14 \text{ feet}}{28 \text{ feet}} = \frac{14}{28} = \frac{14 \cdot 1}{14 \cdot 2} = \frac{1}{2}$

58. $\frac{25 \text{ feet}}{60 \text{ feet}} = \frac{25}{60} = \frac{5 \cdot 5}{5 \cdot 12} = \frac{5}{12}$

59. $\frac{3 \text{ pints}}{81 \text{ pints}} = \frac{3}{81} = \frac{3 \cdot 1}{3 \cdot 27} = \frac{1}{27}$

60. $\frac{6 \text{ pints}}{48 \text{ pints}} = \frac{6}{48} = \frac{6 \cdot 1}{6 \cdot 8} = \frac{1}{8}$

61. $\frac{2 \text{ teachers}}{18 \text{ students}} = \frac{2 \cdot 1 \text{ teachers}}{2 \cdot 9 \text{ students}} = \frac{1 \text{ teacher}}{9 \text{ students}}$

62. $\frac{6 \text{ nurses}}{24 \text{ patients}} = \frac{6 \cdot 1 \text{ nurses}}{6 \cdot 4 \text{ patients}} = \frac{1 \text{ nurse}}{4 \text{ patients}}$

63. $24 \div 6 = 4$, so $\frac{24 \text{ cups}}{6 \text{ people}} = 4$ cups/person.

64. $18 \div 3 = 6$, so $\frac{18 \text{ toys}}{3 \text{ children}} = 6$ toys/child.

65. $136 \div 4 = 34$, so $\frac{136 \text{ miles}}{4 \text{ hours}} = 34$ miles/hour.

66. $12 \div 6 = 2$, so $\frac{12 \text{ gallons}}{6 \text{ cows}} = 2$ gallons/cow.

67. $4.94 \div 4 = 1.235$, so the unit price for the 4-ounce size is $1.235/ounce.
$9.98 \div 8 \approx 1.248$, so the unit price for the 8-ounce size is about $1.248/ounce.
The 4-ounce size is the better buy.

68. $0.65 \div 12 \approx 0.054$, so the unit price for the 12-ounce size is about $0.054/ounce.
$2.98 \div 64 \approx 0.047$, so the unit price for the 64-ounce size is about 0.047/ounce.
The 64-ounce size is the better buy.

69. $\frac{2 \text{ cups cookie dough}}{30 \text{ cookies}} = \frac{4 \text{ cups cookie dough}}{60 \text{ cookies}}$

70. $\frac{5 \text{ nickels}}{3 \text{ dollars}} = \frac{20 \text{ nickels}}{12 \text{ dollars}}$

71. $\frac{3}{4} \overset{?}{=} \frac{87}{116}$
$3 \cdot 116 \overset{?}{=} 4 \cdot 87$
$348 = 348$
The proportion is true.

72. $\frac{2}{3} \overset{?}{=} \frac{4}{9}$
$2 \cdot 9 \overset{?}{=} 3 \cdot 4$
$18 \neq 12$
The proportion is false.

73.
$$\frac{3}{n} = \frac{15}{8}$$
$$3 \cdot 8 = 15n$$
$$24 = 15n$$
$$1.6 = n$$

74.
$$\frac{6}{n} = \frac{30}{24}$$
$$6 \cdot 24 = 30n$$
$$144 = 30n$$
$$4.8 = n$$

75.
$$\frac{42}{5} = \frac{n}{10}$$
$$42 \cdot 10 = 5n$$
$$420 = 5n$$
$$84 = n$$

76.
$$\frac{5}{4} = \frac{n}{20}$$
$$5 \cdot 20 = 4n$$
$$100 = 4n$$
$$25 = n$$

77. Let n = the monthly payment, in dollars.
$$\begin{array}{l} \text{payment} \rightarrow \\ \text{borrowed} \rightarrow \end{array} \frac{n}{23,000} = \frac{39.75}{1500} \begin{array}{l} \leftarrow \text{payment} \\ \leftarrow \text{borrowed} \end{array}$$
$$1500n = 23,000 \cdot 39.75$$
$$1500n = 914,250$$
$$n = \frac{914,250}{1500} = 609.5$$
The monthly payment is $609.50.

78. Let n = the monthly payment, in dollars.
$$\begin{array}{l} \text{payment} \rightarrow \\ \text{borrowed} \rightarrow \end{array} \frac{n}{18,000} = \frac{39.75}{1500} \begin{array}{l} \leftarrow \text{payment} \\ \leftarrow \text{borrowed} \end{array}$$
$$1500n = 18,000 \cdot 39.75$$
$$1500n = 715,500$$
$$n = \frac{715,500}{1500} = 477$$
The monthly payment is $477.

79. Let n = yield per year, in dollars.
$$\begin{array}{l} \text{investment} \rightarrow \\ \text{yield} \rightarrow \end{array} \frac{1350}{n} = \frac{1200}{152} \begin{array}{l} \leftarrow \text{investment} \\ \leftarrow \text{yield} \end{array}$$
$$1350 \cdot 152 = 1200n$$
$$205,200 = 1200n$$
$$171 = n$$
The investment will yield $171 in one year.

80. Let n = yield per year, in dollars.
$$\begin{array}{l} \text{investment} \rightarrow \\ \text{yield} \rightarrow \end{array} \frac{750}{n} = \frac{1200}{152} \begin{array}{l} \leftarrow \text{investment} \\ \leftarrow \text{yield} \end{array}$$
$$750 \cdot 152 = 1200n$$
$$114,000 = 1200n$$
$$95 = n$$
The investment will yield $95 in one year.

Chapter 5 Test

1. $\dfrac{\$75}{\$10} = \dfrac{75}{10} = \dfrac{5 \cdot 15}{5 \cdot 2} = \dfrac{15}{2}$

2. $\dfrac{4500 \text{ trees}}{6500 \text{ trees}} = \dfrac{4500}{6500} = \dfrac{500 \cdot 9}{500 \cdot 13} = \dfrac{9}{13}$

3. $\dfrac{28 \text{ men}}{4 \text{ women}} = \dfrac{4 \cdot 7 \text{ men}}{4 \cdot 1 \text{ women}} = \dfrac{7 \text{ men}}{1 \text{ woman}}$

4. $\dfrac{9 \text{ inches}}{30 \text{ days}} = \dfrac{3 \cdot 3 \text{ inches}}{3 \cdot 10 \text{ days}} = \dfrac{3 \text{ inches}}{10 \text{ days}}$

5. $\dfrac{8.6}{10} = \dfrac{8.6 \cdot 10}{10 \cdot 10} = \dfrac{86}{100} = \dfrac{2 \cdot 43}{2 \cdot 50} = \dfrac{43}{50}$

6.
$$\frac{5\frac{7}{8}}{9\frac{3}{4}} = \frac{\frac{47}{8}}{\frac{39}{4}}$$
$$= \frac{47}{8} \div \frac{39}{4}$$
$$= \frac{47}{8} \cdot \frac{4}{39}$$
$$= \frac{188}{312}$$
$$= \frac{4 \cdot 47}{4 \cdot 78}$$
$$= \frac{47}{78}$$

7. $\dfrac{456 \text{ feet}}{186 \text{ feet}} = \dfrac{456}{186} = \dfrac{6 \cdot 76}{6 \cdot 31} = \dfrac{76}{31}$

8. $\dfrac{650 \text{ kilometers}}{8 \text{ hours}} = \dfrac{650}{8} \text{ km/hr} = 81.25 \text{ km/hr}$

9. $\dfrac{8 \text{ inches}}{12 \text{ hours}} = \dfrac{8}{12} \text{ inch/hour} = \dfrac{2}{3} \text{ inch/hour}$

10. $\dfrac{140 \text{ students}}{5 \text{ teachers}} = \dfrac{140}{5}$ students/teacher

$= 28$ students/teacher

11. $\dfrac{108 \text{ inches}}{12 \text{ seconds}} = \dfrac{108}{12}$ inches/sec $= 9$ inches/sec

12.
$$\begin{array}{r} 0.148 \approx 0.15 \\ 8\overline{)1.190} \\ \underline{-8} \\ 39 \\ \underline{-32} \\ 70 \\ \underline{-64} \\ 6 \end{array}$$

The 8-ounce size costs about \$0.15/ounce.

$$\begin{array}{r} 0.157 \approx 0.16 \\ 12\overline{)1.890} \\ \underline{-1\,2} \\ 69 \\ \underline{-60} \\ 90 \\ \underline{-84} \\ 6 \end{array}$$

The 12-ounce size costs about \$0.16/ounce.
The 8-ounce size is the better buy.

13.
$$\begin{array}{r} 0.0931 \approx 0.093 \\ 16\overline{)1.4900} \\ \underline{-1\,44} \\ 50 \\ \underline{-48} \\ 20 \\ \underline{-16} \\ 4 \end{array}$$

The 16-ounce size costs approximately
\$0.093/ounce.

$$\begin{array}{r} 0.0995 \approx 0.100 \\ 24\overline{)2.3900} \\ \underline{-2\,16} \\ 230 \\ \underline{-216} \\ 140 \\ \underline{-120} \\ 20 \end{array}$$

The 24-ounce size costs approximately
\$0.100/ounce.
The 16-ounce size is the better buy.

14. $\dfrac{28}{16} \overset{?}{=} \dfrac{14}{8}$

$28 \cdot 8 \overset{?}{=} 16 \cdot 14$

$224 = 224$

The proportion is true.

15. $\dfrac{3.6}{2.2} \overset{?}{=} \dfrac{1.9}{1.2}$

$3.6 \cdot 1.2 \overset{?}{=} 2.2 \cdot 1.9$

$4.32 \neq 4.18$

The proportion is false.

16. $\dfrac{25 \text{ computers}}{600 \text{ students}} = \dfrac{1 \text{ computer}}{24 \text{ students}}$

17. $\dfrac{n}{3} = \dfrac{15}{9}$

$9n = 3 \cdot 15$

$9n = 45$

$n = \dfrac{45}{9}$

$n = 5$

18. $\dfrac{8}{n} = \dfrac{11}{6}$

$11n = 8 \cdot 6$

$11n = 48$

$n = \dfrac{48}{11} = 4\dfrac{4}{11}$

19. $\dfrac{\frac{15}{12}}{\frac{3}{7}} = \dfrac{n}{\frac{4}{5}}$

$\dfrac{15}{12} \cdot \dfrac{4}{5} = \dfrac{3}{7}n$

$1 = \dfrac{3}{7}n$

$n = 1 \div \dfrac{3}{7}$

$n = 1 \cdot \dfrac{7}{3}$

$n = \dfrac{7}{3}$

20. $\dfrac{1.5}{5} = \dfrac{2.4}{n}$

$1.5n = 5 \cdot 2.4$

$1.5n = 12$

$n = 12 \div 1.5$

$n = 8$

21.
$$\frac{n}{2\frac{5}{8}} = \frac{1\frac{1}{6}}{3\frac{1}{2}}$$

$$\left(3\frac{1}{2}\right)n = \left(1\frac{1}{6}\right)\left(2\frac{5}{8}\right)$$

$$\frac{7}{2}n = \frac{7}{6} \cdot \frac{21}{8}$$

$$\frac{7}{2}n = \frac{49}{16}$$

$$n = \frac{49}{16} \div \frac{7}{2}$$

$$n = \frac{49}{16} \cdot \frac{2}{7}$$

$$n = \frac{7}{8}$$

22. Let n = length of home, in feet.

$$\begin{array}{l} \text{drawing} \rightarrow \\ \text{reality} \rightarrow \end{array} \frac{11}{n} = \frac{2}{9} \begin{array}{l} \leftarrow \text{drawing} \\ \leftarrow \text{reality} \end{array}$$

$$2n = 11 \cdot 9$$

$$2n = 99$$

$$n = \frac{99}{2}$$

$$n = 49\frac{1}{2}$$

The home is $49\frac{1}{2}$ feet long.

23. Let n = travel time, in hours.

$$\begin{array}{l} \text{distance} \rightarrow \\ \text{time} \rightarrow \end{array} \frac{100}{n} = \frac{80}{3} \begin{array}{l} \leftarrow \text{distance} \\ \leftarrow \text{time} \end{array}$$

$$80n = 100 \cdot 3$$

$$80n = 300$$

$$n = \frac{300}{80}$$

$$n = 3\frac{3}{4}$$

It will take $3\frac{3}{4}$ hours to travel 100 miles.

24. Let n = dose, in grams.

$$\begin{array}{l} \text{dose} \rightarrow \\ \text{body weight} \rightarrow \end{array} \frac{n}{80} = \frac{10}{15} \begin{array}{l} \leftarrow \text{dose} \\ \leftarrow \text{body weight} \end{array}$$

$$15n = 80 \cdot 10$$

$$15n = 800$$

$$n = \frac{800}{15}$$

$$n = 53\frac{1}{3}$$

The standard dose is $53\frac{1}{3}$ grams.

25. Let n = the number of cartons.

$$\begin{array}{l} \text{cartons} \rightarrow \\ \text{hours} \rightarrow \end{array} \frac{n}{8} = \frac{86}{6} \begin{array}{l} \leftarrow \text{cartons} \\ \leftarrow \text{hours} \end{array}$$

$$6n = 8 \cdot 86$$

$$6n = 688$$

$$n = \frac{688}{6}$$

$$n = 114\frac{2}{3}$$

He can pack $114\frac{2}{3}$ cartons in 8 hours.

26. Let n = number of coffee drinkers in the town.

$$\begin{array}{l} \text{drinkers} \rightarrow \\ \text{adults} \rightarrow \end{array} \frac{n}{7900} = \frac{27}{50} \begin{array}{l} \leftarrow \text{drinkers} \\ \leftarrow \text{adults} \end{array}$$

$$50n = 7900 \cdot 27$$

$$50n = 213,300$$

$$n = \frac{213,300}{50}$$

$$n = 4266$$

About 4266 of the adults would be expected to drink coffee every day.

Cumulative Review Chapters 1–5

1. a. $12 - 9 = 3$
   Check: $3 + 9 = 12$

   b. $22 - 7 = 15$
   Check: $15 + 7 = 22$

   c. $35 - 35 = 0$
   Check: $0 + 35 = 35$

   d. $70 - 0 = 70$
   Check: $70 + 0 = 70$

2. a. $20 \cdot 0 = 0$

    b. $20 \cdot 1 = 20$

    c. $0 \cdot 20 = 0$

    d. $1 \cdot 20 = 20$

3. To round 248,982 to the nearest hundred, notice that the digit in the tens place is 8. Since this digit is at least 5, we add 1 to the digit in the hundreds place. 248,982 rounded to the nearest hundred is 249,000.

4. To round 248,982 to the nearest thousand, notice that the digit in the hundreds place is 9. Since this digit is at least 5, we add 1 to the digit in the thousands place. 248,982 rounded to the nearest thousand is 249,000.

5. a.
$$\begin{array}{r} 25 \\ \times\ 8 \\ \hline 200 \end{array}$$

    b.
$$\begin{array}{r} 246 \\ \times\ \ 5 \\ \hline 1230 \end{array}$$

6.
$$\begin{array}{r} 373 \text{ R } 24 \\ 28\overline{)10468} \\ \underline{-84} \\ 206 \\ \underline{-196} \\ 108 \\ \underline{-84} \\ 24 \end{array}$$

7.
$$\begin{array}{r} 187 \\ \times\ 33 \\ \hline 561 \\ 5610 \\ \hline 6171 \end{array}$$
The total cost is \$6171.

8. The total length is
$$\begin{array}{r} \overset{2\,1}{} \\ 4800 \\ 3270 \\ 2761 \\ +\ 5760 \\ \hline 16591 \end{array}$$
16,591 feet

9.
$$\begin{array}{r} 5 \\ 3\overline{)15} \\ 3\overline{)45} \end{array}$$
$$45 = 3 \cdot 3 \cdot 5 = 3^2 \cdot 5$$

10. $\sqrt{64} = 8$, since $8 \cdot 8 = 64$.

11. $\dfrac{12}{20} = \dfrac{4 \cdot 3}{4 \cdot 5} = \dfrac{3}{5}$

12. $9^2 \cdot \sqrt{9} = 9^2 \cdot 3 = 81 \cdot 3 = 243$

13. $\dfrac{3}{4} \cdot \dfrac{8}{5} = \dfrac{3 \cdot 4 \cdot 2}{4 \cdot 5} = \dfrac{3 \cdot 2}{5} = \dfrac{6}{5}$

14. $3\dfrac{3}{8} \cdot 4\dfrac{5}{9} = \dfrac{27}{8} \cdot \dfrac{41}{9} = \dfrac{9 \cdot 3 \cdot 41}{8 \cdot 9} = \dfrac{3 \cdot 41}{8} = \dfrac{123}{8} = 15\dfrac{3}{8}$

15. $\dfrac{6}{13} \cdot \dfrac{26}{30} = \dfrac{6 \cdot 2 \cdot 13}{13 \cdot 6 \cdot 5} = \dfrac{2}{5}$

16. $\dfrac{2}{11} \cdot \dfrac{5}{8} \cdot \dfrac{22}{27} = \dfrac{2 \cdot 5 \cdot 2 \cdot 11}{11 \cdot 2 \cdot 2 \cdot 2 \cdot 27} = \dfrac{5}{2 \cdot 27} = \dfrac{5}{54}$

17. $\dfrac{2}{7} + \dfrac{3}{7} = \dfrac{2+3}{7} = \dfrac{5}{7}$

18. $\dfrac{26}{30} - \dfrac{7}{30} = \dfrac{26-7}{30} = \dfrac{19}{30}$

19. $\dfrac{7}{13} + \dfrac{6}{13} + \dfrac{3}{13} = \dfrac{7+6+3}{13} = \dfrac{16}{13} = 1\dfrac{3}{13}$

20. $\dfrac{7}{10} - \dfrac{3}{10} + \dfrac{4}{10} = \dfrac{7-3+4}{10} = \dfrac{8}{10} = \dfrac{4}{5}$

21. $12 \cdot 1 = 12$ Not a multiple of 9.
$12 \cdot 2 = 24$ Not a multiple of 9.
$12 \cdot 3 = 36$ A multiple of 9.
The LCM of 9 and 12 is 36.

22. The LCD of 25 and 10 is 50.
$$\dfrac{17}{25} + \dfrac{3}{10} = \dfrac{34}{50} + \dfrac{15}{50} = \dfrac{34+15}{50} = \dfrac{49}{50}$$

23. $\dfrac{1}{2} = \dfrac{1 \cdot 12}{2 \cdot 12} = \dfrac{12}{24}$

24. The LCD of $55 = 5 \cdot 11$ and $33 = 3 \cdot 11$ is
$3 \cdot 5 \cdot 11 = 165$.

$$\frac{10}{55} = \frac{10 \cdot 3}{55 \cdot 3} = \frac{30}{165}$$

$$\frac{6}{33} = \frac{6 \cdot 5}{33 \cdot 5} = \frac{30}{165}$$

The fractions are equivalent.

25. The LCD of 11 and 3 is 33.

$$\frac{10}{11} - \frac{2}{3} = \frac{30}{33} - \frac{22}{33} = \frac{30 - 22}{33} = \frac{8}{33}$$

26. The LCD of $24 = 3 \cdot 8$ and $9 = 3 \cdot 3$ is
$3 \cdot 3 \cdot 8 = 72$.

$$\begin{array}{ccc} 17\dfrac{5}{24} & 17\dfrac{15}{72} & 16\dfrac{87}{72} \\[2mm] -9\dfrac{5}{9} & -9\dfrac{40}{72} & -9\dfrac{40}{72} \\[2mm] & & 7\dfrac{47}{72} \end{array}$$

27. The LCD of 12 and 4 is 12.

$$\frac{5}{12} - \frac{1}{4} = \frac{5}{12} - \frac{3}{12} = \frac{2}{12} = \frac{1}{6}$$

$\dfrac{1}{6}$ of an hour remains.

28. $80 \div 8 \cdot 2 + 7 = 10 \cdot 2 + 7 = 20 + 7 = 27$

29. The LCD of 3 and 8 is $3 \cdot 8 = 24$.

$$\begin{array}{cc} 2\dfrac{1}{3} & 2\dfrac{8}{24} \\[2mm] +5\dfrac{3}{8} & +5\dfrac{9}{24} \\[2mm] & 7\dfrac{17}{24} \end{array}$$

30. The average is their sum, divided by 3.
The LCD of 5, 9, and 15 is 45.

$$\left(\frac{3}{5} + \frac{4}{9} + \frac{11}{15}\right) \div 3 = \left(\frac{27}{45} + \frac{20}{45} + \frac{33}{45}\right) \div 3$$

$$= \frac{80}{45} \div \frac{3}{1}$$

$$= \frac{80}{45} \cdot \frac{1}{3}$$

$$= \frac{80}{135}$$

$$= \frac{16}{27}$$

31. The LCD of 10 and 7 is 70.

$$\frac{3}{10} = \frac{21}{70}$$

$$\frac{2}{7} = \frac{20}{70}$$

Since $21 > 20$, $\dfrac{3}{10} > \dfrac{2}{7}$.

32. $28{,}000 \times 500 = 14{,}000{,}000$

33. 1.3 is one and three tenths.

34. Seventy-five thousandths is 0.075.

35. To round 736.2359 to the nearest tenth, notice
that the digit in the hundredths place is 3. Since
this digit is less than 5, we do not add 1 to the
digit in the tenths place, and delete the digits to
the right of the tenths place. 736.2359 rounded
to the nearest tenth is 736.2.

36. To round 736.2359 to the nearest thousandth,
notice that the digit in the ten-thousandths place
is 9. Since this digit is at least 5, we add 1 to the
digit in the thousandths place and delete the digit
to the right of the thousandths place. 736.2359
rounded to the nearest thousandth is 736.236.

37.
$$\begin{array}{r} \overset{1}{23.850} \\ + 1.604 \\ \hline 25.454 \end{array}$$

38.
$$\begin{array}{r} 700.00 \\ - 18.76 \\ \hline 681.24 \end{array}$$

39.
$$\begin{array}{r} 0.283 \\ \times \quad 0.3 \\ \hline 0.0849 \end{array}$$

40.
$$\begin{array}{r} 0.375 \\ 8\overline{)3.000} \\ -2\,4 \\ \hline 60 \\ -56 \\ \hline 40 \\ -40 \\ \hline 0 \end{array}$$

$$\frac{3}{8} = 0.375$$

41.
$$\begin{array}{r} 0.125 \\ 4\overline{)0.500} \\ \underline{-4} \\ 10 \\ \underline{-8} \\ 20 \\ \underline{-20} \\ 0 \end{array}$$

$0.5 \div 4 = 0.125$

Check:
$$\begin{array}{r} 0.125 \\ \times \quad 4 \\ \hline 0.500 \end{array}$$

42. $7.9 = \dfrac{79}{10}$

43. $0.5(8.6 - 1.2) = 0.5(7.4) = 3.7$

44.
$$\frac{n}{4} = \frac{12}{16}$$
$$16n = 4 \cdot 12$$
$$16n = 48$$
$$\frac{16n}{16} = \frac{48}{16}$$
$$n = 3$$

45. Write the fractions as decimals.

$\dfrac{9}{20} = 0.45$

$\dfrac{4}{9} = 0.444....$

Since $0.444... < 0.45 < 0.456$, the numbers in

order from smallest to largest are $\dfrac{4}{9}, \dfrac{9}{20}, 0.456$.

46. $700 \div 5 = 140$, so 700 meters in 5 seconds is
$$\frac{700 \text{ meters}}{5 \text{ seconds}} = \frac{5 \cdot 140 \text{ meters}}{5 \cdot 1 \text{ seconds}}$$
$$= \frac{140 \text{ meters}}{1 \text{ second}}$$
$$= 140 \text{ meters/second}.$$

47. The ratio of 2.6 to 3.1 is $\dfrac{2.6}{3.1}$.

48. The ratio of 7 to 21 is $\dfrac{7}{21} = \dfrac{7 \cdot 1}{7 \cdot 3} = \dfrac{1}{3}$.

49. The ratio of $1\dfrac{1}{2}$ to $7\dfrac{3}{4}$ is $\dfrac{1\frac{1}{2}}{7\frac{3}{4}}$.

50. The ratio of 900 to 9000 is
$$\frac{900}{9000} = \frac{900 \cdot 1}{900 \cdot 10} = \frac{1}{10}.$$

Chapter 6

Section 6.1 Practice

1. $\dfrac{23}{100} = 23\%$

   23% of the students are freshmen.

2. $\dfrac{29}{100} = 29\%$

   29% of executives are in their forties.

3. $89\% = 89(0.01) = 0.89$

4. $2.7\% = 2.7(0.01) = 0.027$

5. $150\% = 150(0.01) = 1.5$

6. $0.69\% = 0.69(0.01) = 0.0069$

7. $800\% = 800(0.01) = 8.00$ or 8

8. $0.19 = 0.19(100\%) = 19\%$

9. $1.75 = 1.75(100\%) = 175\%$

10. $0.044 = 0.044(100\%) = 4.4\%$

11. $0.7 = 0.7(100\%) = 70\%$

Vocabulary and Readiness Check

1. <u>Percent</u> means "per hundred."

2. <u>100%</u> = 1.

3. The % symbol is read as <u>percent</u>.

4. To write a decimal as a *percent*, multiply by 1 in the form of <u>100%</u>.

5. To write a percent as a *decimal*, drop the % symbol and multiply by <u>0.01</u>.

Exercise Set 6.1

1. $\dfrac{96}{100} = 96\%$

   96% of these college students use the Internet.

3. a. $\dfrac{75}{100} = 75\%$

   75% of tart cherries are produced in Michigan.

b. Of every 100 tart cherries,
   $100 - 75 = 25$ are *not* produced in Michigan.

   $\dfrac{25}{100} = 25\%$

   25% of tart cherries are not produced in Michigan.

5. 37 out of 100 adults preferred football.

   $\dfrac{37}{100} = 37\%$

7. 37 of the adults preferred football, while 13 preferred soccer. Thus, $37 + 13 = 50$ preferred football or soccer.

   $\dfrac{50}{100} = 50\%$

9. $41\% = 41(0.01) = 0.41$

11. $6\% = 6(0.01) = 0.06$

13. $100\% = 100(0.01) = 1.00$ or 1

15. $73.6\% = 73.6(0.01) = 0.736$

17. $2.8\% = 2.8(0.01) = 0.028$

19. $0.6\% = 0.6(0.01) = 0.006$

21. $300\% = 300(0.01) = 3.00$ or 3

23. $32.58\% = 32.58(0.01) = 0.3258$

25. $38\% = 38(0.01) = 0.38$

27. $20.2\% = 20.2(0.01) = 0.202$

29. $46.5\% = 46.5(0.01) = 0.465$

31. $0.98 = 0.98(100\%) = 98\%$

33. $3.1 = 3.1(100\%) = 310\%$

35. $29.00 = 29.00(100\%) = 2900\%$

37. $0.003 = 0.003(100\%) = 0.3\%$

39. $0.22 = 0.22(100\%) = 22\%$

41. $5.3 = 5.3(100\%) = 530\%$

43. $0.056 = 0.056(100\%) = 5.6\%$

45. $0.3328 = 0.3328(100\%) = 33.28\%$

47. $3.00 = 3.00(100\%) = 300\%$

49. $0.7 = 0.7(100\%) = 70\%$

51. $0.68 = 0.68(100\%) = 68\%$

53. $0.039 = 0.039(100\%) = 3.9\%$

55. $0.093 = 0.093(100\%) = 9.3\%$

57. $\dfrac{1}{4} = \dfrac{1}{4} \cdot \dfrac{25}{25} = \dfrac{25}{100} = 0.25$

59. $\dfrac{13}{20} = \dfrac{13}{20} \cdot \dfrac{5}{5} = \dfrac{65}{100} = 0.65$

61. $\dfrac{9}{10} = 0.9$

63. a. False: $6.5\% = 6.5 \cdot 0.01 = 0.065 \neq 0.65$

  b. True: $7.8\% = 7.8 \cdot 0.01 = 0.078$

  c. False: $120\% = 120 \cdot 0.01 = 1.2 \neq 0.12$

  d. True: $0.35\% = 0.35 \cdot 0.01 = 0.0035$

65. If the percentages of the four blood types are added, the total will be 100%, since the blood types cover the whole population.
$45\% + 40\% + 11\% = (45 + 40 + 11)\%$
$= 96\%$
$100\% - 96\% = (100 - 96)\% = 4\%$
4% of the U.S. population has the blood type AB.

67. The longest bar is the one for network systems and data communications analysts, so that occupation is predicted to be the fastest growing.

69. $35\% = 35(0.01) = 0.35$

71. answers may vary

Section 6.2 Practice

1. $25\% = 25 \cdot \dfrac{1}{100} = \dfrac{25}{100} = \dfrac{1}{4}$

2. $2.3\% = 2.3 \cdot \dfrac{1}{100} = \dfrac{2.3}{100} \cdot \dfrac{10}{10} = \dfrac{23}{1000}$

3. $225\% = 225 \cdot \dfrac{1}{100} = \dfrac{225}{100} = \dfrac{9}{4} = 2\dfrac{1}{4}$

4. $66\dfrac{2}{3}\% = \dfrac{200}{3} \cdot \dfrac{1}{100} = \dfrac{2}{3}$

5. $8\% = 8 \cdot \dfrac{1}{100} = \dfrac{8}{100} = \dfrac{2}{25}$

6. $\dfrac{1}{2} = \dfrac{1}{2} \cdot 100\% = \dfrac{100}{2}\% = 50\%$

7. $\dfrac{7}{40} = \dfrac{7}{40} \cdot 100\% = \dfrac{700}{40}\% = \dfrac{35}{2}\% = 17\dfrac{1}{2}\%$

8. $2\dfrac{1}{4} = \dfrac{9}{4} = \dfrac{9}{4} \cdot 100\% = \dfrac{900}{4}\% = 225\%$

9. $\dfrac{3}{17} = \dfrac{3}{17} \cdot 100\% = \dfrac{300}{17}\% \approx 17.65\%$

$$
\begin{array}{r}
17.647 \approx 17.65 \\
17\overline{)300.000} \\
\underline{-17} \\
130 \\
\underline{-119} \\
11\,0 \\
\underline{-10\,2} \\
80 \\
\underline{-68} \\
120 \\
\underline{-119} \\
1
\end{array}
$$

10. As a decimal: $22.5\% = 22.5 \cdot 0.01 = 0.225$

As a fraction: $22.5\% = 22.5 \cdot \dfrac{1}{100}$

$= \dfrac{22.5}{100}$

$= \dfrac{22.5}{100} \cdot \dfrac{10}{10}$

$= \dfrac{225}{1000}$

$= \dfrac{9}{40}$

11. $1\dfrac{1}{4} = \dfrac{5}{4} = \dfrac{5}{4} \cdot 100\% = \dfrac{500}{4}\% = 125\%$
This is a 125% increase.

Vocabulary and Readiness Check

1. <u>Percent</u> means "per hundred."

2. <u>100%</u> = 1.

3. To write a decimal or a fraction as a *percent*, multiply by 1 in the form of <u>100%</u>.

4. To write a percent as a *fraction*, drop the % symbol and multiply by $\frac{1}{\underline{100}}$.

5. $\frac{13}{100} = 13\%$

6. $\frac{92}{100} = 92\%$

7. $\frac{87}{100} = 87\%$

8. $\frac{71}{100} = 71\%$

9. $\frac{1}{100} = 1\%$

10. $\frac{2}{100} = 2\%$

Exercise Set 6.2

1. $12\% = 12 \cdot \frac{1}{100} = \frac{12}{100} = \frac{3 \cdot 4}{25 \cdot 4} = \frac{3}{25}$

3. $4\% = 4 \cdot \frac{1}{100} = \frac{4}{100} = \frac{1 \cdot 4}{25 \cdot 4} = \frac{1}{25}$

5. $4.5\% = 4.5 \cdot \frac{1}{100}$
$= \frac{4.5}{100}$
$= \frac{4.5}{100} \cdot \frac{10}{10}$
$= \frac{45}{1000}$
$= \frac{9 \cdot 5}{200 \cdot 5}$
$= \frac{9}{200}$

7. $175\% = 175 \cdot \frac{1}{100} = \frac{175}{100} = \frac{7 \cdot 25}{4 \cdot 25} = \frac{7}{4}$ or $1\frac{3}{4}$

9. $73\% = 73 \cdot \frac{1}{100} = \frac{73}{100}$

11. $12.5\% = 12.5 \cdot \frac{1}{100}$
$= \frac{12.5}{100}$
$= \frac{12.5}{100} \cdot \frac{10}{10}$
$= \frac{125}{1000}$
$= \frac{1 \cdot 125}{8 \cdot 125}$
$= \frac{1}{8}$

13. $6.25\% = 6.25 \cdot \frac{1}{100}$
$= \frac{6.25}{100}$
$= \frac{6.25}{100} \cdot \frac{100}{100}$
$= \frac{625}{10,000}$
$= \frac{1 \cdot 625}{16 \cdot 625}$
$= \frac{1}{16}$

15. $6\% = 6 \cdot \frac{1}{100} = \frac{6}{100} = \frac{3 \cdot 2}{50 \cdot 2} = \frac{3}{50}$

17. $10\frac{1}{3}\% = \frac{31}{3}\% = \frac{31}{3} \cdot \frac{1}{100} = \frac{31}{300}$

19. $22\frac{3}{8}\% = \frac{179}{8}\% = \frac{179}{8} \cdot \frac{1}{100} = \frac{179}{800}$

21. $\frac{3}{4} = \frac{3}{4}(100\%) = \frac{300}{4}\% = 75\%$

23. $\frac{7}{10} = \frac{7}{10} \cdot 100\% = \frac{700}{10}\% = 70\%$

25. $\frac{2}{5} = \frac{2}{5}(100\%) = \frac{200}{5}\% = 40\%$

27. $\dfrac{59}{100} = 59\%$

29. $\dfrac{17}{50} = \dfrac{17}{50}(100\%) = \dfrac{1700}{50}\% = 34\%$

31. $\dfrac{3}{8} = \dfrac{3}{8} \cdot 100\%$

$= \dfrac{300}{8}\%$

$= \dfrac{75 \cdot 4}{2 \cdot 4}\%$

$= \dfrac{75}{2}\%$

$= 37\dfrac{1}{2}\%$

33. $\dfrac{5}{16} = \dfrac{5}{16}(100\%) = \dfrac{500}{16}\% = \dfrac{125}{4}\% = 31\dfrac{1}{4}\%$

35. $1\dfrac{3}{5} = \dfrac{8}{5} = \dfrac{8}{5} \cdot 100\% = \dfrac{800}{5}\% = 160\%$

37. $\dfrac{7}{9} = \dfrac{7}{9}(100\%) = \dfrac{700}{9}\% = 77\dfrac{7}{9}\%$

39. $\dfrac{13}{20} = \dfrac{13}{20} \cdot 100\% = \dfrac{1300}{20}\% = 65\%$

41. $2\dfrac{1}{2} = \dfrac{5}{2} = \dfrac{5}{2}(100\%) = \dfrac{500}{2}\% = 250\%$

43. $1\dfrac{9}{10} = \dfrac{19}{10} = \dfrac{19}{10} \cdot 100\% = \dfrac{1900}{10}\% = 190\%$

45. $\dfrac{7}{11} = \dfrac{7}{11}(100\%) = \dfrac{700}{11}\% \approx 63.64\%$

```
         63.636... ≈ 63.64
    11) 700.000
        −66
        ────
         40
        −33
        ────
         70
        −66
        ────
         40
        −33
        ────
         70
        −66
        ────
          4
```

47. $\dfrac{4}{15} = \dfrac{4}{15} \cdot 100\% = \dfrac{400}{15}\% \approx 26.67\%$

```
         26.666 ≈ 26.67
    15) 400.000
        −30
        ────
        100
        −90
        ────
        10 0
        −9 0
        ────
         1 00
         −90
        ────
          100
          −90
        ────
            0
```

49. $\dfrac{1}{7} = \dfrac{1}{7}(100\%) = \dfrac{100}{7}\% \approx 14.29\%$

```
        14.285 ≈ 14.29
     7) 100.000
        −7
        ────
        30
       −28
        ────
         2 0
        −1 4
        ────
          60
         −56
        ────
          40
         −35
        ────
           5
```

51. $\dfrac{11}{12} = \dfrac{11}{12} \cdot 100\% = \dfrac{1100}{12}\% \approx 91.67\%$

```
          91.666 ≈ 91.67
     12) 1100.000
         −108
         ────
           20
          −12
         ────
            8 0
           −7 2
         ────
             80
            −72
         ────
             80
            −72
         ────
              8
```

53.

Percent	Decimal	Fraction
35%	$35\% = 0.35$	$35\% = \dfrac{35}{100} = \dfrac{7}{20}$
$\dfrac{1}{5} = \dfrac{20}{100} = 20\%$	$\dfrac{1}{5} = \dfrac{2}{10} = 0.2$	$\dfrac{1}{5}$
$0.5 = 0.50 = 50\%$	0.5	$0.5 = \dfrac{5}{10} = \dfrac{1}{2}$
70%	$70\% = 0.7$	$70\% = \dfrac{70}{100} = \dfrac{7}{10}$
$\dfrac{3}{8} = \dfrac{375}{1000} = \dfrac{37.5}{100} = 37.5\%$	$\dfrac{3}{8} = \dfrac{375}{1000} = 0.375$	$\dfrac{3}{8}$

55.

Percent	Decimal	Fraction
40%	$40\% = 0.4$	$40\% = \dfrac{40}{100} = \dfrac{2}{5}$
$0.235 = 23.5\%$	0.235	$0.235 = \dfrac{235}{1000} = \dfrac{47}{200}$
$\dfrac{4}{5} = \dfrac{80}{100} = 80\%$	$\dfrac{4}{5} = \dfrac{8}{10} = 0.8$	$\dfrac{4}{5}$
$33\dfrac{1}{3}\%$	$33\dfrac{1}{3}\% = \dfrac{1}{3} = 0.33\overline{3}$	$33\dfrac{1}{3}\% = \dfrac{\frac{100}{3}}{100} = \dfrac{1}{3}$
$\dfrac{7}{8} = \dfrac{875}{1000} = 87.5\%$	$\dfrac{7}{8} = \dfrac{875}{1000} = 0.875$	$\dfrac{7}{8}$
7.5%	$7.5\% = 0.075$	$7.5\% = \dfrac{7.5}{100} = \dfrac{75}{1000} = \dfrac{3}{40}$

57.

Percent	Decimal	Fraction
200%	$200\% = 2.00 = 2$	$200\% = \dfrac{200}{100} = 2$
$2.8 = \dfrac{28}{10} = \dfrac{280}{100} = 280\%$	2.8	$2.8 = 2\dfrac{8}{10} = 2\dfrac{4}{5}$
705%	$705\% = 7.05$	$705\% = \dfrac{705}{100} = 7\dfrac{5}{100} = 7\dfrac{1}{20}$
$4\dfrac{27}{50} = 4\dfrac{54}{100} = \dfrac{454}{100} = 454\%$	$4\dfrac{27}{50} = 4\dfrac{54}{100} = 4.54$	$4\dfrac{27}{50}$

59. $26.2\% = 26.2(0.01) = 0.262$

$$\frac{26.2}{100} = \frac{262}{1000} = \frac{131}{500}$$

61. $23\% = 0.23$

$$23\% = \frac{23}{100}$$

63. $\frac{483}{1000} = \frac{48.3}{100} = 48.3\%$

65. $8.75\% = 8.75 \cdot 0.01 = 0.0875$

67. $\frac{1}{4} = \frac{25}{100} = 25\%$

69. $0.5\% = 0.5(0.01) = 0.005$

$$0.5\% = 0.5 \cdot \frac{1}{100}$$

$$= \frac{0.5}{100}$$

$$= \frac{0.5}{100} \cdot \frac{10}{10}$$

$$= \frac{5}{1000}$$

$$= \frac{1}{200}$$

71. $14.2\% = 14.2(0.01) = 0.142$

$$14.2\% = 14.2 \cdot \frac{1}{100}$$

$$= \frac{14.2}{100}$$

$$= \frac{14.2 \cdot 10}{100 \cdot 10}$$

$$= \frac{142}{1000}$$

$$= \frac{71}{500}$$

73. $7.9\% = 7.9(0.01) = 0.079$

$$7.9\% = 7.9 \cdot \frac{1}{100} = \frac{7.9}{100} = \frac{79}{1000}$$

75. $3 \cdot n = 45$

$$\frac{3 \cdot n}{3} = \frac{45}{3}$$

$$n = 15$$

77. $8 \cdot n = 80$

$$\frac{8 \cdot n}{8} = \frac{80}{8}$$

$$n = 10$$

79. $6 \cdot n = 72$

$$\frac{6 \cdot n}{6} = \frac{72}{6}$$

$$n = 12$$

81. a. To round 52.8647 to the nearest tenth observe that the digit in the hundredths place is 6. Since this digit is at least 5, we add 1 to the digit in the tenths place. The number 52.8647 rounded to the nearest tenth is 52.9. Thus $52.8647\% \approx 52.9\%$.

b. To round 52.8647 to the nearest hundredth observe that the digit in the thousandths place is 4. Since this digit is less than 5, we do not add 1 to the digit in the hundredths place. The number 52.8647 rounded to the nearest hundredth is 52.86. Thus $52.8647\% \approx 52.86\%$.

83. $1.07835 = 107.835\%$
To round 107.835 to the nearest tenth, observe that the digit in the hundredths place is 3. Since this digit is less than 5, we do not add 1 to the digit in the tenths place. The number 107.835, rounded to the nearest tenth is 107.8. Thus $1.07835 = 107.835\% \approx 107.8\%$.

85. $0.65794 = 65.794\%$
To round 65.794 to the nearest hundredth, observe that the digit in the thousandths place is 4. Since this digit is less than 5, we do not add 1 to the digit in the hundredths place. The number 65.794 rounded to the nearest hundredth is 65.79. Thus, $0.65794 = 65.794\% \approx 65.79\%$.

87. $0.7682 = 76.82\%$
To round 76.82 to the nearest one, observe that the digit in the tenths place is 8. Since this digit is at least 5, we add 1 to the digit in the ones place. The number 76.82, rounded to the nearest one is 77. Thus $0.7682 = 76.82\% \approx 77\%$.

89. 3 of the 4 equal parts are shaded.

$$\frac{3}{4} = \frac{3}{4} \cdot \frac{25}{25} = \frac{75}{100} = 75\%$$

75% of the figure is shaded.

91. 4 of the 5 equal parts are shaded.

$$\frac{4}{5} = \frac{4}{5} \cdot \frac{20}{20} = \frac{80}{100} = 80\%$$

80% of the figure is shaded.

93. A fraction written as a percent is greater than 100% when the numerator is <u>greater</u> than the denominator.

95. answers may vary

97.
$$\begin{array}{r} 0.2658 \approx 0.266 \\ 79\overline{)21.0000} \\ \underline{-15\ 8} \\ 5\ 20 \\ \underline{-4\ 74} \\ 460 \\ \underline{-395} \\ 650 \\ \underline{-632} \\ 18 \end{array}$$

$$\frac{21}{79} \approx 0.266 \text{ or } 26.6\%$$

99.
$$\begin{array}{r} 1.1548 \approx 1.155 \\ 736\overline{)850.0000} \\ \underline{-736} \\ 114\ 0 \\ \underline{-73\ 6} \\ 40\ 40 \\ \underline{-36\ 80} \\ 3\ 600 \\ \underline{-2\ 944} \\ 6560 \\ \underline{-5888} \\ 672 \end{array}$$

$$\frac{850}{736} \approx 1.155 \text{ or } 115.5\%$$

Section 6.3 Practice

1. 6 is <u>what percent</u> of 24?

$$\begin{array}{ccc} \downarrow\downarrow & \downarrow & \downarrow\downarrow \\ 6 = & n & \cdot\ 24 \end{array}$$

2. 1.8 is 20% of <u>what number</u>?

$$\begin{array}{ccccc} \downarrow\downarrow\downarrow & \downarrow & & \downarrow \\ 1.8 = 20\% & \cdot & & n \end{array}$$

3. <u>What number</u> is 40% of 3.6?

$$\begin{array}{ccccc} \downarrow & & \downarrow\downarrow & & \downarrow\downarrow \\ n & & = 40\% & \cdot & 3.6 \end{array}$$

4. 42% of 50 is <u>what number</u>?

$$\begin{array}{ccc} \downarrow\ \downarrow\downarrow\downarrow & & \downarrow \\ 42\%\ \cdot\ 50 = & & n \end{array}$$

5. 15% of <u>what number</u> is 9?

$$\begin{array}{cccc} \downarrow\ \downarrow & \downarrow & \downarrow\downarrow \\ 15\%\ \cdot & n & = 9 \end{array}$$

6. <u>What percent</u> of 150 is 90?

$$\begin{array}{cc} \downarrow & \downarrow\ \downarrow\downarrow \\ n & \cdot\ 150 = 90 \end{array}$$

7. <u>What number</u> is 20% of 85?

$$\begin{array}{ccc} \downarrow & & \downarrow\downarrow\ \downarrow\downarrow \\ n & & = 20\% \cdot 85 \\ n & & = 0.20\ \cdot\ 85 \\ n & & = 17 \end{array}$$

Thus, 17 is 20% of 85.

8. 90% of 150 is <u>what number</u>?

$$\begin{array}{cc} \downarrow\ \downarrow\ \downarrow\ \downarrow & \downarrow \\ 90\%\ \cdot\ 150 = & n \\ 0.90\ \cdot\ 150 = & n \\ 135 = & n \end{array}$$

Thus, 90% of 150 is 135.

9. 15% of <u>what number</u> is 1.2?

$$\begin{array}{ccc} \downarrow\ \downarrow & \downarrow & \downarrow\ \downarrow \\ 15\%\ \cdot & n & = 1.2 \\ 0.15\ \cdot & n & = 1.2 \end{array}$$

$$n = \frac{1.2}{0.15}$$
$$n = 8$$

Thus, 15% of 8 is 1.2.

10. 27 is $4\frac{1}{2}$% of what number?

$\downarrow \downarrow \downarrow \quad \downarrow \qquad \downarrow$

$27 = 4\frac{1}{2}\% \quad \cdot \qquad n$

$27 = 0.045 \quad \cdot \qquad n$

$n = \dfrac{27}{0.045}$

$n = 60$

Thus 27 is $4\frac{1}{2}$% of 60.

11. What percent of 80 is 8?

$\qquad \downarrow \qquad \downarrow \downarrow \downarrow$

$\qquad n \qquad \cdot 80 = 90$

$\qquad\qquad\qquad n = \dfrac{8}{80}$

$\qquad\qquad\qquad n = 0.1$

$\qquad\qquad\qquad n = 10\%$

So, 10% of 80 is 8.

12. 35 is what percent of 25?

$\quad \downarrow \downarrow \qquad \downarrow \qquad \downarrow \downarrow$

$\quad 35 = \qquad n \qquad \cdot 25$

$\dfrac{35}{25} = n$

$1.4 = n$

$140\% = n$

So, 35 is 140% of 25.

Vocabulary and Readiness Check

1. The word <u>is</u> translates to "=."

2. The word <u>of</u> usually translates to "multiplication."

3. In the statement "10% of 90 is 9," the number 9 is called the <u>amount</u>, 90 is called the <u>base</u>, and 10 is called the <u>percent</u>.

4. 100% of a number = <u>the number</u>

5. Any "percent greater than 100%" of "a number" = "a number <u>greater</u> than the original number."

6. Any "percent less than 100%" of "a number" = "a number <u>less</u> than the original number."

7. percent: 42%
base: 50
amount: 21

8. percent: 30%
base: 65
amount: 19.5

9. percent: 125%
base: 86
amount: 107.5

10. percent: 110%
base: 90
amount: 99

Exercise Set 6.3

1. 18% of 81 is what number?

$\downarrow \quad \downarrow \downarrow \downarrow \qquad \downarrow$

$18\% \quad \cdot \;\; 81 = \qquad x$

3. 20% of what number is 105?

$\downarrow \quad \downarrow \qquad \downarrow \qquad \downarrow \downarrow$

$20\% \quad \cdot \qquad x \qquad = 105$

5. 0.6 is 40% of what number?

$\downarrow \quad \downarrow \quad \downarrow \quad \downarrow \qquad \downarrow$

$0.6 = \;\; 40\% \;\cdot \qquad x$

7. What percent of 80 is 3.8?

$\qquad \downarrow \qquad \downarrow \downarrow \downarrow \; \downarrow$

$\qquad x \qquad \cdot 80 = \; 3.8$

9. What number is 9% of 43?

$\qquad \downarrow \qquad \downarrow \downarrow \downarrow \; \downarrow$

$\qquad x \qquad = 9\% \; \cdot \; 43$

11. What percent of 250 is 150?

$\qquad \downarrow \qquad \downarrow \downarrow \; \downarrow \; \downarrow$

$\qquad x \qquad \cdot 250 = \; 150$

13. $10\% \cdot 35 = x$

$0.10 \cdot 35 = x$

$\qquad 3.5 = x$

10% of 35 is 3.5.

15. $x = 14\% \cdot 205$
$x = 0.14 \cdot 205$
$x = 28.7$
28.7 is 14% of 205.

17. $1.2 = 12\% \cdot x$
$1.2 = 0.12x$
$\dfrac{1.2}{0.12} = \dfrac{0.12x}{0.12}$
$10 = x$
1.2 is 12% of 10.

19. $8\dfrac{1}{2}\% \cdot x = 51$
$0.085x = 51$
$\dfrac{0.085x}{0.085} = \dfrac{51}{0.085}$
$x = 600$
$8\dfrac{1}{2}\%$ of 600 is 51.

21. $x \cdot 80 = 88$
$\dfrac{x \cdot 80}{80} = \dfrac{88}{80}$
$x = 1.1$
$x = 110\%$
88 is 110% of 80.

23. $17 = x \cdot 50$
$\dfrac{17}{50} = \dfrac{x \cdot 50}{50}$
$0.34 = x$
$34\% = x$
17 is 34% of 50.

25. $0.1 = 10\% \cdot x$
$0.1 = 0.10x$
$\dfrac{0.1}{0.1} = \dfrac{0.1x}{0.1}$
$1 = x$
0.1 is 10% of 1.

27. $150\% \cdot 430 = x$
$1.5 \cdot 430 = x$
$645 = x$
150% of 430 is 645.

29. $82.5 = 16\dfrac{1}{2}\% \cdot x$
$82.5 = 0.165x$
$\dfrac{82.5}{0.165} = \dfrac{0.165x}{0.165}$
$500 = x$
82.5 is $16\dfrac{1}{2}\%$ of 500.

31. $2.58 = x \cdot 50$
$\dfrac{2.58}{50} = \dfrac{x \cdot 50}{50}$
$0.0516 = x$
$5.16\% = x$
2.58 is 5.16% of 50.

33. $x = 42\% \cdot 60$
$x = 0.42 \cdot 60$
$x = 25.2$
25.2 is 42% of 60.

35. $x \cdot 184 = 64.4$
$\dfrac{x \cdot 184}{184} = \dfrac{64.4}{184}$
$x = 0.35$
$x = 35\%$
35% of 184 is 64.4.

37. $120\% \cdot x = 42$
$1.20 \cdot x = 42$
$\dfrac{1.2x}{1.2} = \dfrac{42}{1.2}$
$x = 35$
120% of 35 is 42.

39. $2.4\% \cdot 26 = x$
$0.024 \cdot 26 = x$
$0.624 = x$
2.4% of 26 is 0.624.

41. $x \cdot 600 = 3$
$\dfrac{x \cdot 600}{600} = \dfrac{3}{600}$
$x = 0.005$
$x = 0.5\%$
0.5% of 600 is 3.

43. $6.67 = 4.6\% \cdot x$
$6.67 = 0.046x$
$\dfrac{6.67}{0.046} = \dfrac{0.046x}{0.046}$
$145 = x$
6.67 is 4.6% of 145.

45. $1575 = x \cdot 2500$

$$\frac{1575}{2500} = \frac{x \cdot 2500}{2500}$$

$0.63 = x$

$63\% = x$

1575 is 63% of 2500.

47. $2 = x \cdot 50$

$$\frac{2}{50} = \frac{x \cdot 50}{50}$$

$0.04 = x$

$4\% = x$

2 is 4% of 50.

49. $\dfrac{27}{n} = \dfrac{9}{10}$

$27 \cdot 10 = 9 \cdot n$

$270 = 9n$

$\dfrac{270}{9} = n$

$30 = n$

51. $\dfrac{n}{5} = \dfrac{8}{11}$

$11n = 5 \cdot 8$

$11n = 40$

$n = \dfrac{40}{11} = 3\dfrac{7}{11}$

53. $\dfrac{17}{12} = \dfrac{n}{20}$

55. $\dfrac{8}{9} = \dfrac{14}{n}$

57. In the equation $5 \cdot n = 32$, the step that should be taken to find the value of n is to divide by 5, obtaining $n = \dfrac{32}{5}$, which is choice **c**.

59. In the equation $0.06 = n \cdot 7$, the next step is to divide by 7, obtaining $n = \dfrac{0.06}{7}$. This is choice **b**.

61. $20\% \cdot n = 18.6$ in words is "twenty percent of some number is eighteen and six-tenths."

63. Since 100% of 20 is 20, and 30 is greater than 20, x must be greater than 100%; b.

65. Since 85 is less than 120, the percent is less than 100%; c.

67. Since 55% is less than 100%, which is 1, 55% of 45 is less than 45; c.

69. Since 100% is 1, 100% of 45 is equal to 45; a.

71. Since 100% is 1, 100% of 45 is equal to 45; a.

73. answers may vary

75. $1.5\% \cdot 45,775 = x$

$0.015 \cdot 45,775 = x$

$686.625 = x$

1.5% of 45,775 is 686.625.

77. $22,113 = 180\% \cdot x$

$22,113 = 1.80 \cdot x$

$\dfrac{22,113}{1.8} = \dfrac{1.8 \cdot x}{1.8}$

$12,285 = x$

22,113 is 180% of 12,285.

Section 6.4 Practice

1. The amount is 55. The base is unknown; call it b. The percent is 15.

$$\frac{55}{b} = \frac{15}{100}$$

2. The amount is 35. The base is 70. The percent is unknown; call it p.

$$\frac{35}{70} = \frac{p}{100}$$

3. The amount is unknown; call it a. The base is 68. The percent is 25.

$$\frac{a}{68} = \frac{25}{100}$$

4. The amount is 520. The base is unknown; call it b. The percent is 65.

$$\frac{520}{b} = \frac{65}{100}$$

5. The amount is 65. The base is 50. The percent is unknown; call it p.

$$\frac{65}{50} = \frac{p}{100}$$

6. The amount is unknown; call it a. The base is 80. The percent is 36.

$$\frac{a}{80} = \frac{36}{100}$$

7. The amount is unknown; call it a. The base is 120. The percent is 8.

$$\frac{a}{120} = \frac{8}{100}$$
$$100a = 120 \cdot 8$$
$$100a = 960$$
$$a = \frac{960}{100}$$
$$a = 9.6$$

9.6 is 8% of 120.

8. The amount is 60. The base is unknown; call it b. The percent is 75.

$$\frac{60}{b} = \frac{75}{100}$$
$$75b = 60 \cdot 100$$
$$75b = 6000$$
$$b = \frac{6000}{75}$$
$$b = 80$$

75% of 80 is 60.

9. The amount is 15.2. The base is unknown; call it b. The percent is 5.

$$\frac{15.2}{b} = \frac{5}{100}$$
$$5b = 15.2 \cdot 100$$
$$5b = 1520$$
$$b = \frac{1520}{5}$$
$$b = 304$$

15.2 is 5% of 304.

10. The amount is 6. The base is 40. The percent is unknown; call it p.

$$\frac{6}{40} = \frac{p}{100}$$
$$40p = 6 \cdot 100$$
$$40p = 600$$
$$p = \frac{600}{40}$$
$$p = 15$$

15% of 40 is 6.

11. The amount is 336. The base is 160. The percent is unknown; call it p.

$$\frac{336}{160} = \frac{p}{100}$$
$$160p = 336 \cdot 100$$
$$160p = 33,600$$
$$p = \frac{33,600}{160}$$
$$p = 210$$

336 is 210% of 160.

Vocabulary and Readiness Check

1. When translating the statement "20% of 15 is 3" to a proportion, the number 3 is called the <u>amount</u>, 15 is the <u>base</u>, and 20 is the <u>percent</u>.

2. In the question "50% of what number is 28?", which part of the percent proportion is unknown? <u>base</u>

3. In the question "What number is 25% of 200?", which part of the percent proportion is unknown? <u>amount</u>

4. In the question "38 is what percent of 380?", which part of the percent proportion is unknown? <u>percent</u>

5. amount = 12.6; base = 42; percent = 30

6. amount = 201; base = 300; percent = 67

7. amount = 102; base = 510; percent = 20

8. amount = 248; base = 620; percent = 40

Exercise Set 6.4

1. 98% of 45 is <u>what number</u>?

$$\downarrow \qquad \downarrow \qquad \downarrow$$

percent   base   amount = a

$$\frac{a}{45} = \frac{98}{100}$$

3. <u>What number</u> is 4% of 150?

$$\downarrow \qquad\qquad \downarrow \qquad \downarrow$$

amount = a percent   base

$$\frac{a}{150} = \frac{4}{100}$$

5.   14.3   is   26% of <u>what number</u>?

    ↓   ↓   ↓

amount   percent   base $= b$

$\dfrac{14.3}{b} = \dfrac{26}{100}$

7.   35% of <u>what number</u> is 84?

    ↓   ↓   ↓

percent   base $= b$ amount

$\dfrac{84}{b} = \dfrac{35}{100}$

9.   <u>What percent</u> of 400 is 70?

       ↓   ↓   ↓

percent $= p$ base   amount

$\dfrac{70}{400} = \dfrac{p}{100}$

11.   8.2 is <u>what percent</u> of 82?

    ↓   ↓   ↓

amount   percent $= p$ base

$\dfrac{8.2}{82} = \dfrac{p}{100}$

13. $\dfrac{a}{65} = \dfrac{40}{100}$ or $\dfrac{a}{65} = \dfrac{2}{5}$

$a \cdot 5 = 65 \cdot 2$

$5a = 130$

$\dfrac{5a}{5} = \dfrac{130}{5}$

$a = 26$

40% of 65 is 26.

15. $\dfrac{a}{105} = \dfrac{18}{100}$

$a \cdot 100 = 105 \cdot 18$

$a \cdot 100 = 1890$

$\dfrac{a \cdot 100}{100} = \dfrac{1890}{100}$

$a = 18.9$

18.9 is 18% of 105.

17. $\dfrac{90}{b} = \dfrac{15}{100}$ or $\dfrac{90}{b} = \dfrac{3}{20}$

$90 \cdot 20 = b \cdot 3$

$1800 = 3b$

$\dfrac{1800}{3} = \dfrac{3b}{3}$

$600 = b$

15% of 600 is 90.

19. $\dfrac{7.8}{b} = \dfrac{78}{100}$

$7.8 \cdot 100 = b \cdot 78$

$780 = 78 \cdot b$

$\dfrac{780}{78} = \dfrac{78 \cdot b}{78}$

$10 = b$

7.8 is 78% of 10.

21. $\dfrac{42}{35} = \dfrac{p}{100}$ or $\dfrac{6}{5} = \dfrac{p}{100}$

$6 \cdot 100 = 5 \cdot p$

$600 = 5p$

$\dfrac{600}{5} = \dfrac{5p}{5}$

$120 = p$

42 is 120% of 35.

23. $\dfrac{14}{50} = \dfrac{p}{100}$ or $\dfrac{7}{25} = \dfrac{p}{100}$

$7 \cdot 100 = 25 \cdot p$

$700 = 25p$

$\dfrac{700}{25} = \dfrac{25p}{25}$

$28 = p$

14 is 28% of 50.

25. $\dfrac{3.7}{b} = \dfrac{10}{100}$ or $\dfrac{3.7}{b} = \dfrac{1}{10}$

$3.7 \cdot 10 = b \cdot 1$

$37 = b$

3.7 is 10% of 37.

27. $\dfrac{a}{70} = \dfrac{2.4}{100}$

$a \cdot 100 = 70 \cdot 2.4$

$100a = 168$

$\dfrac{100a}{100} = \dfrac{168}{100}$

$a = 1.68$

1.68 is 2.4% of 70.

29.
$$\frac{160}{b} = \frac{16}{100} \text{ or } \frac{160}{b} = \frac{4}{25}$$
$$160 \cdot 25 = b \cdot 4$$
$$4000 = 4b$$
$$\frac{4000}{4} = \frac{4b}{4}$$
$$1000 = b$$
160 is 16% of 1000.

31.
$$\frac{394.8}{188} = \frac{p}{100}$$
$$394.8 \cdot 100 = 188 \cdot p$$
$$39,480 = 188p$$
$$\frac{39,480}{188} = \frac{188p}{188}$$
$$210 = p$$
394.8 is 210% of 188.

33.
$$\frac{a}{62} = \frac{89}{100}$$
$$a \cdot 100 = 62 \cdot 89$$
$$100a = 5518$$
$$\frac{100a}{100} = \frac{5518}{100}$$
$$a = 55.18$$
55.18 is 89% of 62.

35.
$$\frac{2.7}{6} = \frac{p}{100}$$
$$2.7 \cdot 100 = 6 \cdot p$$
$$270 = 6p$$
$$\frac{270}{6} = \frac{6p}{6}$$
$$45 = p$$
45% of 6 is 2.7.

37.
$$\frac{105}{b} = \frac{140}{100} \text{ or } \frac{105}{b} = \frac{7}{5}$$
$$105 \cdot 5 = b \cdot 7$$
$$525 = 7b$$
$$\frac{525}{7} = \frac{7b}{7}$$
$$75 = b$$
140% of 75 is 105.

39.
$$\frac{a}{48} = \frac{1.8}{100}$$
$$a \cdot 100 = 48 \cdot 1.8$$
$$100a = 86.4$$
$$\frac{100a}{100} = \frac{86.4}{100}$$
$$a = 0.864$$
1.8% of 48 is 0.864.

41.
$$\frac{4}{800} = \frac{p}{100} \text{ or } \frac{1}{200} = \frac{p}{100}$$
$$1 \cdot 100 = 200 \cdot p$$
$$100 = 200p$$
$$\frac{100}{200} = \frac{200p}{200}$$
$$0.5 = p$$
0.5% of 800 is 4.

43.
$$\frac{3.5}{b} = \frac{2.5}{100}$$
$$3.5 \cdot 100 = b \cdot 2.5$$
$$350 = 2.5b$$
$$\frac{350}{2.5} = \frac{2.5b}{2.5}$$
$$140 = b$$
3.5 is 2.5% of 140.

45.
$$\frac{a}{48} = \frac{20}{100} \text{ or } \frac{a}{48} = \frac{1}{5}$$
$$a \cdot 5 = 48 \cdot 1$$
$$5a = 48$$
$$\frac{5a}{5} = \frac{48}{5}$$
$$a = 9.6$$
20% of 48 is 9.6.

47.
$$\frac{2486}{2200} = \frac{p}{100}$$
$$2486 \cdot 100 = 2200 \cdot p$$
$$248,600 = 2200p$$
$$\frac{248,600}{2200} = \frac{2200p}{2200}$$
$$113 = p$$
2486 is 113% of 2200.

49. $\frac{11}{16} + \frac{3}{16} = \frac{11+3}{16} = \frac{14}{16} = \frac{2 \cdot 7}{2 \cdot 8} = \frac{7}{8}$

51. The LCD is 30.

$$3\frac{1}{2} - \frac{11}{30} = \frac{7}{2} - \frac{11}{30}$$

$$= \frac{7}{2} \cdot \frac{15}{15} - \frac{11}{30}$$

$$= \frac{105}{30} - \frac{11}{30}$$

$$= \frac{94}{30}$$

$$= \frac{47}{15}$$

$$= 3\frac{2}{15}$$

53.
$$\begin{array}{r} 1 \\ 0.41 \\ + 0.29 \\ \hline 0.70 \end{array}$$

55.
$$\begin{array}{r} 2.38 \\ - 0.19 \\ \hline 2.19 \end{array}$$

57. answers may vary

59.
$$\frac{a}{64} = \frac{25}{100}$$

$$\frac{17}{64} \overset{?}{=} \frac{25}{100}$$

$$17 \cdot 100 \overset{?}{=} 64 \cdot 25$$

$$1700 = 1600 \quad \text{False}$$

The amount is not 17.

$$\frac{a}{64} = \frac{25}{100}$$

$$a \cdot 100 = 64 \cdot 25$$

$$100a = 1600$$

$$\frac{100a}{100} = \frac{1600}{100}$$

$$a = 16$$

25% of 64 is 16.

61.
$$\frac{p}{100} = \frac{13}{52}$$

$$\frac{25}{100} \overset{?}{=} \frac{13}{52}$$

$$\frac{1}{4} = \frac{1}{4} \quad \text{True}$$

Yes, the percent is equal to 25 (25%).

63. answers may vary

65.
$$\frac{a}{53,862} = \frac{22.3}{100}$$

$$a \cdot 100 = 53,862 \cdot 22.3$$

$$100a = 1,201,122.6$$

$$\frac{100a}{100} = \frac{1,201,122.6}{100}$$

$$a \approx 12,011.2$$

22.3% of 53,862 is 12,011.2.

67.
$$\frac{8652}{b} = \frac{119}{100}$$

$$8652 \cdot 100 = b \cdot 119$$

$$865,200 = 119b$$

$$\frac{865,200}{119} = \frac{119b}{119}$$

$$7270.6 \approx b$$

8652 is 119% of 7270.6.

Integrated Review

1. $0.12 = 0.12 \cdot 100\% = 12\%$

2. $0.68 = 0.68 \cdot 100\% = 68\%$

3. $\frac{1}{8} = \frac{1}{8} \cdot 100\% = \frac{100}{8}\% = 12.5\%$

4. $\frac{5}{2} = \frac{5}{2} \cdot 100\% = \frac{500}{2}\% = 250\%$

5. $5.2 = 5.2 \cdot 100\% = 520\%$

6. $8 = 8 \cdot 100\% = 800\%$

7. $\frac{3}{50} = \frac{3}{50} \cdot 100\% = \frac{300}{50}\% = 6\%$

8. $\frac{11}{25} = \frac{11}{25} \cdot 100\% = \frac{1100}{25}\% = 44\%$

9. $7\frac{1}{2} = \frac{15}{2} = \frac{15}{2} \cdot 100\% = \frac{1500}{2}\% = 750\%$

10. $3\frac{1}{4} = \frac{13}{4} = \frac{13}{4} \cdot 100\% = \frac{1300}{4}\% = 325\%$

11. $0.03 = 0.03 \cdot 100\% = 3\%$

12. $0.05 = 0.05 \cdot 100\% = 5\%$

13. $65\% = 65 \cdot 0.01 = 0.65$

14. $31\% = 31 \cdot 0.01 = 0.31$

15. $8\% = 8 \cdot 0.01 = 0.08$

16. $7\% = 7 \cdot 0.01 = 0.07$

17. $142\% = 142 \cdot 0.01 = 1.42$

18. $400\% = 400 \cdot 0.01 = 4$

19. $2.9\% = 2.9 \cdot 0.01 = 0.029$

20. $6.6\% = 6.6 \cdot 0.01 = 0.066$

21. $3\% = 3 \cdot 0.01 = 0.03$

$3\% = 3 \cdot \dfrac{1}{100} = \dfrac{3}{100}$

22. $5\% = 5 \cdot 0.01 = 0.05$

$5\% = 5 \cdot \dfrac{1}{100} = \dfrac{5}{100} = \dfrac{1}{20}$

23. $5.25\% = 5.25 \cdot 0.01 = 0.0525$

$5.25\% = 0.0525 = \dfrac{525}{10{,}000} = \dfrac{21 \cdot 25}{400 \cdot 25} = \dfrac{21}{400}$

24. $12.75\% = 12.75 \cdot 0.01 = 0.1275$

$12.75\% = 0.1275 = \dfrac{1275}{10{,}000} = \dfrac{51 \cdot 25}{400 \cdot 25} = \dfrac{51}{400}$

25. $38\% = 38 \cdot 0.01 = 0.38$

$38\% = 38 \cdot \dfrac{1}{100} = \dfrac{38}{100} = \dfrac{19}{50}$

26. $45\% = 45 \cdot 0.01 = 0.45$

$45\% = 45 \cdot \dfrac{1}{100} = \dfrac{45}{100} = \dfrac{9}{20}$

27. $12\dfrac{1}{3}\% = \dfrac{37}{3}\% = \dfrac{37}{3} \cdot \dfrac{1}{100} = \dfrac{37}{300} \approx 0.123$

$12\dfrac{1}{3}\% = \dfrac{37}{300}$

28. $16\dfrac{2}{3}\% = \dfrac{50}{3}\% = \dfrac{50}{3} \cdot \dfrac{1}{100} = \dfrac{50}{300} = \dfrac{1}{6} \approx 0.167$

$16\dfrac{2}{3}\% = \dfrac{1}{6}$

29.
$$\dfrac{a}{70} = \dfrac{12}{100}$$
$$100a = 70 \cdot 12$$
$$100a = 840$$
$$a = \dfrac{840}{100}$$
$$a = 8.4$$
12% of 70 is 8.4.

30.
$$\dfrac{36}{b} = \dfrac{36}{100}$$
$$36b = 36 \cdot 100$$
$$36b = 3600$$
$$b = \dfrac{3600}{36}$$
$$b = 100$$
36 is 36% of 100.

31.
$$\dfrac{212.5}{b} = \dfrac{85}{100}$$
$$85b = 212.5 \cdot 100$$
$$85b = 21{,}250$$
$$b = \dfrac{21{,}250}{85}$$
$$b = 250$$
212.5 is 85% of 250.

32.
$$\dfrac{66}{55} = \dfrac{p}{100}$$
$$55p = 66 \cdot 100$$
$$55p = 6600$$
$$p = \dfrac{6600}{55}$$
$$p = 120$$
66 is 120% of 55.

33.
$$\dfrac{23.8}{85} = \dfrac{p}{100}$$
$$85p = 23.8 \cdot 100$$
$$85p = 2380$$
$$p = \dfrac{2380}{85}$$
$$p = 28$$
23.8 is 28% of 85.

34. $\dfrac{a}{200} = \dfrac{38}{100}$

$100a = 200 \cdot 38$

$100a = 7600$

$a = \dfrac{7600}{100}$

$a = 76$

38% of 200 is 76.

35. $\dfrac{a}{44} = \dfrac{25}{100}$

$100a = 44 \cdot 25$

$100a = 1100$

$a = \dfrac{1100}{100}$

$a = 11$

11 is 25% of 44.

36. $\dfrac{128.7}{99} = \dfrac{p}{100}$

$99p = 128.7 \cdot 100$

$99p = 12,870$

$p = \dfrac{12,870}{99}$

$p = 130$

128.7 is 130% of 99.

37. $\dfrac{215}{250} = \dfrac{p}{100}$

$250p = 215 \cdot 100$

$250p = 21,500$

$p = \dfrac{21,500}{250}$

$p = 86$

215 is 86% of 250.

38. $\dfrac{a}{84} = \dfrac{45}{100}$

$100a = 84 \cdot 45$

$100a = 3780$

$a = \dfrac{3780}{100}$

$a = 37.8$

37.8 is 45% of 84.

39. $\dfrac{63}{b} = \dfrac{42}{100}$

$42b = 63 \cdot 100$

$42b = 6300$

$b = \dfrac{6300}{42}$

$b = 150$

42% of 150 is 63.

40. $\dfrac{58.9}{b} = \dfrac{95}{100}$

$95b = 58.9 \cdot 100$

$95b = 5890$

$b = \dfrac{5890}{95}$

$b = 62$

95% of 62 is 58.9.

Section 6.5 Practice

1. *Method 1*:

What number is 25% of 2174?

$x = 25\% \cdot 2174$

$x = 0.25 \cdot 2174$

$x = 543.5$

We predict 543.5 miles of the trail resides in the state of Virginia.

Method 2:

What number is 25% of 2174?

amount   percent   base

$\dfrac{a}{2174} = \dfrac{25}{100}$

$a \cdot 100 = 2174 \cdot 25$

$100a = 54,350$

$\dfrac{100a}{100} = \dfrac{54,350}{100}$

$a = 543.5$

We predict 543.5 miles of the trail resides in the state of Virginia.

2. *Method 1*:

34,000 is what percent of 130,000?

$34,000 = x \cdot 130,000$

$$34,000 = 130,000x$$
$$\frac{34,000}{130,000} = \frac{130,000x}{130,000}$$
$$0.26 \approx x$$
$$26\% = x$$

In Florida, 34,000 or 26% more new nurses were needed.

Method 2:

34,000 is what percent of 130,000?

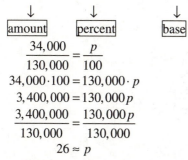

$$\frac{34,000}{130,000} = \frac{p}{100}$$
$$34,000 \cdot 100 = 130,000 \cdot p$$
$$3,400,000 = 130,000p$$
$$\frac{3,400,000}{130,000} = \frac{130,000p}{130,000}$$
$$26 \approx p$$

In Florida, 34,000 or 26% more new nurses were needed.

3. 775 is 31% of what number?
   Method 1
   $$775 = 31\% \cdot n$$
   $$775 = 0.31 \cdot n$$
   $$n = \frac{775}{0.31}$$
   $$n = 2500$$
   Method 2
   $$\frac{775}{b} = \frac{31}{100}$$
   $$31b = 775 \cdot 100$$
   $$31b = 77,500$$
   $$b = \frac{77,500}{31}$$
   $$b = 2500$$
   2500 students go to Euclid University.

4. *Method 1:*
   What number is 3% of 240 million?

   $$x = 3\% \cdot 240$$
   $$x = 0.03 \cdot 240$$
   $$x = 7.2 \text{ million}$$

   a. The increase in the number of vehicles on the road in 2007 is 7.2 million.

   b. The total number of registered vehicles on the road in 2007 was
   240 million + 7.2 million = 247.2 million

   Method 2:
   What number is 3% of 240 million?

   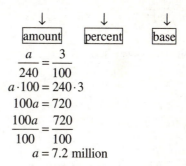

   $$\frac{a}{240} = \frac{3}{100}$$
   $$a \cdot 100 = 240 \cdot 3$$
   $$100a = 720$$
   $$\frac{100a}{100} = \frac{720}{100}$$
   $$a = 7.2 \text{ million}$$

   a. The increase in the number of vehicles on the road in 2007 is 7.2 million.

   b. The total number of registered vehicles on the road in 2007 was
   240 million + 7.2 million = 247.2 million

5. Find the amount of increase by subtracting the original number of attendants from the new number of attendants.
   amount of increase = 333 − 285 = 48
   The amount of increase is 48 attendants.

   $$\text{percent of increase} = \frac{\text{amount of increase}}{\text{original amount}}$$
   $$= \frac{48}{285}$$
   $$\approx 0.168$$
   $$= 16.8\%$$

   The number of attendants to the local play, *Peter Pan*, increased by about 16.8%.

6. Find the amount of decrease by subtracting 18,483 from 20,200.
   Amount of decrease = 20,200 − 18,483 = 1717
   The amount of decrease is 1717.

   $$\text{percent of decrease} = \frac{\text{amount of decrease}}{\text{original amount}}$$
   $$= \frac{1717}{20,200}$$
   $$= 0.085$$
   $$= 8.5\%$$

   The population decreased by 8.5%.

Exercise Set 6.5

1. 24 is 1.5% of what number?
Method 1:
$$24 = 1.5\% \cdot x$$
$$24 = 0.015x$$
$$\frac{24}{0.015} = \frac{0.015x}{0.015}$$
$$1600 = x$$
1600 bolts were inspected.
Method 2:
$$\frac{24}{b} = \frac{1.5}{100}$$
$$24 \cdot 100 = b \cdot 1.5$$
$$2400 = 1.5b$$
$$\frac{2400}{1.5} = \frac{1.5b}{1.5}$$
$$1600 = b$$
1600 bolts were inspected.

3. 4% of 220 is what number?
Method 1:
$$4\% \cdot 220 = x$$
$$0.04 \cdot 220 = x$$
$$8.8 = x$$
The minimum weight resistance is 8.8 pounds.
Method 2:
$$\frac{a}{220} = \frac{4}{100}$$
$$a \cdot 100 = 220 \cdot 4$$
$$100a = 880$$
$$\frac{100a}{100} = \frac{880}{100}$$
$$a = 8.8$$
The minimum weight resistance is 8.8 pounds.

5. 378 is what percent of 2700?
Method 1:
$$378 = x \cdot 2700$$
$$\frac{378}{2700} = \frac{x \cdot 2700}{2700}$$
$$0.14 = x$$
$$14\% = x$$
The student spent 14% of last semester's college cost on books.

Method 2:
$$\frac{378}{2700} = \frac{p}{100}$$
$$378 \cdot 100 = 2700 \cdot p$$
$$37,800 = 2700p$$
$$\frac{37,800}{2700} = \frac{2700p}{2700}$$
$$14 = p$$
The student spent 14% of last semester's college cost on books.

7. 32% of 725 is what number?
Method 1:
$$32\% \cdot 725 = x$$
$$0.32 \cdot 725 = x$$
$$232 = x$$
232 films were rated R.
Method 2:
$$\frac{a}{725} = \frac{32}{100}$$
$$a \cdot 100 = 725 \cdot 32$$
$$100a = 23,200$$
$$\frac{100a}{100} = \frac{23,200}{100}$$
$$a = 232$$
232 films were rated R.

9. 160,650 is what percent of 945,000?
Method 1:
$$160,650 = x \cdot 945,000$$
$$\frac{160,650}{945,000} = \frac{x \cdot 945,000}{945,000}$$
$$0.17 = x$$
$$17\% = x$$
17% of America's restaurants are pizza restaurants.
Method 2:
$$\frac{160,650}{945,000} = \frac{p}{100}$$
$$160,650 \cdot 100 = 945,000 \cdot p$$
$$16,065,000 = 945,000p$$
$$\frac{16,065,000}{945,000} = \frac{945,000p}{945,000}$$
$$17 = p$$
17% of America's restaurants are pizza restaurants.

11. 8% of 6200 is what number?
Method 1:
$$8\% \cdot 6200 = x$$
$$0.08 \cdot 6200 = x$$
$$496 = x$$

The decrease in the number of chairs produced is
496. The new number of chairs produced each
month is 6200 − 496 = 5704 chairs.
Method 2:

$$\frac{a}{6200} = \frac{8}{100}$$
$$100 \cdot a = 6200 \cdot 8$$
$$100a = 49,600$$
$$\frac{100a}{100} = \frac{49,600}{100}$$
$$a = 496$$

The decrease in the number of chairs produced is
496. The new number of chairs produced each
month is 6200 − 496 = 5704 chairs.

13. What number is 27% of 66,000?
Method 1:
$$x = 27\% \cdot 66,000$$
$$x = 0.27 \cdot 66,000$$
$$x = 17,820$$
The number of people employed as physician
assistants is expected to be
66,000 + 17,820 = 83,820.
Method 2:
$$\frac{a}{66,000} = \frac{27}{100}$$
$$a \cdot 100 = 66,000 \cdot 27$$
$$100a = 1,782,000$$
$$\frac{100a}{100} = \frac{1,782,000}{100}$$
$$a = 17,820$$
The number of people employed as physician
assistants is expected to be
66,000 + 17,820 = 83,820.

15. 0.8% of 642 thousand is what number?
Method 1:
$$0.8\% \cdot 642 = x$$
$$0.008 \cdot 642 = x$$
$$5.136 = x$$
The population of North Dakota in 2009 was
642 − 5.136 = 636.864 thousand or
approximately 637 thousand.
Method 2:
$$\frac{a}{642} = \frac{0.8}{100}$$
$$100 \cdot a = 642 \cdot 0.8$$
$$100a = 513.6$$
$$\frac{100a}{100} = \frac{513.6}{100}$$
$$a = 5.136$$
The population of North Dakota in 2009 was

642 − 5.136 = 636.864 thousand or
approximately 637 thousand.

17. 41 is what percent of 135?
Method 1:
$$41 = x \cdot 135$$
$$\frac{41}{135} = \frac{x \cdot 135}{135}$$
$$0.30 \approx x$$
$$30\% \approx x$$
30% of the runs are intermediate.
Method 2:
$$\frac{41}{135} = \frac{p}{100}$$
$$41 \cdot 100 = 135 \cdot p$$
$$4100 = 135p$$
$$\frac{4100}{135} = \frac{135p}{135}$$
$$30 \approx p$$

30% of the runs are intermediate.

19. 20 is what percent of 40?
Method 1:
$$20 = x \cdot 40$$
$$\frac{20}{40} = \frac{40x}{40}$$
$$0.5 = x$$
$$50\% = x$$
50% of the total calories come from fat.
Method 2:
$$\frac{20}{40} = \frac{p}{100}$$
$$20 \cdot 100 = 40 \cdot p$$
$$2000 = 40p$$
$$\frac{2000}{40} = \frac{40p}{40}$$
$$50 = p$$
50% of the total calories come from fat.

21. 10 is what percent of 80?
Method 1:
$$10 = x \cdot 80$$
$$\frac{10}{80} = \frac{x \cdot 80}{80}$$
$$0.125 = x$$
$$12.5\% = x$$
12.5% of the total calories come from fat.

Method 2:
$$\frac{10}{80} = \frac{p}{100}$$
$$10 \cdot 100 = 80 \cdot p$$
$$1000 = 80p$$
$$\frac{1000}{80} = \frac{80p}{80}$$
$$12.5 = p$$
12.5% of the total calories come from fat.

23. 35 is what percent of 120?
Method 1:
$$35 = x \cdot 120$$
$$\frac{35}{120} = \frac{120x}{120}$$
$$0.292 \approx x$$
$$29.2\% \approx x$$
29.2% of the total calories come from fat.
Method 2:
$$\frac{35}{120} = \frac{p}{100}$$
$$35 \cdot 100 = 120 \cdot p$$
$$3500 = 120p$$
$$\frac{3500}{120} = \frac{120p}{120}$$
$$29.2 \approx p$$
29.2% of the total calories come from fat.

25. 26,250 is 15% of what number?
Method 1:
$$26,250 = 15\% \cdot x$$
$$26,250 = 0.15 \cdot x$$
$$\frac{26,250}{0.15} = \frac{0.15 \cdot x}{0.15}$$
$$175,000 = x$$
The price of the home was $175,000.
Method 2:
$$\frac{26,250}{b} = \frac{15}{100}$$
$$26,250 \cdot 100 = b \cdot 15$$
$$2,625,000 = 15b$$
$$\frac{2,625,000}{15} = \frac{15b}{15}$$
$$175,000 = b$$
The price of the home was $175,000.

27. What number is 78% of 40?
Method 1:
$$x = 78\% \cdot 40$$
$$x = 0.78 \cdot 40$$
$$x = 31.2$$
The owner can bill 31.2 hours each week for a

repairman.
Method 2:
$$\frac{a}{40} = \frac{78}{100}$$
$$a \cdot 100 = 40 \cdot 78$$
$$100a = 3120$$
$$\frac{100a}{100} = \frac{3120}{100}$$
$$a = 31.2$$
The owner can bill 31.2 hours each week for a repairman.

29. What number is 4.5% of 19,286?
Method 1:
$$x = 4.5\% \cdot 19,286$$
$$x = 0.045 \cdot 19,286$$
$$x = 867.87$$
The price of the car will increase by $867.87.
The new price of that model will be
$19,286 + $867.87 = $20,153.87.
Method 2:
$$\frac{a}{19,286} = \frac{4.5}{100}$$
$$a \cdot 100 = 19,286 \cdot 4.5$$
$$100a = 86,787$$
$$\frac{100a}{100} = \frac{86,787}{100}$$
$$a = 867.87$$
The price of the car will increase by $867.87.
The new price of that model will be
$19,286 + $867.87 = $20,153.87.

31. 21 is 60% of what number?
Method 1:
$$21 = 60\% \cdot x$$
$$21 = 0.60x$$
$$\frac{21}{0.60} = \frac{0.60x}{0.60}$$
$$35 = x$$
60% of the tower's total height is 35 feet.
Method 2:
$$\frac{21}{b} = \frac{60}{100}$$
$$21 \cdot 100 = b \cdot 60$$
$$2100 = 60b$$
$$\frac{2100}{60} = \frac{60b}{60}$$
$$35 = b$$
60% of the tower's total height is 35 feet.

33. 82.3% of 4761 is what number?

Method 1:

$82.3\% \cdot 4761 = x$

$0.823 \cdot 4761 = x$

$3918 \approx x$

$4761 + 3918 = 8679$

The increase is $3918 and the tuition in 2007–2008 was $8679.

Method 2:

$$\frac{a}{4761} = \frac{82.3}{100}$$

$$a \cdot 100 = 4761 \cdot 82.3$$

$$100a = 391,830.3$$

$$\frac{100a}{100} = \frac{391,830.3}{100}$$

$$a \approx 3918$$

$4761 + 3918 = 8679$

The increase is $3918 and the tuition in 2007–2008 was $8679.

35. What number is 5.7% of 731,000?

Method 1:

$x = 5.7\% \cdot 731,000$

$x = 0.057 \cdot 731,000$

$x = 41,667$

The increase is projected to be 41,667. The number of associate degrees awarded in 2017–2018 is projected to be $731,000 + 41,667 = 772,667$.

Method 2:

$$\frac{a}{731,000} = \frac{5.7}{100}$$

$$a \cdot 100 = 731,000 \cdot 5.7$$

$$100a = 4,166,700$$

$$\frac{100a}{100} = \frac{4,166,700}{100}$$

$$a = 41,667$$

The increase is projected to be 41,667. The number of associate degrees awarded in 2017–2018 is projected to be $731,000 + 41,667 = 772,667$.

	Original Amount	New Amount	Amount of Increase	Percent Increase
37.	50	80	$80 - 50 = 30$	$\dfrac{30}{50} = 0.6 = 60\%$
39.	65	117	$117 - 65 = 52$	$\dfrac{52}{65} = 0.8 = 80\%$

	Original Amount	New Amount	Amount of Decrease	Percent Decrease
41.	8	6	$8 - 6 = 2$	$\dfrac{2}{8} = 0.25 = 25\%$
43.	160	40	$160 - 40 = 120$	$\dfrac{120}{160} = 0.75 = 75\%$

45. $\text{percent decrease} = \dfrac{\text{amount of decrease}}{\text{original amount}}$

$= \dfrac{150 - 84}{150}$

$= \dfrac{66}{150}$

$= 0.44$

The decrease in calories is 44%.

47. $\text{percent decrease} = \dfrac{\text{amount of decrease}}{\text{original amount}}$

$= \dfrac{10,845 - 10,700}{10,845}$

$= \dfrac{145}{10,845}$

≈ 0.013

The decrease in cable TV systems was 1.3%.

49. $\text{percent increase} = \dfrac{\text{amount of increase}}{\text{original amount}}$

$= \dfrac{449 - 174}{174}$

$= \dfrac{275}{174}$

≈ 1.580

The increase in acres was 158.0.

51. $\text{percent decrease} = \dfrac{\text{amount of decrease}}{\text{original amount}}$

$= \dfrac{21.50 - 14.88}{21.50}$

$= \dfrac{6.62}{21.50}$

≈ 0.308

The decrease in the list price is 30.8%.

53. $\text{percent increase} = \dfrac{\text{amount of increase}}{\text{original amount}}$

$= \dfrac{3769 - 3570}{3570}$

$= \dfrac{199}{3570}$

≈ 0.056

The increase in elementary and secondary teachers is expected to be 5.6%.

55. $\text{percent decrease} = \dfrac{\text{amount of decrease}}{\text{original amount}}$

$= \dfrac{6903 - 5545}{6903}$

$= \dfrac{1358}{6903}$

≈ 0.197

The decrease in cinema sites was 19.7%.

57. $\text{percent increase} = \dfrac{\text{amount of increase}}{\text{original amount}}$

$= \dfrac{650,000 - 504,000}{504,000}$

$= \dfrac{146,000}{504,000}$

≈ 0.290

The percent increase in computer systems analysts is expected to be 29.0.

59. $\text{percent increase} = \dfrac{\text{amount of increase}}{\text{original amount}}$

$= \dfrac{247,081 - 178,025}{178,025}$

$= \dfrac{69,056}{178,025}$

≈ 0.388

The percent increase of cell sites was 38.8%.

61.
$$\begin{array}{r} 0.12 \\ \times\, 38 \\ \hline 96 \\ 360 \\ \hline 4.56 \end{array}$$

63.
$$\begin{array}{r} {}^{1\,1} \\ 9.20 \\ +\, 1.98 \\ \hline 11.18 \end{array}$$

65.
$$\begin{array}{r} 78.00 \\ -\, 19.46 \\ \hline 58.54 \end{array}$$

67. The increased number is double the original number.

69. To find the percent increase, she should have divided by the original amount, which is 150.

$\text{percent increase} = \dfrac{30}{150} = 0.2 = 20\%$

71. False; the percents are different.
percent increase from 1980 to 1990 = 20%

percent decrease from 1990 to 2000 = $16\dfrac{2}{3}\%$

Section 6.6 Practice

1. sales tax = tax rate · purchase price

$\qquad\qquad\quad \downarrow \qquad\quad \downarrow$

$\qquad = \quad 8.5\% \cdot \quad \$59.90$

$\qquad = 0.085 \cdot \$59.90$

$\qquad \approx \$5.09$

The sales tax is \$5.09.

Total Price = purchase price + sales tax

$\qquad\qquad\qquad\quad \downarrow \qquad\qquad\quad \downarrow$

$\qquad = \quad \$59.90 \quad + \ \$5.09$

$\qquad = \$64.99$

The sales tax on \$59.90 is \$5.09, and the total price is \$64.99.

2. sales tax = tax rate · purchase price

$\quad\ \downarrow \qquad\quad \downarrow \qquad\qquad \downarrow$

$\$1665 \ = \quad r \quad \cdot \quad \$18,500$

$\dfrac{1665}{18,500} = \dfrac{r \cdot 18,500}{18,500}$

$\qquad 0.09 = r$

The sales tax rate is 9%.

3. commission = commission rate · sales

$\qquad\qquad\qquad\quad \downarrow \qquad\qquad \downarrow$

$\qquad = \quad 6.6\% \quad \cdot \ \$47,632$

$\qquad = 0.066 \cdot \$47,632$

$\qquad \approx \$3143.712$

The sales representative's commission for the month is \$3143.71.

4. commission = commission rate · sales

$$\downarrow \qquad\qquad \downarrow \qquad\quad \downarrow$$

$$\$645 \quad = \quad\quad r \quad\cdot\quad \$4300$$

$$\frac{645}{4300}=r$$

$$0.15=r$$

$$15\%=r$$

The commission rate is 15%.

5. amount of discount
= discount rate · original price

$$\downarrow \qquad\qquad \downarrow$$

$$= \quad 35\% \quad\cdot\quad \$700$$

$$=0.35\cdot\$700$$

$$=\$245$$

The discount is $245.

sale price = original price − discount

$$\downarrow \qquad\qquad \downarrow$$

$$= \quad \$700 \quad - \quad \$245$$

$$=\$455$$

The sale price is $455.

Vocabulary and Readiness Check

1. <u>sales tax</u> = tax rate · purchase price.

2. <u>total price</u> = purchase price + sales tax.

3. <u>commission</u> = commission rate · sales.

4. <u>amount of discount</u>
= discount rate · original price.

5. <u>sale price</u> = original price − amount of discount.

6. sale price = original price − <u>amount of discount</u>

Exercise Set 6.6

1. sales tax = 5% · $150 = 0.05 · $150 = $7.50
The sales tax is $7.50.

3. sales tax = 7.5% · $799 = 0.075 · $799 ≈ $59.93
total price = $799 + $59.93 = $858.93
The total price of the camcorder is $858.93.

5. $335.30 = r · $4790

$$\frac{335.30}{4790}=r$$

$$0.07=r$$

The sales tax rate is 7%.

7. a. $10.20 = 8.5% · p
$10.2 = 0.085p

$$\frac{\$10.2}{0.085}=p$$

$$\$120=p$$

The purchase price of the table saw is $120.

b. total price = $120 + $10.20 = $130.20
The total price of the table saw is $130.20.

9. sales tax = 6.5% · $1800 = 0.065 · $1800 = $117
total price = $1800 + $117 = $1917
The sales tax was $117 and the total price of the bracelet is $1917.

11. $24.25 = 5% · p
$24.25 = 0.05p

$$\frac{\$24.25}{0.05}=p$$

$$\$485=p$$

The purchase price of the futon is $485.

13. $98.70 = r · $1645

$$\frac{98.70}{1645}=r$$

$$0.06=r$$

The sales tax rate is 6%.

15. Purchase price = $210 + $15 + $5 = $230
Sales tax = 7% · $230 = 0.07 · $230 = $16.10
Total price = $230 + $16.10 = $246.10
The sales tax is $16.10 and the total price of the items is $246.10.

17. commission = 4% · $1,329,401
= 0.04 · $1,329,401
= $53,176.04
Her commission was $53,176.04.

19. $1380.40 = r · $9860

$$\frac{\$1380.40}{\$9860}=r$$

$$0.14=r$$

The commission rate is 14%.

21. commission = 1.5% · $325,900
= 0.015 · $325,900
= $4888.50
His commission will be $4888.50.

23.
$$\$5565 = 3\% \cdot \text{sales}$$
$$\$5565 = 0.03 \cdot s$$
$$\frac{\$5565}{0.03} = s$$
$$\$185,500 = s$$

The selling price of the house was $185,500.

	Original Price	Discount Rate	Amount of Discount	Sale Price
25.	$89	10%	10% · $89 = $8.90	$89 − $8.90 = $80.10
27.	$196.50	50%	50% · $196.50 = $98.25	$196.50 − $98.25 = $98.25
29.	$410	35%	35% · $410 = $143.50	$410 − $143.50 = $266.50
31.	$21,700	15%	15% · $21,700 = $3255	$21,700 − $3255 = $18,445

33. discount = 15% · $300 = 0.15 · $300 = $45
sale price = $300 − $45 = $255
The discount is $45 and the sale price is $255.

	Purchase Price	Tax Rate	Sales Tax	Total Price
35.	$305	9%	9% · $305 = $27.45	$305 + $27.45 = $332.45
37.	$56	5.5%	5.5% · $56 = $3.08	$56 + $3.08 = $59.08

	Sale	Commission Rate	Commission
39.	$235,800	3%	$235,800 · 3% = $7074
41.	$17,900	$\dfrac{\$1432}{\$17,900} = 0.08 = 8\%$	$1432

43. $2000 \cdot \dfrac{3}{10} \cdot 2 = 600 \cdot 2 = 1200$

45. $400 \cdot \dfrac{3}{100} \cdot 11 = 12 \cdot 11 = 132$

47. $600 \cdot 0.04 \cdot \dfrac{2}{3} = 24 \cdot \dfrac{2}{3} = 16$

49. Round $68 to $70 and 9.5% to 10%.
10% · $70 = 0.10 · $70 = $7
$70 + $7 = $77
The best estimate of the total price is $77; d.

	Bill Amount	10%	15%	20%
51.	$40.21 \approx $40	$4.00	$4 + \dfrac{1}{2}($4) = $4 + $2 = $6.00	2($4) = $8.00
53.	$72.17 \approx $72.00	$7.20	$7.20 + \dfrac{1}{2}($7.20) = $7.20 + 3.60 $= 10.80	2($7.20) = $14.40

55. A discount of 60% is better; answers may vary.

57. $7.5\% \cdot \$24{,}966 = 0.075 \cdot \$24{,}966 = \$1872.45$
$\$24{,}966 + \$1872.45 = \$26{,}838.45$
The total price of the necklace is $26,838.45.

Section 6.7 Practice

1. $I = P \cdot R \cdot T$
$I = \$875 \cdot 7\% \cdot 5$
$\quad = \$875 \cdot (0.07) \cdot 5$
$\quad = \$306.25$
The simple interest is $306.25.

2. $I = P \cdot R \cdot T$
$I = \$1500 \cdot 20\% \cdot \dfrac{9}{12}$
$\quad = \$1500 \cdot (0.20) \cdot \dfrac{9}{12}$
$\quad = \$225$
She paid $225 in interest.

3. $I = P \cdot R \cdot T$
$\quad = \$2100 \cdot 13\% \cdot \dfrac{6}{12}$
$\quad = \$2100 \cdot (0.13) \cdot \dfrac{6}{12}$
$\quad = \$136.50$
The interest is $136.50.
$\begin{aligned} \text{total amount} &= \text{principal} + \text{interest} \\ &= \$2100 + \$136.50 \\ &= \$2236.50 \end{aligned}$
After 6 months, the total amount paid will be $2236.50.

4. $P = \$3000,\ r = 4\% = 0.04,\ n = 1,\ t = 6$ years
$A = P\left(1 + \dfrac{r}{n}\right)^{n \cdot t}$
$\quad = \$3000\left(1 + \dfrac{0.04}{1}\right)^{1 \cdot 6}$
$\quad = \$3000(1.04)^6$
$\quad \approx \$3795.96$
The total amount after 6 years is $3795.96.

5. $P = \$5500,\ r = 6\frac{1}{4}\% = 0.0625,\ n = 365,\ t = 5$

$$A = P\left(1 + \frac{r}{n}\right)^{n \cdot t}$$

$$= \$5500\left(1 + \frac{0.0625}{365}\right)^{365 \cdot 5}$$

$$\approx \$7517.41$$

The total amount after 5 years is $7517.41.

6. Since there are 12 months per year, the number of payments is $3 \cdot 12 = 36$.

$$\text{monthly payment} = \frac{\text{principal} + \text{interest}}{\text{total number of payments}}$$

$$= \frac{\$3000 + \$1123.58}{36}$$

$$= \frac{\$4123.58}{36}$$

$$\approx \$114.54$$

The monthly payment is about $114.54.

Calculator Explorations

1. $\text{compound interest factor} = \left(1 + \frac{0.09}{4}\right)^{4 \cdot 5}$

$$\approx 1.56051$$

2. $\text{compound interest factor} = \left(1 + \frac{0.14}{365}\right)^{365 \cdot 15}$

$$\approx 8.16288$$

3. $\text{compound interest factor} = \left(1 + \frac{0.11}{1}\right)^{1 \cdot 20}$

$$\approx 8.06231$$

4. $\text{compound interest factor} = \left(1 + \frac{0.07}{2}\right)^{2 \cdot 1}$

$$\approx 1.07123$$

5. $\text{compound interest factor} = \left(1 + \frac{0.06}{4}\right)^{4 \cdot 4}$

$$\text{total amount} = \$500 \cdot \left(1 + \frac{0.06}{4}\right)^{4 \cdot 4} \approx \$634.49$$

6. $\text{compound interest factor} = \left(1 + \frac{0.05}{365}\right)^{365 \cdot 19}$

$$\text{total amount} = \$2500 \cdot \left(1 + \frac{0.05}{365}\right)^{365 \cdot 19}$$

$$\approx \$6463.85$$

Vocabulary and Readiness Check

1. To calculate <u>simple</u> interest, use $I = P \cdot R \cdot T$.

2. To calculate <u>compound</u> interest, use
$$A = P\left(1 + \frac{r}{n}\right)^{n \cdot t}.$$

3. <u>Compound</u> interest is computed on not only the original principal, but on interest already earned in previous compounding periods.

4. When interest is computed on the original principal only, it is called <u>simple</u> interest.

5. <u>Total amount</u> (paid or received) = principal + interest.

6. The <u>principal amount</u> is the money borrowed, loaned, or invested.

Exercise Set 6.7

1. $\text{simple interest} = \text{principal} \cdot \text{rate} \cdot \text{time}$
$$= \$200 \cdot 8\% \cdot 2$$
$$= \$200 \cdot 0.08 \cdot 2$$
$$= \$32$$

3. $\text{simple interest} = \text{principal} \cdot \text{rate} \cdot \text{time}$
$$= \$160 \cdot 11.5\% \cdot 4$$
$$= \$160 \cdot 0.115 \cdot 4$$
$$= \$73.60$$

5. $\text{simple interest} = \text{principal} \cdot \text{rate} \cdot \text{time}$
$$= \$5000 \cdot 10\% \cdot 1\frac{1}{2}$$
$$= \$5000 \cdot 0.10 \cdot 1.5$$
$$= \$750$$

7. $\text{simple interest} = \text{principal} \cdot \text{rate} \cdot \text{time}$
$$= \$375 \cdot 18\% \cdot \frac{6}{12}$$
$$= \$375 \cdot 0.18 \cdot 0.5$$
$$= \$33.75$$

9. simple interest = principal · rate · time

$$= \$2500 \cdot 16\% \cdot \frac{21}{12}$$
$$= \$2500 \cdot 0.16 \cdot 1.75$$
$$= \$700$$

11. simple interest = principal · rate · time

$$= \$162,500 \cdot 12.5\% \cdot 5$$
$$= \$162,500 \cdot 0.125 \cdot 5$$
$$= \$101,562.50$$

$\$162,500 + \$101,562.50 = \$264,062.50$
The amount of interest is $\$101,562.50$.
The total amount paid back is $\$264,062.50$.

13. simple interest = principal · rate · time

$$= \$5000 \cdot 9\% \cdot \frac{15}{12}$$
$$= \$5000 \cdot 0.09 \cdot 1.25$$
$$= \$562.50$$

Total = $\$5000 + \$562.50 = \$5562.50$

15. Simple interest = principal · rate · time

$$= \$8500 \cdot 17\% \cdot 4$$
$$= \$8500 \cdot 0.17 \cdot 4$$
$$= \$5780$$

Total amount = $\$8500 + \$5780 = \$14,280$

17. $A = P\left(1 + \dfrac{r}{n}\right)^{n \cdot t}$

$$= 6150\left(1 + \frac{0.14}{2}\right)^{2 \cdot 15}$$
$$= 6150(1.07)^{30}$$
$$\approx 46,815.37$$

The total amount is $\$46,815.37$.

19. $A = P\left(1 + \dfrac{r}{n}\right)^{n \cdot t}$

$$= 1560\left(1 + \frac{0.08}{365}\right)^{365 \cdot 5}$$
$$= 1560\left(1 + \frac{0.08}{365}\right)^{1825}$$
$$\approx 2327.14$$

The total amount is $\$2327.14$.

21. $A = P\left(1 + \dfrac{r}{n}\right)^{n \cdot t}$

$$= 10,000\left(1 + \frac{0.09}{2}\right)^{2 \cdot 20}$$
$$= 10,000(1.045)^{40}$$
$$\approx 58,163.65$$

The total amount is $\$58,163.65$.

23. $A = P\left(1 + \dfrac{r}{n}\right)^{n \cdot t}$

$$= 2675\left(1 + \frac{0.09}{1}\right)^{1 \cdot 1}$$
$$= 2675(1.09)$$
$$= 2915.75$$

The total amount is $\$2915.75$.

25. $A = P\left(1 + \dfrac{r}{n}\right)^{n \cdot t}$

$$= 2000\left(1 + \frac{0.08}{1}\right)^{1 \cdot 5}$$
$$= 2000(1.08)^{5}$$
$$\approx 2938.66$$

The total amount is $\$2938.66$.

27. $A = P\left(1 + \dfrac{r}{n}\right)^{n \cdot t}$

$$= 2000\left(1 + \frac{0.08}{4}\right)^{4 \cdot 5}$$
$$= 2000(1.02)^{20}$$
$$\approx 2971.89$$

The total amount is $\$2971.89$.

29. monthly payment

$$= \frac{\text{principal} + \text{simple interest}}{\text{number of payments}}$$
$$= \frac{\$1500 + \$61.88}{6}$$
$$= \frac{\$1561.88}{6}$$
$$\approx \$260.31$$

The monthly payment is $\$260.31$.

31. monthly payment

$$= \frac{\text{principal} + \text{simple interest}}{\text{number of payments}}$$

$$= \frac{\$20,000 + \$10,588.70}{4 \cdot 12}$$

$$= \frac{\$30,588.70}{48}$$

$$\approx \$637.26$$

The monthly payment is $637.26.

33. perimeter = 10 + 6 + 10 + 6 = 32
The perimeter is 32 yards.

35. perimeter = 5 · 7 = 35
The perimeter is 35 meters.

37. answers may vary

39. Account 2 earns more; answers may vary

Chapter 6 Vocabulary Check

1. In a mathematical statement, <u>of</u> usually means "multiplication."

2. In a mathematical statement, <u>is</u> means "equals."

3. <u>Percent</u> means "per hundred."

4. <u>Compound interest</u> is computed not only on the principal, but also on interest already earned in previous compounding periods.

5. In the percent proportion $\dfrac{\text{amount}}{\text{base}} = \dfrac{\text{percent}}{100}$.

6. To write a decimal or fraction as a percent, multiply by <u>100%</u>.

7. The decimal equivalent of the % symbol is <u>0.01</u>.

8. The fraction equivalent of the % symbol is $\dfrac{1}{100}$.

9. The percent equation is <u>base</u> · percent = <u>amount</u>.

10. <u>Percent of decrease</u> = $\dfrac{\text{amount of decrease}}{\text{original amount}}$.

11. <u>Percent of increase</u> = $\dfrac{\text{amount of increase}}{\text{original amount}}$.

12. <u>Sales tax</u> = tax rate · purchase price.

13. <u>Total price</u> = purchase price + sales tax.

14. <u>Commission</u> = commission rate · sales.

15. <u>Amount of discount</u> = discount rate · original price.

16. <u>Sale price</u> = original price − amount of discount.

Chapter 6 Review

1. $\dfrac{37}{100} = 37\%$ of the adults preferred pepperoni.

2. $\dfrac{77}{100} = 77\%$ of the free throws were made.

3. $83\% = 83 \cdot 0.01 = 0.83$

4. $75\% = 75 \cdot 0.01 = 0.75$

5. $73.5\% = 73.5 \cdot 0.01 = 0.735$

6. $1.5\% = 1.5 \cdot 0.01 = 0.015$

7. $125\% = 125 \cdot 0.01 = 1.25$

8. $145\% = 145 \cdot 0.01 = 1.45$

9. $0.5\% = 0.5 \cdot 0.01 = 0.005$

10. $0.7\% = 0.7 \cdot 0.01 = 0.007$

11. $200\% = 200 \cdot 0.01 = 2.00$ or 2

12. $400\% = 400 \cdot 0.01 = 4.00$ or 4

13. $26.25\% = 26.25 \cdot 0.01 = 0.2625$

14. $85.34\% = 85.34 \cdot 0.01 = 0.8534$

15. $2.6 = 2.6 \cdot 100\% = 260\%$

16. $1.02 = 1.02 \cdot 100\% = 102\%$

17. $0.35 = 0.35 \cdot 100\% = 35\%$

18. $0.055 = 0.055 \cdot 100\% = 5.5\%$

19. $0.725 = 0.725 \cdot 100\% = 72.5\%$

20. $0.252 = 0.252 \cdot 100\% = 25.2\%$

21. $0.076 = 0.076 \cdot 100\% = 7.6\%$

22. $0.085 = 0.085 \cdot 100\% = 8.5\%$

23. $0.71 = 0.71 \cdot 100\% = 71\%$

24. $0.65 = 0.65 \cdot 100\% = 65\%$

25. $4 = 4 \cdot 100\% = 400\%$

26. $9 = 9 \cdot 100\% = 900\%$

27. $1\% = \dfrac{1}{100}$

28. $10\% = \dfrac{10}{100} = \dfrac{1}{10}$

29. $25\% = \dfrac{25}{100} = \dfrac{1}{4}$

30. $8.5\% = \dfrac{8.5}{100} = \dfrac{85}{1000} = \dfrac{17}{200}$

31. $10.2\% = \dfrac{10.2}{100} = \dfrac{102}{1000} = \dfrac{51}{500}$

32. $16\dfrac{2}{3}\% = \dfrac{50}{3} = \dfrac{50}{3} \cdot \dfrac{1}{100} = \dfrac{50}{300} = \dfrac{1}{6}$

33. $33\dfrac{1}{3}\% = \dfrac{100}{3} = \dfrac{100}{3} \cdot \dfrac{1}{100} = \dfrac{1}{3}$

34. $110\% = \dfrac{110}{100} = \dfrac{11}{10}$ or $1\dfrac{1}{10}$

35. $\dfrac{1}{5} = \dfrac{1}{5} \cdot 100\% = \dfrac{100}{5}\% = 20\%$

36. $\dfrac{7}{10} = \dfrac{7}{10} \cdot 100\% = \dfrac{700}{10}\% = 70\%$

37. $\dfrac{5}{6} = \dfrac{5}{6} \cdot 100\% = \dfrac{500}{6}\% = \dfrac{250}{3}\% = 83\dfrac{1}{3}\%$

38. $\dfrac{3}{5} = \dfrac{3}{5} \cdot 100\% = \dfrac{300}{5}\% = 60\%$

39. $1\dfrac{1}{4} = \dfrac{5}{4} = \dfrac{5}{4} \cdot 100\% = \dfrac{500}{4}\% = 125\%$

40. $1\dfrac{2}{3} = \dfrac{5}{3} = \dfrac{5}{3} \cdot 100\% = \dfrac{500}{3}\% = 166\dfrac{2}{3}\%$

41. $\dfrac{1}{16} = \dfrac{1}{16} \cdot 100\% = \dfrac{100}{16}\% = \dfrac{25}{4}\% = 6.25\%$

42. $\dfrac{5}{8} = \dfrac{5}{8} \cdot 100\%$

$= \dfrac{500}{8}\%$

$= \dfrac{125}{2}\%$

$= 62.5\%$

43. $1250 = 1.25\% \cdot n$
$1250 = 0.0125 \cdot n$
$n = \dfrac{1250}{0.0125}$
$n = 100,000$
1250 is 1.25% of 100,000.

44. $n = 33\dfrac{1}{3}\% \cdot 24,000$

$n = \dfrac{100}{3}\% \cdot 24,000$

$n = \dfrac{100}{3} \cdot \dfrac{1}{100} \cdot 24,000$

$n = 8000$

8000 is $33\dfrac{1}{3}\%$ of 24,000.

45. $124.2 = n \cdot 540$
$n = \dfrac{124.2}{540}$
$n = 0.23$
$n = 23\%$
124.2 is 23% of 540.

46. $22.9 = 20\% \cdot n$
$22.9 = 0.20 \cdot n$
$n = \dfrac{22.9}{0.20}$
$n = 114.5$
22.9 is 20% of 114.5.

47. $n = 40\% \cdot 7500$
$n = 0.40 \cdot 7500 = 3000$
3000 is 40% of 7500.

48. $693 = n \cdot 462$
$n = \dfrac{693}{462}$
$n = 1.5$
$n = 150\%$
693 is 150% of 462.

49. $\dfrac{104.5}{b} = \dfrac{25}{100}$

$25b = 104.5 \cdot 100$

$25b = 10{,}450$

$b = \dfrac{10{,}450}{25}$

$b = 418$

104.5 is 25% of 418.

50. $\dfrac{16.5}{b} = \dfrac{5.5}{100}$

$5.5b = 16.5 \cdot 100$

$5.5b = 1650$

$b = \dfrac{1650}{5.5}$

$b = 300$

16.5 is 5.5% of 300.

51. $\dfrac{a}{180} = \dfrac{36}{100}$

$100a = 180 \cdot 36$

$100a = 6480$

$a = \dfrac{6480}{100}$

$a = 64.8$

64.8 is 36% of 180.

52. $\dfrac{63}{35} = \dfrac{p}{100}$

$35p = 63 \cdot 100$

$35p = 6300$

$p = \dfrac{6300}{35}$

$p = 180$

63 is 180% of 35.

53. $\dfrac{93.5}{85} = \dfrac{p}{100}$

$85p = 93.5 \cdot 100$

$85p = 9350$

$p = \dfrac{9350}{85}$

$p = 110$

93.5 is 110% of 85.

54. $\dfrac{a}{500} = \dfrac{33}{100}$

$100a = 500 \cdot 33$

$100a = 16{,}500$

$a = \dfrac{16{,}500}{100}$

$a = 165$

165 is 33% of 500.

55. 1320 is what percent of 2000?
Method 1

$1320 = n \cdot 2000$

$\dfrac{1320}{2000} = n$

$0.66 = n$

$66\% = n$

66% of people own microwaves.
Method 2

$\dfrac{1320}{2000} = \dfrac{p}{100}$

$1320 \cdot 100 = 2000p$

$132{,}000 = 2000p$

$\dfrac{132{,}000}{2000} = p$

$66 = p$

66% of people own microwaves.

56. 2000 is what percent of 12,360?
Method 1

$2000 = n \cdot 12{,}360$

$\dfrac{2000}{12{,}360} = n$

$0.1618 \approx n$

$16\% \approx n$

16% of the entering freshmen are enrolled in basic college mathematics.
Method 2

$\dfrac{2000}{12{,}360} = \dfrac{p}{100}$

$2000 \cdot 100 = 12{,}360p$

$200{,}000 = 12{,}360p$

$\dfrac{200{,}000}{12{,}360} = p$

$16 \approx p$

16% of the entering freshmen are enrolled in basic college mathematics.

57. percent decrease $= \dfrac{\text{amount of decrease}}{\text{original amount}}$

$\quad = \dfrac{675 - 534}{675}$

$\quad = \dfrac{141}{675}$

$\quad \approx 0.209$

$\quad \approx 20.9\%$

58. percent increase $= \dfrac{\text{amount of increase}}{\text{original amount}}$

$\quad = \dfrac{33 - 16}{16}$

$\quad = \dfrac{17}{16}$

$\quad = 1.0625$

$\quad = 106.25\%$

59. amount of decrease $= 4\% \cdot \$215,000$

$\quad = 0.04 \cdot \$215,000$

$\quad = \$8600$

expected amount $= \$215,000 - \$8600$

$\quad = \$206,400$

60. amount of increase $= 15\% \cdot \$11.50$

$\quad = 0.15 \cdot \$11.50$

$\quad = \$1.725$

$\quad \approx \$1.73$

new hourly rate $= \$11.50 + \$1.73 = \$13.23$

61. sales tax $= 5.5\% \cdot \$250 = 0.055 \cdot \$250 = \$13.75$

total amount $= \$250 + \$13.75 = \$263.75$

62. sales tax $= 4.5\% \cdot \$25.50$

$\quad = 0.045 \cdot \$25.50$

$\quad = \$1.1475$

$\quad \approx \$1.15$

63. commission $= 5\% \cdot \$100,000$

$\quad = 0.05 \cdot \$100,000$

$\quad = \$5000$

64. commission $= 7.5\% \cdot \$4005$

$\quad = 0.075 \cdot \$4005$

$\quad = \$300.375$

$\quad \approx \$300.38$

65. discount $= 30\% \cdot \$3000 = 0.30 \cdot \$3000 = \$900$

sale price $= \$3000 - \$900 = \$2100$

66. discount $= 10\% \cdot \$90 = 0.10 \cdot \$90 = \$9$

sale price $= \$90 - \$9 = \$81$

67. simple interest $= \text{principal} \cdot \text{rate} \cdot \text{time}$

$\quad = \$4000 \cdot 12\% \cdot \dfrac{4}{12}$

$\quad = \$4000 \cdot 0.12 \cdot \dfrac{1}{3}$

$\quad = \$160$

68. simple interest $= \text{principal} \cdot \text{rate} \cdot \text{time}$

$\quad = \$6500 \cdot 20\% \cdot \dfrac{3}{12}$

$\quad = \$6500 \cdot 0.20 \cdot \dfrac{1}{4}$

$\quad = \$325$

69. total amount

$= \text{original principal} \cdot \text{compound interest factor}$

$= \$5500 \cdot 5.47357$

$\approx \$30,104.64$

70. total amount

$= \text{original principal} \cdot \text{compound interest factor}$

$= \$6000 \cdot 2.91776$

$= \$17,506.56$

71. total amount

$= \text{original principal} \cdot \text{compound interest factor}$

$= \$100 \cdot 1.80611$

$\approx \$180.61$

compound interest

$= \text{total amount} - \text{original principal}$

$= \$180.61 - \$100$

$= \$80.61$

72. total amount

$= \text{original principal} \cdot \text{compound interest factor}$

$= \$1000 \cdot 33.83010$

$= \$33,830.10$

compound interest

$= \text{total amount} - \text{original principal}$

$= \$33,830.10 - \$1000$

$\approx \$32,830.10$

73. $3.8\% = 3.8 \cdot 0.01 = 0.038$

74. $24.5\% = 24.5 \cdot 0.01 = 0.245$

75. $0.9\% = 0.9 \cdot 0.01 = 0.009$

76. $0.54 = 0.54 \cdot 100\% = 54\%$

77. $95.2 = 95.2 \cdot 100\% = 9520\%$

78. $0.3 = 0.3 \cdot 100\% = 30\%$

79. $47\% = 47 \cdot \dfrac{1}{100} = \dfrac{47}{100}$

80. $6\dfrac{2}{5}\% = \dfrac{32}{5} \cdot \dfrac{1}{100} = \dfrac{32}{500} = \dfrac{8}{125}$

81. $5.6\% = 5.6 \cdot \dfrac{1}{100} = \dfrac{5.6}{100} = \dfrac{56}{1000} = \dfrac{7}{125}$

82. $\dfrac{3}{8} = \dfrac{3}{8} \cdot 100\% = \dfrac{300}{8}\% = \dfrac{75}{2}\% = 37\dfrac{1}{2}\%$

83. $\dfrac{2}{13} = \dfrac{2}{13} \cdot 100\% = \dfrac{200}{13}\% = 15\dfrac{5}{13}\%$

84. $\dfrac{6}{5} = \dfrac{6}{5} \cdot 100\% = \dfrac{600}{5}\% = 120\%$

85. $43 = 16\% \cdot n$
$43 = 0.16 \cdot n$
$n = \dfrac{43}{0.16}$
$n = 268.75$
43 is 16% of 268.75.

86. $27.5 = n \cdot 25$
$n = \dfrac{27.5}{25}$
$n = 1.1$
$n = 110\%$
27.5 is 110% of 25.

87. $n = 36\% \cdot 1968$
$n = 0.36 \cdot 1968$
$n = 708.48$
708.48 is 36% of 1968.

88. $67 = n \cdot 50$
$n = \dfrac{67}{50}$
$n = 1.34$
$n = 134\%$
67 is 134% of 50.

89. $\dfrac{75}{25} = \dfrac{p}{100}$
$25p = 75 \cdot 100$
$25p = 7500$
$p = \dfrac{7500}{25}$
$p = 300$
75 is 300% of 25.

90. $\dfrac{a}{240} = \dfrac{16}{100}$
$100a = 240 \cdot 16$
$100a = 3840$
$a = \dfrac{3840}{100}$
$a = 38.4$
38.4 is 16% of 240.

91. $\dfrac{28}{b} = \dfrac{5}{100}$
$5b = 28 \cdot 100$
$5b = 2800$
$b = \dfrac{2800}{5}$
$b = 560$
28 is 5% of 560.

92. $\dfrac{52}{16} = \dfrac{p}{100}$
$16p = 52 \cdot 100$
$16p = 5200$
$p = \dfrac{5200}{16}$
$p = 325$
52 is 325% of 16.

93. 78 is what percent of 300?
Method 1
$78 = n \cdot 300$
$\dfrac{78}{300} = n$
$0.26 = n$
$26\% = n$
26% of the soft drink cans have been sold.

Method 2

$$\frac{78}{300} = \frac{p}{100}$$

$$78 \cdot 100 = 300\,p$$

$$7800 = 300\,p$$

$$\frac{7800}{300} = p$$

$$26 = p$$

26% of the soft drink cans have been sold.

94. What number is 7% of 96,950?

Method 1

$n = 7\% \cdot 96{,}950$

$n = 0.07 \cdot 96{,}950$

$n = 6786.5$

The loss of value was $6786.50.

Method 2

$$\frac{a}{96{,}950} = \frac{7}{100}$$

$$100a = 7 \cdot 96{,}950$$

$$100a = 678{,}650$$

$$a = \frac{678{,}650}{100}$$

$$a = 6786.5$$

The loss of value was $6786.50.

95. sales tax $= 8.75\% \cdot \$568$

$\qquad\quad = 0.0875 \cdot \$568$

$\qquad\quad = \$49.70$

purchase price $= \$568 + \$49.70 = \$617.70$

96. discount $= 15\% \cdot \$23 = 0.15 \ \$23 = \$3.45$

97. commission rate $= \dfrac{\$1.60}{\$12.80} = 0.125 = 12.5\%$

98. simple interest $=$ principal $\cdot$ rate $\cdot$ time

$$= \$1400 \cdot 13\% \cdot \frac{6}{12}$$

$$= \$1400 \cdot 0.13 \cdot \frac{1}{2}$$

$$= \$91$$

total amount $= \$1400 + \$91 = \$1491$

99. $A = P\left(1 + \dfrac{r}{n}\right)^{n \cdot t}$

$$= 8800\left(1 + \frac{0.08}{4}\right)^{4 \cdot 9}$$

$$= 8800(2.0399)$$

$$= 17{,}951.01$$

total amount $= \$17{,}951.01$

100. simple interest $=$ principal $\cdot$ rate $\cdot$ time

$$= \$5500 \cdot 12.5\% \cdot 9$$

$$= \$5500 \cdot 0.125 \cdot 9$$

$$= \$6187.50$$

total amount $= \$5500 + \$6187.50 = \$11{,}687.50$

Chapter 6 Test

1. $85\% = 85(0.01) = 0.85$

2. $500\% = 500 \cdot 0.01 = 5.00$ or 5

3. $0.8\% = 0.8 \cdot 0.01 = 0.008$

4. $0.056 = 0.056 \cdot 100\% = 5.6\%$

5. $6.1 = 6.1 \cdot 100\% = 610\%$

6. $0.39 = 0.39 \cdot 100\% = 39\%$

7. $120\% = 120 \cdot \dfrac{1}{100} = \dfrac{120}{100} = \dfrac{6}{5}$

8. $38.5\% = 38.5 \cdot \dfrac{1}{100} = \dfrac{38.5}{100} = \dfrac{385}{1000} = \dfrac{77}{200}$

9. $0.2\% = 0.2 \cdot \dfrac{1}{100} = \dfrac{0.2}{100} = \dfrac{2}{1000} = \dfrac{1}{500}$

10. $\dfrac{11}{20} = \dfrac{11}{20} \cdot 100\% = \dfrac{1100}{20}\% = 55\%$

11. $\dfrac{3}{8} = \dfrac{3}{8} \cdot 100\% = \dfrac{300}{8}\% = 37.5\%$

12. $1\dfrac{5}{9} = \dfrac{14}{9} = \dfrac{14}{9} \cdot 100\% = \dfrac{1400}{9}\% = 155\dfrac{5}{9}\%$

13. $n = 42\% \cdot 80$

$n = 0.42 \cdot 80$

$n = 33.6$

42% of 80 is 33.6.

14. $7.5 = 0.6\% \cdot n$

$7.5 = 0.006 \cdot n$

$n = \dfrac{7.5}{0.006}$

$n = 1250$

0.6% of 1250 is 7.5.

15. $567 = n \cdot 756$

$$\frac{567}{756} = n$$

$$0.75 = n$$

$$75\% = n$$

567 is 75% of 756.

16. $12\% \cdot 320 = 0.12 \cdot 320 = 38.4$
The alloy contains 38.4 pounds of copper.

17.   20% of what number is $11,350?

$$0.20n = \$11,350$$

$$n = \frac{\$11,350}{0.20}$$

$$n = \$56,750$$

The value is $56,750.

18.   sales tax $= 1.25\% \cdot \$354$

$$= 0.0125 \cdot \$354$$

$$= \$4.425$$

$$\approx \$4.43$$

total price $= \$354 + \$4.43 = \$358.43$

19.   percent increase $= \dfrac{\text{amount of increase}}{\text{original amount}}$

$$= \frac{26,460 - 25,200}{25,200}$$

$$= \frac{1260}{25,200}$$

$$= 0.05$$

$$= 5\%$$

20.   discount $= 15\% \cdot \$120 = 0.15 \cdot \$120 = \$18$
sale price $= \$120 - \$18 = \$102$

21.   commission $= 4\% \cdot \$9875 = 0.04 \cdot \$9875 = \$395$
His commission is $395.

22.   tax rate $= \dfrac{\$1.53}{\$152.99} \approx 0.01$
The tax rate is 1%.

23.   simple interest $=$ principal $\cdot$ rate $\cdot$ time

$$= \$2000 \cdot 9.25\% \cdot 3\frac{1}{2}$$

$$= \$2000 \cdot 0.0925 \cdot 3.5$$

$$= \$647.50$$

24.   From Appendix A.7, the compound interest factor for 5 years at 8% compounded annually is 1.46933.
total amount $= \$1365 \cdot 1.46933 \approx \$2005.64$

25.   simple interest $=$ principal $\cdot$ rate $\cdot$ time

$$= \$400 \cdot 13.5\% \cdot \frac{6}{12}$$

$$= \$400 \cdot 0.135 \cdot 0.5$$

$$= \$27.00$$

Total amount due the bank $= \$400 + \$27 = \$427$

Cumulative Review Chapters 1–6

1.
$$58\overline{)9900} \quad \frac{206 \text{ R } 12}{}$$
$$\underline{-96}$$
$$30$$
$$\underline{-0}$$
$$300$$
$$\underline{-288}$$
$$12$$

The cans will fill 206 cases with 12 cans left over. There will be enough cans to fill the order.

2.
$$\begin{array}{r} 409 \\ \times\ 76 \\ \hline 2\ 454 \\ 28\ 630 \\ \hline 31,084 \end{array}$$

3. a. $\dfrac{30}{7} = 4\dfrac{2}{7}$

    b. $\dfrac{16}{15} = 1\dfrac{1}{15}$

    c. $\dfrac{84}{6} = 14$

4. a. $2\dfrac{5}{7} = \dfrac{2 \cdot 7 + 5}{7} = \dfrac{14 + 5}{7} = \dfrac{19}{7}$

    b. $10\dfrac{1}{10} = \dfrac{10 \cdot 10 + 1}{10} = \dfrac{100 + 1}{10} = \dfrac{101}{10}$

    c. $5\dfrac{3}{8} = \dfrac{5 \cdot 8 + 3}{8} = \dfrac{40 + 3}{8} = \dfrac{43}{8}$

5. Other factor trees are possible.

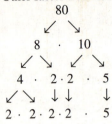

The factorization is $80 = 2 \cdot 2 \cdot 2 \cdot 2 \cdot 5$ or $2^4 \cdot 5$.

6. area $= 17$ miles $\cdot 7$ mile $= 119$ square miles

7. Since 10 and 27 have no common factors, $\dfrac{10}{27}$ is already in simplest form.

8. The average is the sum, divided by 3.
$$\frac{28 + 34 + 70}{3} = \frac{132}{3} = 44$$
The average is 44.

9. $\dfrac{23}{32} \cdot \dfrac{4}{7} = \dfrac{23 \cdot 4}{4 \cdot 8 \cdot 7} = \dfrac{23}{8 \cdot 7} = \dfrac{23}{56}$

10. To round 76,498 to the nearest ten, observe that the digit in the ones place is 8. Since this digit is at least 5, we add 1 to the digit in the tens place. The number 76,498, rounded to the nearest ten is 76,500.

11. The reciprocal of $\dfrac{11}{8}$ is $\dfrac{8}{11}$.

12. Each part is $\dfrac{1}{4}$ of a whole and there are 15 parts shaded, or 3 wholes and 3 more parts.
improper fraction: $\dfrac{15}{4}$; mixed number: $3\dfrac{3}{4}$

13. perimeter $= 2 \cdot$ length $+ 2 \cdot$ width
$$= 2 \cdot \frac{4}{15} \text{ in.} + 2 \cdot \frac{2}{15} \text{ in.}$$
$$= \frac{8}{15} \text{ in.} + \frac{4}{15} \text{ in.}$$
$$= \frac{12}{15} \text{ in.}$$
$$= \frac{4}{5} \text{ in.}$$

14. $2 \cdot 5^2 = 2 \cdot 25 = 50$

15. The first few multiples of 20 are:
$20 \cdot 1 = 20$ (not a multiple of 12)
$20 \cdot 2 = 40$ (not a multiple of 12)
$20 \cdot 3 = 60 = 5 \cdot 12$ (a multiple of 12)
The LCM of 12 and 20 is 60.

16. $\dfrac{10}{9} - \dfrac{7}{9} = \dfrac{10 - 7}{9} = \dfrac{3}{9} = \dfrac{1}{3}$

17. The LCD of 5 and 15 is 15.
$$\frac{2}{5} + \frac{4}{15} = \frac{6}{15} + \frac{4}{15} = \frac{6 + 4}{15} = \frac{10}{15} = \frac{2}{3}$$

18. $\dfrac{2}{3} \cdot 510 = \dfrac{2 \cdot 510}{3} = \dfrac{1020}{3} = 340$

19. The LCD of 14 and 7 is 14.
$$\begin{array}{ccc} 7\dfrac{3}{14} & 7\dfrac{3}{14} & 6\dfrac{17}{14} \\[2mm] -3\dfrac{6}{7} & -3\dfrac{12}{14} & -3\dfrac{12}{14} \\ \hline & & 3\dfrac{5}{14} \end{array}$$

20. $9 \cdot \sqrt{25} - 6 \cdot \sqrt{4} = 9 \cdot 5 - 6 \cdot 2 = 45 - 12 = 33$

21. $\dfrac{1}{2} \div \dfrac{8}{7} = \dfrac{1}{2} \cdot \dfrac{7}{8} = \dfrac{7}{16}$

22. The LCD of 5 and 8 is 40.
$$\begin{array}{cc} 20\dfrac{4}{5} & 20\dfrac{32}{40} \\[2mm] +12\dfrac{7}{8} & +12\dfrac{35}{40} \\ \hline & 32\dfrac{67}{40} = 32 + 1\dfrac{27}{40} = 33\dfrac{27}{40} \end{array}$$

23. $\dfrac{2}{9} \cdot \dfrac{3}{11} = \dfrac{2 \cdot 3}{3 \cdot 3 \cdot 11} = \dfrac{2}{33}$

24. $1\dfrac{7}{8} \cdot 3\dfrac{2}{5} = \dfrac{15}{8} \cdot \dfrac{17}{5} = \dfrac{3 \cdot 5 \cdot 17}{8 \cdot 5} = \dfrac{51}{8}$ or $6\dfrac{3}{8}$

25. $\dfrac{8}{10} = 0.8$

26. $\dfrac{9}{100} = 0.09$

27. $\dfrac{87}{10} = 8.7$

28. $\dfrac{48}{10,000} = 0.0048$

29. To round 3.1779 to the nearest hundredth, observe that the digit in the thousandths place is 7. Since this digit is at least 5, we add 1 to the digit in the hundredths place. The number 3.1779, rounded to the nearest hundredth, is 3.18. Thus $3.1779 \approx 3.18$.

30.
$$\begin{array}{r} 38.00 \\ -\,10.06 \\ \hline 27.94 \end{array}$$

31.
$$\begin{array}{r} {}^{1\ 1\ 1} \\ 763.7651 \\ 22.0010 \\ +\,43.8900 \\ \hline 829.6561 \end{array}$$

32. $12.483 \times 100 = 1248.3$

33.
$$\begin{array}{r} 23.6 \\ \times\,0.78 \\ \hline 1\,888 \\ 16\,520 \\ \hline 18.408 \end{array}$$

34. $76.3 \times 1000 = 76{,}300$

35. $\dfrac{786.1}{1000} = 0.7861$

36. $0.5\overline{)0.638}$ becomes
$$\begin{array}{r} 1.276 \\ 5\overline{)6.380} \\ \underline{-5} \\ 13 \\ \underline{-10} \\ 38 \\ \underline{-35} \\ 30 \\ \underline{-30} \\ 0 \end{array}$$

37. $\dfrac{0.12}{10} = 0.012$

38. $0.23\overline{)11.6495}$ becomes
$$\begin{array}{r} 50.65 \\ 23\overline{)1164.95} \\ \underline{-115} \\ 14 \\ \underline{-0} \\ 14\,9 \\ \underline{-13\,8} \\ 1\,15 \\ \underline{-1\,15} \\ 0 \end{array}$$

39. $723.6 \div 1000 \times 10 = 0.7236 \times 10 = 7.236$

40. $\dfrac{3.19 - 0.707}{13} = \dfrac{2.483}{13} = 0.191$

41. $\dfrac{1}{4} = \dfrac{1}{4} \cdot \dfrac{25}{25} = \dfrac{25}{100} = 0.25$

42.
$$\begin{array}{r} 0.5555\ldots \\ 9\overline{)5.000} \\ \underline{-4\,5} \\ 50 \\ \underline{-45} \\ 50 \\ \underline{-45} \\ 5 \end{array}$$

$\dfrac{5}{9} = 0.\overline{5} \approx 0.556$

43. $\dfrac{10 \text{ nails}}{6 \text{ feet}} = \dfrac{5 \text{ nails}}{3 \text{ feet}}$

44. $\dfrac{115 \text{ miles}}{5 \text{ gallons}} = \dfrac{23 \text{ miles}}{1 \text{ gallon}} = 23 \text{ miles/gallons}$

45. The cross products are $4.1 \cdot 5 = 20.5$ and $7 \cdot 2.9 = 20.3$. Since these are not equal, the proportion is not true.

46.
$$\begin{array}{r} 0.0516 \approx 0.052 \\ 18\overline{)0.9300} \\ \underline{-90} \\ 30 \\ \underline{-18} \\ 120 \\ \underline{-108} \\ 12 \end{array}$$

$0.93 for 18 flour tortillas is $0.052/tortilla.

$$\begin{array}{r} 0.0583 \approx 0.058 \\ 24\overline{)1.4000} \\ \underline{-1\ 20} \\ 200 \\ \underline{-192} \\ 80 \\ \underline{-72} \\ 8 \end{array}$$

$1.40 for 24 flour tortillas is $0.058/tortilla.
$0.93 for 18 tortillas is a better buy.

47. n miles is to 7 inches as 5 miles is to 2 inches.

$$\frac{n}{5} = \frac{7}{2}$$
$$2n = 5 \cdot 7$$
$$2n = 35$$
$$n = \frac{35}{2}$$
$$n = 17.5$$

17.5 miles correspond to 7 inches.

48. a. $7\% = 7 \cdot 0.01 = 0.07$

b. $200\% = 200 \cdot 0.01 = 2.00$ or 2

c. $0.5\% = 0.5 \cdot 0.01 = 0.005$

49. $n = 25\% \cdot 0.008$

50. $\dfrac{3}{8} = \dfrac{3}{8} \cdot 100\% = \dfrac{300}{8}\% = 37.5\%$ or $37\dfrac{1}{2}\%$

Chapter 7

Section 7.1 Practice

1. $6 \text{ ft} = \frac{6 \text{ ft}}{1} \cdot 1 = \frac{6 \text{ ft}}{1} \cdot \frac{12 \text{ in.}}{1 \text{ ft}} = 6 \cdot 12 \text{ in.} = 72 \text{ in.}$

2. $8 \text{ yd} = \frac{8 \text{ yd}}{1} \cdot 1 = \frac{8 \text{ yd}}{1} \cdot \frac{3 \text{ ft}}{1 \text{ yd}} = 8 \cdot 3 \text{ ft} = 24 \text{ ft}$

3. $18 \text{ in.} = \frac{18 \text{ in.}}{1} \cdot \frac{1 \text{ ft}}{12 \text{ in.}} = \frac{18}{12} \text{ ft} = 1.5 \text{ ft}$

4. $68 \text{ in.} = \frac{68 \text{ in.}}{1} \cdot \frac{1 \text{ ft}}{12 \text{ in.}} = \frac{68}{12} \text{ ft}$

$$12 \overline{) \begin{array}{r} 5 \\ 68 \\ -60 \\ \hline 8 \end{array}}$$

Thus, 68 in. = 5 ft 8 in.

5. $5 \text{ yd} = \frac{5 \text{ yd}}{1} \cdot \frac{3 \text{ ft}}{1 \text{ yd}} = 15 \text{ ft}$

$5 \text{ yd } 2 \text{ ft} = 15 \text{ ft} + 2 \text{ ft} = 17 \text{ ft}$

6. $\begin{array}{r} 4 \text{ ft} \quad 8 \text{ in.} \\ + 8 \text{ ft } 11 \text{ in.} \\ \hline 12 \text{ ft } 19 \text{ in.} \end{array}$

Since 19 inches is the same as 1 ft 7 in., we have
12 ft 19 in. = 12 ft + 1 ft 7 in. = 13 ft 7 in.

7. $\begin{array}{r} 4 \text{ ft} \quad 7 \text{ in.} \\ \times \qquad\qquad 4 \\ \hline 16 \text{ ft } 28 \text{ in.} \end{array}$

Since 28 in. is the same as 2 ft 4 in., we simplify
as 16 ft 28 in. = 16 ft + 2 ft 4 in. = 18 ft 4 in.

8. $\begin{array}{rr} 5 \text{ ft } 8 \text{ in.} \rightarrow & 4 \text{ ft } 20 \text{ in.} \\ -1 \text{ ft } 9 \text{ in.} & -1 \text{ ft} \quad 9 \text{ in.} \\ \hline & 3 \text{ ft } 11 \text{ in.} \end{array}$

The remaining board length is 3 ft 11 in.

9. $2.5 \text{ m} = \frac{2.5 \text{ m}}{1} \cdot \frac{1000 \text{ mm}}{1 \text{ m}} = 2500 \text{ mm}$

10. $3500 \text{ m} = 3.500 \text{ km or } 3.5 \text{ km}$

11. $640 \text{ m} = 0.64 \text{ km}$

$\begin{array}{r} 2.10 \text{ km} \\ - 0.64 \text{ km} \\ \hline 1.46 \text{ km} \end{array}$

$2.1 \text{ km} = 2100 \text{ m}$

$\begin{array}{r} 2100 \text{ m} \\ - 640 \text{ m} \\ \hline 1460 \text{ m} \end{array}$

12. $\begin{array}{r} 18.3 \text{ hm} \\ \times \quad 5 \\ \hline 91.5 \text{ hm} \end{array}$

13. $1.72 \text{ m} = 172 \text{ cm}$
$172 \text{ cm} - 55 \text{ cm} = 117 \text{ cm}$
She grew 117 cm or 1.17 m.

14. $0.8 \text{ m} = 80 \text{ cm}$
$80 \text{ cm} + 45 \text{ cm} = 125 \text{ cm}$
The scarf will be 125 cm or 1.25 m.

Vocabulary and Readiness Check

1. The basic unit of length in the metric system is the <u>meter</u>.

2. The expression $\frac{1 \text{ foot}}{12 \text{ inches}}$ is an example of a <u>unit fraction</u>.

3. A meter is slightly longer than a <u>yard</u>.

4. One foot equals 12 <u>inches</u>.

5. One yard equals 3 <u>feet</u>.

6. One yard equals 36 <u>inches</u>.

7. One mile equals 5280 <u>feet</u>.

8. One foot equals $\frac{1}{3}$ <u>yard</u>.

Exercise Set 7.1

1. $60 \text{ in.} = \frac{60 \text{ in.}}{1} \cdot \frac{1 \text{ ft}}{12 \text{ in.}} = \frac{60}{12} \text{ ft} = 5 \text{ ft}$

3. $12 \text{ yd} = \frac{12 \text{ yd}}{1} \cdot \frac{3 \text{ ft}}{1 \text{ yd}} = 12 \cdot 3 \text{ ft} = 36 \text{ ft}$

5. $42,240 \text{ ft} = \dfrac{42,240 \text{ ft}}{1} \cdot \dfrac{1 \text{ mi}}{5280 \text{ ft}}$

$\qquad\qquad = \dfrac{42,240}{5280} \text{ mi}$

$\qquad\qquad = 8 \text{ mi}$

7. $8\dfrac{1}{2} \text{ ft} = \dfrac{8.5 \text{ ft}}{1} \cdot \dfrac{12 \text{ in.}}{1 \text{ ft}} = 8.5 \cdot 12 \text{ in.} = 102 \text{ in.}$

9. $10 \text{ ft} = \dfrac{10 \text{ ft}}{1} \cdot \dfrac{1 \text{ yd}}{3 \text{ ft}} = \dfrac{10}{3} \text{ yd} = 3\dfrac{1}{3} \text{ yd}$

11. $6.4 \text{ mi} = \dfrac{6.4 \text{ mi}}{1} \cdot \dfrac{5280 \text{ ft}}{1 \text{ mi}}$

$\qquad\qquad = 6.4 \cdot 5280 \text{ ft}$

$\qquad\qquad = 33,792 \text{ ft}$

13. $162 \text{ in.} = \dfrac{162 \text{ in.}}{1} \cdot \dfrac{1 \text{ ft}}{12 \text{ in.}} \cdot \dfrac{1 \text{ yd}}{3 \text{ ft}}$

$\qquad\qquad = \dfrac{162}{36} \text{ yd}$

$\qquad\qquad = 4.5 \text{ yd}$

15. $3 \text{ in.} = \dfrac{3 \text{ in.}}{1} \cdot \dfrac{1 \text{ ft}}{12 \text{ in.}} = \dfrac{3}{12} \text{ ft} = 0.25 \text{ ft}$

17. $40 \text{ ft} = \dfrac{40 \text{ ft}}{1} \cdot \dfrac{1 \text{ yd}}{3 \text{ ft}} = \dfrac{40}{3} \text{ yd}$

$$\begin{array}{r} 13 \text{ yd } 1 \text{ ft} \\ 3 \overline{)\, 40} \\ \underline{-3} \\ 10 \\ \underline{-9} \\ 1 \end{array}$$

19. $85 \text{ in.} = \dfrac{85 \text{ in.}}{1} \cdot \dfrac{1 \text{ ft}}{12 \text{ in.}} = \dfrac{85}{12} \text{ ft}$

$$\begin{array}{r} 7 \text{ ft } 1 \text{ in.} \\ 12 \overline{)\, 85} \\ \underline{-84} \\ 1 \end{array}$$

21. $10,000 \text{ ft} = \dfrac{10,000 \text{ ft}}{1} \cdot \dfrac{1 \text{ mi}}{5280 \text{ ft}} = \dfrac{10,000}{5280} \text{ mi}$

$$\begin{array}{r} 1 \text{ mi } 4720 \text{ ft} \\ 5280 \overline{)\, 10,000} \\ \underline{-5280} \\ 4720 \end{array}$$

23. $5 \text{ ft } 2 \text{ in.} = \dfrac{5 \text{ ft}}{1} \cdot \dfrac{12 \text{ in.}}{1 \text{ ft}} + 2 \text{ in.}$

$\qquad\qquad = 60 \text{ in.} + 2 \text{ in.}$

$\qquad\qquad = 62 \text{ in.}$

25. $8 \text{ yd } 2 \text{ ft} = \dfrac{8 \text{ yd}}{1} \cdot \dfrac{3 \text{ ft}}{1 \text{ yd}} + 2 \text{ ft}$

$\qquad\qquad = 24 \text{ ft} + 2 \text{ ft}$

$\qquad\qquad = 26 \text{ ft}$

27. $2 \text{ yd } 1 \text{ ft} = \dfrac{2 \text{ yd}}{1} \cdot \dfrac{3 \text{ ft}}{1 \text{ yd}} + 1 \text{ ft} = 6 \text{ ft} + 1 \text{ ft} = 7 \text{ ft}$

$7 \text{ ft} = \dfrac{7 \text{ ft}}{1} \cdot \dfrac{12 \text{ in.}}{1 \text{ ft}} = 7 \cdot 12 \text{ in.} = 84 \text{ in.}$

29. $3 \text{ ft } 10 \text{ in.} + 7 \text{ ft } 4 \text{ in.} = 10 \text{ ft } 14 \text{ in.}$

$\qquad\qquad\qquad\qquad\qquad = 10 \text{ ft} + 1 \text{ ft } 2 \text{ in.}$

$\qquad\qquad\qquad\qquad\qquad = 11 \text{ ft } 2 \text{ in.}$

31. $12 \text{ yd } 2 \text{ ft} + 9 \text{ yd } 2 \text{ ft} = 21 \text{ yd } 4 \text{ ft}$

$\qquad\qquad\qquad\qquad\qquad = 21 \text{ yd} + 1 \text{ yd } 1 \text{ ft}$

$\qquad\qquad\qquad\qquad\qquad = 22 \text{ yd } 1 \text{ ft}$

33.
$$\begin{array}{r} 22 \text{ ft } 8 \text{ in.} \\ -\ 16 \text{ ft } 3 \text{ in.} \\ \hline 6 \text{ ft } 5 \text{ in.} \end{array}$$

35.
$$\begin{array}{rcl} 18 \text{ ft } 3 \text{ in.} & \rightarrow & 17 \text{ ft } 15 \text{ in.} \\ -\ 10 \text{ ft } 9 \text{ in.} & & -\ 10 \text{ ft }\ \ 9 \text{ in.} \\ \hline & & 7 \text{ ft }\ \ 6 \text{ in.} \end{array}$$

37. $28 \text{ ft } 8 \text{ in.} \div 2 = 14 \text{ ft } 4 \text{ in.}$

39.
$$\begin{array}{r} 16 \text{ yd }\ \ 2 \text{ ft} \\ \times \qquad\quad 5 \\ \hline 80 \text{ yd } 10 \text{ ft} = 80 \text{ yd} + 3 \text{ yd } 1 \text{ ft} = 83 \text{ yd } 1 \text{ ft} \end{array}$$

41. $60 \text{ m} = \dfrac{60 \text{ m}}{1} \cdot \dfrac{100 \text{ cm}}{1 \text{ m}} = 6000 \text{ cm}$

43. $40 \text{ mm} = \dfrac{40 \text{ mm}}{1} \cdot \dfrac{1 \text{ cm}}{10 \text{ mm}} = 4 \text{ cm}$

45. $500 \text{ m} = \dfrac{500 \text{ m}}{1} \cdot \dfrac{1 \text{ km}}{1000 \text{ m}} = \dfrac{500}{1000} \text{ km} = 0.5 \text{ km}$

47. $1700 \text{ mm} = \dfrac{1700 \text{ mm}}{1} \cdot \dfrac{1 \text{ m}}{1000 \text{ mm}} = 1.7 \text{ m}$

49. $1500 \text{ cm} = \dfrac{1500 \text{ cm}}{1} \cdot \dfrac{1 \text{ m}}{100 \text{ cm}} = \dfrac{1500}{100} \text{ m} = 15 \text{ m}$

51. $0.42 \text{ km} = \dfrac{0.42 \text{ km}}{1} \cdot \dfrac{100,000 \text{ cm}}{1 \text{ km}} = 42,000 \text{ cm}$

53. $7 \text{ km} = \dfrac{7 \text{ km}}{1} \cdot \dfrac{1000 \text{ m}}{1 \text{ km}} = 7000 \text{ m}$

55. $8.3 \text{ cm} = \dfrac{8.3 \text{ cm}}{1} \cdot \dfrac{10 \text{ mm}}{1 \text{ cm}} = 83 \text{ mm}$

57. $\begin{aligned} 20.1 \text{ mm} &= \dfrac{20.1 \text{ mm}}{1} \cdot \dfrac{1 \text{ dm}}{100 \text{ mm}} \\ &= \dfrac{20.1}{100} \text{ dm} \\ &= 0.201 \text{ dm} \end{aligned}$

59. $0.04 \text{ m} = \dfrac{0.04 \text{ m}}{1} \cdot \dfrac{1000 \text{ mm}}{1 \text{ m}} = 40 \text{ mm}$

61.
$$\begin{array}{r} 8.60 \text{ m} \\ + 0.34 \text{ m} \\ \hline 8.94 \text{ m} \end{array}$$

63.
$$\begin{array}{r} 2.9 \text{ m} \\ + 40.0 \text{ mm} \\ \hline \phantom{2.94 \text{ m}} \end{array} \qquad \begin{array}{r} 2.90 \text{ m} \\ + 0.04 \text{ m} \\ \hline 2.94 \text{ m} \end{array} \text{ or } \begin{array}{r} 2900 \text{ mm} \\ + 40 \text{ mm} \\ \hline 2940 \text{ mm} \end{array}$$

65.
$$\begin{array}{r} 24.8 \text{ mm} \\ - 1.19 \text{ cm} \\ \hline \phantom{12.9 \text{ mm}} \end{array} \qquad \begin{array}{r} 24.8 \text{ mm} \\ - 11.9 \text{ mm} \\ \hline 12.9 \text{ mm} \end{array} \text{ or } \begin{array}{r} 2.48 \text{ cm} \\ - 1.19 \text{ cm} \\ \hline 1.29 \text{ cm} \end{array}$$

67.
$$\begin{array}{r} 15 \text{ km} \\ - 2360 \text{ m} \\ \hline \phantom{12.64 \text{ km}} \end{array} \qquad \begin{array}{r} 15.00 \text{ km} \\ - 2.36 \text{ km} \\ \hline 12.64 \text{ km} \end{array} \text{ or } \begin{array}{r} 15,000 \text{ m} \\ - 2360 \text{ m} \\ \hline 12,640 \text{ m} \end{array}$$

69. $18.3 \text{ m} \times 3 = 54.9 \text{ m}$

71. $6.2 \text{ km} \div 4 = 1.55 \text{ km}$

		Yards	Feet	Inches
73.	Chrysler Building in New York City	$348\frac{2}{3}$	1046	12,552
75.	Python length	$11\frac{2}{3}$	35	420

		Meters	Millimeters	Kilometers	Centimeters
77.	Length of elephant	5	5000	0.005	500
79.	Tennis ball diameter	0.065	65	0.000065	6.5
81.	Distance from London to Paris	342,000	342,000,000	342	34,200,000

83. a. $641 \text{ ft} = \dfrac{641 \text{ ft}}{1} \cdot \dfrac{1 \text{ yd}}{3 \text{ ft}}$

$= \dfrac{641}{3} \text{ yd}$

$= 213\dfrac{2}{3} \text{ yd}$

b. $641 \text{ ft} = \dfrac{641 \text{ ft}}{1} \cdot \dfrac{12 \text{ in.}}{1 \text{ ft}}$
$= 641 \cdot 12 \text{ in.}$
$= 7692 \text{ in.}$

85. 6 ft 10 in.
  + 3 ft 8 in.
  $\overline{\text{9 ft 18 in.}}$ = 9 ft + 1 ft 6 in. = 10 ft 6 in.
The bamboo is 10 ft 6 in. now.

87. 6000 ft
  − 900 ft
  $\overline{\text{5100 ft}}$
The Grand Canyon of the Colorado River is 5100 feet deeper than the Grand Canyon of the Yellowstone River.

89. 8 ft 11 in. ÷ 22.5 in. = 107 in. ÷ 22.5 in. ≈ 4.8
Robert is 4.8 times as tall as Gul.

91. 16 ft 5 in.   15 ft 17 in.
  − 2 ft 6 in.   − 2 ft 6 in.
            $\overline{\text{13 ft 11 in.}}$

A hand of the Statue of Liberty is 13 ft 11 in. longer than the width of an eye.

93. 80 mm   80.0 mm
  − 5.33 cm   − 53.3 mm
          $\overline{\text{26.7 mm}}$

The ice must be 26.7 mm thicker before skating is allowed.

95. 1 ft 9 in. $\times 9 = 9$ ft 81 in.
$\qquad\qquad = 9$ ft $+ 6$ ft 9 in.
$\qquad\qquad = 15$ ft 9 in.
The stacks extend 15 ft 9 in. from the wall.

97. 9 ft 3 in. $\div 3 = 3$ ft 1 in.
Each cut piece will be 3 ft 1 in. long.

99. 1 m 65 cm $= 165$ cm
165 cm $\times 25 = 4125$ cm
or
1 m 65 cm $= 1.65$ m
1.65 m $\times 25 = 41.25$ m
The total length is 4125 cm or 41.25 m.

101.
$$\begin{array}{r} 3.35 \\ 20\overline{)67.00} \\ \underline{-60} \\ 70 \\ \underline{-60} \\ 1\,00 \\ \underline{-1\,00} \\ 0 \end{array}$$

Each piece will be 3.35 meters long.

103.
$$\begin{array}{r} 2.15 \text{ m} \\ -\ \ \ 2 \text{ cm} \\ \hline \end{array} \qquad \begin{array}{r} 2.15 \text{ m} \\ -\ 0.02 \text{ m} \\ \hline 2.13 \text{ m} \end{array}$$

The remaining cord is 2.13 m.

105. 182 ft $\times 2 = 364$ ft
Two trucks are 364 feet long.
$$364 \text{ ft} = \frac{364 \text{ ft}}{1} \cdot \frac{1 \text{ yd}}{3 \text{ ft}} = 121\frac{1}{3} \text{ yd}$$
Two trucks are $121\frac{1}{3}$ yards long.

107. 3.429 m $= 342.9$ cm

$$22.86\overline{)342.90} \text{ becomes } \begin{array}{r} 15 \\ 2286\overline{)34290} \\ \underline{-2286} \\ 11430 \\ \underline{-11430} \\ 0 \end{array}$$

15 tiles are needed to cross the room.

109. $0.21 = \dfrac{21}{100}$

111. $\dfrac{13}{100} = 0.13$

113. $\dfrac{1}{4} = \dfrac{1}{4} \cdot \dfrac{25}{25} = \dfrac{25}{100} = 0.25$

115. No, the width of a twin-size bed being 20 meters is not reasonable.

117. Yes, glass for a drinking glass being 2 millimeters thick is reasonable.

119. No, the distance across the Colorado River being 50 kilometers is not reasonable.

121. 5 yd 2 in. is close to 5 yd. 7 yd 30 in. is close to 7 yd 36 in. $= 8$ yd.
Estimate: 5 yd $+ 8$ yd $= 13$ yd.

123. answers may vary; for example,
$$4 \text{ ft} = \frac{4 \text{ ft}}{1} \cdot \frac{1 \text{ yd}}{3 \text{ ft}} = \frac{4}{3} \text{ yd} = 1\frac{1}{3} \text{ yd}$$
$$4 \text{ ft} = \frac{4 \text{ ft}}{1} \cdot \frac{12 \text{ in.}}{1 \text{ ft}} = 48 \text{ in.}$$

125. answers may vary

127. 18.3 m $\times 18.3$ m $= 334.89$ sq m
The area of the sign is 334.89 square meters.

Section 7.2 Practice

1. $6500 \text{ lb} = \dfrac{6500 \text{ lb}}{1} \cdot \dfrac{1 \text{ ton}}{2000 \text{ lb}}$
$\qquad\qquad = \dfrac{6500}{2000} \text{ tons}$
$\qquad\qquad = \dfrac{13}{4} \text{ tons or } 3\dfrac{1}{4} \text{ tons}$

2. $72 \text{ oz} = \dfrac{72 \text{ oz}}{1} \cdot \dfrac{1 \text{ lb}}{16 \text{ oz}} = \dfrac{72}{16} \text{ lb} = \dfrac{9}{2} \text{ lb or } 4\dfrac{1}{2} \text{ lb}$

3. $47 \text{ oz} = 47 \text{ oz} \cdot \dfrac{1 \text{ lb}}{16 \text{ oz}} = \dfrac{47}{16} \text{ lb}$

$$\begin{array}{r} 2 \text{ lb 15 oz} \\ 16\overline{)47} \\ \underline{-32} \\ 15 \end{array}$$

Thus, 47 oz $= 2$ lb 15 oz

4.
$$\begin{array}{r} 8 \text{ tons } 100 \text{ lb} \\ -\ 5 \text{ tons } 1200 \text{ lb} \\ \hline \end{array} \rightarrow \begin{array}{r} 7 \text{ tons } 2100 \text{ lb} \\ -\ 5 \text{ tons } 1200 \text{ lb} \\ \hline 2 \text{ tons } 900 \text{ lb} \end{array}$$

5.

$$
\begin{array}{r}
1\text{ lb}\quad 6\text{ oz} \\
4\overline{)\ 5\text{ lb}\quad 8\text{ oz}} \\
\underline{-\ 4\text{ lb}} \\
1\text{ lb} = \underline{16\text{ oz}} \\
24\text{ oz}
\end{array}
$$

6.

$$
\begin{array}{ll}
\text{batch weight} & 5\text{ lb }14\text{ oz} \\
\underline{+\text{ container weight}} \rightarrow \underline{+\qquad 6\text{ oz}} \\
\text{total weight} & 5\text{ lb }20\text{ oz}
\end{array}
$$

5 lb 20 oz = 5 lb + 1 lb 4 oz = 6 lb 4 oz
The total weight is 6 lb 4 oz.

7. $3.41\text{ g} = \dfrac{3.41\text{ g}}{1} \cdot \dfrac{1000\text{ mg}}{1\text{ g}} = 3410\text{ mg}$

8. $56.2\text{ cg} = \underline{56}.2\text{ cg} = 0.562\text{ g}$

9. $3.1\text{ dg} = 0.31\text{ g} \quad$ or $\quad 2.5\text{ g} = 25\text{ dg}$

$$
\begin{array}{ll}
2.50\text{ g} & 25.0\text{ dg} \\
\underline{-\ 0.31\text{ g}} & \underline{-\ 3.1\text{ dg}} \\
2.19\text{ g} & 21.9\text{ dg}
\end{array}
$$

10.

$$
\begin{array}{r}
22.9\quad \approx 23 \\
24\overline{)\ 550.0\text{ kg}} \\
\underline{-48} \\
70 \\
\underline{-48} \\
22\ 0 \\
\underline{-21\ 6} \\
4
\end{array}
$$

Each bag weighs about 23 kg.

Vocabulary and Readiness Check

1. Mass is a measure of the amount of substance in an object. This measure does not change.

2. Weight is the measure of the pull of gravity.

3. The basic unit of mass in the metric system is the gram.

4. One pound equals 16 ounces.

5. One ton equals 2000 pounds.

6. 3 tons = 6000 lb (1 ton = 2000 lb)

7. 32 oz = 2 lb (16 oz = 1 lb)

8. 3 lb = 48 oz (1 lb = 16 oz)

9. 4000 lb = 2 tons (1 ton = 2000 lb)

10. 1 ton = 2000 lb (1 ton = 2000 lb)

Exercise Set 7.2

1. $2\text{ lb} = \dfrac{2\text{ lb}}{1} \cdot \dfrac{16\text{ oz}}{1\text{ lb}} = 2 \cdot 16\text{ oz} = 32\text{ oz}$

3. $5\text{ tons} = \dfrac{5\text{ tons}}{1} \cdot \dfrac{2000\text{ lb}}{1\text{ ton}}$
$= 5 \cdot 2000\text{ lb}$
$= 10{,}000\text{ lb}$

5. $18{,}000\text{ lb} = \dfrac{18{,}000\text{ lb}}{1} \cdot \dfrac{1\text{ ton}}{2000\text{ lb}}$
$= \dfrac{18{,}000}{2000}\text{ tons}$
$= 9\text{ tons}$

7. $60\text{ oz} = \dfrac{60\text{ oz}}{1} \cdot \dfrac{1\text{ lb}}{16\text{ oz}} = \dfrac{60}{16}\text{ lb} = \dfrac{15}{4}\text{ lb} = 3\dfrac{3}{4}\text{ lb}$

9. $3500\text{ lb} = \dfrac{3500\text{ lb}}{1} \cdot \dfrac{1\text{ ton}}{2000\text{ lb}}$
$= \dfrac{3500}{2000}\text{ tons}$
$= \dfrac{7}{4}\text{ tons}$
$= 1\dfrac{3}{4}\text{ tons}$

11. $12.75\text{ lb} = \dfrac{12.75\text{ lb}}{1} \cdot \dfrac{16\text{ oz}}{1\text{ lb}}$
$= 12.75 \cdot 16\text{ oz}$
$= 204\text{ oz}$

13. $4.9\text{ tons} = \dfrac{4.9\text{ tons}}{1} \cdot \dfrac{2000\text{ lb}}{1\text{ ton}}$
$= 4.9 \cdot 2000\text{ lb}$
$= 9800\text{ lb}$

15. $4\dfrac{3}{4}\text{ lb} = \dfrac{19}{4}\text{ lb}$
$= \dfrac{\frac{19}{4}\text{ lb}}{1} \cdot \dfrac{16\text{ oz}}{1\text{ lb}}$
$= \dfrac{19}{4} \cdot 16\text{ oz}$
$= 76\text{ oz}$

17. $2950 \text{ lb} = \dfrac{2950 \text{ lb}}{1} \cdot \dfrac{1 \text{ ton}}{2000 \text{ lb}}$

$= \dfrac{2950}{2000} \text{ tons}$

$= \dfrac{59}{40} \text{ tons}$

$= 1.475 \text{ tons}$

$\approx 1.5 \text{ tons}$

19. $\dfrac{4}{5} \text{ oz} = \dfrac{\frac{4}{5}}{1} \cdot \dfrac{1 \text{ lb}}{16 \text{ oz}} = \dfrac{4}{5} \cdot \dfrac{1}{16} \text{ lb} = \dfrac{1}{20} \text{ lb}$

21. $5\dfrac{3}{4} \text{ lb} = \dfrac{23}{4} \text{ lb}$

$= \dfrac{\frac{23}{4} \text{ lb}}{1} \cdot \dfrac{16 \text{ oz}}{1 \text{ lb}}$

$= \dfrac{23}{4} \cdot 16 \text{ oz}$

$= 92 \text{ oz}$

23. $10 \text{ lb } 1 \text{ oz} = 10 \cdot 16 \text{ oz} + 1 \text{ oz}$

$= 160 \text{ oz} + 1 \text{ oz}$

$= 161 \text{ oz}$

25. $89 \text{ oz} = \dfrac{89 \text{ oz}}{1} \cdot \dfrac{1 \text{ lb}}{16 \text{ oz}} = \dfrac{89}{16} \text{ lb}$

$\begin{array}{r} 5 \text{ lb } 9 \text{ oz} \\ 16\overline{)\ 89\ } \\ -80 \\ \hline 9 \end{array}$

$89 \text{ oz} = 5 \text{ lb } 9 \text{ oz}$

27. $34 \text{ lb } 12 \text{ oz} + 18 \text{ lb } 14 \text{ oz} = 52 \text{ lb } 26 \text{ oz}$

$= 52 \text{ lb} + 1 \text{ lb } 10 \text{ oz}$

$= 53 \text{ lb } 10 \text{ oz}$

29. $3 \text{ tons } 1820 \text{ lb} + 4 \text{ tons } 930 \text{ lb}$

$= 7 \text{ tons } 2750 \text{ lb}$

$= 7 \text{ tons} + 1 \text{ ton } 750 \text{ lb}$

$= 8 \text{ tons } 750 \text{ lb}$

31. $\begin{array}{r} 5 \text{ tons } 1050 \text{ lb} \\ -\ 2 \text{ tons }\ \ 875 \text{ lb} \\ \hline 3 \text{ tons }\ \ 175 \text{ lb} \end{array}$

33. $\begin{array}{r} 12 \text{ lb } 4 \text{ oz} \\ -\ 3 \text{ lb } 9 \text{ oz} \end{array} \quad \begin{array}{r} 11 \text{ lb } 20 \text{ oz} \\ -\ 3 \text{ lb }\ \ 9 \text{ oz} \\ \hline 8 \text{ lb } 11 \text{ oz} \end{array}$

35. $5 \text{ lb } 3 \text{ oz} \times 6 = 30 \text{ lb } 18 \text{ oz}$

$= 30 \text{ lb} + 1 \text{ lb } 2 \text{ oz}$

$= 31 \text{ lb } 2 \text{ oz}$

37. $6 \text{ tons } 1500 \text{ lb} \div 5 = \dfrac{6}{5} \text{ tons } 300 \text{ lb}$

$= 1\dfrac{1}{5} \text{ tons } 300 \text{ lb}$

$= 1 \text{ ton} + \dfrac{2000 \text{ lb}}{5} + 300 \text{ lb}$

$= 1 \text{ ton} + 400 \text{ lb} + 300 \text{ lb}$

$= 1 \text{ ton } 700 \text{ lb}$

39. $500 \text{ g} = \dfrac{500 \text{ g}}{1} \cdot \dfrac{1 \text{ kg}}{1000 \text{ g}} = \dfrac{500}{1000} \text{ kg} = 0.5 \text{ kg}$

41. $4 \text{ g} = \dfrac{4 \text{ g}}{1} \cdot \dfrac{1000 \text{ mg}}{1 \text{ g}} = 4 \cdot 1000 \text{ mg} = 4000 \text{ mg}$

43. $25 \text{ kg} = \dfrac{25 \text{ kg}}{1} \cdot \dfrac{1000 \text{ g}}{1 \text{ kg}} = 25 \cdot 1000 \text{ g} = 25{,}000 \text{ g}$

45. $48 \text{ mg} = \dfrac{48 \text{ mg}}{1} \cdot \dfrac{1 \text{ g}}{1000 \text{ mg}} = \dfrac{48}{1000} \text{ g} = 0.048 \text{ g}$

47. $6.3 \text{ g} = \dfrac{6.3 \text{ g}}{1} \cdot \dfrac{1 \text{ kg}}{1000 \text{ g}} = \dfrac{6.3}{1000} \text{ kg} = 0.0063 \text{ kg}$

49. $15.14 \text{ g} = \dfrac{15.14 \text{ g}}{1} \cdot \dfrac{1000 \text{ mg}}{1 \text{ g}}$

$= 15.14 \cdot 1000 \text{ mg}$

$= 15{,}140 \text{ mg}$

51. $6.25 \text{ kg} = \dfrac{6.25 \text{ kg}}{1} \cdot \dfrac{1000 \text{ g}}{1 \text{ kg}}$

$= 6.25 \cdot 1000 \text{ g}$

$= 6250 \text{ g}$

53. $35 \text{ hg} = \dfrac{35 \text{ hg}}{1} \cdot \dfrac{10{,}000 \text{ cg}}{1 \text{ hg}}$

$= 35 \cdot 10{,}000 \text{ cg}$

$= 350{,}000 \text{ cg}$

55. $\begin{array}{r} 3.8 \text{ mg} \\ +\ 9.7 \text{ mg} \\ \hline 13.5 \text{ mg} \end{array}$

57. $205 \text{ mg} + 5.61 \text{ g} = 0.205 \text{ g} + 5.61 \text{ g} = 5.815 \text{ g}$
or
$205 \text{ mg} + 5.61 \text{ g} = 205 \text{ mg} + 5610 \text{ mg}$
$= 5815 \text{ mg}$

59. $9 \text{ g} - 7150 \text{ mg}$

$$
\begin{array}{r}
9000 \text{ mg} \\
- 7150 \text{ mg} \\
\hline
1850 \text{ mg}
\end{array}
\quad \text{or} \quad
\begin{array}{r}
9.000 \text{ g} \\
- 7.150 \text{ g} \\
\hline
1.850 \text{ g}
\end{array}
\text{ or } 1.85 \text{ g}
$$

61. $1.61 \text{ kg} - 250 \text{ g} = 1.61 \text{ kg} - 0.250 \text{ kg} = 1.36 \text{ kg}$
or
$1.61 \text{ kg} - 250 \text{ g} = 1610 \text{ g} - 250 \text{ g} = 1360 \text{ g}$

63.
$$
\begin{array}{r}
5.2 \text{ kg} \\
\times 2.6 \\
\hline
13.52 \text{ kg}
\end{array}
$$

65. $17 \text{ kg} \div 8 = \dfrac{17}{8} \text{ kg}$

$$
\begin{array}{r}
2.125 \\
8 \overline{)\ 17.000} \\
\underline{-16} \\
1\ 0 \\
\underline{-8} \\
20 \\
\underline{-16} \\
40 \\
\underline{-40} \\
0
\end{array}
$$

$17 \text{ kg} \div 8 = 2.125 \text{ kg}$

	Object	Tons	Pounds	Ounces
67.	Statue of Liberty—weight of copper sheeting	100	200,000	3,200,000
69.	A 12-inch cube of osmium (heaviest metal)	$\frac{269}{400}$ or 0.6725	1345	21,520

	Object	Grams	Kilograms	Milligrams	Centigrams
71.	Capsule of Amoxicillin (antibiotic)	0.5	0.0005	500	50
73.	A six-year-old boy	21,000	21	21,000,000	2,100,000

75.
$$
\begin{array}{r}
336 \\
\times 24 \\
\hline
1344 \\
6720 \\
\hline
8064
\end{array}
$$

$8064 \text{ g} = \dfrac{8064 \text{ g}}{1} \cdot \dfrac{1 \text{ kg}}{1000 \text{ g}} = \dfrac{8064}{1000} \text{ kg} = 8.064 \text{ kg}$

24 cans weigh 8.064 kg.

77. $0.09 \text{ g} = \dfrac{0.09 \text{ g}}{1} \cdot \dfrac{1000 \text{ mg}}{1 \text{ g}}$

$\qquad\qquad = 0.09 \cdot 1000 \text{ mg}$

$\qquad\qquad = 90 \text{ mg}$

$90 \text{ mg} - 60 \text{ mg} = 30 \text{ mg}$

The extra-strength tablet contains 30 mg more medication.

79.
$$
\begin{array}{r}
1 \text{ lb } 10 \text{ oz} \\
+ \ 3 \text{ lb } 14 \text{ oz} \\
\hline
4 \text{ lb } 24 \text{ oz}
\end{array}
$$
$4 \text{ lb } 24 \text{ oz} = 4 \text{ lb} + 1 \text{ lb } 8 \text{ oz} = 5 \text{ lb } 8 \text{ oz}$

The total amount of rice is 5 lb 8 oz.

81.
$$
\begin{array}{r}
64 \text{ lb } \ \ 8 \text{ oz} \\
- \ 28 \text{ lb } 10 \text{ oz} \\
\hline
\end{array}
\qquad
\begin{array}{r}
63 \text{ lb } 24 \text{ oz} \\
- \ 28 \text{ lb } 10 \text{ oz} \\
\hline
35 \text{ lb } 14 \text{ oz}
\end{array}
$$

Carla's zucchini was 35 lb 14 oz lighter than the record weight.

83.
$$
\begin{array}{r}
7 \text{ lb } \ \ 8 \text{ oz} \\
- \qquad 8.6 \text{ oz} \\
\hline
\end{array}
\qquad
\begin{array}{r}
6 \text{ lb } \ \ 24 \text{ oz} \\
- \qquad 8.6 \text{ oz} \\
\hline
6 \text{ lb } 15.4 \text{ oz}
\end{array}
$$

This is 6 lb 15.4 oz lighter.

85. $3 \times 16 = 48$

3 cartons contain 48 boxes of fruit.

$3 \text{ mg} \times 48 = 144 \text{ mg}$

3 cartons contain 144 mg of preservatives.

87. $26 \text{ g} \times 12 = 312 \text{ g}$

$\dfrac{312 \text{ g}}{1} \cdot \dfrac{1 \text{ kg}}{1000 \text{ g}} = 0.312 \text{ kg}$ of packaging in a carton

$6.432 \text{ kg} - 0.312 \text{ kg} = 6.12 \text{ kg}$

The actual weight of the oatmeal is 6.12 kg.

89. $3 \text{ lb } 4 \text{ oz} \times 10 = 30 \text{ lb } 40 \text{ oz}$

$\qquad\qquad\qquad\quad = 30 \text{ lb} + 2 \text{ lb } 8 \text{ oz}$

$\qquad\qquad\qquad\quad = 32 \text{ lb } 8 \text{ oz}$

Each box weighs 32 lb 8 oz.

$32 \text{ lb } 8 \text{ oz} \times 4 = 128 \text{ lb } 32 \text{ oz}$

$\qquad\qquad\qquad\quad = 128 \text{ lb} + 2 \text{ lb}$

$\qquad\qquad\qquad\quad = 130 \text{ lb}$

4 boxes of meat weigh 130 lb.

91.
$$
\begin{array}{r}
55 \text{ lb } 4 \text{ oz} \\
- \ 2 \text{ lb } 8 \text{ oz} \\
\hline
\end{array}
\qquad
\begin{array}{r}
54 \text{ lb } 20 \text{ oz} \\
- \ 2 \text{ lb } \ \ 8 \text{ oz} \\
\hline
52 \text{ lb } 12 \text{ oz}
\end{array}
$$

$$
\begin{array}{r}
52 \text{ lb } 12 \text{ oz} \\
\times \qquad\qquad 4 \\
\hline
208 \text{ lb } 48 \text{ oz}
\end{array}
$$
$= 208 \text{ lb} + 3 \text{ lb} = 211 \text{ lb}$

4 cartons contain 211 lb of pineapple.

93. $\dfrac{4}{25} = \dfrac{4}{25} \cdot \dfrac{4}{4} = \dfrac{16}{100} = 0.16$

95. $\dfrac{7}{8} = \dfrac{7}{8} \cdot \dfrac{125}{125} = \dfrac{875}{1000} = 0.875$

97. No, a pill containing 2 kg of medication is not reasonable.

99. Yes, a bag of flour weighing 4.5 kg is reasonable.

101. No, a professor weighing less than 150 g is not reasonable.

103. answers may vary; for example 250 mg or 0.25 g

105. True, a kilogram is 1000 grams.

107. answers may vary

Section 7.3 Practice

1. $43 \text{ pt} = \dfrac{43 \text{ pt}}{1} \cdot \dfrac{1 \text{ qt}}{2 \text{ pt}} = \dfrac{43}{2} \text{ qt} = 21\dfrac{1}{2} \text{ qt}$

2. $26 \text{ qt} = \dfrac{26 \text{ qt}}{1} \cdot \dfrac{4 \text{ c}}{1 \text{ qt}} = 26 \cdot 4 \text{ c} = 104 \text{ c}$

3.
$$
\begin{array}{r}
1 \text{ gal } 1 \text{ qt} \\
- \qquad 2 \text{ qt} \\
\hline
\end{array}
\quad \rightarrow \quad
\begin{array}{r}
5 \text{ qt} \\
- \ 2 \text{ qt} \\
\hline
3 \text{ qt}
\end{array}
$$

4.
$$
\begin{array}{r}
3 \text{ gal } 1 \text{ qt} \\
2\overline{)\ 6 \text{ gal } 3 \text{ qt}} \\
\underline{-6} \qquad\qquad \\
0 \text{ gal } 3 \text{ qt} \\
\underline{- \quad 2} \qquad \\
1 \text{ qt}
\end{array}
$$

$6 \text{ gal } 3 \text{ qt} \div 2 = 3 \text{ gal } 1 \text{ qt} + \dfrac{1}{2} \text{ qt}$

$\qquad\qquad\qquad\quad = 3 \text{ gal } 1 \text{ qt} + \dfrac{\frac{1}{2} \text{ qt}}{1} \cdot \dfrac{2 \text{ pt}}{1 \text{ qt}}$

$\qquad\qquad\qquad\quad = 3 \text{ gal } 1 \text{ qt } 1 \text{ pt}$

5.
$$
\begin{array}{r}
15 \text{ gal } 3 \text{ qt} \\
+ \ 4 \text{ gal } 3 \text{ qt} \\
\hline
19 \text{ gal } 6 \text{ qt}
\end{array}
$$
$= 19 \text{ gal} + 1 \text{ gal } 2 \text{ qt} = 20 \text{ gal } 2 \text{ qt}$

The total amount of oil will be 20 gal 2 qt.

6. $2100 \text{ ml} = \dfrac{2100 \text{ ml}}{1} \cdot \dfrac{1 \text{ L}}{1000 \text{ ml}} = \dfrac{2100}{1000} \text{ L} = 2.1 \text{ L}$

7. $2.13 \text{ dal} = \dfrac{2.13 \text{ dal}}{1} \cdot \dfrac{10 \text{ L}}{1 \text{ dal}} = 2.13 \cdot 10 \text{ L} = 21.3 \text{ L}$

8. $1250 \text{ ml} = 1.250 \text{ L}$　　　$2.9 \text{ L} = 2900 \text{ ml}$

$$\begin{array}{r} 1.25 \text{ L} \\ +\ 2.9\ \text{ L} \\ \hline 4.15 \text{ L} \end{array} \qquad \begin{array}{r} 1250 \text{ ml} \\ +\ 2900 \text{ ml} \\ \hline 4150 \text{ ml} \end{array}$$

The total is 4.15 L or 4150 ml.

9.
$$\begin{array}{r} 28.6 \text{ L} \\ \times\ \ 85 \\ \hline 143\ 0 \\ 2288\ 0 \\ \hline 2431.0 \text{ L} \end{array}$$
Thus, 2431 L can be pumped in 85 minutes.

Vocabulary and Readiness Check

1. Units of <u>capacity</u> are generally used to measure liquids.

2. The basic unit of capacity in the metric system is the <u>liter</u>.

3. One cup equals 8 <u>fluid ounces</u>.

4. One quart equals 2 <u>pints</u>.

5. One pint equals 2 <u>cups</u>.

6. One quart equals 4 <u>cups</u>.

7. One gallon equals 4 <u>quarts</u>.

8. 2 c = 1 pt (1 pt = 2 c)

9. 4 c = 2 pt (1 pt = 2 c)

10. 4 qt = 1 gal (1 gal = 4 qt)

11. 8 qt = 2 gal (1 gal = 4 qt)

12. 2 pt = 1 qt (1 qt = 2 pt)

13. 6 pt = 3 qt (1 qt = 2 pt)

14. 8 fl oz = 1 c (1 c = 8 fl oz)

15. 24 fl oz = 3 c (1 c = 8 fl oz)

16. 3 pt = 6 c (2 c = 1 pt)

Exercise Set 7.3

1. $32 \text{ fl oz} = \dfrac{32 \text{ fl oz}}{1} \cdot \dfrac{1 \text{ c}}{8 \text{ fl oz}} = \dfrac{32}{8} \text{ c} = 4 \text{ c}$

3. $8 \text{ qt} = \dfrac{8 \text{ qt}}{1} \cdot \dfrac{2 \text{ pt}}{1 \text{ qt}} = 8 \cdot 2 \text{ pt} = 16 \text{ pt}$

5. $14 \text{ qt} = \dfrac{14 \text{ qt}}{1} \cdot \dfrac{1 \text{ gal}}{4 \text{ qt}} = \dfrac{14}{4} \text{ gal} = 3\dfrac{1}{2} \text{ gal}$

7. $80 \text{ fl oz} = \dfrac{80 \text{ fl oz}}{1} \cdot \dfrac{1 \text{ c}}{8 \text{ fl oz}} = \dfrac{80}{8} \text{ c} = 10 \text{ c}$

$10 \text{ c} = \dfrac{10 \text{ c}}{1} \cdot \dfrac{1 \text{ pt}}{2 \text{ c}} = \dfrac{10}{2} \text{ pt} = 5 \text{ pt}$

9. $2 \text{ qt} = \dfrac{2 \text{ qt}}{1} \cdot \dfrac{2 \text{ pt}}{1 \text{ qt}} \cdot \dfrac{2 \text{ c}}{1 \text{ pt}} = 2 \cdot 2 \cdot 2 \text{ c} = 8 \text{ c}$

11. $120 \text{ fl oz} = \dfrac{120 \text{ fl oz}}{1} \cdot \dfrac{1 \text{ c}}{8 \text{ fl oz}} \cdot \dfrac{1 \text{ pt}}{2 \text{ c}} \cdot \dfrac{1 \text{ qt}}{2 \text{ pt}}$

$\qquad = \dfrac{120}{8 \cdot 2 \cdot 2} \text{ qt}$

$\qquad = \dfrac{15}{4} \text{ qt}$

$\qquad = 3\dfrac{3}{4} \text{ qt}$

13. $42 \text{ c} = \dfrac{42 \text{ c}}{1} \cdot \dfrac{1 \text{ qt}}{4 \text{ c}} = \dfrac{42}{4} \text{ qt} = 10\dfrac{1}{2} \text{ qt}$

15. $4\dfrac{1}{2} \text{ pt} = \dfrac{9}{2} \text{ pt} = \dfrac{\frac{9}{2} \text{ pt}}{1} \cdot \dfrac{2 \text{ c}}{1 \text{ pt}} = \dfrac{9}{2} \cdot 2 \text{ c} = 9 \text{ c}$

17. $5 \text{ gal } 3 \text{ qt} = \dfrac{5 \text{ gal}}{1} \cdot \dfrac{4 \text{ qt}}{1 \text{ gal}} + 3 \text{ qt}$

$\qquad = 5 \cdot 4 \text{ qt} + 3 \text{ qt}$

$\qquad = 20 \text{ qt} + 3 \text{ qt}$

$\qquad = 23 \text{ qt}$

19. $\dfrac{1}{2} \text{ c} = \dfrac{\frac{1}{2} \text{ c}}{1} \cdot \dfrac{1 \text{ pt}}{2 \text{ c}} = \dfrac{1}{2} \cdot \dfrac{1}{2} \text{ pt} = \dfrac{1}{4} \text{ pt}$

21. $58 \text{ qt} = 56 \text{ qt} + 2 \text{ qt}$

$= \dfrac{56 \text{ qt}}{1} \cdot \dfrac{1 \text{ gal}}{4 \text{ qt}} + 2 \text{ qt}$

$= \dfrac{56}{4} \text{ gal} + 2 \text{ qt}$

$= 14 \text{ gal } 2 \text{ qt}$

23. $39 \text{ pt} = 38 \text{ pt} + 1 \text{ pt}$

$= \dfrac{38 \text{ pt}}{1} \cdot \dfrac{1 \text{ qt}}{2 \text{ pt}} + 1 \text{ pt}$

$= 19 \text{ qt} + 1 \text{ pt}$

$= 16 \text{ qt} + 3 \text{ qt} + 1 \text{ pt}$

$= \dfrac{16 \text{ qt}}{1} \cdot \dfrac{1 \text{ gal}}{4 \text{ qt}} + 3 \text{ qt} + 1 \text{ pt}$

$= 4 \text{ gal} + 3 \text{ qt} + 1 \text{ pt}$

$= 4 \text{ gal } 3 \text{ qt } 1 \text{ pt}$

25. $2\dfrac{3}{4} \text{ gal} = \dfrac{11}{4} \text{ gal}$

$= \dfrac{\frac{11}{4} \text{ gal}}{1} \cdot \dfrac{4 \text{ qt}}{1 \text{ gal}} \cdot \dfrac{2 \text{ pt}}{1 \text{ qt}}$

$= \dfrac{11}{4} \cdot 4 \cdot 2 \text{ pt}$

$= 22 \text{ pt}$

27.
$$\begin{array}{r} 5 \text{ gal } 3 \text{ qt} \\ + \ 7 \text{ gal } 3 \text{ qt} \\ \hline 12 \text{ gal } 6 \text{ qt} \end{array} = 12 \text{ gal} + 1 \text{ gal } 2 \text{ qt} = 13 \text{ gal } 2 \text{ qt}$$

29. $1 \text{ c } 5 \text{ fl oz} + 2 \text{ c } 7 \text{ fl oz} = 3 \text{ c } 12 \text{ fl oz}$

$= 3 \text{ c} + 1 \text{ c } 4 \text{ fl oz}$

$= 4 \text{ c } 4 \text{ fl oz}$

31.
$$\begin{array}{r} 3 \text{ gal} \\ - \ 1 \text{ gal } 3 \text{ qt} \\ \hline \end{array} \qquad \begin{array}{r} 2 \text{ gal } 4 \text{ qt} \\ - \ 1 \text{ gal } 3 \text{ qt} \\ \hline 1 \text{ gal } 1 \text{ qt} \end{array}$$

33.
$$\begin{array}{r} 3 \text{ gal } 1 \text{ qt} \\ - \ 1 \text{ qt } 1 \text{ pt} \\ \hline \end{array} \qquad \begin{array}{r} 2 \text{ gal } 5 \text{ qt} \\ - \ 1 \text{ qt } 1 \text{ pt} \\ \hline \end{array} \qquad \begin{array}{r} 2 \text{ gal } 4 \text{ qt } 2 \text{ pt} \\ - \ 1 \text{ qt } 1 \text{ pt} \\ \hline 2 \text{ gal } 3 \text{ qt } 1 \text{ pt} \end{array}$$

35. $8 \text{ gal } 2 \text{ qt} \times 2 = 16 \text{ gal } 4 \text{ qt}$

$= 16 \text{ gal} + 1 \text{ gal}$

$= 17 \text{ gal}$

37. $9 \text{ gal } 2 \text{ qt} \div 2 = (8 \text{ gal } 4 \text{ qt} + 2 \text{ qt}) \div 2$

$= 8 \text{ gal } 6 \text{ qt} \div 2$

$= 4 \text{ gal } 3 \text{ qt}$

39. $5 \text{L} = \dfrac{5 \text{L}}{1} \cdot \dfrac{1000 \text{ ml}}{1 \text{ L}} = 5000 \text{ ml}$

41. $0.16 \text{ L} = \dfrac{0.16 \text{ L}}{1} \cdot \dfrac{1 \text{ kl}}{1000 \text{ L}}$

$= \dfrac{0.16}{1000} \text{ kl}$

$= 0.00016 \text{ kl}$

43. $5600 \text{ ml} = \dfrac{5600 \text{ ml}}{1} \cdot \dfrac{1 \text{ L}}{1000 \text{ ml}} = \dfrac{5600}{1000} \text{ L} = 5.6 \text{ L}$

45. $3.2 \text{ L} = \dfrac{3.2 \text{ L}}{1} \cdot \dfrac{100 \text{ cl}}{1 \text{ L}} = 3.2 \cdot 100 \text{ cl} = 320 \text{ cl}$

47. $410 \text{ L} = \dfrac{410 \text{ L}}{1} \cdot \dfrac{1 \text{ kl}}{1000 \text{ L}} = \dfrac{410}{1000} \text{ kl} = 0.41 \text{ kl}$

49. $64 \text{ ml} = \dfrac{64 \text{ ml}}{1} \cdot \dfrac{1 \text{ L}}{1000 \text{ ml}} = \dfrac{64}{1000} \text{ L} = 0.064 \text{ L}$

51. $0.16 \text{ kl} = \dfrac{0.16 \text{ kl}}{1} \cdot \dfrac{1000 \text{ L}}{1 \text{ kl}}$

$= 0.16 \cdot 1000 \text{ L}$

$= 160 \text{ L}$

53. $3.6 \text{ L} = \dfrac{3.6 \text{ L}}{1} \cdot \dfrac{1000 \text{ ml}}{1 \text{ L}}$

$= 3.6 \cdot 1000 \text{ ml}$

$= 3600 \text{ ml}$

55. $3.4 \text{ L} + 15.9 \text{ L} = 19.3 \text{ L}$

57. $2700 \text{ ml} + 1.8 \text{ L} = 2.7 \text{ L} + 1.8 \text{ L} = 4.5 \text{ L}$
or
$2700 \text{ ml} + 1.8 \text{ L} = 2700 \text{ ml} + 1800 \text{ ml} = 4500 \text{ ml}$

59.
$$\begin{array}{r} 8.6 \text{ L} \\ - \ 190 \text{ ml} \\ \hline \end{array} \quad \begin{array}{r} 8600 \text{ ml} \\ - \ 190 \text{ ml} \\ \hline 8410 \text{ ml} \end{array} \ \text{or} \ \begin{array}{r} 8.60 \text{ L} \\ - \ 0.19 \text{ L} \\ \hline 8.41 \text{ L} \end{array}$$

61. $17,500 \text{ ml} - 0.9 \text{ L} = 17,500 \text{ ml} - 900 \text{ ml}$

$= 16,600 \text{ ml}$

or

$17,500 \text{ ml} - 0.9 \text{ L} = 17.5 \text{ L} - 0.9 \text{ L} = 16.6 \text{ L}$

63. $480 \text{ ml} \times 8 = 3840 \text{ ml}$

65. $81.2 \text{ L} \div 0.5 = 81.2 \text{ L} \div \dfrac{1}{2}$

$= 81.2 \text{ L} \cdot 2$

$= 162.4 \text{ L}$

	Capacity	Cups	Gallons	Quarts	Pints
67.	An average-size bath of water	336	21	84	168
69.	Your kidneys filter about this amount of blood every minute	4	$\frac{1}{4}$	1	2

71. $\begin{array}{cc} 2\text{ L} & 2.000\text{ L} \\ -\,410\text{ ml} & -\,0.410\text{ L} \\ \hline & 1.590\text{ L} \end{array}$

There was 1.59 L left in the bottle.

73. 354 ml + 18.6 L = 0.354 L + 18.6 L = 18.954 L
There were 18.954 liters of gasoline in her tank.

75. $\dfrac{1}{30}$ gal $= \dfrac{\frac{1}{30}\text{ gal}}{1}\cdot\dfrac{4\text{ qt}}{1\text{ gal}}\cdot\dfrac{2\text{ pt}}{1\text{ qt}}\cdot\dfrac{2\text{ c}}{1\text{ pt}}\cdot\dfrac{8\text{ fl oz}}{1\text{ c}}$

$\qquad = \dfrac{1}{30}\cdot 128\text{ fl oz}$

$\qquad \approx 4.3\text{ fl oz}$

$\dfrac{1}{30}$ gal is about 4.3 fluid ounces.

77. 5 pt 1 c + 2 pt 1 c = 7 pt 2 c
$\qquad = 7\text{ pt} + 1\text{ pt}$
$\qquad = 8\text{ pt}$
$\qquad = \dfrac{8\text{ pt}}{1}\cdot\dfrac{1\text{ qt}}{2\text{ pt}}$
$\qquad = \dfrac{8}{2}\text{ qt}$
$\qquad = \dfrac{4\text{ qt}}{1}\cdot\dfrac{1\text{ gal}}{4\text{ qt}}$
$\qquad = \dfrac{4}{4}\text{ gal}$
$\qquad = 1\text{ gal}$

Yes, the liquid can be poured into the container without causing it to overflow.

79. $44.3\overline{)14.0}$ becomes $\begin{array}{r} 0.3160 \approx 0.316 \\ 443\overline{)140.0000} \\ -132\ 9 \\ \hline 7\ 10 \\ -4\ 43 \\ \hline 2\ 670 \\ -2\ 658 \\ \hline 120 \\ -0 \\ \hline 120 \end{array}$

$\dfrac{\$14}{44.3\text{ L}} \approx \dfrac{\$0.316}{1\text{ L}}$
The price was \$0.316 per liter.

81. $\dfrac{20}{25} = \dfrac{4 \cdot 5}{5 \cdot 5} = \dfrac{4}{5}$

83. $\dfrac{27}{45} = \dfrac{3 \cdot 9}{5 \cdot 9} = \dfrac{3}{5}$

85. $\dfrac{72}{80} = \dfrac{8 \cdot 9}{8 \cdot 10} = \dfrac{9}{10}$

87. No, a 2 L dose of cough medicine is not reasonable.

89. No, a tub filled with 3000 ml of hot water is not reasonable.

91. less than; answers may vary

93. answers may vary

95. $1 \text{ gal} = \dfrac{1 \text{ gal}}{1} \cdot \dfrac{4 \text{ qt}}{1 \text{ gal}} \cdot \dfrac{2 \text{ pt}}{1 \text{ qt}} \cdot \dfrac{2 \text{ c}}{1 \text{ pt}} \cdot \dfrac{8 \text{ fl oz}}{1 \text{ c}}$
$= 1 \cdot 4 \cdot 2 \cdot 2 \cdot 8 \text{ fl oz}$
$= 128 \text{ fl oz}$
There are 128 fl oz in 1 gallon.

97. B indicates 1.5 cc.

99. D indicates 2.7 cc.

101. B indicates 54 u or 0.54 cc.

103. D indicates 86 u or 0.86 cc.

Integrated Review

1. $36 \text{ in.} = \dfrac{36 \text{ in.}}{1} \cdot \dfrac{1 \text{ ft}}{12 \text{ in.}} = \dfrac{36}{12} \text{ ft} = 3 \text{ ft}$

2. $10,560 \text{ ft} = \dfrac{10,560 \text{ ft}}{1} \cdot \dfrac{1 \text{ mi}}{5280 \text{ ft}}$
$= \dfrac{10,560}{5280} \text{ mi}$
$= 2 \text{ mi}$

3. $20 \text{ ft} = \dfrac{20 \text{ ft}}{1} \cdot \dfrac{1 \text{ yd}}{3 \text{ ft}} = \dfrac{20}{3} \text{ yd} = 6\dfrac{2}{3} \text{ yd}$

4. $6\dfrac{1}{3} \text{ yd} = \dfrac{19}{3} \text{ yd} = \dfrac{\frac{19}{3} \text{ yd}}{1} \cdot \dfrac{3 \text{ ft}}{1 \text{ yd}} = \dfrac{19}{3} \cdot 3 \text{ ft} = 19 \text{ ft}$

5. $2.1 \text{ mi} = \dfrac{2.1 \text{ mi}}{1} \cdot \dfrac{5280 \text{ ft}}{1 \text{ mi}}$
$= 2.1 \cdot 5280 \text{ ft}$
$= 11,088 \text{ ft}$

6. $3.2 \text{ ft} = \dfrac{3.2 \text{ ft}}{1} \cdot \dfrac{12 \text{ in.}}{1 \text{ ft}} = 3.2 \cdot 12 \text{ in.} = 38.4 \text{ in.}$

7. $30 \text{ m} = \dfrac{30 \text{ m}}{1} \cdot \dfrac{100 \text{ cm}}{1 \text{ m}} = 30 \cdot 100 \text{ cm} = 3000 \text{ cm}$

8. $24 \text{ mm} = \dfrac{24 \text{ mm}}{1} \cdot \dfrac{1 \text{ cm}}{10 \text{ mm}} = \dfrac{24}{10} \text{ cm} = 2.4 \text{ cm}$

9. $2000 \text{ mm} = \dfrac{2000 \text{ mm}}{1} \cdot \dfrac{1 \text{ m}}{1000 \text{ mm}}$
$= \dfrac{2000}{1000} \text{ m}$
$= 2 \text{ m}$

10. $1800 \text{ cm} = \dfrac{1800 \text{ cm}}{1} \cdot \dfrac{1 \text{ m}}{100 \text{ cm}} = \dfrac{1800}{100} \text{ m} = 18 \text{ m}$

11. $7.2 \text{ cm} = \dfrac{7.2 \text{ cm}}{1} \cdot \dfrac{10 \text{ mm}}{1 \text{ cm}} = 7.2 \cdot 10 \text{ mm} = 72 \text{ mm}$

12. $600 \text{ m} = \dfrac{600 \text{ m}}{1} \cdot \dfrac{1 \text{ km}}{1000 \text{ m}} = \dfrac{600}{1000} \text{ km} = 0.6 \text{ km}$

13. $7\dfrac{1}{2} \text{ tons} = \dfrac{15}{2} \text{ tons}$
$= \dfrac{\frac{15}{2} \text{ tons}}{1} \cdot \dfrac{2000 \text{ lb}}{1 \text{ ton}}$
$= \dfrac{15}{2} \cdot 2000 \text{ lb}$
$= 15,000 \text{ lb}$

14. $11,000 \text{ lb} = \dfrac{11,000 \text{ lb}}{1} \cdot \dfrac{1 \text{ ton}}{2000 \text{ lb}}$
$= \dfrac{11,000}{2000} \text{ tons}$
$= 5.5 \text{ tons}$

15. $8.5 \text{ lb} = \dfrac{8.5 \text{ lb}}{1} \cdot \dfrac{16 \text{ oz}}{1 \text{ lb}} = 8.5 \cdot 16 \text{ oz} = 136 \text{ oz}$

16. $72 \text{ oz} = \dfrac{72 \text{ oz}}{1} \cdot \dfrac{1 \text{ lb}}{16 \text{ oz}} = \dfrac{72}{16} \text{ lb} = 4.5 \text{ lb}$

17. $104 \text{ oz} = \dfrac{104 \text{ oz}}{1} \cdot \dfrac{1 \text{ lb}}{16 \text{ oz}} = \dfrac{104}{16} \text{ lb} = 6.5 \text{ lb}$

18. $5 \text{ lb} = \dfrac{5 \text{ lb}}{1} \cdot \dfrac{16 \text{ oz}}{1 \text{ lb}} = 5 \cdot 16 \text{ oz} = 80 \text{ oz}$

19. $28 \text{ kg} = \dfrac{28 \text{ kg}}{1} \cdot \dfrac{1000 \text{ g}}{1 \text{ kg}} = 28 \cdot 1000 \text{ g} = 28,000 \text{ g}$

20. $1400 \text{ mg} = \dfrac{1400 \text{ mg}}{1} \cdot \dfrac{1 \text{ g}}{1000 \text{ mg}} = \dfrac{1400}{1000} \text{ g} = 1.4 \text{ g}$

21. $5.6 \text{ g} = \dfrac{5.6 \text{ g}}{1} \cdot \dfrac{1 \text{ kg}}{1000 \text{ g}} = \dfrac{5.6}{1000} \text{ kg} = 0.0056 \text{ kg}$

22. $6 \text{ kg} = \dfrac{6 \text{ kg}}{1} \cdot \dfrac{1000 \text{ g}}{1 \text{ kg}} = 6 \cdot 1000 \text{ g} = 6000 \text{ g}$

23. $670 \text{ mg} = \dfrac{670 \text{ mg}}{1} \cdot \dfrac{1 \text{ g}}{1000 \text{ mg}} = \dfrac{670}{1000} \text{ g} = 0.67 \text{ g}$

24. $3.6 \text{ g} = \dfrac{3.6 \text{ g}}{1} \cdot \dfrac{1 \text{ kg}}{1000 \text{ g}} = \dfrac{3.6}{1000} \text{ kg} = 0.0036 \text{ kg}$

25. $6 \text{ qt} = \dfrac{6 \text{ qt}}{1} \cdot \dfrac{2 \text{ pt}}{1 \text{ qt}} = 6 \cdot 2 \text{ pt} = 12 \text{ pt}$

26. $5 \text{ pt} = \dfrac{5 \text{ pt}}{1} \cdot \dfrac{1 \text{ qt}}{2 \text{ pt}} = \dfrac{5}{2} \text{ qt} = 2.5 \text{ qt}$

27. $14 \text{ qt} = \dfrac{14 \text{ qt}}{1} \cdot \dfrac{1 \text{ gal}}{4 \text{ qt}} = \dfrac{14}{4} \text{ gal} = 3.5 \text{ gal}$

28. $17 \text{ c} = \dfrac{17 \text{ c}}{1} \cdot \dfrac{1 \text{ pt}}{2 \text{ c}} = \dfrac{17}{2} \text{ pt} = 8.5 \text{ pt}$

29. $3\dfrac{1}{2} \text{ pt} = \dfrac{7}{2} \text{ pt} = \dfrac{\frac{7}{2} \text{ pt}}{1} \cdot \dfrac{2 \text{ c}}{1 \text{ pt}} = \dfrac{7}{2} \cdot 2 \text{ c} = 7 \text{ c}$

30. $26 \text{ qt} = \dfrac{26 \text{ qt}}{1} \cdot \dfrac{1 \text{ gal}}{4 \text{ qt}} = \dfrac{26}{4} \text{ gal} = 6.5 \text{ gal}$

31. $7 \text{ L} = \dfrac{7 \text{ L}}{1} \cdot \dfrac{1000 \text{ ml}}{1 \text{ L}} = 7 \cdot 1000 \text{ ml} = 7000 \text{ ml}$

32. $350 \text{ L} = \dfrac{350 \text{ L}}{1} \cdot \dfrac{1 \text{ kl}}{1000 \text{ L}} = \dfrac{350}{1000} \text{ kl} = 0.35 \text{ kl}$

33. $47 \text{ ml} = \dfrac{47 \text{ ml}}{1} \cdot \dfrac{1 \text{ L}}{1000 \text{ ml}} = \dfrac{47}{1000} \text{ L} = 0.047 \text{ L}$

34. $0.97 \text{ kl} = \dfrac{0.97 \text{ kl}}{1} \cdot \dfrac{1000 \text{ L}}{1 \text{ kl}}$
$= 0.97 \cdot 1000 \text{ L}$
$= 970 \text{ L}$

35. $0.126 \text{ kl} = \dfrac{0.126 \text{ kl}}{1} \cdot \dfrac{1000 \text{ L}}{1 \text{ kl}}$
$= 0.126 \cdot 1000 \text{ L}$
$= 126 \text{ L}$

36. $75 \text{ ml} = \dfrac{75 \text{ ml}}{1} \cdot \dfrac{1 \text{ L}}{1000 \text{ ml}} = \dfrac{75}{1000} \text{ L} = 0.075 \text{ L}$

37. $\dfrac{1}{2} \text{ c} = \dfrac{\frac{1}{2} \text{ c}}{1} \cdot \dfrac{8 \text{ fl oz}}{1 \text{ c}} = \dfrac{1}{2} \cdot 8 \text{ fl oz} = 4 \text{ fl oz}$

38. $\dfrac{3}{4} \text{ gal} = \dfrac{\frac{3}{4} \text{ gal}}{1} \cdot \dfrac{4 \text{ qt}}{1 \text{ gal}} \cdot \dfrac{2 \text{ pt}}{1 \text{ qt}} \cdot \dfrac{2 \text{ c}}{1 \text{ pt}}$
$= \dfrac{3}{4} \cdot 4 \cdot 2 \cdot 2 \text{ c}$
$= 12 \text{ c}$

Section 7.4 Practice

1. $1.5 \text{ cm} = \dfrac{1.5 \text{ cm}}{1} \cdot \dfrac{1 \text{ in.}}{2.54 \text{ cm}} = \dfrac{1.5}{2.54} \text{ in.} \approx 0.59 \text{ in.}$

2. $8 \text{ oz} \approx \dfrac{8 \text{ oz}}{1} \cdot \dfrac{28.35 \text{ g}}{1 \text{ oz}} = 8 \cdot 28.35 \text{ g} = 226.8 \text{ g}$

3. $237 \text{ ml} \approx \dfrac{237 \text{ ml}}{1} \cdot \dfrac{1 \text{ fl oz}}{29.57 \text{ ml}}$
$= \dfrac{237}{29.57} \text{ fl oz}$
$\approx 8 \text{ fl oz}$

Exercise Set 7.4

1. $756 \text{ ml} \approx \dfrac{756 \text{ ml}}{1} \cdot \dfrac{1 \text{ fl oz}}{29.57 \text{ ml}}$
$= \dfrac{756}{29.57} \text{ fl oz}$
$\approx 25.57 \text{ fl oz}$

3. $86 \text{ in.} = \dfrac{86 \text{ in.}}{1} \cdot \dfrac{2.54 \text{ cm}}{1 \text{ in.}}$
$= 86 \cdot 2.54 \text{ cm}$
$= 218.44 \text{ cm}$

5. $1000 \text{ g} \approx \dfrac{1000 \text{ g}}{1} \cdot \dfrac{0.04 \text{ oz}}{1 \text{ g}}$
$= 1000 \cdot 0.04 \text{ oz}$
$= 40 \text{ oz}$

7. $93 \text{ km} \approx \dfrac{93 \text{ km}}{1} \cdot \dfrac{0.62 \text{ mi}}{1 \text{ km}}$
$= 93 \cdot 0.62 \text{ mi}$
$= 57.66 \text{ mi}$

9. $14.5 \text{ L} \approx \dfrac{14.5 \text{ L}}{1} \cdot \dfrac{0.26 \text{ gal}}{1 \text{ L}} \approx 3.77 \text{ gal}$

11. $30 \text{ lb} \approx \dfrac{30 \text{ lb}}{1} \cdot \dfrac{0.45 \text{ kg}}{1 \text{ lb}} = 30 \cdot 0.45 \text{ kg} = 13.5 \text{ kg}$

		Meters	Yards	Centimeters	Feet	Inches
13.	The Height of a Woman	1.5	$1\frac{2}{3}$	150	5	60
15.	Leaning Tower of Pisa	55	60	5500	180	2160

17. $10 \text{ cm} = \dfrac{10 \text{ cm}}{1} \cdot \dfrac{1 \text{ in.}}{2.54 \text{ cm}} \approx 3.94 \text{ in.}$
The balance beam is approximately 3.94 inches wide.

19. $50 \text{ mph} \approx \dfrac{50 \text{ mph}}{1} \cdot \dfrac{1.61 \text{ km}}{1 \text{ mi}} = 80.5 \text{ kph}$
The speed limit is approximately 80.5 kilometers per hour.

21. $200 \text{ mg} = 0.2 \text{ g} \approx \dfrac{0.2 \text{ g}}{1} \cdot \dfrac{0.04 \text{ oz}}{1 \text{ g}} = 0.008 \text{ oz}$

23. $100 \text{ kg} \approx \dfrac{100 \text{ kg}}{1} \cdot \dfrac{2.2 \text{ lb}}{1 \text{ kg}} = 100 \cdot 2.2 \text{ lb} = 220 \text{ lb}$

$15 \text{ stone } 10 \text{ lb} = \dfrac{15 \text{ stone}}{1} \cdot \dfrac{14 \text{ lb}}{1 \text{ stone}} + 10 \text{ lb}$
$= 15 \cdot 14 \text{ lb} + 10 \text{ lb}$
$= 210 \text{ lb} + 10 \text{ lb}$
$= 220 \text{ lb}$
Yes; the stamp is approximately correct.

25. $4500 \text{ km} = \dfrac{4500 \text{ km}}{1} \cdot \dfrac{0.62 \text{ mi}}{1 \text{ km}} = 2790 \text{ mi}$
The trip is about 2790 miles.

27. $3\frac{1}{2}$ in. $= \dfrac{3\frac{1}{2}\text{ in.}}{1} \cdot \dfrac{2.54\text{ cm}}{1\text{ in.}} = 8.89$ cm

8.89 cm $= \dfrac{8.89\text{ cm}}{1} \cdot \dfrac{10\text{ mm}}{1\text{ cm}} = 88.9$ mm ≈ 90 mm

The width is approximately 90 mm.

29. 1.5 lb $- 1.25$ lb $= 0.25$ lb

0.25 lb $\approx \dfrac{0.25\text{ lb}}{1} \cdot \dfrac{0.45\text{ kg}}{1\text{ lb}} \cdot \dfrac{1000\text{ g}}{1\text{ kg}} \approx 112.5$ g

The difference is approximately 112.5 g.

31. 167 kmh $\approx \dfrac{167\text{ kmh}}{1} \cdot \dfrac{0.62\text{ mi}}{1\text{ km}} \approx 104$ mph

The sneeze is approximately 104 miles per hour.

33. 8 m $\approx \dfrac{8\text{ m}}{1} \cdot \dfrac{3.28\text{ ft}}{1\text{ m}} \approx 26.24$ ft

The base diameter is approximately 26.24 ft.

35. 4.5 km $\approx \dfrac{4.5\text{ km}}{1} \cdot \dfrac{0.62\text{ mi}}{1\text{ km}} \approx 3$ mi

The track is approximately 3 mi.

37. One dose every 4 hours results in $\dfrac{24}{4} = 6$ doses

per day and $6 \times 7 = 42$ doses per week.

5 ml $\times 42 = 210$ ml

210 ml $\approx \dfrac{210\text{ ml}}{1} \cdot \dfrac{1\text{ fl oz}}{29.57\text{ ml}} \approx 7.1$ fl oz

8 fluid ounces of medicine should be purchased.

39. This math book has a height of about 28 cm; b.

41. A liter has greater capacity than a quart; b.

43. A kilogram weighs greater than a pound; c

45. An $8\frac{1}{2}$-ounce glass of water has a capacity of

about 250 ml $\left(\dfrac{1}{4}\text{ L}\right)$; d.

47. The weight of an average man is about 70 kg
(70 kg $\approx 2.2 \cdot 70$ lb $= 154$ lb); d

49. $6 \cdot 4 + 5 \div 1 = 24 + 5 \div 1 = 24 + 5 = 29$

51. $\dfrac{10+8}{10-8} = \dfrac{18}{2} = 9$

53. $3 + 5(19 - 17) - 8 = 3 + 5(2) - 8$
$\qquad\qquad\qquad\quad = 3 + 10 - 8$
$\qquad\qquad\qquad\quad = 13 - 8$
$\qquad\qquad\qquad\quad = 5$

55. $3[(1 + 5) \cdot (8 - 6)] = 3(6 \cdot 2) = 3(12) = 36$

57. BSA $= \sqrt{\dfrac{90 \times 182}{3600}} \approx 2.13$

The BSA is approximately 2.13 sq m.

59. 40 in. $= \dfrac{40\text{ in.}}{1} \cdot \dfrac{2.54\text{ cm}}{1\text{ in.}} = 101.6$ cm

BSA $= \sqrt{\dfrac{50 \times 101.6}{3600}} \approx 1.19$

The BSA is approximately 1.19 sq m.

61. 60 in. $= \dfrac{60\text{ in.}}{1} \cdot \dfrac{2.54\text{ cm}}{1\text{ in.}} = 152.4$ cm

150 lb $\approx \dfrac{150\text{ lb}}{1} \cdot \dfrac{0.45\text{ kg}}{1\text{ lb}} \approx 67.5$ kg

BSA $\approx \sqrt{\dfrac{67.5 \times 152.4}{3600}} \approx 1.69$

The BSA is approximately 1.69 sq m.

63. The adult's BSA is 2.13 sq m.
$2.13 \times 10 = 21.3$
$2.13 \times 12 = 25.56$
The dosage range is 21.3 mg to 25.56 mg.

65. 20 m $\times 40$ m $= 800$ sq m

20 m $\approx \dfrac{20\text{ m}}{1} \cdot \dfrac{3.28\text{ ft}}{1\text{ m}} \approx 65.6$ ft

20 m $\times 40$ m ≈ 65.6 ft $\times 131.2$ ft ≈ 8606.72 sq ft
The area is 800 sq m or approximately
8606.72 sq ft.

Section 7.5 Practice

1. $F = \dfrac{9}{5} \cdot C + 32 = \dfrac{9}{5} \cdot 50 + 32 = 90 + 32 = 122$

Thus, 50°C is equivalent to 122°F.

2. $F = 1.8 \cdot C + 32 = 1.8 \cdot 18 + 32 = 32.4 + 32 = 64.4$
Therefore, 18°C is the same as 64.4°F.

3. $C = \dfrac{5}{9} \cdot (F - 32) = \dfrac{5}{9} \cdot (68 - 32) = \dfrac{5}{9} \cdot (36) = 20$

Therefore, 68°F is the same temperature as 20°C.

4. $C = \dfrac{5}{9} \cdot (F - 32) = \dfrac{5}{9} \cdot (113 - 32) = \dfrac{5}{9} \cdot (81) = 45$

Therefore, 113°F is 45°C.

5. $C = \dfrac{5}{9} \cdot (F - 32)$

$= \dfrac{5}{9} \cdot (102.8 - 32)$

$= \dfrac{5}{9} \cdot (70.8)$

$= 39.3$

Albert's temperature is 39.3°C.

Exercise Set 7.5

1. $C = \dfrac{5}{9}(F - 32) = \dfrac{5}{9}(41 - 32) = \dfrac{5}{9}(9) = 5$

41°F is 5°C.

3. $C = \dfrac{5}{9}(F - 32) = \dfrac{5}{9}(104 - 32) = \dfrac{5}{9}(72) = 40$

104°F is 40°C.

5. $F = \dfrac{9}{5}C + 32 = \dfrac{9}{5}(60) + 32 = 108 + 32 = 140$

60°C is 140°F.

7. $F = \dfrac{9}{5}C + 32 = \dfrac{9}{5}(115) + 32 = 207 + 32 = 239$

115°C is 239°F.

9. $C = \dfrac{5}{9}(F - 32) = \dfrac{5}{9}(62 - 32) = \dfrac{5}{9}(30) \approx 16.7$

62°F is 16.7°C.

11. $C = \dfrac{5}{9}(F - 32)$

$= \dfrac{5}{9}(142.1 - 32)$

$= \dfrac{5}{9}(110.1)$

≈ 61.2

142.1°F is 61.2°C.

13. $F = 1.8C + 32$

$= 1.8(92) + 32$

$= 165.6 + 32$

$= 197.6$

92°C is 197.6°F.

15. $F = 1.8C + 32$

$= 1.8(16.3) + 32$

$= 29.34 + 32$

≈ 61.3

16.3°C is 61.3°F.

17. $C = \dfrac{5}{9}(F - 32)$

$= \dfrac{5}{9}(122 - 32)$

$= \dfrac{5}{9}(90)$

$= 50$

122°F is 50°C.

19. $F = 1.8C + 32$

$= 1.8(27) + 32$

$= 48.6 + 32$

$= 80.6$

27°C is 80.6°F.

21. $C = \dfrac{5}{9}(F - 32) = \dfrac{5}{9}(212 - 32) = \dfrac{5}{9}(180) = 100$

212°F is 100°C.

23. $C = \dfrac{5}{9}(F - 32)$

$= \dfrac{5}{9}(100.2 - 32)$

$= \dfrac{5}{9}(68.2)$

≈ 37.9

100.2°F is 37.9°C.

25. $F = 1.8C + 32$

$= 1.8(118) + 32$

$= 212.4 + 32$

$= 244.4$

118°C is 244.4°F.

27. $C = \dfrac{5}{9}(F - 32)$

$= \dfrac{5}{9}(864 - 32)$

$= \dfrac{5}{9}(832)$

≈ 462.2

864°F is 462.2°C.

29. $C = \dfrac{5}{9}(F-32)$

$\quad = \dfrac{5}{9}(70-32)$

$\quad = \dfrac{5}{9}(38)$

$\quad \approx 21.1$

70°F is 21.1°C.

31. 4×3 in. = 12 in.

or 3 in. + 3in. + 3 in. + 3 in. = 12 in.

The perimeter is 12 in.

33. 4 cm + 3 cm + 5 cm = (4 + 3 + 5) cm

$\qquad\qquad\qquad\qquad = 12$ cm

The perimeter is 12 cm.

35. $2 \times (2$ ft 8 in. $+1$ ft 6 in.$) = 2 \times (3$ ft 14 in.$)$

$\qquad\qquad\qquad\qquad\quad = 6$ ft 28 in.

$\qquad\qquad\qquad\qquad\quad = 6$ ft $+2$ ft 4 in.

$\qquad\qquad\qquad\qquad\quad = 8$ ft 4 in.

The perimeter is 8 ft 4 in.

37. False; the freezing point of water is 32°F.

39. True; the boiling point of water is 100°C.

41. $C = \dfrac{5}{9}(F-32)$

$\quad = \dfrac{5}{9}(936,000,000-32)$

$\quad = \dfrac{5}{9}(935,999,968)$

$\quad \approx 510,000,00$

936,000,000°F is approximately 520,000,000°C.

43. Yes, a 72°F room feels comfortable.

45. No, a fever of 40°F is not reasonable.

47. No, an overcoat is not needed when the temperature is 30°C.

49. Yes, a fever of 40°C is reasonable.

51. answers may vary

Section 7.6 Practice

1. total weight = 63 lb $\times$ 3 = 189 lb

energy = 189 lb $\times$ 340 ft = 64,260 ft-lb

Thus, 64,260 ft-lb of energy are required.

2. $17,065$ BTU $= 17,065$ BTU $\cdot \dfrac{778 \text{ ft-lb}}{1 \text{ BTU}}$

$\qquad\qquad\quad = 13,276,570$ ft-lb

Thus, 17,065 BTU is equivalent to 13,276,570 ft-lb.

3. total calories = 30 · 2 = 60 cal

Therefore, Alan uses 60 calories to fly his kite.

4. total calories = 200 · 5 = 1000

Therefore, Melanie uses 1000 calories.

5. calories used each day = 300(1.25) = 375

calories used for 4 days = 375 · 4 = 1500

Thus, Martha uses 1500 calories.

Vocabulary and Readiness Check

1. Energy is defined as "the capacity to do work."

2. The amount of energy needed to lift a 1-pound object a distance of 1 foot is called a foot-pound.

3. The amount of heat required to raise the temperature of 1 pound of water 1 degree Fahrenheit is called a British Thermal Unit.

4. The amount of heat required to raise the temperature of 1 kilogram of water 1 degree Celsius is called a calorie.

5. $6 \times 5 = 30$ ft-lb

6. $10 \times 4 = 40$ ft-lb

7. $3 \times 20 = 60$ ft-lb

8. $5 \times 9 = 45$ ft-lb

9. $30 \times 3 = 90$ cal

10. $15 \times 2 = 30$ cal

11. $20 \times \dfrac{1}{4} = 5$ cal

12. $50 \times \dfrac{1}{2} = 25$ cal

Exercise Set 7.6

1. energy = 3 pounds · 380 feet = 1140 ft-lb

3. energy = 168 pounds · 22 feet = 3696 ft-lb

5. energy $= \dfrac{2.5 \text{ tons}}{1} \cdot \dfrac{2000 \text{ pounds}}{1 \text{ ton}} \cdot 85 \text{ feet}$
$= 425,000 \text{ ft-lb}$

7. $30 \text{ BTU} = 30 \text{ BTU} \cdot \dfrac{778 \text{ ft-lb}}{1 \text{ BTU}} = 23,340 \text{ ft-lb}$

9. $1000 \text{ BTU} = 1000 \text{ BTU} \cdot \dfrac{778 \text{ ft-lb}}{1 \text{ BTU}}$
$= 778,000 \text{ ft-lb}$

11. $20,000 \text{ BTU} = 20,000 \text{ BTU} \cdot \dfrac{778 \text{ ft-lb}}{1 \text{ BTU}}$
$= 15,560,000 \text{ ft-lb}$

13. $34,130 \text{ BTU} = 34,130 \text{ BTU} \cdot \dfrac{778 \text{ ft-lb}}{1 \text{ BTU}}$
$= 26,553,140 \text{ ft-lb}$

15. $8,000,000 \text{ ft-lb} = 8,000,000 \text{ ft-lb} \cdot \dfrac{1 \text{ BTU}}{778 \text{ ft-lb}}$
$\approx 10,283 \text{ BTU}$

17. total hours $= 1 \cdot 7 = 7$ hours
calories $= 7 \cdot 115 = 805$ calories

19. total hours $= \dfrac{1}{2} \cdot 5 = 2.5$ hours
calories $= 2.5 \cdot 300 = 750$ calories

21. total hours $= \dfrac{1}{3} \cdot 6 = 2$ hours
calories $= 2 \cdot 720 = 1440$ calories

23. hours $= \dfrac{425}{165} \approx 2.6$ hours

25. miles $= \dfrac{3500}{200} = 17.5$ miles

27.
$$\begin{array}{r} 90 \\ -\ 17 \\ \hline 73 \end{array}$$

29.
$$\begin{array}{r} 180 \\ -\ 17 \\ \hline 163 \end{array}$$

31.
$$\begin{array}{r} 180 \\ -\ 161 \\ \hline 19 \end{array}$$

33. energy $= 123.9 \text{ pounds} \cdot \dfrac{9 \text{ inches}}{1} \cdot \dfrac{1 \text{ foot}}{12 \text{ inches}}$
$= 92.925 \text{ ft-lb}$

35. weight $= \dfrac{6400 \text{ ft-lb}}{25 \text{ ft}} = 256 \text{ lb}$

37. answers may vary

Chapter 7 Vocabulary Check

1. <u>Weight</u> is a measure of the pull of gravity.

2. <u>Mass</u> is a measure of the amount of substance in an object. This measure does not change.

3. The basic unit of length in the metric system is the <u>meter</u>.

4. To convert from one unit of length to another, <u>unit fractions</u> may be used.

5. A <u>gram</u> is the basic unit of mass in the metric system.

6. <u>Energy</u> is the capacity to do work.

7. In the U.S. system of measurement, a <u>British Thermal Unit</u> is the amount of heat required to raise the temperature of 1 pound of water 1 degree Fahrenheit.

8. The <u>liter</u> is the basic unit of capacity in the metric system.

9. In the metric system, a <u>calorie</u> is the amount of heat required to raise the temperature of 1 kilogram of water 1 degree Celsius.

Chapter 7 Review

1. $108 \text{ in.} = \dfrac{108 \text{ in.}}{1} \cdot \dfrac{1 \text{ ft}}{12 \text{ in.}} = \dfrac{108}{12} \text{ ft} = 9 \text{ ft}$

2. $72 \text{ ft} = \dfrac{72 \text{ ft}}{1} \cdot \dfrac{1 \text{ yd}}{3 \text{ ft}} = \dfrac{72}{3} \text{ yd} = 24 \text{ yd}$

3. $1.5 \text{ mi} = \dfrac{1.5 \text{ mi}}{1} \cdot \dfrac{5280 \text{ ft}}{1 \text{ mi}} = 1.5 \cdot 5280 \text{ ft} = 7920 \text{ ft}$

4. $\frac{1}{2} \text{ yd} = \frac{\frac{1}{2} \text{ yd}}{1} \cdot \frac{3 \text{ ft}}{1 \text{ yd}} \cdot \frac{12 \text{ in.}}{1 \text{ ft}} = \frac{1}{2} \cdot 3 \cdot 12 \text{ in.} = 18 \text{ in.}$

5. $52 \text{ ft} = 51 \text{ ft} + 1 \text{ ft}$
$= \frac{51 \text{ ft}}{1} \cdot \frac{1 \text{ yd}}{3 \text{ ft}} + 1 \text{ ft}$
$= \frac{51}{3} \text{ yd} + 1 \text{ ft}$
$= 17 \text{ yd} \ 1 \text{ ft}$

6. $46 \text{ in.} = 36 \text{ in.} + 10 \text{ in.}$
$= \frac{36 \text{ in.}}{1} \cdot \frac{1 \text{ ft}}{12 \text{ in.}} + 10 \text{ in.}$
$= \frac{36}{12} \text{ ft} + 10 \text{ in.}$
$= 3 \text{ ft} \ 10 \text{ in.}$

7. $42 \text{ m} = \frac{42 \text{ m}}{1} \cdot \frac{100 \text{ cm}}{1 \text{ m}} = 42 \cdot 100 \text{ cm} = 4200 \text{ cm}$

8. $82 \text{ cm} = \frac{82 \text{ cm}}{1} \cdot \frac{10 \text{ mm}}{1 \text{ cm}} = 82 \cdot 10 \text{ mm} = 820 \text{ mm}$

9. $12.18 \text{ mm} = \frac{12.18 \text{ mm}}{1} \cdot \frac{1 \text{ m}}{1000 \text{ mm}}$
$= \frac{12.18}{1000} \text{ m}$
$= 0.01218 \text{ m}$

10. $2.31 \text{ m} = \frac{2.31 \text{ m}}{1} \cdot \frac{1 \text{ km}}{1000 \text{ m}}$
$= \frac{2.31}{1000} \text{ km}$
$= 0.00231 \text{ km}$

11. $\begin{array}{r} 4 \text{ yd} \ 2 \text{ ft} \\ + 16 \text{ yd} \ 2 \text{ ft} \\ \hline 20 \text{ yd} \ 4 \text{ ft} \end{array} = 20 \text{ yd} + 1 \text{ yd} \ 1 \text{ ft} = 21 \text{ yd} \ 1 \text{ ft}$

12. $\begin{array}{r} 12 \text{ ft} \ 1 \text{ in.} \\ - \ 4 \text{ ft} \ 8 \text{ in.} \\ \hline \end{array}$ $\begin{array}{r} 11 \text{ ft} \ 13 \text{ in.} \\ - \ 4 \text{ ft} \ \ 8 \text{ in.} \\ \hline 7 \text{ ft} \ \ 5 \text{ in.} \end{array}$

13. $\begin{array}{r} 8 \text{ ft} \ \ 3 \text{ in.} \\ \times \ \ \ \ \ \ \ \ 5 \\ \hline 40 \text{ ft} \ 15 \text{ in.} \end{array} = 40 \text{ ft} + 1 \text{ ft} \ 3 \text{ in.} = 41 \text{ ft} \ 3 \text{ in.}$

14. $7 \text{ ft} \ 4 \text{ in.} \div 2 = 6 \text{ ft} + 1 \text{ ft} \ 4 \text{ in.} \div 2$
$= 6 \text{ ft} + 16 \text{ in.} \div 2$
$= \frac{6}{2} \text{ ft} + \frac{16}{2} \text{ in.}$
$= 3 \text{ ft} \ 8 \text{ in.}$

15. $8 \text{ cm} = 80 \text{ mm}$ $15 \text{ mm} = 1.5 \text{ cm}$
$\begin{array}{r} 80 \text{ mm} \\ + \ 15 \text{ mm} \\ \hline 95 \text{ mm} \end{array}$ or $\begin{array}{r} 8.0 \text{ cm} \\ + \ 1.5 \text{ cm} \\ \hline 9.5 \text{ cm} \end{array}$

16. $4 \text{ m} = 400 \text{ cm}$ $126 \text{ cm} = 1.26 \text{ m}$
$\begin{array}{r} 400 \text{ cm} \\ - \ 126 \text{ cm} \\ \hline 274 \text{ cm} \end{array}$ or $\begin{array}{r} 4.00 \text{ m} \\ - \ 1.26 \text{ m} \\ \hline 2.74 \text{ m} \end{array}$

17. $\begin{array}{r} 8.62 \text{ m} \\ \times \ \ \ \ \ 4 \\ \hline 34.48 \text{ m} \end{array}$

18. $19.6 \text{ km} \div 8 = \frac{19.6}{8} \text{ km} = 2.45 \text{ km}$

19. $\begin{array}{r} 333 \text{ yd} \ 1 \text{ ft} \\ - 163 \text{ yd} \ 2 \text{ ft} \\ \hline \end{array}$ $\begin{array}{r} 332 \text{ yd} \ 4 \text{ ft} \\ - 163 \text{ yd} \ 2 \text{ ft} \\ \hline 169 \text{ yd} \ 2 \text{ ft} \end{array}$

The amount of material that remains is 169 yd 2 ft.

20. $\begin{array}{r} 6 \text{ ft} \ 4 \text{ in.} \\ \times \ \ \ \ \ \ \ \ 20 \\ \hline 120 \text{ ft} \ 80 \text{ in.} \end{array} = 120 \text{ ft} + 6 \text{ ft} \ 8 \text{ in.} = 126 \text{ ft} \ 8 \text{ in.}$
The frames require 126 ft 8 in. of material.

21. $\begin{array}{r} 217 \text{ km} \\ \times \ \ \ \ \ 2 \\ \hline 434 \text{ km} \end{array}$

$434 \text{ km} \div 4 = \frac{434}{4} \text{ km} = 108.5 \text{ km}$

Each must drive 108.5 km.

22. $\begin{array}{r} 0.8 \text{ m} \\ \times \ 30 \text{ cm} \\ \hline \end{array}$ $\begin{array}{r} 0.8 \text{ m} \\ \times \ \ 0.3 \text{ m} \\ \hline 0.24 \text{ sq m} \end{array}$

The area is 0.24 sq m.

23. $66 \text{ oz} = \frac{66 \text{ oz}}{1} \cdot \frac{1 \text{ lb}}{16 \text{ oz}} = \frac{66}{16} \text{ lb} = 4.125 \text{ lb}$

24. $2.3 \text{ tons} = \dfrac{2.3 \text{ tons}}{1} \cdot \dfrac{2000 \text{ lb}}{1 \text{ ton}}$

$\qquad = 2.3 \cdot 2000 \text{ lb}$

$\qquad = 4600 \text{ lb}$

25. $52 \text{ oz} = 48 \text{ oz} + 4 \text{ oz}$

$\qquad = \dfrac{48 \text{ oz}}{1} \cdot \dfrac{1 \text{ lb}}{16 \text{ oz}} + 4 \text{ oz}$

$\qquad = \dfrac{48}{16} \text{ lb} + 4 \text{ oz}$

$\qquad = 3 \text{ lb } 4 \text{ oz}$

26. $10,300 \text{ lb} = 10,000 \text{ lb} + 300 \text{ lb}$

$\qquad = \dfrac{10,000 \text{ lb}}{1} \cdot \dfrac{1 \text{ ton}}{2000 \text{ lb}} + 300 \text{ lb}$

$\qquad = \dfrac{10,000}{200} \text{ tons} + 300 \text{ lb}$

$\qquad = 5 \text{ tons } 300 \text{ lb}$

27. $27 \text{ mg} = \dfrac{27 \text{ mg}}{1} \cdot \dfrac{1 \text{ g}}{1000 \text{ mg}} = \dfrac{27}{1000} \text{ g} = 0.027 \text{ g}$

28. $40 \text{ kg} = \dfrac{40 \text{ kg}}{1} \cdot \dfrac{1000 \text{ g}}{1 \text{ kg}} = 40 \cdot 1000 \text{ g} = 40,000 \text{ g}$

29. $2.1 \text{ hg} = \dfrac{2.1 \text{ hg}}{1} \cdot \dfrac{10 \text{ dag}}{1 \text{ hg}} = 2.1 \cdot 10 \text{ dag} = 21 \text{ dag}$

30. $0.03 \text{ mg} = \dfrac{0.03 \text{ mg}}{1} \cdot \dfrac{1 \text{ dg}}{100 \text{ mg}}$

$\qquad = \dfrac{0.03}{100} \text{ dg}$

$\qquad = 0.0003 \text{ dg}$

31.
$$\begin{array}{r} 6 \text{ lb } 5 \text{ oz} \\ -\ 2 \text{ lb } 12 \text{ oz} \\ \hline \end{array} \qquad \begin{array}{r} 5 \text{ lb } 21 \text{ oz} \\ -\ 2 \text{ lb } 12 \text{ oz} \\ \hline 3 \text{ lb } 9 \text{ oz} \end{array}$$

32.
$$\begin{array}{r} 5 \text{ tons } 1600 \text{ lb} \\ +\ 4 \text{ tons } 1200 \text{ lb} \\ \hline 9 \text{ tons } 2800 \text{ lb} \end{array} \quad \begin{array}{l} = 9 \text{ tons} + 1 \text{ ton } 800 \text{ lb} \\ = 10 \text{ tons } 800 \text{ lb} \end{array}$$

33. $6 \text{ tons } 2250 \text{ lb} \div 3 = \dfrac{6}{3} \text{ tons } \dfrac{2250}{3} \text{ lb}$

$\qquad\qquad = 2 \text{ tons } 750 \text{ lb}$

34.
$$\begin{array}{r} 8 \text{ lb } 6 \text{ oz} \\ \times \qquad\quad 4 \\ \hline 32 \text{ lb } 24 \text{ oz} \end{array} = 32 \text{ lb} + 1 \text{ lb } 8 \text{ oz} = 33 \text{ lb } 8 \text{ oz}$$

35.
$$\begin{array}{r} 4.3 \text{ mg} \\ \times \quad 5 \\ \hline 21.5 \text{ mg} \end{array}$$

36. $4.8 \text{ kg} = 4800 \text{ g} \qquad\qquad 4200 \text{ g} = 4.2 \text{ kg}$

$$\begin{array}{r} 4800 \text{ g} \\ -\ 4200 \text{ g} \\ \hline 600 \text{ g} \end{array} \quad \text{or} \quad \begin{array}{r} 4.8 \text{ kg} \\ -\ 4.2 \text{ kg} \\ \hline 0.6 \text{ kg} \end{array}$$

37.
$$\begin{array}{r} 1 \text{ lb } 12 \text{ oz} \\ +\ 2 \text{ lb } 8 \text{ oz} \\ \hline 3 \text{ lb } 20 \text{ oz} \end{array} = 3 \text{ lb} + 1 \text{ lb } 4 \text{ oz} = 4 \text{ lb } 4 \text{ oz}$$

The total weight was 4 lb 4 oz.

38. $38 \text{ tons } 300 \text{ lb} \div 4 = \dfrac{38}{4} \text{ tons } \dfrac{300}{4} \text{ lb}$

$\qquad\qquad = 9\dfrac{1}{2} \text{ tons } 75 \text{ lb}$

$\qquad\qquad = 9 \text{ tons} + \dfrac{1}{2} \text{ ton} + 75 \text{ lb}$

$\qquad\qquad = 9 \text{ tons} + 1000 \text{ lb} + 75 \text{ lb}$

$\qquad\qquad = 9 \text{ tons } 1075 \text{ lb}$

They each receive 9 tons 1075 lb.

39. $450 \text{ g} = 0.450 \text{ kg}$

$8.3 \text{ kg} - 0.450 \text{ kg} = 7.85 \text{ kg}$

She actually received 7.85 kg.

40. $9.3 \text{ kg} \div 8 = \dfrac{9.3}{8} \text{ kg} = 1.1625 \text{ kg}$

Each received 1.1625 kg.

41. $16 \text{ pt} = \dfrac{16 \text{ pt}}{1} \cdot \dfrac{1 \text{ qt}}{2 \text{ pt}} = \dfrac{16}{2} \text{ qt} = 8 \text{ qt}$

42. $40 \text{ fl oz} = \dfrac{40 \text{ fl oz}}{1} \cdot \dfrac{1 \text{ c}}{8 \text{ fl oz}} = \dfrac{40}{8} \text{ c} = 5 \text{ c}$

43. $3 \text{ qt } 1 \text{ pt} = \dfrac{3 \text{ qt}}{1} \cdot \dfrac{2 \text{ pt}}{1 \text{ qt}} + 1 \text{ pt}$

$\qquad\qquad = 3 \cdot 2 \text{ pt} + 1 \text{ pt}$

$\qquad\qquad = 6 \text{ pt} + 1 \text{ pt}$

$\qquad\qquad = 7 \text{ pt}$

44. $18 \text{ qt} = \dfrac{18 \text{ qt}}{1} \cdot \dfrac{2 \text{ pt}}{1 \text{ qt}} \cdot \dfrac{2 \text{ c}}{1 \text{ pt}} = 18 \cdot 2 \cdot 2 \text{ c} = 72 \text{ c}$

45. $9 \text{ pt} = 8 \text{ pt} + 1 \text{ pt}$

$= \dfrac{8 \text{ pt}}{1} \cdot \dfrac{1 \text{ qt}}{2 \text{ pt}} + 1 \text{ pt}$

$= \dfrac{8}{2} \text{ qt} + 1 \text{ pt}$

$= 4 \text{ qt } 1 \text{ pt}$

46. $15 \text{ qt} = 12 \text{ qt} + 3 \text{ qt}$

$= \dfrac{12 \text{ qt}}{1} \cdot \dfrac{1 \text{ gal}}{4 \text{ qt}} + 3 \text{ qt}$

$= \dfrac{12}{4} \text{ gal} + 3 \text{ qt}$

$= 3 \text{ gal } 3 \text{ qt}$

47. $3.8 \text{ L} = \dfrac{3.8 \text{ L}}{1} \cdot \dfrac{1000 \text{ ml}}{1 \text{ L}}$

$= 3.8 \cdot 1000 \text{ ml}$

$= 3800 \text{ ml}$

48. $4.2 \text{ ml} = \dfrac{4.2 \text{ ml}}{1} \cdot \dfrac{1 \text{ dl}}{100 \text{ ml}} = \dfrac{4.2}{100} \text{ dl} = 0.042 \text{ dl}$

49. $14 \text{ hl} = \dfrac{14 \text{ hl}}{1} \cdot \dfrac{1 \text{ kl}}{10 \text{ hl}} = \dfrac{14}{10} \text{ kl} = 1.4 \text{ kl}$

50. $30.6 \text{ L} = \dfrac{30.6 \text{ L}}{1} \cdot \dfrac{100 \text{ cl}}{1 \text{ L}} = 30.6 \cdot 100 \text{ cl} = 3060 \text{ cl}$

51.
$$\begin{array}{r} 1 \text{ qt } 1 \text{ pt} \\ + \, 3 \text{ qt } 1 \text{ pt} \\ \hline 4 \text{ qt } 2 \text{ pt} \end{array} = 4 \text{ qt} + 1 \text{ qt} = 1 \text{ gal } 1 \text{ qt}$$

52.
$$\begin{array}{r} 3 \text{ gal } 2 \text{ qt} \\ \times \qquad 2 \\ \hline 6 \text{ gal } 4 \text{ qt} \end{array} = 6 \text{ gal} + 1 \text{ gal} = 7 \text{ gal}$$

53. $0.946 \text{ L} = 946 \text{ ml} \qquad 210 \text{ ml} = 0.21 \text{ L}$

$$\begin{array}{r} 946 \text{ ml} \\ - \, 210 \text{ ml} \\ \hline 736 \text{ ml} \end{array} \quad \text{or} \quad \begin{array}{r} 0.946 \text{ L} \\ - \, 0.210 \text{ L} \\ \hline 0.736 \text{ L} \end{array}$$

54. $6.1 \text{ L} = 6100 \text{ ml} \qquad 9400 \text{ ml} = 9.4 \text{ L}$

$$\begin{array}{r} 6100 \text{ ml} \\ + \, 9400 \text{ ml} \\ \hline 15,500 \text{ ml} \end{array} \quad \text{or} \quad \begin{array}{r} 6.1 \text{ L} \\ + \, 9.4 \text{ L} \\ \hline 15.5 \text{ L} \end{array}$$

55.
$$\begin{array}{r} 4 \text{ gal } 2 \text{ qt} \\ - \, 1 \text{ gal } 3 \text{ qt} \end{array} \qquad \begin{array}{r} 3 \text{ gal } 6 \text{ qt} \\ - \, 1 \text{ gal } 3 \text{ qt} \\ \hline 2 \text{ gal } 3 \text{ qt} \end{array}$$

There are 2 gal 3 qt of tea remaining.

56. $1 \text{ c } 4 \text{ fl oz} \div 2 = (8 \text{ fl oz} + 4 \text{ fl oz}) \div 2$

$= 12 \text{ fl oz} \div 2$

$= 6 \text{ fl oz}$

Use 6 fl oz of stock for half of a recipe.

57. $85 \text{ ml} \times 8 \times 16 = 10,880 \text{ ml}$

$\dfrac{10,880 \text{ ml}}{1} \cdot \dfrac{1 \text{ L}}{1000 \text{ ml}} = \dfrac{10,880}{1000} \text{ L} = 10.88 \text{ L}$

58. $6 \text{ L} + 1300 \text{ ml} + 2.6 \text{ L} = 6 \text{ L} + 1.3 \text{ L} + 2.6 \text{ L}$

$= 9.9 \text{ L}$

Since 9.9 L is less than 10 L, yes it will fit.

59. $7 \text{ m} = \dfrac{7 \text{ m}}{1} \cdot \dfrac{3.28 \text{ ft}}{1 \text{ m}} \approx 22.96 \text{ ft}$

60. $11.5 \text{ yd} = \dfrac{11.5 \text{ yd}}{1} \cdot \dfrac{1 \text{ m}}{1.09 \text{ yd}} \approx 10.55 \text{ m}$

61. $17.5 \text{ L} = \dfrac{17.5 \text{ L}}{1} \cdot \dfrac{0.26 \text{ gal}}{1 \text{ L}} \approx 4.55 \text{ gal}$

62. $7.8 \text{ L} = \dfrac{7.8 \text{ L}}{1} \cdot \dfrac{1.06 \text{ qt}}{1 \text{ L}} \approx 8.27 \text{ qt}$

63. $15 \text{ oz} = \dfrac{15 \text{ oz}}{1} \cdot \dfrac{28.35 \text{ g}}{1 \text{ oz}} \approx 425.25 \text{ g}$

64. $23 \text{ lb} = \dfrac{23 \text{ lb}}{1} \cdot \dfrac{0.45 \text{ kg}}{1 \text{ lb}} \approx 10.35 \text{ kg}$

65. $100 \text{ m} = \dfrac{100 \text{ m}}{1} \cdot \dfrac{1.09 \text{ yd}}{1 \text{ m}} \approx 109 \text{ yd}$

The race is about 109 yd.

66. $82 \text{ kg} = \dfrac{82 \text{ kg}}{1} \cdot \dfrac{2.20 \text{ lb}}{1 \text{ kg}} \approx 180.4$

The person weighs approximately 180.4 lb.

67. $3 \text{ L} = \dfrac{3 \text{ L}}{1} \cdot \dfrac{1.06 \text{ qt}}{1 \text{ L}} \approx 3.18 \text{ qt}$

The bottle contains approximately 3.18 qt.

68. $1.2 \text{ mm} \times 50 = 60 \text{ mm}$

$60 \text{ mm} = \dfrac{60 \text{ mm}}{1} \cdot \dfrac{1 \text{ cm}}{10 \text{ mm}} = 6 \text{ cm}$

$6 \text{ cm} = \dfrac{6 \text{ cm}}{1} \cdot \dfrac{1 \text{ in.}}{2.54 \text{ cm}} \approx 2.36 \text{ in.}$

The height of the stack is approximately 2.36 in.

69. F = 1.8C + 32 = 1.8(245) + 32 = 441 + 32 = 473
245°C is 473°F.

70. F = 1.8C + 32 = 1.8(160) + 32 = 288 + 32 = 320
160°C is 320°F.

71. $F = 1.8C + 32$
$= 1.8(42) + 32$
$= 75.6 + 32$
$= 107.6$
42°C is 107.6°F.

72. $C = \dfrac{5}{9}(F - 32) = \dfrac{5}{9}(93.2 - 32) = \dfrac{5}{9}(61.2) = 34$
93.2°F is 34°C.

73. $C = \dfrac{5}{9}(F - 32) = \dfrac{5}{9}(41.3 - 32) = \dfrac{5}{9}(9.3) \approx 5.2$
41.3°F is 5.2°C.

74. $C = \dfrac{5}{9}(F - 32) = \dfrac{5}{9}(80 - 32) = \dfrac{5}{9}(48) \approx 26.7$
80°F is 26.7°C.

75. $C = \dfrac{5}{9}(F - 32) = \dfrac{5}{9}(35 - 32) = \dfrac{5}{9}(3) \approx 1.7$
35°F is 1.7°C.

76. $F = 1.8C + 32 = 1.8(165) + 32 = 297 + 32 = 329$
165°C is 329°F.

77. energy = 5.6 lb · 12 ft = 67.2 ft-lb
It requires 67.2 ft-lb of energy.

78. energy = 21 lb · 6.5 ft = 136.5 ft-lb
It requires 136.5 ft-lb of energy.

79. energy $= \dfrac{1.2 \text{ tons}}{1} \cdot \dfrac{2000 \text{ lb}}{1 \text{ ton}} \cdot \dfrac{15 \text{ yd}}{1} \cdot \dfrac{3 \text{ ft}}{1 \text{ yd}}$
$= 108,000 \text{ ft-lb}$
It requires 108,000 ft-lb of energy.

80. 12,000 BTU $= 12,000 \text{ BTU} \cdot \dfrac{778 \text{ ft-lb}}{1 \text{ BTU}}$
$= 9,336,000 \text{ ft-lb}$
12,000 BTU is equivalent to 9,336,000 ft-lb.

81. 2,000,000 ft-lb $= 2,000,000 \text{ ft-lb} \cdot \dfrac{1 \text{ BTU}}{778 \text{ ft-lb}}$
$\approx 2600 \text{ BTU}$

82. calories $= 450 \times 2\dfrac{1}{2} = 1125 \text{ cal}$
He uses 1125 calories.

83. total hours = 3 × 24 = 72 hr
total calories = 72 × 210 = 15,120 cal
She used a total of 15,120 calories.

84. hours $= \dfrac{420}{180} = \dfrac{7}{3} = 2\dfrac{1}{3}$ hr
She needs to walk $2\dfrac{1}{3}$ hours to burn of the calories.

85. 2.5 mi $= \dfrac{2.5 \text{ mi}}{1} \cdot \dfrac{5280 \text{ ft}}{1 \text{ mi}} = 13,200 \text{ ft}$

86. 6.25 ft $= \dfrac{6.25 \text{ ft}}{1} \cdot \dfrac{12 \text{ in.}}{1 \text{ ft}} = 75 \text{ in.}$

87. 23,760 ft $= \dfrac{23,760 \text{ ft}}{1} \cdot \dfrac{1 \text{ mi}}{5280 \text{ ft}} = 4.5 \text{ mi}$

88. 129 in. $= \dfrac{129 \text{ in.}}{1} \cdot \dfrac{1 \text{ ft}}{12 \text{ in.}} = 10.75 \text{ ft}$

89. 8200 lb = 8000 lb + 200 lb
$= \dfrac{8000 \text{ lb}}{1} \cdot \dfrac{1 \text{ ton}}{2000 \text{ lb}} + 200 \text{ lb}$
$= 4 \text{ tons } 200 \text{ lb}$

90. 4300 lb = 4000 lb + 300 lb
$= \dfrac{4000 \text{ lb}}{1} \cdot \dfrac{1 \text{ ton}}{2000 \text{ lb}} + 300 \text{ lb}$
$= 2 \text{ tons } 300 \text{ lb}$

91. 5 m $= \dfrac{5 \text{ m}}{1} \cdot \dfrac{100 \text{ cm}}{1 \text{ m}} = 500 \text{ cm}$

92. 286 mm $= \dfrac{286 \text{ mm}}{1} \cdot \dfrac{1 \text{ km}}{1,000,000 \text{ mm}}$
$= 0.000286 \text{ km}$

93. 1400 mg $= \dfrac{1400 \text{ mg}}{1} \cdot \dfrac{1 \text{ g}}{1000 \text{ mg}} = 1.4 \text{ g}$

94. 240 mg $= \dfrac{240 \text{ mg}}{1} \cdot \dfrac{1 \text{ g}}{1000 \text{ mg}} = 0.24 \text{ g}$

95. $6.75 \text{ gal} = \dfrac{6.75 \text{ gal}}{1} \cdot \dfrac{4 \text{ qt}}{1 \text{ gal}} = 27 \text{ qt}$

96. $5.25 \text{ gal} = \dfrac{5.25 \text{ gal}}{1} \cdot \dfrac{4 \text{ qt}}{1 \text{ gal}} = 21 \text{ qt}$

97. $8.5 \text{ pt} = \dfrac{8.5 \text{ pt}}{1} \cdot \dfrac{2 \text{ c}}{1 \text{ pt}} = 17 \text{ c}$

98. $6.25 \text{ pt} = \dfrac{6.25 \text{ pt}}{1} \cdot \dfrac{2 \text{ c}}{1 \text{ pt}} = 12.5 \text{ c}$

99. $F = \dfrac{9}{5}C + 32 = \dfrac{9}{5}(86) + 32 = 154.8 + 32 = 186.8$

$86°$ is $186.8°$F.

100. $F = \dfrac{9}{5}C + 32 = \dfrac{9}{5}(15) + 32 = 27 + 32 = 59$

$15°$C is $59°$F.

101. $C = \dfrac{5}{9}(F - 32) = \dfrac{5}{9}(51.8 - 32) = \dfrac{5}{9}(19.8) = 11$

$51.8°$F is $11°$C.

102. $C = \dfrac{5}{9}(F - 32) = \dfrac{5}{9}(82.4 - 32) = \dfrac{5}{9}(50.4) = 28$

$82.4°$F is $28°$C.

103. 9.3 km = 9300 m   183 m = 0.183 km

9300 m		9.300 km
− 183 m	or	− 0.183 km
9117 m		9.117 km

104. 8.6 km = 8600 m   247 m = 0.247 km

8600 m		8.600 km
− 247 m	or	− 0.247 km
8353 m		8.353 km

105. 7.4 L = 7400 ml   6500 ml = 6.5 L

7400 ml		7.4 L
+ 6500 ml	or	+ 6.5 L
13,900 ml		13.9 L

106. 35 L = 35,000 ml   700 ml = 0.7 L

35,000 ml		35.0 L
+ 700 ml	or	+ 0.7 L
35,700 ml		35.7 L

107. 9.3 g = 9300 mg   1200 mg = 1.2 g

9300 mg		9.3 g
− 1200 mg	or	− 1.2 g
8100 mg		8.1 g

108. 3.4 g = 3400 mg   1800 mg = 1.8 g

3400 mg		3.4 g
− 1800 mg	or	− 1.8 g
1600 mg		1.6 g

109.

6.3 kg
× 8
50.4 kg

110.

3.2 kg
× 4
12.8 kg

111.

3 gal 1 qt
+ 4 gal 2 qt
7 gal 3 qt

112.

6 gal 1 qt
+ 2 gal 1 qt
8 gal 2 qt

113. 4100 mm = 41 dm   3 dm = 300 mm

41 dm		4100 mm
− 3 dm	or	− 300 mm
38 dm		3800 mm

114. 6300 mm = 63 dm   5 dm = 500 mm

63 dm		6300 mm
− 5 dm	or	− 500 mm
58 dm		5800 mm

115. $4.5 \text{ tons} \div 2 = \dfrac{4.5}{2} \text{ tons} = 2.25 \text{ tons}$

116. $6.75 \text{ tons} \div 3 = \dfrac{6.75}{3} \text{ tons} = 2.25 \text{ tons}$

Chapter 7 Test

1.

$$12)\overline{280}$$
$$\underline{-24}$$
$$40$$
$$\underline{-36}$$
$$4$$

280 inches = 23 ft 4 in.

2. $2\dfrac{1}{2}$ gal $= \dfrac{2\frac{1}{2}\text{ gal}}{1} \cdot \dfrac{4\text{ qt}}{1\text{ gal}} = 10$ qt

3. 30 oz $= \dfrac{30\text{ oz}}{1} \cdot \dfrac{1\text{ lb}}{16\text{ oz}} = 1.875$ lb

4. 2.8 tons $= \dfrac{2.8\text{ tons}}{1} \cdot \dfrac{2000\text{ lb}}{1\text{ ton}} = 5600$ lb

5. 38 pt $= \dfrac{38\text{ pt}}{1} \cdot \dfrac{1\text{ qt}}{2\text{ pt}} \cdot \dfrac{1\text{ gal}}{4\text{ qt}}$
$= \dfrac{38}{8}$ gal
$= \dfrac{19}{4}$ gal
$= 4\dfrac{3}{4}$ gal

6. 40 mg $= \dfrac{40\text{ mg}}{1} \cdot \dfrac{1\text{ g}}{1000\text{ mg}} = 0.04$ g

7. 2.4 kg $= \dfrac{2.4\text{ kg}}{1} \cdot \dfrac{1000\text{ g}}{1\text{ kg}} = 2400$ g

8. 3.6 cm $= \dfrac{3.6\text{ cm}}{1} \cdot \dfrac{10\text{ mm}}{1\text{ cm}} = 36$ mm

9. 4.3 dg $= \dfrac{4.3\text{ dg}}{1} \cdot \dfrac{1\text{ g}}{10\text{ dg}} = \dfrac{4.3}{10}$ g $= 0.43$ g

10. 0.83 L $= \dfrac{0.83\text{ L}}{1} \cdot \dfrac{1000\text{ ml}}{1\text{ L}} = 830$ ml

11.

$$\begin{array}{r} 3\text{ qt }1\text{ pt} \\ +\ 2\text{ qt }1\text{ pt} \\ \hline \end{array}$$

5 qt 2 pt $= 4$ qt $+ 1$ qt $+ 2$ pt
$= 1$ gal $+ 1$ qt $+ 1$ qt
$= 1$ gal $+ 1$ qt $+ 1$ qt
$= 1$ gal 2 qt

12.

$$\begin{array}{r} 8\text{ lb }6\text{ oz} \\ -\ 4\text{ lb }9\text{ oz} \\ \hline \end{array} \rightarrow \begin{array}{r} 7\text{ lb }22\text{ oz} \\ -\ 4\text{ lb }\ \ 9\text{ oz} \\ \hline 3\text{ lb }13\text{ oz} \end{array}$$

13. 2 ft 9 in. $\times 3 = 6$ ft 27 in.
$= 6$ ft $+ 2$ ft 3 in.
$= 8$ ft 3 in.

14. 5 gal 2 qt $\div 2 = 4$ gal 6 qt $\div 2$
$= \dfrac{4}{2}$ gal $\dfrac{6}{2}$ qt
$= 2$ gal 3 qt

15.

8 cm = 80 mm   14 mm = 1.4 cm

$$\begin{array}{r} 80\text{ mm} \\ -\ 14\text{ mm} \\ \hline 66\text{ mm} \end{array} \quad \text{or} \quad \begin{array}{r} 8.0\text{ cm} \\ -\ 1.4\text{ cm} \\ \hline 6.6\text{ cm} \end{array}$$

16.

1.8 km = 1800 m   456 m = 0.456 km

$$\begin{array}{r} 1800\text{ m} \\ +\ 456\text{ m} \\ \hline 2256\text{ m} \end{array} \quad \text{or} \quad \begin{array}{r} 1.800\text{ km} \\ +\ 0.456\text{ km} \\ \hline 2.256\text{ km} \end{array}$$

17. $C = \dfrac{5}{9}(F - 32)$
$= \dfrac{5}{9}(84 - 32)$
$= \dfrac{5}{9}(52)$
≈ 28.9
84°F is 28.9°C.

18. $F = \dfrac{9}{5}C + 32 = \dfrac{9}{5}(12.6) + 32 = 22.68 + 32 \approx 54.7$

12.6°C is 54.7°F.

19. 8.4 m $\cdot \dfrac{2}{3} = \dfrac{8.4}{1} \cdot \dfrac{2}{3}$ m $= 5.6$ m

The trees will be 5.6 m tall.

20.

$$\begin{array}{r} 20\text{ gal} \\ -\ 15\text{ gal }1\text{ qt} \\ \hline \end{array} \qquad \begin{array}{r} 19\text{ gal }4\text{ qt} \\ -\ 15\text{ gal }1\text{ qt} \\ \hline 4\text{ gal }3\text{ qt} \end{array}$$

Thus, 4 gal 3 qt remains in the container.

21. 88 m $+ 340$ cm $= 88$ m $+ 3.40$ m $= 91.4$ m
The span is 91.4 meters

22.

$$\begin{array}{r} 2 \text{ ft } 9 \text{ in.} \\ \times \qquad 6 \\ \hline 12 \text{ ft } 54 \text{ in.} \end{array} = 12 \text{ ft} + 4 \text{ ft } 6 \text{ in.} = 16 \text{ ft } 6 \text{ in.}$$

Thus, 16 ft 6 in. of material is needed.

23.

$$\begin{array}{r} 246 \text{ ft} \quad 9 \text{ in.} \\ \times \qquad 2 \\ \hline 492 \text{ ft } 18 \text{ in.} \end{array} = 492 \text{ ft} + 1 \text{ ft } 6 \text{ in.} = 493 \text{ ft } 6 \text{ in.}$$

24. $101.6 \text{ cm} \times 148 = 15,036.8 \text{ cm}$

$$\frac{15,036.8 \text{ cm}}{1} \cdot \frac{1 \text{ m}}{100 \text{ cm}} = 150.368 \text{ m}$$

The total length is 150.368 m.

25. $C = \dfrac{5}{9}(F - 32)$

$$= \frac{5}{9}(136 - 32)$$

$$= \frac{5}{9}(104)$$

$$\approx 57.8$$

136°F is 57.8°C.

26. $F = \dfrac{9}{5}C + 32 = \dfrac{9}{5}(41) + 32 = 73.8 + 32 = 105.8$

41°C is 105.8°F.

27. $4667 \text{ gal} = \dfrac{4667 \text{ gal}}{1} \cdot \dfrac{4 \text{ qt}}{1 \text{ gal}} \cdot \dfrac{2 \text{ pt}}{1 \text{ qt}}$

$$= 4667 \cdot 4 \cdot 2 \text{ pt}$$

$$= 37,336 \text{ pt}$$

There were 37,336 pints of ice cream used.

28. $5 \text{ g} \approx \dfrac{5 \text{ g}}{1} \cdot \dfrac{0.04 \text{ oz}}{1 \text{ g}} \approx 0.2 \text{ oz}$

The candy weighs about 0.2 oz.

29. $5 \text{ km} \approx \dfrac{5 \text{ km}}{1} \cdot \dfrac{1 \text{ mi}}{1.61 \text{ km}} \approx 3.1 \text{ mi}$

5 km is about 3.1 mi.

30. $5 \text{ gal} \approx \dfrac{5 \text{ gal}}{1} \cdot \dfrac{3.79 \text{ L}}{1 \text{ gal}} \approx 18.95 \text{ L}$

The container holds about 18.95 L.

31. energy $= 48.5 \text{ lb} \cdot 14 \text{ ft} = 679 \text{ ft-lb}$
The energy required is 679 ft-lb.

32. $26,000 \text{ BTU} = \dfrac{26,000 \text{ BTU}}{1} \cdot \dfrac{778 \text{ ft-lb}}{1 \text{ BTU}}$

$$= 20,228,000 \text{ ft-lb}$$

The heater requires 20,228,000 ft-lb.

33. $\dfrac{180 \text{ cal}}{1 \text{ hour}} \cdot \dfrac{1 \text{ hour}}{1 \text{ day}} \cdot \dfrac{5 \text{ days}}{1} = 180 \cdot 5 \text{ cal} = 900 \text{ cal}$

She burns 900 calories.

Cumulative Review Chapters 1–7

1.

$$\begin{array}{r} {\scriptstyle 1\,2\,2} \\ 1647 \\ 246 \\ 32 \\ + \quad 85 \\ \hline 2010 \end{array}$$

2.

$$\begin{array}{r} 2000 \\ - \quad 469 \\ \hline 1531 \end{array}$$

3.

$$\begin{array}{r} 7 \\ 5\overline{)35} \\ 3\overline{)105} \\ 3\overline{)315} \\ 3\overline{)945} \end{array}$$

The prime factorization of 945 is $3 \cdot 3 \cdot 3 \cdot 5 \cdot 7$ or $3^3 \cdot 5 \cdot 7$.

4. area $= $ length $\cdot$ width $= 17 \text{ in.} \cdot 9 \text{ in.} = 153 \text{ sq in.}$
The area is 153 sq in.

5. $11 = 11$
$33 = 3 \cdot 11$
The LCM is $3 \cdot 11$ or 33.

6. The LCD of 21 and 9 is 63.

$$\frac{8}{21} - \frac{2}{9} = \frac{8}{21} \cdot \frac{3}{3} - \frac{2}{9} \cdot \frac{7}{7}$$

$$= \frac{24}{63} - \frac{14}{63}$$

$$= \frac{24 - 14}{63}$$

$$= \frac{10}{63}$$

7. $3\dfrac{4}{5} = 3\dfrac{12}{15}$

$+1\dfrac{4}{15} = 1\dfrac{4}{15}$

$4\dfrac{16}{15}$

$4\dfrac{16}{15} = 4 + 1\dfrac{1}{15} = 5\dfrac{1}{15}$

8. $2\dfrac{1}{2} \cdot 4\dfrac{2}{15} = \dfrac{5}{2} \cdot \dfrac{62}{15} = \dfrac{5 \cdot 2 \cdot 31}{2 \cdot 5 \cdot 3} = \dfrac{31}{3}$ or $10\dfrac{1}{3}$

9. $0.125 = \dfrac{125}{1000} = \dfrac{1}{8}$

10. $1.2 = 1\dfrac{2}{10} = 1\dfrac{1}{5}$

11. $105.083 = 105\dfrac{83}{1000}$

12. $\left(\dfrac{2}{3}\right)^3 = \dfrac{2}{3} \cdot \dfrac{2}{3} \cdot \dfrac{2}{3} = \dfrac{8}{27}$

13. $\begin{array}{cc} 0.052 & 0.236 \\ \downarrow & \downarrow \\ 0 & < \quad 2, \text{ so} \end{array}$

$0.052 < 0.236$

14. $30 \div 6 \cdot 5 = 5 \cdot 5 = 25$

15.
$$\begin{array}{r} 85.00 \\ -17.31 \\ \hline 67.69 \end{array}$$

Check: $\begin{array}{r} 1\ 1\ 1 \\ 67.69 \\ +17.31 \\ \hline 85.00 \end{array}$

16.
$$\begin{array}{r} {\scriptstyle 1\ 1\ \ 1} \\ 27.900 \\ 8.070 \\ +103.261 \\ \hline 139.231 \end{array}$$

17. $42.1 \times 0.1 = 4.21$

18. $186.04 \times 1000 = 186{,}040$

19. $9.2 \times 0.001 = 0.0092$

20. The average is the sum divided by 3.

$$\dfrac{6.8 + 9.7 + 0.9}{3} = \dfrac{17.4}{3} = 5.8$$

21.
$$\begin{array}{r} 0.26 \\ 32\overline{)\,8.32} \\ -6\ 4 \\ \hline 1\ 92 \\ -1\ 92 \\ \hline 0 \end{array}$$

Check: $\begin{array}{r} 0.26 \\ \times\ 32 \\ \hline 0.52 \\ 7.80 \\ \hline 8.32 \end{array}$

22. The LCD of 10 and 4 is 20.

$$\dfrac{3}{10} + \dfrac{3}{4} = \dfrac{3}{10} \cdot \dfrac{2}{2} + \dfrac{3}{4} \cdot \dfrac{5}{5} = \dfrac{6}{20} + \dfrac{15}{20} = \dfrac{21}{20} \text{ or } 1\dfrac{1}{20}$$

23. $\dfrac{3}{16} \rightarrow$
$$\begin{array}{r} 0.1875 \\ 16\overline{)\,3.0000} \\ -1\ 6 \\ \hline 1\ 40 \\ -1\ 28 \\ \hline 120 \\ -112 \\ \hline 80 \\ -80 \\ \hline 0 \end{array}$$

Thus $2\dfrac{3}{16} = 2.1875$.

24. To round 7.2846 to the nearest tenth, observe that the digit in the hundredths place is 8. Since this digit is at least 5, we add 1 to the digit in the tenths place. The number 7.2846, rounded to the nearest tenth is 7.3.

25.
$$\begin{array}{r} 0.666\ldots \\ 3\overline{)\,2.000} \\ -1\ 8 \\ \hline 20 \\ -18 \\ \hline 20 \\ -18 \\ \hline 2 \end{array}$$

$\dfrac{2}{3} = 0.666\ldots = 0.\overline{6}$

26. $\dfrac{0.12 + 0.96}{0.5} = \dfrac{1.08}{0.5} = 2.16$

27. The ratio of 12 to 17 is $\dfrac{12}{17}$.

28. The ratio of $2\dfrac{2}{3}$ to $5\dfrac{1}{9}$ is

$$\dfrac{2\frac{2}{3}}{5\frac{1}{9}} = \dfrac{\frac{8}{3}}{\frac{46}{9}} = \dfrac{8}{3} \div \dfrac{46}{9} = \dfrac{8}{3} \cdot \dfrac{9}{46} = \dfrac{12}{23}.$$

29.
$$
\begin{array}{r}
22.5 \\
15\overline{)\,337.5} \\
\underline{-30} \\
37 \\
\underline{-30} \\
7\,5 \\
\underline{-7\,5} \\
0
\end{array}
$$
The unit rate is 22.5 miles/gallon.

30. $\dfrac{\text{side}}{\text{perimeter}} = \dfrac{9 \text{ inches}}{4 \times 9 \text{ inches}} = \dfrac{9}{36} = \dfrac{1}{4}$

31.
$$\dfrac{7}{n} = \dfrac{6}{5}$$
$$7 \cdot 5 = n \cdot 6$$
$$35 = n \cdot 6$$
$$\dfrac{35}{6} = n$$
$$5\dfrac{5}{6} = n$$

32. $\dfrac{3 \text{ cups}}{2 \text{ crusts}} = \dfrac{n \text{ cups}}{5 \text{ crusts}}$
$$3 \cdot 5 = 2 \cdot n$$
$$15 = 2 \cdot n$$
$$\dfrac{15}{2} = n$$
$$7.5 = n$$
Thus, 7.5 cups of flour are needed.

33.
$$\dfrac{1}{2400} = \dfrac{n}{15,360}$$
$$1 \cdot 15,360 = 2400 \cdot n$$
$$15,360 = 2400, n$$
$$\dfrac{15,360}{2400} = n$$
$$6.4 = n$$
Therefore, 7 bags of fertilizer are needed.

34. $23\% = \dfrac{23}{100}$

35. $23\% = 23(0.01) = 0.23$

36. $\dfrac{7}{8} = \dfrac{7}{8}(100\%) = \dfrac{700}{8}\% = 87\dfrac{1}{2}\%$ or 87.5%

37. $\dfrac{1}{12} = \dfrac{1}{12} \cdot 100\% = \dfrac{1}{12} \cdot \dfrac{100\%}{1} = \dfrac{100}{12}\% \approx 8.33\%$

$$
\begin{array}{r}
8.333... \\
12\overline{)\,100.000} \\
\underline{-96} \\
4\,0 \\
\underline{-3\,6} \\
40 \\
\underline{-36} \\
40 \\
\underline{-36} \\
4
\end{array}
$$

Thus, $\dfrac{1}{12}$ is approximately 8.33%.

38.
$$\dfrac{108}{450} = \dfrac{p}{100}$$
$$108 \cdot 100 = 450p$$
$$10,800 = 450p$$
$$\dfrac{10,800}{450} = p$$
$$24 = p$$
Thus, 108 is 24% of 450.

39. $n = 35\% \cdot 40$
$$n = 0.35 \cdot 40$$
$$n = 14$$
Thus, 14 is 35% of 40.

40. $4 \text{ gal} = \dfrac{4 \text{ gal}}{1} \cdot \dfrac{4 \text{ qt}}{1 \text{ gal}} \cdot \dfrac{2 \text{ pt}}{1 \text{ qt}} = 4 \cdot 4 \cdot 2 \text{ pt} = 32 \text{ pt}$

41. percent = p, base = 30, amount = 75
$$\dfrac{\text{amount}}{\text{base}} = \dfrac{\text{percent}}{100}$$
$$\dfrac{75}{30} = \dfrac{p}{100}$$

42. $8.6 \text{ m} = \dfrac{8.6 \text{ m}}{1} \cdot \dfrac{100 \text{ cm}}{1 \text{ m}} = 8.6 \cdot 100 \text{ cm} = 860 \text{ cm}$

43. percent decrease $= \dfrac{\text{amount of decrease}}{\text{original amount}}$

$= \dfrac{1500 - 1230}{1500}$

$= \dfrac{270}{1500}$

$= 0.18$

$= 18\%$

Thus, this is an 18% decrease.

44. $13,000 \text{ lb} = \dfrac{13,000 \text{ lb}}{1} \cdot \dfrac{1 \text{ ton}}{2000 \text{ lb}}$

$= \dfrac{13,000}{2000} \text{ tons}$

$= 6.5 \text{ tons}$

45. amount of discount $= 25\% \cdot \$65$

$= 0.25 \cdot \$65$

$= \$16.25$

The discount is \$16.25.

sale price $= \$65 - \$16.25 = \$48.75$

The sale price is \$48.75.

46. $3\dfrac{1}{4} \text{ lb} = \dfrac{3\frac{1}{4} \text{ lb}}{1} \cdot \dfrac{16 \text{ oz}}{1 \text{ lb}} = \dfrac{13}{4} \cdot 16 \text{ oz} = 52 \text{ oz}$

47. Simple interest $= \text{principal} \cdot \text{rate} \cdot \text{time}$

$= \$500 \cdot 12\% \cdot 2$

$= \$500 \cdot 0.12 \cdot 2$

$= \$120$

The simple interest is \$120.

48. percent increase $= \dfrac{\text{amount of increase}}{\text{original amount}}$

$= \dfrac{276 - 240}{240}$

$= \dfrac{36}{240}$

$= 0.15$

$= 15\%$

The percent increase is 15%.

49. $9000 \text{ lb} = \dfrac{9000 \text{ lb}}{1} \cdot \dfrac{1 \text{ ton}}{2000 \text{ lb}}$

$= \dfrac{9000}{2000} \text{ tons}$

$= 4\dfrac{1}{2} \text{ tons}$

50. $F = \dfrac{9}{5}C + 32 = \dfrac{9}{5}(25) + 32 = 45 + 32 = 77$

25°C is 77°F.

Chapter 8

Section 8.1 Practice

1. Figure (a) is part of a line with one endpoint, so it is a ray. It is ray AB or $\overrightarrow{AB}$.
   Figure (b) has two endpoints, so it is a line segment. It is line segment RS or $\overline{RS}$.
   Figure (c) extends indefinitely in two directions, so it is a line. It is line EF or $\overleftrightarrow{EF}$.
   Figure (d) has two rays with a common endpoint, so it is an angle. It is $\angle HVT$ or $\angle TVH$ or $\angle V$.

2. Two other ways to name $\angle z$ are $\angle RTS$ and $\angle STR$.

3. a. $\angle R$ is an obtuse angle. It measures between $90°$ and $180°$.

   b. $\angle N$ is a straight angle. It measures $180°$.

   c. $\angle M$ is an acute angle. It measures between $0°$ and $90°$.

   d. $\angle Q$ is a right angle. It measures $90°$.

4. The complement of a $29°$ angle is an angle that measures $90° - 29° = 61°$.

5. The supplement of a $67°$ angle is an angle that measures $180° - 67° = 113°$.

6. a. $m\angle y = m\angle ADC - m\angle BDC$
      $= 141° - 97°$
      $= 44°$

   b. $m\angle x = 79° - 51° = 28°$

   c. Since the measures of both $\angle x$ and $\angle y$ are between $0°$ and $90°$, they are acute angles.

7. Since $\angle a$ and the angle marked $109°$ are vertical angles, they have the same measure; so $m\angle a = 109°$.
   Since $\angle a$ and $\angle b$ are adjacent angles, their measures have a sum of $180°$. So $m\angle b = 180° - 109° = 71°$.
   Since $\angle b$ and $\angle c$ are vertical angles, they have the same measure; so $m\angle c = 71°$.

8. $\angle w$ and $\angle x$ are vertical angles. $\angle w$ and $\angle y$ are corresponding angles, as are $\angle x$ and $\angle d$. So all of these angles have the same measure: $m\angle x = m\angle y = m\angle d = m\angle w = 45°$.
   $\angle w$ and $\angle a$ are adjacent angles, as are $\angle w$ and $\angle b$, so $m\angle a = m\angle b = 180° - 45° = 135°$.
   $\angle a$ and $\angle c$ are corresponding angles so $m\angle c = m\angle a = 135°$. $\angle c$ and $\angle z$ are vertical angles, so $m\angle z = m\angle c = 135°$.

Vocabulary and Readiness Check

1. A <u>plane</u> is a flat surface that extends indefinitely.

2. A <u>point</u> has no length, no width, and no height.

3. <u>Space</u> extends in all directions indefinitely.

4. A <u>line</u> is a set of points extending indefinitely in two directions.

5. A <u>ray</u> is part of a line with one endpoint.

6. An <u>angle</u> is made up of two rays that share a common endpoint. The common endpoint is called the <u>vertex</u>.

7. A <u>straight</u> angle measures $180°$.

8. A <u>right</u> angle measures $90°$.

9. An <u>acute</u> angle measures between $0°$ and $90°$.

10. An <u>obtuse</u> angle measures between $90°$ and $180°$.

11. <u>Parallel</u> lines never meet and <u>intersecting</u> lines meet at a point.

12. Two intersecting lines are <u>perpendicular</u> if they form right angles when they intersect.

13. An angle can be measured in <u>degrees</u>.

14. A line that intersects two or more lines at different points is called a <u>transversal</u>.

15. When two lines intersect, four angles are formed, called <u>vertical</u> angles.

16. Two angles that share a common side are called <u>adjacent</u> angles.

Exercise Set 8.1

1. The figure extends indefinitely in two directions, so it is a line. It is line *CD*, line *l*, or $\overleftrightarrow{CD}$

3. The figure has two end points, so it is a line segment. It is line segment *MN* or $\overline{MN}$.

5. The figure has two rays with a common end point. It is an angle, which can be named $\angle GHI$, $\angle IHG$, or $\angle H$.

7. The figure has one end point and extends indefinitely in one direction, so it is a ray. It is ray *UW* or $\overrightarrow{UW}$.

9. Two other ways to name $\angle x$ are $\angle CPR$ and $\angle RPC$.

11. Two other ways to name $\angle z$ are $\angle TPM$ and $\angle MPT$.

13. $\angle S$ is a straight angle. It measures 180°.

15. $\angle R$ is a right angle. It measures 90°.

17. $\angle Q$ is an obtuse angle. It measures between 90° and 180°.

19. $\angle P$ is an acute angle. It measures between 0° and 90°.

21. The complement of an angle that measures 23° is an angle that measures 90° − 23° = 67°.

23. The supplement of an angle that measures 17° is an angle that measures 180° − 17° = 163°.

25. The complement of an angle that measures 58° is an angle that measures 90° − 58° = 32°.

27. The supplement of an angle that measures 150° is an angle that measures 180° − 150° = 30°.

29. 52° + 38° = 90°, so $\angle PNQ$ and $\angle QNR$ are complementary. 60° + 30° = 90°, so $\angle MNP$ and $\angle RNO$ are complementary.

31. 45° + 135° = 180°, so there are 4 pairs of supplementary angles: $\angle SPT$ and $\angle RPS$, $\angle SPT$ and $\angle QPT$, $\angle QPR$ and $\angle RPS$, $\angle QPR$ and $\angle QPT$.

33. $m\angle x = 74° - 47° = 27°$

35. $m\angle x = 42° + 90° = 132°$

37. $\angle x$ and the angle marked 150° are supplementary, so $m\angle x = 180° - 150° = 30°$. $\angle y$ and the angle marked 150° are vertical angles, so $m\angle y = 150°$. $\angle z$ and $\angle x$ are vertical angles so $m\angle z = m\angle x = 30°$.

39. $\angle x$ and the angle marked 103° are supplementary, so $m\angle x = 180° - 103° = 77°$. $\angle y$ and the angle marked 103° are vertical angles, so $m\angle y = 103°$. $\angle x$ and $\angle z$ are vertical angles, so $m\angle z = m\angle x = 77°$.

41. $\angle x$ and the angle marked 80° are supplementary, so $m\angle x = 180° - 80° = 100°$. $\angle y$ and the angle marked 80° are alternate interior angles, so $m\angle y = 80°$. $\angle x$ and $\angle z$ are corresponding angles, so $m\angle z = m\angle x = 100°$.

43. $\angle x$ and the angle marked 46° are supplementary, so $m\angle x = 180° - 46° = 134°$. $\angle y$ and the angle marked 46° are corresponding angles, so $m\angle y = 46°$. $\angle x$ and $\angle z$ are corresponding angles, so $m\angle z = m\angle x = 134°$.

45. $\angle x$ can also be named $\angle ABC$ or $\angle CBA$.

47. $\angle z$ can also be named $\angle DBE$ or $\angle EBD$.

49. $m\angle ABC = 15°$

51. $m\angle CBD = 50°$

53. $m\angle DBA = m\angle DBC + m\angle CBA$
$= 50° + 15°$
$= 65°$

55. $m\angle CBE = m\angle CBD + m\angle DBE$
$= 50° + 45°$
$= 95°$

57. $\dfrac{7}{8} + \dfrac{1}{4} = \dfrac{7}{8} + \dfrac{2}{8} = \dfrac{9}{8}$ or $1\dfrac{1}{8}$

59. $\dfrac{7}{8} \cdot \dfrac{1}{4} = \dfrac{7 \cdot 1}{8 \cdot 4} = \dfrac{7}{32}$

61.
$$3\dfrac{1}{3} - 2\dfrac{1}{2} = \dfrac{10}{3} - \dfrac{5}{2}$$
$$= \dfrac{10 \cdot 2}{3 \cdot 2} - \dfrac{5 \cdot 3}{2 \cdot 3}$$
$$= \dfrac{20}{6} - \dfrac{15}{6}$$
$$= \dfrac{5}{6}$$

63. $3\dfrac{1}{3} \div 2\dfrac{1}{2} = \dfrac{10}{3} \div \dfrac{5}{2} = \dfrac{10}{3} \cdot \dfrac{2}{5} = \dfrac{5 \cdot 2 \cdot 2}{3 \cdot 5} = \dfrac{4}{3}$ or $1\dfrac{1}{3}$

65. The supplement of an angle that measures 125.2° is an angle with measure $180° - 125.2° = 54.8°$.

67. False; answers may vary

69. True

71. $\angle a$ and the angle marked 60° are alternate interior angles, so $m\angle a = 60°$. The sum of $\angle a$, $\angle b$, and the angle marked 70° is a straight angle, so $m\angle b = 180° - 60° - 70° = 50°$. $\angle d$ and the angle marked 70° are alternate interior angles, so $m\angle d = 70°$. $\angle c$ and $\angle d$ are supplementary, so $m\angle c = 180° - 70° = 110°$. $\angle e$ and the angle marked 60° are supplementary, so $m\angle e = 180° - 60° = 120°$.

73. no; answers may vary

75. Let x be the measure of each of the angles, in degrees. We are given that $x° + x° = 90°$, so $2x = 90$ and $x = 45$. The angles both measure 45°.

Section 8.2 Practice

1. $m\angle x = 180° - 25° - 110° = 45°$

2. $m\angle y = 180° - 90° - 25° = 65°$

3. The radius is half the diameter.
$r = d \div 2 = 16$ in. $\div 2 = 8$ in.

4. The diameter is twice the radius.
$d = 2 \cdot r = 2 \cdot 7$ mi $= 14$ mi

Exercise Set 8.2

1. The figure has five sides, so it is a pentagon.

3. The figure has six sides, so it is a hexagon.

5. The figure has four sides, so it is a quadrilateral.

7. The figure has five sides, so it is a pentagon.

9. All three sides of the triangle have the same length, therefore the triangle is equilateral.

11. No two sides of the triangle have the same length. Also, one of the angles is a right angle. Therefore the triangle is a scalene right triangle.

13. Two sides of the triangle have the same length, therefore the triangle is isosceles.

15. $m\angle x = 180° - 70° - 85° = 25°$

17. $m\angle x = 180° - 95° - 72° = 13°$

19. $m\angle x = 180° - 90° - 50° = 40°$

21. Twice the radius of a circle is its diameter.

23. A parallelogram with four right angles is a rectangle.

25. A quadrilateral with opposite sides parallel is a parallelogram.

27. The side opposite the right angle of a right triangle is called the hypotenuse.

29. $d = 2 \cdot r = 2 \cdot 7 = 14$ m

31. $r = d \div 2 = 29 \div 2 = 14.5$ cm

33. $d = 2 \cdot r = 2 \cdot 20.3 = 40.6$ cm

35. $r = d \div 2 = 168 \div 2 = 84$ in.

37. The solid is a cylinder.

39. The solid is a rectangular solid.

41. The solid is a cone.

43. The object has the shape of a cube.

45. The object has the shape of a rectangular solid.

47. The object has the shape of a sphere.

49. The object has the shape of a pyramid.

51. $d = 2 \cdot r = 2 \cdot 7.4 = 14.8$ in.

53. $r = \frac{1}{2}d = \frac{1}{2} \cdot 26 = 13$ miles

55. $d = 2 \cdot r \approx 2 \cdot 36{,}184 = 72{,}368$ mi

57. $2(18) + 2(36) = 36 + 72 = 108$

59. $4(3.14) = 12.56$

61. True; since all four sides of a square are equal in length, a square is also a rhombus.

63. True; since opposite sides of a rectangle are parallel, a rectangle is also a parallelogram.

65. False; a pentagon has five sides, so it cannot be a quadrilateral, which has four sides.

67. yes; answers may vary

69. answers may vary

Section 8.3 Practice

1. a. Perimeter $= 12 \text{ m} + 12 \text{ m} + 15 \text{ m} + 15 \text{ m}$
$= 54$ meters

b. Perimeter $= 80 \text{ ft} + 80 \text{ ft} + 60 \text{ ft} + 60 \text{ ft}$
$= 280$ feet

2. $P = 2 \cdot l + 2 \cdot w$
$= 2 \cdot 22 \text{ cm} + 2 \cdot 10 \text{ cm}$
$= 44 \text{ cm} + 20 \text{ cm}$
$= 64$ centimeters

3. $P = 4 \cdot s = 4 \cdot 5$ feet $= 20$ feet

4. $P = a + b + c$
$= 5 \text{ cm} + 9 \text{ cm} + 7 \text{ cm}$
$= 21$ centimeters

5. Perimeter $= 4 \text{ km} + 6 \text{ km} + 4 \text{ km} + 9 \text{ km}$
$= 23$ kilometers

6. The unmarked horizontal side has length $20 \text{ m} - 15 \text{ m} = 5 \text{ m}$. The unmarked vertical side has length $26 \text{ m} - 7 \text{ m} = 19 \text{ m}$.
$P = 15 \text{ m} + 26 \text{ m} + 20 \text{ m} + 7 \text{ m} + 5 \text{ m} + 19 \text{ m}$
$= 92$ meters

7. $P = 2 \cdot l + 2 \cdot w$
$= 2 \cdot 120 \text{ feet} + 2 \cdot 60 \text{ feet}$
$= 240 \text{ feet} + 120 \text{ feet}$
$= 360$ feet
cost $= \$1.90$ per foot $\cdot$ 360 feet $= \$684$
The cost of the fencing is $684.

8. a. $C = \pi \cdot d = \pi \cdot 20 \text{ yd} = 20\pi \text{ yd} \approx 62.8 \text{ yd}$
The exact circumference of the watered region is 20π yards, which is approximately 62.8 yards.

b. $C = \pi \cdot d = \pi \cdot 12 \text{ in.} = 12\pi \text{ in.} \approx 37.68 \text{ in.}$
The exact circumference of the clock face is 12π inches, which is approximately 37.68 inches.

Vocabulary and Readiness Check

1. The <u>perimeter</u> of a polygon is the sum of the lengths of its sides.

2. The distance around a circle is called the <u>circumference</u>.

3. The exact ratio of circumference to diameter is <u>π</u>.

4. The diameter of a circle is double its <u>radius</u>.

5. Both $\underline{\frac{22}{7} \text{ (or 3.14)}}$ and $\underline{3.14 \left(\text{or } \frac{22}{7} \right)}$ are approximations for π.

6. The radius of a circle is half its <u>diameter</u>.

Exercise Set 8.3

1. $P = 2 \cdot l + 2 \cdot w$
$= 2 \cdot 17 \text{ ft} + 2 \cdot 15 \text{ ft}$
$= 34 \text{ ft} + 30 \text{ ft}$
$= 64$ ft
The perimeter is 64 feet.

3. $P = 35 \text{ cm} + 25 \text{ cm} + 35 \text{ cm} + 25 \text{ cm}$
$= 120$ cm
The perimeter is 120 centimeters.

5. $P = a + b + c$
$= 5 \text{ in.} + 7 \text{ in.} + 9 \text{ in.}$
$= 21$ in.
The perimeter is 21 inches.

7. Sum the lengths of the sides.
$P = 10 \text{ ft} + 8 \text{ ft} + 8 \text{ ft} + 15 \text{ ft} + 7 \text{ ft}$
$\quad = 48 \text{ ft}$
The perimeter is 48 feet.

9. All sides of a regular polygon have the same length, so the perimeter is the number of sides multiplied by the length of a side.
$P = 3 \cdot 14 \text{ in.} = 42 \text{ in.}$
The perimeter is 42 inches.

11. All sides of a regular polygon have the same length, so the perimeter is the number of sides multiplied by the length of a side.
$P = 5 \cdot 31 \text{ cm} = 155 \text{ cm}$
The perimeter is 155 centimeters.

13. Sum the lengths of the sides.
$P = 5 \text{ ft} + 3 \text{ ft} + 2 \text{ ft} + 7 \text{ ft} + 4 \text{ ft}$
$\quad = 21 \text{ ft}$
The perimeter is 21 feet.

15. total distance $= 2(312) = 624$
624 feet of lime powder will be deposited.

17. $P = 2 \cdot l + 2 \cdot w$
$\quad = 2 \cdot 120 \text{ yd} + 2 \cdot 53 \text{ yd}$
$\quad = 240 \text{ yd} + 106 \text{ yd}$
$\quad = 346 \text{ yd}$
The perimeter of the football field is 346 yards.

19. $P = 2 \cdot l + 2 \cdot w$
$\quad = 2 \cdot 8 \text{ ft} + 2 \cdot 3 \text{ ft}$
$\quad = 16 \text{ ft} + 6 \text{ ft}$
$\quad = 22 \text{ ft}$
22 feet of stripping is needed for this project.

21. The amount of stripping needed is 22 feet.
22 feet $\cdot$ \$2.50 per foot $=$ \$55
The total cost of the stripping is \$55.

23. All sides of a regular polygon have the same length, so the perimeter is the number of sides multiplied by the length of a side.
$P = 8 \cdot 9 \text{ in.} = 72 \text{ in.}$
The perimeter is 72 inches.

25. $P = 4 \cdot s = 4 \cdot 7 \text{ in.} = 28 \text{ in.}$
The perimeter is 28 inches.

27. $P = 2 \cdot l + 2 \cdot w$
$\quad = 2 \cdot 11 \text{ ft} + 2 \cdot 10 \text{ ft}$
$\quad = 22 \text{ ft} + 20 \text{ ft}$
$\quad = 42 \text{ ft}$
42 ft $\cdot$ \$0.86 per foot $=$ \$36.12
The cost is \$36.12.

29. The unmarked vertical side has length
$28 \text{ m} - 20 \text{ m} = 8 \text{ m}$.
The unmarked horizontal side has length
$20 \text{ m} - 17 \text{ m} = 3 \text{m}$.
$P = 17 \text{ m} + 8 \text{ m} + 3 \text{ m} + 20 \text{ m} + 20 \text{ m} + 28 \text{ m}$
$\quad = 96 \text{ m}$
The perimeter is 96 meters.

31. The unmarked horizontal side has length
$(3 + 6 + 4) \text{ ft} = 13 \text{ ft}$.
$P = (3 + 5 + 6 + 5 + 4 + 15 + 13 + 15) \text{ ft} = 66 \text{ ft}$
The perimeter is 66 feet.

33. The unmarked vertical side has length
$5 \text{ cm} + 14 \text{ cm} = 19 \text{ cm}$.
The unmarked horizontal side has length
$18 \text{ cm} - 9 \text{ cm} = 9 \text{ cm}$.
$P = 18 \text{ cm} + 19 \text{ cm} + 9 \text{ cm} + 14 \text{ cm} + 9 \text{ cm} + 5 \text{ cm}$
$\quad = 74 \text{ cm}$
The perimeter is 74 centimeters.

35. $C = \pi \cdot d = \pi \cdot 17 \text{ cm} = 17\pi \text{ cm} \approx 53.38 \text{ cm}$
The circumference is exactly 17π centimeters or approximately 53.38 centimeters.

37. $C = 2 \cdot \pi \cdot r$
$\quad = 2 \cdot \pi \cdot 8 \text{ mi}$
$\quad = 16\pi \text{ mi}$
$\quad \approx 50.24 \text{ mi}$
The circumference is exactly 16π miles, or approximately 50.24 miles.

39. $C = \pi \cdot d = \pi \cdot 26 \text{ m} = 26\pi \text{ m} \approx 81.64 \text{ m}$
The circumference is exactly 26π meters or approximately 81.64 meters.

41. $\pi \cdot d = \pi \cdot 15 \text{ ft} = 15\pi \text{ ft} \approx 47.1$
He needs 15π feet of netting or 47.1 feet.

43. $C = \pi \cdot d = \pi \cdot 4000 \text{ ft} = 4000\pi \text{ ft} \approx 12,560 \text{ ft}$
The distance around is about 12,560 feet.

45. Sum the lengths of the sides.
$P = 9 \text{ mi} + 6 \text{ mi} + 11 \text{ mi} + 4.7 \text{ mi}$
$\quad = 30.7 \text{ mi}$
The perimeter is 30.7 miles.

47. $C = \pi \cdot d = \pi \cdot 14$ cm $= 14\pi$ cm ≈ 43.96 cm
The circumference is
14π centimeters ≈ 43.96 centimeters.

49. $P = 5 \cdot 8$ mm $= 40$ mm
The perimeter is 40 millimeters.

51. The unmarked vertical side has length
$(22 - 8)$ ft $= 14$ ft.
The unmarked horizontal side has length
$(20 - 7)$ ft $= 13$ ft.
$P = (7 + 8 + 13 + 14 + 20 + 22)$ ft $= 84$ ft
The perimeter is 84 feet.

53. $5 + 6 \cdot 3 = 5 + 18 = 23$

55. $(20 - 16) \div 4 = 4 \div 4 = 1$

57. $72 \div (2 \cdot 6) = 72 \div 12 = 6$

59. $(18 + 8) - (12 + 4) = 26 - 16 = 10$

61. a. The first age category that 8-year-old children fit into is "Under 9," thus the minimum width is 30 yards and the minimum length is 40 yards.

   b. $\begin{aligned} P &= 2 \cdot l + 2 \cdot w \\ &= 2 \cdot 40 \text{ yd} + 2 \cdot 30 \text{ yd} \\ &= 80 \text{ yd} + 60 \text{ yd} \\ &= 140 \text{ yd} \end{aligned}$

     The perimeter of the field is 140 yards.

63. The square's perimeter is $4 \cdot 3$ in. $= 12$ in.
The circle's circumference is $\pi \cdot 4$ in. ≈ 12.56 in.
So the circle has the greater distance around; b.

65. a. Smaller circle:
$\begin{aligned} C &= 2 \cdot \pi \cdot r \\ &= 2 \cdot \pi \cdot 10 \text{ m} \\ &= 20\pi \text{ m} \\ &\approx 62.8 \text{ m} \end{aligned}$
Larger circle:
$\begin{aligned} C &= 2 \cdot \pi \cdot r \\ &= 2 \cdot \pi \cdot 20 \text{ m} \\ &= 40\pi \text{ m} \\ &\approx 125.6 \text{ m} \end{aligned}$

   b. Yes, when the radius of a circle is doubled, the circumference is also doubled.

67. answers may vary

69. The length of the curved section at the top is half of the circumference of a circle of diameter 6 meters.
$\frac{1}{2} \cdot C = \frac{1}{2} \cdot \pi \cdot d = \frac{1}{2} \cdot \pi \cdot 6$ m $= 3\pi$ m ≈ 9.4 meters
The total length of the straight sides is
$3 \cdot 6$ m $= 18$ m.
The perimeter is the sum of these.
9.4 m $+ 18$ m $= 27.4$ m
The perimeter of the figure is 27.4 meters.

71. The total length of the two straight sections is $2 \cdot 22$ m $= 44$ m. The total length of the two curved sections is the circumference of a circle of radius 5 m.
$C = 2 \cdot \pi \cdot r = 2 \cdot \pi \cdot 5$ m $= 10\pi$ m
The perimeter of the track is the sum of these.
$P = 44$ m $+ 10\pi$ m $= (44 + 10\pi)$ m ≈ 75.4 m

73. $\begin{aligned} P &= 2 \cdot l + 2 \cdot w \\ 31 \text{ ft} &= 2 \cdot 9 \text{ ft} + 2 \cdot w \\ 31 \text{ ft} &= 18 \text{ ft} + 2 \cdot w \\ 13 \text{ ft} &= 2 \cdot w \\ 6.5 \text{ ft} &= w \end{aligned}$
The width is 6.5 feet.

Section 8.4 Practice

1. $\begin{aligned} A &= \frac{1}{2} \cdot b \cdot h \\ &= \frac{1}{2} \cdot 8 \text{ in.} \cdot 6\frac{1}{4} \text{ in.} \\ &= \frac{1}{2} \cdot 8 \text{ in.} \cdot \frac{25}{4} \text{ in.} \\ &= 25 \text{ sq in.} \end{aligned}$
The area is 25 square inches.

2. $A = s \cdot s = 4.2$ yd $\cdot 4.2$ yd $= 17.64$ sq yd
The area is 17.64 square yards.

3. Split the rectangle into two pieces, a top rectangle with dimensions 12 m by 24 m, and a bottom rectangle with dimensions 6 m by 18 m. The area of the figure is the sum of the areas of these.
$\begin{aligned} A &= 12 \text{ m} \cdot 24 \text{ m} + 6 \text{ m} \cdot 18 \text{ m} \\ &= 288 \text{ sq m} + 108 \text{ sq m} \\ &= 396 \text{ sq m} \end{aligned}$
The area is 396 square meters.

4. $A = \pi \cdot r^2$

$= \pi \cdot (7 \text{ cm})^2$

$= 49\pi \text{ sq cm}$

$\approx 153.86 \text{ sq cm}$

The area is 49π square centimeters, which is approximately 153.86 square centimeters.

Exercise Set 8.4

1. $A = l \cdot w = 3.5 \text{ m} \cdot 2 \text{ m} = 7 \text{ sq m}$

3. $A = \dfrac{1}{2} \cdot b \cdot h$

$= \dfrac{1}{2} \cdot 6\dfrac{1}{2} \text{ yd} \cdot 3 \text{ yd}$

$= \dfrac{1}{2} \cdot \dfrac{13}{2} \text{ yd} \cdot 3 \text{ yd}$

$= \dfrac{39}{4} \text{ sq yd}$

$= 9\dfrac{3}{4} \text{ sq yd}$

5. $A = \dfrac{1}{2} \cdot b \cdot h = \dfrac{1}{2} \cdot 6 \text{ yd} \cdot 5 \text{ yd} = 15 \text{ sq yd}$

7. $r = d \div 2 = (3 \text{ in.}) \div 2 = 1.5 \text{ in.}$

$A = \pi r^2$

$= \pi (1.5 \text{ in.})^2$

$= 2.25\pi \text{ sq in.}$

$\approx 7.065 \text{ sq in.}$

9. $A = s^2 = (4.2 \text{ ft})^2 = 17.64 \text{ sq ft}$

11. $A = \dfrac{1}{2}(b + B) \cdot h$

$= \dfrac{1}{2}(5 \text{ m} + 9 \text{ m}) \cdot 4 \text{ m}$

$= \dfrac{1}{2} \cdot 14 \text{ m} \cdot 4 \text{ m}$

$= 28 \text{ sq m}$

13. $A = \dfrac{1}{2}(b + B) \cdot h$

$= \dfrac{1}{2}(7 \text{ yd} + 4 \text{ yd}) \cdot 4 \text{ yd}$

$= \dfrac{1}{2}(11 \text{ yd}) \cdot 4 \text{ yd}$

$= 22 \text{ sq yd}$

15. $A = b \cdot h$

$= 7 \text{ ft} \cdot 5\dfrac{1}{4} \text{ ft}$

$= 7 \text{ ft} \cdot \dfrac{21}{4} \text{ ft}$

$= \dfrac{147}{4} \text{ sq ft}$

$= 36\dfrac{3}{4} \text{ sq ft}$

17. $A = b \cdot h$

$= 5 \text{ in.} \cdot 4\dfrac{1}{2} \text{ in.}$

$= 5 \text{ in.} \cdot \dfrac{9}{2} \text{ in.}$

$= \dfrac{45}{2} \text{ sq in.}$

$= 22\dfrac{1}{2} \text{ sq in.}$

19. The base of the triangle is
$7 \text{ cm} - 1\dfrac{1}{2} \text{ cm} - 1\dfrac{1}{2} \text{ cm} = 4 \text{ cm},$ so its area is

$\dfrac{1}{2} \cdot 4 \text{ cm} \cdot 2 \text{ cm} = 4 \text{ sq cm}.$

The area of the rectangle is
$7 \text{ cm} \cdot 3 \text{ cm} = 21 \text{ sq cm}.$
The area of the figure is the sum of these.
$A = 4 \text{ sq cm} + 21 \text{ sq cm} = 25 \text{ sq cm}$

21. The figure can be divided into two rectangles, one measuring 10 mi by 5 mi and one measuring 12 mi by 3 mi. The area of the figure is the sum of the areas of the two rectangles.
$A = 10 \text{ mi} \cdot 5 \text{ mi} + 12 \text{ mi} \cdot 3 \text{ mi}$
$= 50 \text{ sq mi} + 36 \text{ sq mi}$
$= 86 \text{ sq mi}$

23. The top of the figure is a square with sides of length 3 cm, so its area is
$s^2 = (3 \text{ cm})^2 = 9 \text{ sq cm}.$
The bottom of the figure is a parallelogram with area $b \cdot h = 3 \text{ cm} \cdot 5 \text{ cm} = 15 \text{ sq cm}.$
The area of the figure is the sum of these.
$A = 9 \text{ sq cm} + 15 \text{ sq cm} = 24 \text{ sq cm}$

25. $A = \pi r^2$

$= \pi (6 \text{ in.})^2$

$= 36\pi$ sq in.

$\approx 36 \cdot \dfrac{22}{7}$ sq in.

$= 113\dfrac{1}{7}$ sq in.

27. $A = l \cdot w$

$= 16 \text{ ft} \cdot 10\dfrac{1}{2}$ ft

$= 16 \text{ ft} \cdot \dfrac{21}{2}$ ft

$= 168$ sq ft

The area of the wall is 168 square feet.

29. $A = l \cdot w = 505 \text{ ft} \cdot 225 \text{ ft} = 113{,}625$ sq ft

The area of the flag is 113,625 square feet.

31. $r = d \div 2 = 4 \text{ ft} \div 2 = 2$ ft

$A = \pi r^2$

$= \pi \cdot (2 \text{ ft})^2$

$= 4\pi$ sq ft

≈ 12.56 sq ft

The area of the top of the pizza is 4π square feet; about 12.56 square feet.

33. $A = l \cdot w = 16 \text{ in.} \cdot 8 \text{ in.} = 128$ sq in.

$128 \text{ sq in.} \cdot \dfrac{1 \text{ sq ft}}{144 \text{ sq in.}} = \dfrac{128}{144} \text{ sq ft} = \dfrac{8}{9}$ sq ft

The side has area 128 square inches, which is $\dfrac{8}{9}$ square foot.

35. $A = l \cdot w$

$= 25\dfrac{1}{2} \text{ in.} \cdot 20$ in.

$= \dfrac{51}{2} \text{ in.} \cdot 20$ in.

$= 510$ sq in.

The frame requires 510 square inches of glass.

37. $A = l \cdot w = 7 \text{ ft} \cdot 6 \text{ ft} = 42$ sq ft

$4 \cdot 42 \text{ sq ft} = 168$ sq ft

Four panels have an area of 168 square feet.

39. The land is in the shape of a trapezoid.

$A = \dfrac{1}{2}(b + B) \cdot h$

$= \dfrac{1}{2}(90 \text{ ft} + 140 \text{ ft}) \cdot 80$ ft

$= \dfrac{1}{2} \cdot 230 \text{ ft} \cdot 80$ ft

$= 9200$ sq ft

There are 9200 square feet of land in the plot.

41. a. $A = \dfrac{1}{2}(b + B) \cdot h$

$= \dfrac{1}{2}(25 \text{ ft} + 36 \text{ ft}) \cdot 12\dfrac{1}{2}$ ft

$= \dfrac{1}{2} \cdot 61 \text{ ft} \cdot 12\dfrac{1}{2}$ ft

$= 381\dfrac{1}{4}$ sq ft

To the nearest square foot, the area is 381 square feet.

b. Divide the area of the roof by the area covered by one "square."

$\dfrac{381}{100} = 3.81$

Since you cannot purchase a part of a square, a total of 4 squares needs to be purchased.

43. $C = \pi \cdot d = \pi \cdot 14 \text{ in.} = 14\pi \text{ in.} \approx 43.96$ in.

45. Sum the lengths of the sides.

$3 \text{ ft} + 3\dfrac{1}{2} \text{ ft} + 6 \text{ ft} + 8\dfrac{1}{2} \text{ ft} + 4 \text{ ft} = 25$ ft

47. $P = 6 \cdot 2\dfrac{1}{8} \text{ ft} = 6 \cdot \dfrac{17}{8} \text{ ft} = \dfrac{51}{4} \text{ ft} = 12\dfrac{3}{4}$ ft

49. A fence goes around the edge of a yard, thus the situation involves perimeter.

51. Carpet covers the entire floor of a room, so the situation involves area.

53. Paint covers the surface of the wall, thus the situation involves area.

55. A wallpaper border goes around the edge of a room, so the situation involves perimeter.

$d = \dfrac{1}{2} \cdot 12$ in. $= 6$ in.

$A = \pi \cdot r^2 = \pi(6 \text{ in.})^2 = 36\pi$ sq in.

Price per square inch $= \dfrac{\$10}{36\pi \text{ sq in.}} \approx \$0.0884$

8-inch pizzas:

$r = \dfrac{1}{2} \cdot d = \dfrac{1}{2} \cdot 8$ in. $= 4$ in.

$A = \pi \cdot r^2 = \pi(4 \text{ in.})^2 = 16\pi$ sq in.

$2 \cdot A = 2 \cdot 16\pi$ sq in. $= 32\pi$ sq in.

Price per square inch: $\dfrac{\$9}{32\pi \text{ sq in.}} \approx \$0.0895$

Since the price per square inch for the 12-inch pizza is less, the 12-inch pizza is the better deal.

59. 2 feet $= 2 \cdot 12$ inches $= 24$ inches
$A = l \cdot w = 24$ in. $\cdot 8$ in. $= 192$ sq in.

8 inches $= \dfrac{8}{12}$ foot $= \dfrac{2}{3}$ foot

$A = l \cdot w = 2$ ft $\cdot \dfrac{2}{3}$ ft $= \dfrac{4}{3}$

61. The area of the shaded region is the area of the square minus the area of the circle.
Square:

$A = s^2 = (6 \text{ in.})^2 = 36$ sq in.

Circle:

$r = \dfrac{1}{2} \cdot d = \dfrac{1}{2}(6 \text{ in.}) = 3$ in.

$A = \pi \cdot r^2 = \pi(3 \text{ in.})^2 = 9\pi$ sq in. ≈ 28.26 sq in. 3

6 sq in. $- 28.26$ sq in. $= 7.74$ sq in.
The shaded region has area of approximately 7.74 square inches.

63. $r = d \div 2 = 168$ in. $\div 2 = 84$ in.

$A = \pi r^2$

$= \pi \cdot (84 \text{ in.})^2$

$= 7056\pi$ sq in.

$\approx 22,155.84$ sq in.

The area of the top of the pie was 7056π sq in., or about 22,155.84 sq in.

65. The skating area is a rectangle with a half circle on each end.
Rectangle:
$A = l \cdot w = 22$ m $\cdot 10$ m $= 220$ sq m

Half circles:

$A = 2 \cdot \dfrac{1}{2} \cdot \pi \cdot r^2$

$= \pi(5 \text{ m})^2$

$= 25\pi$ sq m ≈ 78.5 sq m

220 sq m $+ 78.5$ sq m $= 298.5$ sq m
The skating surface has area of about 298.5 square meters.

67. a. The first age category that 9-year-old children fit into is "Under 10," thus the minimum width is 40 yards and the minimum length is 60 yards.

   b. $A = l \cdot w = 40$ yd $\cdot 60$ yd $= 2400$ sq yd
   The area of the field is 2400 square yards.

69. no; answers may vary

Section 8.5 Practice

1. $V = l \cdot w \cdot h = 5$ ft $\cdot 2$ ft $\cdot 4$ ft $= 40$ cubic ft
The volume of the rectangular box is 40 cubic feet.

2. $V = \dfrac{4}{3} \cdot \pi \cdot r^3$

$= \dfrac{4}{3} \cdot \pi \cdot \left(\dfrac{1}{2} \text{ cm}\right)^3$

$= \dfrac{4}{3} \cdot \pi \cdot \dfrac{1}{8}$ cu cm

$= \dfrac{1}{6} \pi$ cu cm

$\approx \dfrac{1}{6} \cdot \dfrac{22}{7}$ cu cm

$= \dfrac{11}{21}$ cu cm

The volume of the ball is $\dfrac{1}{6}\pi$ cubic centimeter,

approximately $\dfrac{11}{12}$ cubic centimeter.

3. $V = \pi \cdot r^2 \cdot h$

$= \pi \cdot (5 \text{ in.})^2 \cdot 7$ in.

$= \pi \cdot 25$ sq in. $\cdot 7$ in.

$= 175\pi$ cu in.

≈ 549.5 cu in.

The volume of the cylinder is 175π cubic inches, approximately 549.5 cubic inches.

4. $V = \frac{1}{3} \cdot s^2 \cdot h$

$\quad = \frac{1}{3} \cdot (3 \text{ m})^2 \cdot 5.1 \text{ m}$

$\quad = \frac{1}{3} \cdot 9 \text{ sq m} \cdot 5.1 \text{ m}$

$\quad = 15.3 \text{ cu m}$

The volume of the pyramid is 15.3 cubic meters.

Vocabulary and Readiness Check

1. The measure of the amount of space inside a solid is its <u>volume</u>.

2. <u>Area</u> measures the amount of surface enclosed by a region.

3. Volume is measured in <u>cubic</u> units.

4. Area is measured in <u>square</u> units.

5. The <u>perimeter</u> of a polygon is the sum of the lengths of its sides.

6. Perimeter is measured in <u>units</u>.

Exercise Set 8.5

1. $V = l \cdot w \cdot h = 6 \text{ in.} \cdot 4 \text{ in.} \cdot 3 \text{ in.} = 72 \text{ cu in.}$

3. $V = s^3 = (8 \text{ cm})^3 = 512 \text{ cu cm}$

5. $V = \frac{1}{3} \pi \cdot r^2 \cdot h$

$\quad = \frac{1}{3} \cdot (2 \text{ yd})^2 \cdot 3 \text{ yd}$

$\quad = 4\pi \text{ cu yd}$

$\quad \approx 4 \cdot \frac{22}{7} \text{ cu yd}$

$\quad = \frac{88}{7} \text{ cu yd}$

$\quad = 12\frac{4}{7} \text{ cu yd}$

7. $r = \frac{1}{2} \cdot d = \frac{1}{2} \cdot 10 \text{ in.} = 5 \text{ in.}$

$V = \frac{4}{3} \pi \cdot r^3$

$\quad = \frac{4}{3} \pi \cdot (5 \text{ in.})^3$

$\quad = \frac{500}{3} \pi \text{ cu in.}$

$\quad \approx \frac{500}{3} \cdot \frac{22}{7} \text{ cu in.}$

$\quad = \frac{11,000}{21} \text{ cu in.}$

$\quad = 523\frac{17}{21} \text{ cu in.}$

9. $r = \frac{1}{2}d = \frac{1}{2} \cdot 2 \text{ in.} = 1 \text{ in.}$

$V = \pi \cdot r^2 \cdot h$

$\quad = \pi(1 \text{ in.})^2 \cdot 9 \text{ in}$

$\quad = 9\pi \text{ cu in.}$

$\quad \approx 9 \cdot \frac{22}{7} \text{ cu in.}$

$\quad = \frac{198}{7} \text{ cu in.}$

$\quad = 28\frac{2}{7} \text{ cu in.}$

11. $V = \frac{1}{3} \cdot s^2 \cdot h = \frac{1}{3} \cdot (5 \text{ cm})^2 \cdot 9 \text{ cm} = 75 \text{ cu cm}$

13. $V = s^3 = \left(1\frac{1}{3} \text{ in.}\right)^3$

$\quad = \left(\frac{4}{3} \text{ in.}\right)^3$

$\quad = \frac{64}{27} \text{ cu in.}$

$\quad = 2\frac{10}{27} \text{ cu in.}$

15. $V = l \cdot w \cdot h = 2 \text{ ft} \cdot 1.4 \text{ ft} \cdot 3 \text{ ft} = 8.4 \text{ cu ft}$

17. $V = \frac{1}{3} \cdot s^2 \cdot h$

$\quad = \frac{1}{3} (5 \text{ in.})^2 \cdot \frac{13}{10} \text{ in.}$

$\quad = \frac{1}{3} \cdot 25 \cdot \frac{13}{10} \text{ cu in.}$

$\quad = 10\frac{5}{6} \text{ cu in.}$

$$\frac{-}{3} \cdot (12 \text{ cm})^2 \cdot 20 \text{ cm}$$
$$= 960 \text{ cu cm}$$

21. $V = \frac{4}{3} \cdot \pi \cdot r^3$

$$= \frac{4}{3} \pi (7 \text{ in.})^3$$

$$= \frac{1372}{3} \pi \text{ cu in. or } 457\frac{1}{3} \pi \text{ cu in.}$$

23. $V = l \cdot w \cdot h$

$$= 2 \text{ ft} \cdot 2\frac{1}{2} \text{ ft} \cdot 1\frac{1}{2} \text{ ft}$$

$$= 2 \text{ ft} \cdot \frac{5}{2} \text{ ft} \cdot \frac{3}{2} \text{ ft}$$

$$= \frac{15}{2} \text{ cu ft}$$

$$= 7\frac{1}{2} \text{ cu ft}$$

25. $r = \frac{1}{2} \cdot d = \frac{1}{2} \cdot 12 \text{ yd} = 6 \text{ yd}$

$$V = \frac{4}{3} \pi r^3$$

$$= \frac{4}{3} \pi (6 \text{ yd})^3$$

$$= 288\pi \text{ cu yd}$$

27. $r = \frac{1}{2} \cdot d = \frac{1}{2} \cdot 3 \text{ in.} = 1.5 \text{ in.}$

$$V = \frac{1}{3} \pi \cdot r^2 \cdot h$$

$$= \frac{1}{3} \pi \cdot (1.5 \text{ in.})^2 \cdot 7 \text{ in.}$$

$$= 5.25\pi \text{ cu in.}$$

The volume is 5.25π cubic inches.

29. $r = \frac{1}{2} \cdot d = \frac{1}{2} \cdot 2.6 \text{ m} = 1.3 \text{ m}$

$$V = \pi \cdot r^2 \cdot h$$

$$= \pi (1.3 \text{ m})^2 \cdot 1.5 \text{ m}$$

$$= 2.535\pi \text{ cu m}$$

$$\approx 7.96 \text{ cu m}$$

31. $r = \frac{1}{2} \cdot d = \frac{1}{2} \cdot 4 \text{ cm} = 2 \text{ cm}$

$$V = \frac{1}{3} \pi \cdot r^2 \cdot h$$

$$= \frac{1}{3} \pi \cdot (2 \text{ cm})^2 \cdot 3 \text{ cm}$$

$$= 4\pi \text{ cu cm}$$

$$\approx 4 \cdot \frac{22}{7} \text{ cu cm}$$

$$= 12\frac{4}{7} \text{ cu cm}$$

There are approximately $12\frac{4}{7}$ cubic centimeters of ice cream in the cone.

33. $V = lwh = (2 \text{ in.})(2 \text{ in.})(2.2 \text{ in.}) = 8.8 \text{ cu in.}$
The volume is 8.8 cubic inches.

35. $V = s^3 = (2.2 \text{ in.})^3 = 10.648 \text{ cu in.}$

37. $5^2 = 5 \cdot 5 = 25$

39. $3^2 = 3 \cdot 3 = 9$

41. $1^2 + 2^2 = 1 \cdot 1 + 2 \cdot 2 = 1 + 4 = 5$

43. $4^2 + 2^2 = 4 \cdot 4 + 2 \cdot 2 = 16 + 4 = 20$

45. $r = \frac{d}{2} = \frac{20 \text{ m}}{2} = 10 \text{ m}$
The volume is half the volume of a sphere.

$$V = \frac{1}{2} \cdot \frac{4}{3} \pi r^3$$

$$= \frac{1}{2} \cdot \frac{4}{3} (3.14)(10 \text{ m})^3$$

$$\approx 2093.33 \text{ cu m}$$

The volume of the hemisphere is about 2093.33 cu m.

47. no; answers may vary

49. Kennel (a):

$$2\frac{1}{12} \text{ ft} \cdot 1\frac{8}{12} \text{ ft} \cdot 1\frac{7}{12} \text{ ft} = \frac{2375}{432} \text{ cu ft} \approx 5.5 \text{ cu ft}$$

Kennel (b):

$$1\frac{1}{12} \text{ ft} \cdot 2 \text{ ft} \cdot 2\frac{8}{12} \text{ ft} = \frac{52}{9} \text{ cu ft} \approx 5.8 \text{ cu ft}$$

Kennel (b) is larger.

51. The radius of the hemisphere (and cylinder) is
$$\frac{1}{2} \cdot d = \frac{1}{2} \cdot 2 \text{ in.} = 1 \text{ in.}$$
hemisphere volume:
$$\frac{1}{2} \cdot \frac{4}{3} \cdot \pi \cdot r^3 = \frac{2}{3} \pi \cdot (1 \text{ in.})^3 = \frac{2}{3} \pi \text{ cu in.}$$
cylinder volume:
$$\pi \cdot r^2 \cdot h = \pi \cdot (1 \text{ in.})^2 \cdot 6 \text{ in.} = 6\pi \text{ cu in.}$$
The total volume is the sum of these.
$$V = \frac{2}{3} \pi \text{ cu in.} + 6\pi \text{ cu in.}$$
$$= 6\frac{2}{3} \cdot \pi \text{ cu in.}$$
$$\approx 21 \text{ cu in.}$$

Integrated Review

1. The supplement is $180° - 27° = 153°$.
The complement is $90° - 27° = 63°$.

2. $\angle x$ and the angle marked $105°$ are adjacent angles, so $m\angle x = 180° - 105° = 75°$. $\angle y$ and the angle marked $105°$ are vertical angles, so $m\angle y = 105°$. $\angle z$ and $\angle x$ are vertical angles, so $m\angle z = 75°$.

3. $\angle x$ and the angle marked $52°$ are adjacent angles, so $m\angle x = 180° - 52° = 128°$. $\angle y$ and the angle marked $52°$ are corresponding angles, so $m\angle y = 52°$. $\angle y$ and $\angle z$ are adjacent angles, so $m\angle z = 180° - 52° = 128°$.

4. The sum of the interior angles of a triangle is $180°$, so $m\angle x = 180° - 90° - 38° = 52°$.

5. $d = 2 \cdot r = 2 \cdot 2.3 \text{ in.} = 4.6 \text{ in.}$

6. $r = d \div 2 = 8\frac{1}{2} \text{ in.} \div 2 = 4\frac{1}{4} \text{ in.}$

7. $P = 4 \cdot s = 4 \cdot 5 \text{ m} = 20 \text{ m}$
$A = s^2 = (5 \text{ m})^2 = 25 \text{ sq m}$

8. $P = a + b + c = 4 \text{ ft} + 3 \text{ ft} + 5 \text{ ft} = 12 \text{ ft}$
$A = \frac{1}{2} \cdot b \cdot h = \frac{1}{2} \cdot 4 \text{ ft} \cdot 3 \text{ ft} = 6 \text{ sq ft}$

9. $C = 2 \cdot \pi \cdot r = 2 \cdot \pi \cdot 5 \text{ cm} = 10\pi \text{ cm} \approx 31.4 \text{ cm}$
$A = \pi \cdot r^2$
$= \pi \cdot (5 \text{ cm})^2$
$= 25\pi \text{ sq cm}$
$\approx 78.5 \text{ sq cm}$

10. $P = 11 \text{ mi} + 5 \text{ mi} + 11 \text{ mi} + 5 \text{ mi} = 32 \text{ mi}$
$A = b \cdot h = 11 \text{ mi} \cdot 4 \text{ mi} = 44 \text{ sq mi}$

11. The unmarked vertical side has length $3 \text{ cm} + 7 \text{ cm} = 10 \text{ cm}$.
The unmarked horizontal side has length $17 \text{ cm} - 8 \text{ cm} = 9 \text{ cm}$.
$P = (8 + 3 + 9 + 7 + 17 + 10) \text{ cm} = 54 \text{ cm}$
$A = 10 \text{ cm} \cdot 8 \text{ cm} + 7 \text{ cm} \cdot 9 \text{ cm}$
$= 80 \text{ sq cm} + 63 \text{ sq cm}$
$= 143 \text{ sq cm}$

12. $P = 2 \cdot l + 2 \cdot w = 2 \cdot 17 \text{ ft} + 2 \cdot 14 \text{ ft} = 62 \text{ ft}$
$A = l \cdot w = 17 \text{ ft} \cdot 14 \text{ ft} = 238 \text{ sq ft}$

13. $V = s^3 = (4 \text{ in.})^3 = 64 \text{ cu in.}$

14. $V = l \cdot w \cdot h = 2 \text{ ft} \cdot 3 \text{ ft} \cdot 5.1 \text{ ft} = 30.6 \text{ cu ft}$

15. $V = \frac{1}{3} \cdot s^2 \cdot h = \frac{1}{3} \cdot (10 \text{ cm})^2 \cdot 12 \text{ cm} = 400 \text{ cu cm}$

16. $r = d \div 2 = 3 \text{ mi} \div 2 = \frac{3}{2} \text{ mi}$
$$V = \frac{4}{3} \cdot \pi \cdot r^3$$
$$= \frac{4}{3} \pi \cdot \left(\frac{3}{2} \text{ mi}\right)^3$$
$$= \frac{9}{2} \pi \text{ cu mi}$$
$$= 4\frac{1}{2} \pi \text{ cu mi}$$
$$\approx \frac{99}{7} \text{ cu mi}$$
$$= 14\frac{1}{7} \text{ cu mi}$$

Section 8.6 Practice

1. a. $\sqrt{100} = 10$ because $10^2 = 100$.

b. $\sqrt{64} = 8$ because $8^2 = 64$.

c. $\sqrt{121} = 11$ because $11^2 = 121$.

d. $\sqrt{0} = 0$ because $0^2 = 0$.

2. $\sqrt{\dfrac{1}{4}} = \dfrac{1}{2}$ because $\left(\dfrac{1}{2}\right)^2 = \dfrac{1}{4}$.

3. $\sqrt{\dfrac{9}{16}} = \dfrac{3}{4}$ because $\left(\dfrac{3}{4}\right)^2 = \dfrac{9}{16}$.

4. a. $\sqrt{21} \approx 4.583$

   b. $\sqrt{62} \approx 7.874$

5. Recall that $\sqrt{49} = 7$ and $\sqrt{64} = 8$. Since 62 is between 49 and 64, then $\sqrt{62}$ is between $\sqrt{49}$ and $\sqrt{64}$. Thus, $\sqrt{62}$ is between 7 and 8. Since 62 is closer to 64, then $\sqrt{62}$ is closer to $\sqrt{64}$, or 8.

6. hypotenuse $= \sqrt{(12)^2 + (16)^2}$
$= \sqrt{144 + 256}$
$= \sqrt{400}$
$= 20$
The hypotenuse is 20 feet long.

7. hypotenuse $= \sqrt{(7)^2 + (9)^2}$
$= \sqrt{49 + 81}$
$= \sqrt{130}$
≈ 11
The hypotenuse is $\sqrt{130}$ kilometers, or approximately 11 kilometers.

8. leg $= \sqrt{(11)^2 - (7)^2}$
$= \sqrt{121 - 49}$
$= \sqrt{72}$
≈ 8.49
The length of the leg is $\sqrt{72}$ feet, or approximately 8.49 feet.

9.

100 yd

53 yd

Let $a = 53$ and $b = 100$.

$a^2 + b^2 = c^2$
$53^2 + 100^2 = c^2$
$2809 + 10,000 = c^2$
$12,809 = c^2$
$\sqrt{12,809} = c$
$113 \approx c$
The diagonal is approximately 113 yards.

Calculator Explorations

1. $\sqrt{1024} = 32$

2. $\sqrt{676} = 26$

3. $\sqrt{31} \approx 5.568$

4. $\sqrt{19} \approx 4.359$

5. $\sqrt{97} \approx 9.849$

6. $\sqrt{56} \approx 7.483$

Vocabulary and Readiness Check

1. $\sqrt{100} = \underline{10}$ only because $10 \cdot 10 = 100$.

2. The <u>radical</u> sign is used to denote the square root of a number.

3. The reverse process of <u>squaring</u> a number is finding a square root of a number.

4. The numbers 9, 1, and $\dfrac{1}{25}$ are called <u>perfect squares</u>.

5. Label the parts of the right triangle.

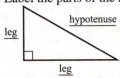

6. The <u>Pythagorean theorem</u> can be used for right triangles.

Exercise Set 8.6

1. $\sqrt{4} = 2$ because $2^2 = 4$.

3. $\sqrt{121} = 11$ because $11^2 = 121$.

5. $\sqrt{\dfrac{1}{81}} = \dfrac{1}{9}$ because $\dfrac{1}{9} \cdot \dfrac{1}{9} = \dfrac{1}{81}$.

7. $\sqrt{\dfrac{16}{64}} = \dfrac{4}{8} = \dfrac{1}{2}$ because $\dfrac{4}{8} \cdot \dfrac{4}{8} = \dfrac{16}{64}$.

9. $\sqrt{3} \approx 1.732$

11. $\sqrt{15} \approx 3.873$

13. $\sqrt{47} \approx 6.856$

15. $\sqrt{26} \approx 5.099$

17. Since 38 is between $36 = 6 \cdot 6$ and $49 = 7 \cdot 7$, $\sqrt{38}$ is between 6 and 7; $\sqrt{38} \approx 6.16$.

19. Since 101 is between $100 = 10 \cdot 10$ and $121 = 11 \cdot 11$, $\sqrt{101}$ is between 10 and 11; $\sqrt{101} \approx 10.05$.

21. $\sqrt{256} = 16$ because $16^2 = 256$.

23. $\sqrt{92} \approx 9.592$

25. $\sqrt{\dfrac{49}{144}} = \dfrac{7}{12}$ because $\left(\dfrac{7}{12}\right)^2 = \dfrac{7}{12} \cdot \dfrac{7}{12} = \dfrac{49}{144}$.

27. $\sqrt{71} \approx 8.426$

29. Let $a = 5$ and $b = 12$.
$$a^2 + b^2 = c^2$$
$$5^2 + 12^2 = c^2$$
$$25 + 144 = c^2$$
$$169 = c^2$$
$$\sqrt{169} = c$$
$$13 = c$$
The missing length is 13 inches.

31. Let $a = 10$ and $c = 12$.
$$a^2 + b^2 = c^2$$
$$10^2 + b^2 = 12^2$$
$$100 + b^2 = 144$$
$$b^2 = 44$$
$$b = \sqrt{44}$$
$$b \approx 6.633$$
The missing length is approximately 6.633 centimeters.

33. Let $a = 22$ and $b = 48$.
$$c^2 = a^2 + b^2$$
$$c^2 = 22^2 + 48^2$$
$$c^2 = 484 + 2304$$
$$c^2 = 2788$$
$$c = \sqrt{2788}$$
$$c \approx 52.802$$
The missing length is approximately 52.802 meters.

35. Let $a = 108$ and $b = 45$.
$$a^2 + b^2 = c^2$$
$$108^2 + 45^2 = c^2$$
$$11,664 + 2025 = c^2$$
$$13,689 = c^2$$
$$\sqrt{13,689} = c$$
$$117 = c$$
The missing length is 117 millimeters.

37.

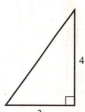

$$\text{hypotenuse} = \sqrt{(\text{leg})^2 + (\text{other leg})^2}$$
$$= \sqrt{(3)^2 + (4)^2}$$
$$= \sqrt{9 + 16}$$
$$= \sqrt{25}$$
$$= 5$$
The hypotenuse has length 5 units.

39.

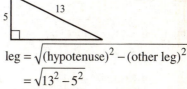

$$\text{leg} = \sqrt{(\text{hypotenuse})^2 - (\text{other leg})^2}$$
$$= \sqrt{13^2 - 5^2}$$
$$= \sqrt{169 - 25}$$
$$= \sqrt{144}$$
$$= 12$$
The leg has length 12 units.

41.

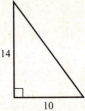

$$\text{hypotenuse} = \sqrt{(\text{leg})^2 + (\text{other leg})^2}$$
$$= \sqrt{(10)^2 + (14)^2}$$
$$= \sqrt{100 + 196}$$
$$= \sqrt{296}$$
$$\approx 17.205$$

The hypotenuse has length of about 17.205 units.

43.

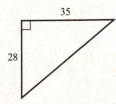

$$\text{hypotenuse} = \sqrt{(\text{leg})^2 + (\text{other leg})^2}$$
$$= \sqrt{35^2 + 28^2}$$
$$= \sqrt{1225 + 784}$$
$$= \sqrt{2009}$$
$$\approx 44.822$$

The hypotenuse has length of about 44.822 units.

45.

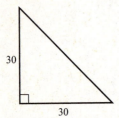

$$\text{hypotenuse} = \sqrt{(\text{leg})^2 + (\text{other leg})^2}$$
$$= \sqrt{(30)^2 + (30)^2}$$
$$= \sqrt{900 + 900}$$
$$= \sqrt{1800}$$
$$\approx 42.426$$

The hypotenuse has length of about 42.426 units.

47.

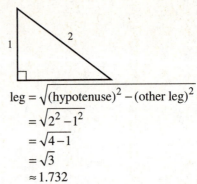

$$\text{leg} = \sqrt{(\text{hypotenuse})^2 - (\text{other leg})^2}$$
$$= \sqrt{2^2 - 1^2}$$
$$= \sqrt{4 - 1}$$
$$= \sqrt{3}$$
$$\approx 1.732$$

The leg has length of about 1.732 units.

49.

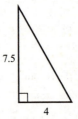

$$\text{hypotenuse} = \sqrt{(\text{leg})^2 + (\text{other leg})^2}$$
$$= \sqrt{(7.5)^2 + (4)^2}$$
$$= \sqrt{56.25 + 16}$$
$$= \sqrt{72.25}$$
$$= 8.5$$

The hypotenuse has length 8.5 units.

51.
$$\text{hypotenuse} = \sqrt{(\text{leg})^2 + (\text{other leg})^2}$$
$$= \sqrt{100^2 + 100^2}$$
$$= \sqrt{10,000 + 10,000}$$
$$= \sqrt{20,000}$$
$$\approx 141.42$$

The length of the diagonal is about 141.42 yards.

53.
$$\text{leg} = \sqrt{(\text{hypotenuse})^2 - (\text{other leg})^2}$$
$$= \sqrt{(32)^2 - (20)^2}$$
$$= \sqrt{1024 - 400}$$
$$= \sqrt{624}$$
$$\approx 25.0$$

The tree is about 25.0 feet tall.

55.
$$\text{hypotenuse} = \sqrt{(\text{leg})^2 + (\text{other leg})^2}$$
$$= \sqrt{160^2 + 300^2}$$
$$= \sqrt{25,600 + 90,000}$$
$$= \sqrt{115,600}$$
$$= 340$$

The length of the run was 340 feet.

57.
$$\frac{n}{6} = \frac{2}{3}$$
$$3 \cdot n = 2 \cdot 6$$
$$3n = 12$$
$$n = \frac{12}{3}$$
$$n = 4$$

59.
$$\frac{9}{11} = \frac{n}{55}$$
$$9 \cdot 55 = 11 \cdot n$$
$$495 = 11n$$
$$\frac{495}{11} = n$$
$$45 = n$$

61.
$$\frac{3}{n} = \frac{7}{14}$$
$$7n = 3 \cdot 14$$
$$7n = 42$$
$$n = \frac{42}{7}$$
$$n = 6$$

63. Recall that $\sqrt{36} = 6$ and $\sqrt{49} = 7$. Since 38 is between 36 and 49, then $\sqrt{38}$ is between $\sqrt{36}$ and $\sqrt{49}$. Thus, $\sqrt{38}$ is between 6 and 7. Since 38 is closer to 36, then $\sqrt{38}$ is closer to $\sqrt{36}$, or 6. Check: $\sqrt{38} \approx 6.16$

65. Recall that $\sqrt{100} = 10$ and $\sqrt{121} = 11$. Since 101 is between 100 and 121, then $\sqrt{101}$ is between $\sqrt{100}$ and $\sqrt{121}$. Thus, $\sqrt{101}$ is between 10 and 11. Since 101 is closer to 100, then $\sqrt{101}$ is closer to $\sqrt{100}$, or 10. Check: $\sqrt{101} \approx 10.05$

67. answers may vary

69.
$$a^2 + b^2 = c^2$$
$$25^2 + 60^2 \stackrel{?}{=} 65^2$$
$$625 + 3600 \stackrel{?}{=} 4225$$
$$4225 = 4225$$

Yes, the set forms the lengths of the sides of a right triangle.

71. Find the missing length in the large right triangle by letting $a = 8$ and $c = 12$.
$$a^2 + b^2 = c^2$$
$$8^2 + b^2 = 12^2$$
$$64 + b^2 = 144$$
$$b^2 = 80$$
$$b = \sqrt{80}$$

Find the unlabeled length by letting $a = 8$ and $c = 10$.
$$a^2 + b^2 = c^2$$
$$8^2 + b^2 = 10^2$$
$$64 + b^2 = 100$$
$$b^2 = 36$$
$$b^2 = \sqrt{36}$$
$$b = 6$$

The unlabeled length is 6 inches. Thus
$$6 + x = \sqrt{80} \quad \text{or} \quad x = \sqrt{80} - 6 \approx 2.94 \text{ inches.}$$

Section 8.7 Practice

1. a. The triangles are congruent by Side-Angle-Side.

   b. The triangles are not congruent.

2.
$$\frac{9 \text{ meters}}{13 \text{ meters}} = \frac{9}{13}$$

The ratio of corresponding sides is $\frac{9}{13}$.

3. a. The ratios of corresponding sides in the two triangles are equal, so:
$$\frac{n}{4} = \frac{10}{5}$$
$$5n = 4 \cdot 10$$
$$5n = 40$$
$$n = \frac{40}{5}$$
$$n = 8$$

The missing length is 8 units.

b.
$$\frac{5}{n} = \frac{9}{6}$$
$$9n = 5 \cdot 6$$
$$9n = 30$$
$$n = \frac{30}{9}$$
$$n = \frac{10}{3} \text{ or } 3\frac{1}{3}$$

The missing length is $\frac{10}{3}$ or $3\frac{1}{3}$ units.

4. The triangle formed by the sun's rays, Tammy, and her shadow is similar to the triangle formed by the sun's rays, the building, and its shadow.
$$\frac{n}{5} = \frac{60}{8}$$
$$8n = 5 \cdot 60$$
$$8n = 300$$
$$n = \frac{300}{8}$$
$$n = 37.5$$
The building is approximately 37.5 feet tall.

Vocabulary and Readiness Check

1. Two triangles that have the same shape, but not necessarily the same size are congruent. <u>false</u>

2. Two triangles are congruent if they have the same shape and size. <u>true</u>

3. Congruent triangles are also similar. <u>true</u>

4. Similar triangles are also congruent. <u>false</u>

5. For the two similar triangles, the ratio of corresponding sides is $\frac{5}{6}$. <u>false</u>

6. $\angle A$ and $\angle D$ are congruent, $\angle B$ and $\angle E$ are congruent, $\angle C$ and $\angle F$ are congruent, and
$$\frac{a}{d} = \frac{b}{e} = \frac{c}{f}.$$

7. $\angle M$ and $\angle Y$ are congruent, $\angle N$ and $\angle X$ are congruent, $\angle P$ and $\angle Z$ are congruent, and
$$\frac{m}{y} = \frac{n}{x} = \frac{p}{z}.$$

Exercise Set 8.7

1. The triangles are congruent by Side-Side-Side.

3. The triangles are not congruent.

5. The triangles are congruent by Angle-Side-Angle.

7. The triangles are congruent by Side-Angle-Side.

9. $\frac{22}{11} = \frac{14}{7} = \frac{12}{6} = \frac{2}{1}$

The ratio of corresponding sides is $\frac{2}{1}$.

11. $\frac{10.5}{7} = \frac{9}{6} = \frac{12}{8} = \frac{3}{2}$

13.
$$\frac{n}{3} = \frac{9}{6}$$
$$n \cdot 6 = 3 \cdot 9$$
$$6n = 27$$
$$\frac{6n}{6} = \frac{27}{6}$$
$$n = 4.5$$

15.
$$\frac{n}{18} = \frac{4}{12}$$
$$n \cdot 12 = 18 \cdot 4$$
$$12n = 72$$
$$\frac{12n}{12} = \frac{72}{12}$$
$$n = 6$$

17.
$$\frac{n}{3.75} = \frac{12}{9}$$
$$n \cdot 9 = 12 \cdot 3.75$$
$$9n = 45$$
$$\frac{9n}{9} = \frac{45}{9}$$
$$n = 5$$

19.
$$\frac{n}{18} = \frac{30}{40}$$
$$n \cdot 40 = 18 \cdot 30$$
$$40n = 540$$
$$\frac{40n}{40} = \frac{540}{40}$$
$$n = 13.5$$

21.
$$\frac{n}{3.25} = \frac{17.5}{3.25}$$
$$n \cdot 3.25 = 3.25 \cdot 17.5$$
$$3.25n = 56.875$$
$$\frac{3.25n}{3.25} = \frac{56.875}{3.25}$$
$$n = 17.5$$

23.
$$\frac{n}{2} = \frac{18\frac{1}{3}}{3\frac{2}{3}}$$
$$n \cdot 3\frac{2}{3} = 2 \cdot 18\frac{1}{3}$$
$$n \cdot \frac{11}{3} = 2 \cdot \frac{55}{3}$$
$$\frac{11}{3} \cdot n = \frac{110}{3}$$
$$\frac{3}{11} \cdot \frac{11}{3} n = \frac{3}{11} \cdot \frac{110}{3}$$
$$n = 10$$

25.
$$\frac{n}{60} = \frac{15}{32}$$
$$n \cdot 32 = 60 \cdot 15$$
$$32n = 900$$
$$\frac{32n}{32} = \frac{900}{32}$$
$$n = 28.125$$

27.
$$\frac{n}{7} = \frac{15}{10\frac{1}{2}}$$
$$n \cdot 10\frac{1}{2} = 7 \cdot 15$$
$$10.5n = 105$$
$$\frac{10.5n}{10.5} = \frac{105}{10.5}$$
$$n = 10$$

29.
$$n = \frac{80}{2}$$
$$n \cdot 2 = 13 \cdot 80$$
$$2n = 1040$$
$$\frac{2n}{2} = \frac{1040}{2}$$
$$n = 520$$
The observation deck is 520 feet high.

31.
$$\frac{x}{25} = \frac{40}{2}$$
$$x \cdot 2 = 40 \cdot 25$$
$$2x = 1000$$
$$\frac{2x}{2} = \frac{1000}{2}$$
$$x = 500$$
The building is 500 feet tall.

33. The triangle formed by the sun's rays, Samantha, and her shadow is similar to the triangle formed by the sun's rays, the tree, and its shadow. Let x be the height of the tree.
$$\frac{x}{5} = \frac{48}{4}$$
$$4x = 5 \cdot 48$$
$$4x = 240$$
$$x = \frac{240}{4}$$
$$x = 60$$
The tree is about 60 feet tall.

35.
$$\frac{n}{18} = \frac{24}{30}$$
$$30 \cdot n = 24 \cdot 18$$
$$30n = 432$$
$$n = \frac{432}{30}$$
$$n = 14.4$$
The shadow of the tree is 14.4 feet long.

37.
$$\frac{x}{55} = \frac{19}{20}$$
$$x \cdot 20 = 55 \cdot 19$$
$$20x = 1045$$
$$\frac{20x}{20} = \frac{1045}{20}$$
$$x = 52.25$$
$$x \approx 52$$
Pete can place 52 neon tetras in the tank.

39. Let $a = 200$ and $c = 430$.
$$a^2 + b^2 = c^2$$
$$200^2 + b^2 = 430^2$$
$$40,000 + b^2 = 184,900$$
$$b^2 = 144,900$$
$$b = \sqrt{144,900}$$
$$b \approx 381$$
The gantry is approximately 381 feet tall.

41.
$$\begin{array}{r} 3.60 \\ + \ 0.41 \\ \hline 4.01 \end{array}$$

43. $(0.41)(3) = 1.23$

45. Let x be the new width.

$$\frac{x}{7} = \frac{5}{9}$$
$$x \cdot 9 = 7 \cdot 5$$
$$9x = 35$$
$$\frac{9x}{9} = \frac{35}{9}$$
$$x = 3\frac{8}{9}$$

The new width is $3\frac{8}{9}$ inches. No, it will not fit

on a 3-by-5-inch card because $3\frac{8}{9} > 3$.

47.
$$\frac{n}{5.2} = \frac{12.6}{7.8}$$
$$n \cdot 7.8 = 5.2 \cdot 12.6$$
$$7.8n = 65.52$$
$$\frac{7.8n}{7.8} = \frac{65.52}{7.8}$$
$$n = 8.4$$

49. answers may vary

51.
$$\frac{x}{5} = \frac{10}{\frac{1}{4}}$$
$$x \cdot \frac{1}{4} = 5 \cdot 10$$
$$\frac{1}{4}x = 50$$
$$x = 200$$
$$\frac{y}{7\frac{1}{2}} = \frac{10}{\frac{1}{4}}$$
$$y \cdot \frac{1}{4} = 7\frac{1}{2} \cdot 10$$
$$\frac{1}{4}y = 75$$
$$y = 300$$

$$\frac{z}{10\frac{5}{8}} = \frac{10}{\frac{1}{4}}$$
$$z \cdot \frac{1}{4} = 10\frac{5}{8} \cdot 10$$
$$\frac{1}{4}z = 106.25$$
$$z = 425$$

The actual proposed dimensions are 200 feet by 300 feet by 425 feet.

Chapter 8 Vocabulary Check

1. A <u>right triangle</u> is a triangle with a right angle. The side opposite the right angle is called the <u>hypotenuse</u>, and the other two sides are called <u>legs</u>.

2. A <u>line segment</u> is a piece of a line with two endpoints.

3. Two angles that have a sum of 90° are called <u>complementary</u> angles.

4. A <u>line</u> is a set of points extending indefinitely in two directions.

5. The <u>perimeter</u> of a polygon is the distance around the polygon.

6. An <u>angle</u> is made up of two rays that share the same endpoint. The common endpoint is called the <u>vertex</u>.

7. <u>Congruent</u> triangles have the same shape and the same size.

8. <u>Area</u> measures the amount of surface of a region.

9. A <u>ray</u> is a part of a line with one endpoint. A ray extends indefinitely in one direction.

10. A <u>square root</u> of a number a is a number b whose square is a.

11. A line that intersects two or more lines at different points is called a <u>transversal</u>.

12. An angle that measures 180° is called a <u>straight</u> angle.

13. The measure of the space of a solid is called its <u>volume</u>.

14. When two lines intersect, four angles are formed. Two of these angles that are opposite each other are called <u>vertical</u> angles.

15. Two of these angles from Exercise 14 that share a common side are called <u>adjacent</u> angles.

16. An angle whose measure is between 90° and 180° is called an <u>obtuse</u> angle.

17. An angle that measures 90° is called a <u>right</u> angle.

18. An angle whose measure is between 0° and 90° is called an <u>acute</u> angle.

19. Two angles that have a sum of 180° are called <u>supplementary</u> angles.

20. <u>Similar</u> triangles have exactly the same shape but not necessarily the same size.

Chapter 8 Review

1. $\angle A$ is a right angle.

2. $\angle B$ is a straight angle.

3. $\angle C$ is an acute angle.

4. $\angle D$ is an obtuse angle.

5. The complement of a 25° angle has measure $90° - 25° = 65°$.

6. The supplement of a 105° angle has measure $180° - 105° = 75°$.

7. $m\angle x = 90° - 32° = 58°$

8. $m\angle x = 180° - 82° = 98°$

9. $m\angle x = 105° - 15° = 90°$

10. $m\angle x = 45° - 20° = 25°$

11. $47° + 133° = 180°$, so $\angle a$ and $\angle b$ are supplementary. So are $\angle b$ and $\angle c$, $\angle c$ and $\angle d$, and $\angle d$ and $\angle a$.

12. $47° + 43° = 90°$, so $\angle x$ and $\angle w$ are complementary. Also, $58° + 32° = 90°$, so $\angle y$ and $\angle z$ are complementary.

13. $\angle x$ and the angle marked 100° are vertical angles, so $m\angle x = 100°$. $\angle x$ and $\angle y$ are adjacent angles, so $m\angle y = 180° - 100° = 80°$. $\angle y$ and $\angle z$ are vertical angles, so $m\angle z = 80°$.

14. $\angle x$ and the angle marked 25° are adjacent angles, so $m\angle x = 180° - 25° = 155°$. $\angle x$ and $\angle y$ are vertical angles, so $m\angle y = 155°$. $\angle z$ and the angle marked 25° are vertical angles, so $m\angle z = 25°$.

15. $\angle x$ and the angle marked 53° are vertical angles, so $m\angle x = 53°$. $\angle x$ and $\angle y$ are alternate interior angles, so $m\angle y = 53°$. $\angle y$ and $\angle z$ are adjacent angles, so $m\angle z = 180° - 53° = 127°$.

16. $\angle x$ and the angle marked 42° are vertical angles, so $m\angle x = 42°$. $\angle x$ and $\angle y$ are alternate interior angles, so $m\angle y = 42°$. $\angle y$ and $\angle z$ are adjacent angles, so $m\angle z = 180° - 42° = 138°$.

17. $m\angle x = 180° - 32° - 45° = 103°$

18. $m\angle x = 180° - 62° - 58° = 60°$

19. $m\angle x = 180° - 90° - 30° = 60°$

20. $m\angle x = 180° - 25° - 90° = 65°$

21. $d = 2 \cdot r = 2 \cdot 2\frac{1}{10}$ m $= 4\frac{1}{5}$ m

22. $r = d \div 2 = 14$ ft $\div 2 = 7$ ft

23. $r = d \div 2 = 19$ m $\div 2 = 9.5$ m

24. $d = 2 \cdot r = 2 \cdot 7\frac{3}{5}$ cm $= 15\frac{1}{5}$ cm

25. The solid is a cube.

26. The solid is a cylinder.

27. The solid is a square-based pyramid.

28. The solid is a rectangular solid.

29. $d = 2 \cdot r = 2 \cdot 9$ in. $= 18$ in.

30. $r = d \div 2 = 4.7$ m $\div 2 = 2.35$ m

31. The polygon has 5 sides, so it is a pentagon.

32. The polygon has 6 sides, so it is a hexagon.

33. The triangle has 3 equal sides, so it is equilateral.

34. The triangle has a right angle and 2 equal sides, so it is an isosceles right triangle.

35. $P = 27 \text{ m} + 17\frac{1}{2} \text{ m} + 27 \text{ m} + 17\frac{1}{2} \text{ m} = 89 \text{ m}$

36. $P = 11 \text{ cm} + 7.6 \text{ cm} + 12 \text{ cm} = 30.6 \text{ cm}$

37. The unmarked vertical side has length
$8 \text{ m} - 5 \text{ m} = 3 \text{ m}$. The unmarked horizontal side has length $10 \text{ m} - 7 \text{ m} = 3 \text{ m}$.
$P = (7 + 3 + 3 + 5 + 10 + 8) \text{ m} = 36 \text{ m}$

38. The unmarked vertical side has length
$5 \text{ ft} + 4 \text{ ft} + 11 \text{ ft} = 20 \text{ ft}$.
$P = (22 + 20 + 22 + 11 + 3 + 4 + 3 + 5) \text{ ft} = 90 \text{ ft}$

39. $P = 2 \cdot l + 2 \cdot w = 2 \cdot 10 \text{ ft} + 2 \cdot 6 \text{ ft} = 32 \text{ ft}$

40. $P = 4 \cdot s = 4 \cdot 110 \text{ ft} = 440 \text{ ft}$

41. $C = \pi \cdot d = \pi \cdot 1.7 \text{ in.} \approx 3.14 \cdot 1.7 \text{ in.} = 5.338 \text{ in.}$

42. $C = 2 \cdot \pi \cdot r$
$= 2 \cdot \pi \cdot 5 \text{ yd}$
$= \pi \cdot 10 \text{ yd}$
$\approx 3.14 \cdot 10 \text{ yd}$
$= 31.4 \text{ yd}$

43. $A = \frac{1}{2} \cdot (b + B) \cdot h$
$= \frac{1}{2} \cdot (12 \text{ ft} + 36 \text{ ft}) \cdot 10 \text{ ft}$
$= \frac{1}{2} \cdot 48 \text{ ft} \cdot 10 \text{ ft}$
$= 240 \text{ sq ft}$

44. $A = \frac{1}{2} \cdot b \cdot h = \frac{1}{2} \cdot 20 \text{ m} \cdot 14 \text{ m} = 140 \text{ sq m}$

45. $A = l \cdot w = 40 \text{ cm} \cdot 15 \text{ cm} = 600 \text{ sq cm}$

46. $A = b \cdot h = 21 \text{ yd} \cdot 9 \text{ yd} = 189 \text{ sq yd}$

47. $A = \pi \cdot r^2 = \pi \cdot (7 \text{ ft})^2 = 49\pi \text{ sq ft} \approx 153.86 \text{ sq ft}$

48. $A = s^2 = (9.1 \text{ m})^2 = 82.81 \text{ sq m}$

49. $A = \frac{1}{2} \cdot b \cdot h = \frac{1}{2} \cdot 34 \text{ in.} \cdot 7 \text{ in.} = 119 \text{ sq in.}$

50. $A = \frac{1}{2} \cdot (b + B) \cdot h$
$= \frac{1}{2} \cdot (64 \text{ cm} + 32 \text{ cm}) \cdot 26 \text{ cm}$
$= \frac{1}{2} \cdot 96 \text{ cm} \cdot 26 \text{ cm}$
$= 1248 \text{ sq cm}$

51. The unmarked horizontal side has length
$13 \text{ m} - 3 \text{ m} = 10 \text{ m}$. The unmarked vertical side has length $12 \text{ m} - 4 \text{ m} = 8 \text{ m}$. The area is the sum of the areas of the two rectangles.
$A = 12 \text{ m} \cdot 10 \text{ m} + 8 \text{ m} \cdot 3 \text{ m}$
$= 120 \text{ sq m} + 24 \text{ sq m}$
$= 144 \text{ sq m}$

52. $A = l \cdot w = 36 \text{ ft} \cdot 12 \text{ ft} = 432 \text{ sq ft}$
The area of the driveway is 432 square feet.

53. $A = 10 \text{ ft} \cdot 13 \text{ ft} = 130 \text{ sq ft}$
130 square feet of carpet are needed.

54. $V = s^3$
$= \left(2\frac{1}{2} \text{ in.}\right)^3$
$= \left(\frac{5}{2} \text{ in.}\right)^3$
$= \frac{125}{8} \text{ cu in.}$
$= 15\frac{5}{8} \text{ cu in.}$

55. $V = l \cdot w \cdot h = 2 \text{ ft} \cdot 7 \text{ ft} \cdot 6 \text{ ft} = 84 \text{ cu ft}$

56. $V = \pi \cdot r^2 \cdot h$
$= \pi \cdot (20 \text{ cm})^2 \cdot 50 \text{ cm}$
$= 20,000\pi \text{ cu cm}$
$\approx 62,800 \text{ cu cm}$

57. $V = \frac{4}{3} \cdot \pi \cdot r^3$
$= \frac{4}{3} \cdot \pi \cdot \left(\frac{1}{2} \text{ km}\right)^3$
$= \frac{1}{6} \pi \text{ cu km}$
$\approx \frac{11}{21} \text{ cu km}$

58. $V = \frac{1}{3} \cdot s^2 \cdot h$

$= \frac{1}{3} \cdot (2 \text{ ft})^2 \cdot 2 \text{ ft}$

$= \frac{8}{3} \text{ cu ft}$

$= 2\frac{2}{3} \text{ cu ft}$

The volume of the pyramid is $2\frac{2}{3}$ cubic feet.

59. $V = \pi \cdot r^2 \cdot h$

$= \pi \cdot (3.5 \text{ in.})^2 \cdot 8 \text{ in.}$

$= 98\pi \text{ cu in.}$

$\approx 307.72 \text{ cu in.}$

The volume of the can is about 307.72 cubic inches.

60. Find the volume of each drawer.

$V = l \cdot w \cdot h$

$= \left(2\frac{1}{2} \text{ ft}\right) \cdot \left(1\frac{1}{2} \text{ ft}\right) \cdot \left(\frac{2}{3} \text{ ft}\right)$

$= \frac{5}{2} \cdot \frac{3}{2} \cdot \frac{2}{3} \text{ cu ft}$

$= \frac{5}{2} \text{ cu ft}$

The three drawers have volume

$3 \cdot \frac{5}{5} = \frac{15}{2} = 7\frac{1}{2}$ cubic feet.

61. $r = d \div 2 = 1 \text{ ft} \div 2 = 0.5 \text{ ft}$

$V = \pi \cdot r^2 \cdot h = \pi \cdot (0.5 \text{ ft})^2 \cdot 2 \text{ ft} = 0.5\pi \text{ cu ft}$

62. $\sqrt{64} = 8$ because $8^2 = 64$.

63. $\sqrt{144} = 12$ because $12^2 = 144$.

64. $\sqrt{\frac{4}{25}} = \frac{2}{5}$ because $\frac{2}{5} \cdot \frac{2}{5} = \frac{4}{25}$.

65. $\sqrt{\frac{1}{100}} = \frac{1}{10}$ because $\frac{1}{10} \cdot \frac{1}{10} = \frac{1}{100}$.

66. $\text{hypotenuse} = \sqrt{(\text{leg})^2 + (\text{other leg})^2}$

$= \sqrt{12^2 + 5^2}$

$= \sqrt{144 + 25}$

$= \sqrt{169}$

$= 13$

The hypotenuse is 13 units.

67. $\text{hypotenuse} = \sqrt{(\text{leg})^2 + (\text{other leg})^2}$

$= \sqrt{20^2 + 21^2}$

$= \sqrt{400 + 441}$

$= \sqrt{841}$

$= 29$

The hypotenuse is 29 units.

68. $\text{leg} = \sqrt{(\text{hypotenuse})^2 - (\text{other leg})^2}$

$= \sqrt{14^2 - 9^2}$

$= \sqrt{196 - 81}$

$= \sqrt{115}$

≈ 10.7

The leg is about 10.7 units.

69. $\text{leg} = \sqrt{(\text{hypotenuse})^2 - (\text{other leg})^2}$

$= \sqrt{155^2 - 124^2}$

$= \sqrt{24,025 - 15,376}$

$= \sqrt{8649}$

$= 93$

The leg is 93 units.

70. The distance is the hypotenuse of a right triangle whose legs have length 90 feet.

$\text{hypotenuse} = \sqrt{(\text{leg})^2 + (\text{other leg})^2}$

$= \sqrt{90^2 + 90^2}$

$= \sqrt{8100 + 8100}$

$= \sqrt{16,200}$

≈ 127.3

The distance is about 127.3 feet.

71. The height is the leg of a right triangle whose hypotenuse is 126 feet and whose other leg is 90 feet.

$$\text{leg} = \sqrt{(\text{hypotenuse})^2 - (\text{other leg})^2}$$
$$= \sqrt{126^2 - 90^2}$$
$$= \sqrt{15,876 - 8100}$$
$$= \sqrt{7776}$$
$$\approx 88.2$$

The height of the building is about 88.2 feet.

72. $\dfrac{n}{15} = \dfrac{20}{8}$

$8n = 15 \cdot 20$
$8n = 300$
$n = \dfrac{300}{8}$
$n = 37.5$

73. $\dfrac{n}{20} = \dfrac{20}{30}$

$30n = 20 \cdot 20$
$30n = 400$
$n = \dfrac{400}{30}$
$n = \dfrac{40}{3}$
$n = 13\dfrac{1}{3}$

74. $\dfrac{n}{5.8} = \dfrac{24}{8}$

$8n = 5.8 \cdot 24$
$8n = 139.2$
$n = \dfrac{139.2}{8}$
$n = 17.4$

75. The triangle formed by the sun's rays, the painter, and his shadow is similar to the triangle formed by the sun's rays, the building, and its shadow. Let x be the height of the building, in feet.

$$\dfrac{x}{5.5} = \dfrac{42}{7}$$
$$7x = 5.5 \cdot 42$$
$$7x = 231$$
$$x = \dfrac{231}{7}$$
$$x = 33$$

The building is about 33 feet tall.

76. $\dfrac{x}{10} = \dfrac{2}{24}$

$24 \cdot x = 10 \cdot 2$
$24x = 20$
$x = \dfrac{20}{24}$
$x = \dfrac{5}{6}$

$\dfrac{y}{26} = \dfrac{2}{24}$

$24 \cdot y = 26 \cdot 2$
$24y = 52$
$y = \dfrac{52}{24}$
$y = \dfrac{13}{6}$
$y = 2\dfrac{1}{6}$

The lengths are $x = \dfrac{5}{6}$ inch and $y = 2\dfrac{1}{6}$ inches.

77. The supplement of a 72° angle is an angle that measures $180° - 72° = 108°$.

78. The complement of a 1° angle is an angle that measures $90° - 1° = 89°$.

79. $m\angle x = 116° - 34° = 82°$

80. $m\angle x = 180° - 58° - 44° = 78°$

81. $\angle x$ and the angle marked 85° are adjacent angles, so $m\angle x = 180° - 85° = 95°$.

82. Let $\angle y$ be the angle corresponding to $\angle x$ at the bottom intersection. Then $\angle y$ and the angle marked 123° are adjacent angles, so $m\angle x = m\angle y = 180° - 123° = 57°$.

83. $r = d \div 2 = 26 \text{ m} \div 2 = 13 \text{ m}$

84. $d = 2 \cdot r = 2 \cdot 6.3 \text{ cm} = 12.6 \text{ cm}$

85. $P = 4 \cdot s = 4 \cdot 5\dfrac{1}{2} \text{ dm} = 22 \text{ dm}$

86. $P = 7 \text{ in.} + 11.2 \text{ in.} + 9.1 \text{ in.} = 27.3 \text{ in.}$

87. The unmarked horizontal side has length
40 ft − 22 ft − 11 ft = 7 ft.
$$P = (22+15+7+15+11+42+40+42)\text{ ft}$$
$$= 194\text{ ft}$$

88. The unmarked horizontal side has length
43 m − 13 m = 30 m. The unmarked vertical side
has length 42 m − 14 m = 28 m. The area is the
sum of the areas of the two rectangles.
$$A = 28\text{ m}\cdot 13\text{ m}+42\text{ m}\cdot 30\text{ m}$$
$$= 364\text{ sq m}+1260\text{ sq m}$$
$$= 1624\text{ sq m}$$

89. $A = \pi\cdot r^2 = \pi\cdot(3\text{ m})^2 = 9\pi$ sq m ≈ 28.26 sq m

90. $V = \dfrac{1}{3}\cdot\pi\cdot r^2\cdot h$
$$= \frac{1}{3}\cdot\pi\cdot\left(5\frac{1}{4}\text{ in.}\right)^2\cdot 12\text{ in.}$$
$$= \frac{1}{3}\cdot\pi\cdot\left(\frac{21}{4}\text{ in.}\right)^2\cdot 12\text{ cu in.}$$
$$= \frac{441}{4}\pi\text{ cu in.}$$
$$\approx \frac{441}{4}\cdot\frac{22}{7}\text{ cu in.}$$
$$= 346\frac{1}{2}\text{ cu in.}$$

91. $V = l\cdot w\cdot h = 5$ in. $\cdot$ 4 in. $\cdot$ 7 in. $= 140$ cu in.

92. $V = l\cdot w\cdot h = 15$ ft $\cdot$ 12 ft $\cdot$ 7 ft $= 1260$ cu ft
The volume of air in the room is 1260 cubic feet.

93. $V = (3\text{ ft})^3 + (1.2\text{ ft})^3$
$$= 27\text{ cu ft}+1.728\text{ cu ft}$$
$$= 28.728\text{ cu ft}$$
The combined volume of the boxes is
28.728 cubic feet.

94. $\sqrt{1} = 1$ because $1^2 = 1$.

95. $\sqrt{36} = 6$ because $6^2 = 36$.

96. $\sqrt{\dfrac{16}{81}} = \dfrac{4}{9}$ because $\dfrac{4}{9}\cdot\dfrac{4}{9} = \dfrac{16}{81}$.

97. hypotenuse $= \sqrt{(\text{leg})^2 + (\text{other leg})^2}$
$$= \sqrt{66^2 + 56^2}$$
$$= \sqrt{4356+3136}$$
$$= \sqrt{7492}$$
$$\approx 86.6$$
The hypotenuse is about 86.6 units.

98. leg $= \sqrt{(\text{hypotenuse})^2 - (\text{other leg})^2}$
$$= \sqrt{24^2 - 12^2}$$
$$= \sqrt{576-144}$$
$$= \sqrt{432}$$
$$\approx 20.8$$
The leg is about 20.8 units.

99. leg $= \sqrt{(\text{hypotenuse})^2 - (\text{other leg})^2}$
$$= \sqrt{51^2 - 17^2}$$
$$= \sqrt{2601-289}$$
$$= \sqrt{2312}$$
$$\approx 48.1$$
The leg is about 48.1 units.

100. hypotenuse $= \sqrt{(\text{leg})^2 + (\text{other leg})^2}$
$$= \sqrt{10^2 + 17^2}$$
$$= \sqrt{100+289}$$
$$= \sqrt{389}$$
$$\approx 19.7$$
The hypotenuse is about 19.7 units.

101.
$$\frac{n}{8\frac{2}{3}} = \frac{9\frac{3}{8}}{12\frac{1}{2}}$$
$$12\frac{1}{2}n = 8\frac{2}{3}\cdot 9\frac{3}{8}$$
$$\frac{25}{2}n = \frac{26}{3}\cdot\frac{75}{8}$$
$$\frac{25}{2}n = \frac{325}{4}$$
$$n = \frac{325}{4}\div\frac{25}{2}$$
$$n = \frac{325}{4}\cdot\frac{2}{25}$$
$$n = \frac{13}{2}$$
$$n = 6\frac{1}{2}$$

102.
$$\frac{n}{6} = \frac{10}{5}$$
$$5n = 6 \cdot 10$$
$$5n = 60$$
$$n = \frac{60}{5}$$
$$n = 12$$

Chapter 8 Test

1. The complement of an angle that measures 78° is an angle that measures 90° − 78° = 12°.

2. The supplement of a 124° angle is an angle that measures 180° − 124° = 56°.

3. $m\angle x = 97° - 40° = 57°$

4. $\angle x$ and the angle marked 62° are adjacent angles, so $m\angle x = 180° - 62° = 118°$. $\angle y$ and the angle marked 62° are vertical angles, so $m\angle y = 62°$. $\angle x$ and $\angle z$ are vertical angles, so $m\angle z = m\angle x = 118°$.

5. $\angle x$ and the angle marked 73° are vertical angles, so $m\angle x = 73°$. $\angle x$ and $\angle y$ are alternate interior angles, so $m\angle y = m\angle x = 73°$. $\angle x$ and $\angle z$ are corresponding angles, so $m\angle z = m\angle x = 73°$.

6. $d = 2 \cdot r = 2 \cdot 3.1 \text{ m} = 6.2 \text{ m}$

7. $r = d \div 2 = 20\frac{1}{2} \text{ in.} \div 2 = 10\frac{1}{4} \text{ in.}$

8. $m\angle x = 180° - 92° - 62° = 26°$

9. Circumference:
$$C = 2 \cdot \pi \cdot r$$
$$= 2 \cdot \pi \cdot 9 \text{ in.}$$
$$= 18\pi \text{ in.}$$
$$\approx 56.52 \text{ in.}$$
Area:
$$A = \pi r^2$$
$$= \pi(9 \text{ in.})^2$$
$$= 81\pi \text{ sq in.}$$
$$\approx 254.34 \text{ sq in.}$$

10. $P = 2 \cdot l + 2 \cdot w$
$$= 2 \cdot 7 \text{ yd} + 2 \cdot 5.3 \text{ yd}$$
$$= 14 \text{ yd} + 10.6 \text{ yd}$$
$$= 24.6 \text{ yd}$$
$A = l \cdot w = 7 \text{ yd} \cdot 5.3 \text{ yd} = 37.1 \text{ sq yd}$

11. The unmarked vertical side has length 11 in. − 7 in. = 4 in. The unmarked horizontal side has length 23 in. − 6 in. = 17 in. $P = (6 + 4 + 17 + 7 + 23 + 11) \text{ in.} = 68 \text{ in.}$ Extending the unmarked vertical side downward divides the region into two rectangles. The region's area is the sum of the areas of these:
$$A = 11 \text{ in.} \cdot 6 \text{ in.} + 7 \text{ in.} \cdot 17 \text{ in.}$$
$$= 66 \text{ sq in.} + 119 \text{ sq in.}$$
$$= 185 \text{ sq in.}$$

12. $V = \pi \cdot r^2 \cdot h$
$$= \pi \cdot (2 \text{ in.})^2 \cdot 5 \text{ in.}$$
$$= 20\pi \text{ cu in.}$$
$$\approx 20 \cdot \frac{22}{7} \text{ cu in.}$$
$$= 62\frac{6}{7} \text{ cu in.}$$

13. $V = l \cdot w \cdot h = 5 \text{ ft} \cdot 3 \text{ ft} \cdot 2 \text{ ft} = 30 \text{ cu ft}$

14. $\sqrt{49} = 7$ because $7^2 = 49$.

15. $\sqrt{79} \approx 8.888$

16. $\sqrt{\dfrac{64}{100}} = \dfrac{8}{10} = \dfrac{4}{5}$ because $\dfrac{8}{10} \cdot \dfrac{8}{10} = \dfrac{64}{100}$.

17. $P = 4 \cdot s = 4 \cdot 4 \text{ in.} = 16 \text{ in.}$
The perimeter of the photo is 16 inches.

18. $V = l \cdot w \cdot h = 3 \text{ ft} \cdot 3 \text{ ft} \cdot 2 \text{ ft} = 18 \text{ cu ft}$
18 cubic feet of soil are needed.

19. $P = 2 \cdot l + 2 \cdot w$
$$= 2 \cdot 18 \text{ ft} + 2 \cdot 13 \text{ ft}$$
$$= 36 \text{ ft} + 26 \text{ ft}$$
$$= 62 \text{ ft}$$
$\text{cost} = P \cdot \$1.87 \text{ per ft}$
$$= 62 \text{ ft} \cdot \$1.87 \text{ per ft}$$
$$= \$115.94$$
62 feet of baseboard are needed, at a total cost of $115.94.

20. $\text{hypotenuse} = \sqrt{(\text{leg})^2 + (\text{other leg})^2}$

$$= \sqrt{4^2 + 4^2}$$
$$= \sqrt{16 + 16}$$
$$= \sqrt{32}$$
$$\approx 5.66$$

The hypotenuse is about 5.66 cm long.

21. First find the area of the lawn.

$A = l \cdot w = 123.8 \text{ ft} \cdot 80 \text{ ft} = 9904 \text{ sq ft}$

0.02 ounce per square foot is $\dfrac{0.02 \text{ oz}}{1 \text{ sq ft}}$.

$\dfrac{0.02 \text{ oz}}{1 \text{ sq ft}} \cdot 9904 \text{ sq ft} = 198.08 \text{ oz}$

Vivian needs to purchase 198.08 ounces of insecticide.

22. $\dfrac{n}{12} = \dfrac{5}{8}$

$8n = 12 \cdot 5$

$8n = 60$

$n = \dfrac{60}{8}$

$n = 7.5$

23. The triangle formed by the sun's rays, Tamara, and her shadow is similar to the triangle formed by the sun's rays, the tower, and its shadow. Let x be the height of the tower, in feet.

$\dfrac{x}{5.75} = \dfrac{48}{4}$

$4x = 5.75 \cdot 48$

$4x = 276$

$x = \dfrac{276}{4}$

$x = 69$

The tower is about 69 feet tall.

Cumulative Review Chapters 1–8

1. 19.5023 is nineteen and five thousand, twenty-three ten-thousandths.

2. The LCD of 11 and 6 is 66.

$\dfrac{7}{11} + \dfrac{1}{6} = \dfrac{42}{66} + \dfrac{11}{66} = \dfrac{53}{66}$

3. To round 736.2359 to the nearest tenth, observe that the digit in the hundredths place is 3. Since this digit is less than 5, we do not add 1 to the digit in the tenths place. The number 736.2359, rounded to the nearest tenth is 736.2

4. To round 736.2359 to the nearest hundred, observe that the digit in the tens place is 3. Since this digit is less than 5, we do not add 1 to the digit in the hundreds place. The number 736.2359, rounded to the nearest hundred is 700.

5. $\begin{array}{r} 45.00 \\ + \ 2.06 \\ \hline 47.06 \end{array}$

6. $3\dfrac{1}{3} \div 1\dfrac{5}{6} = \dfrac{10}{3} \div \dfrac{11}{6} = \dfrac{10}{3} \cdot \dfrac{6}{11} = \dfrac{60}{33} = \dfrac{20}{11} \text{ or } 1\dfrac{9}{11}$

7. $7.68 \times 10 = 76.8$

8. $\dfrac{7}{11} \cdot \dfrac{1}{6} = \dfrac{7}{66}$

9. $76.3 \times 1000 = 76{,}300$

10. $5\dfrac{1}{2} \cdot 2\dfrac{1}{11} = \dfrac{11}{2} \cdot \dfrac{23}{11} = \dfrac{23}{2} \text{ or } 11\dfrac{1}{2}$

11. $\begin{array}{r} 38.6 \\ 7\overline{)\ 270.2} \\ \underline{-21} \\ 60 \\ \underline{-56} \\ 4\ 2 \\ \underline{-4\ 2} \\ 0 \end{array}$

Check: $\begin{array}{r} 38.6 \\ \times \ \ \ 7 \\ \hline 270.2 \end{array}$

12. $\dfrac{56.7}{100} = 0.567$

13. $0.5(8.6 - 1.2) = 0.5 \cdot 7.4 = 3.7$

14. $\dfrac{5 + 2(8 - 3)}{30 \div 6 \cdot 5} = \dfrac{5 + 2 \cdot 5}{5 \cdot 5} = \dfrac{5 + 10}{25} = \dfrac{15}{25} = \dfrac{3}{5}$

15. The decimal form of $\dfrac{1}{8}$ is 0.125, so $\dfrac{1}{8} > 0.12$.

16. $\dfrac{3}{4} = \dfrac{12}{16}$, so $\dfrac{3}{4} < \dfrac{13}{16}$.

17. $\dfrac{2.6}{3.1} = \dfrac{2.6 \cdot 10}{3.1 \cdot 10} = \dfrac{26}{31}$

18. The LCD of 9, 15, and 3 is 45.

$$\frac{2}{9}+\frac{7}{15}-\frac{1}{3}=\frac{10}{45}+\frac{21}{45}-\frac{15}{45}=\frac{31}{45}-\frac{15}{45}=\frac{16}{45}$$

19. The cross products are $2 \cdot 6 = 12$ and $3 \cdot 4 = 12$. Since these are equal, the proportion is true.

20.
$$\frac{7}{8}=\frac{n}{20}$$
$$8n = 7 \cdot 20$$
$$8n = 140$$
$$n = \frac{140}{8}$$
$$n = \frac{35}{2} \text{ or } 17\frac{1}{2}$$

21. The percent of people who drive silver cars is
$$\frac{25}{100} = 25\%.$$

22. The percent of people who prefer digital cameras is $\frac{34}{50} = \frac{68}{100} = 68\%.$

23. $1.9\% = \frac{1.9}{100} = \frac{19}{1000}$

24. $26\% = \frac{26}{100} = \frac{13}{50}$

25. $125\% = \frac{125}{100} = \frac{5}{4} = 1\frac{1}{4}$

26. $560\% = \frac{560}{100} = \frac{28}{5} = 5\frac{3}{5}$

27.
$$85\% \cdot 300 = n$$
$$0.85 \cdot 300 = n$$
$$255 = n$$
85% of 300 is 255.

28.
$$\frac{2.4}{16}=\frac{p}{100}$$
$$2.4 \cdot 100 = 16p$$
$$240 = 16p$$
$$\frac{240}{16} = p$$
$$15 = p$$
2.4 is 15% of 16.

29.
$$20.8 = 40\% \cdot n$$
$$20.8 = 0.40n$$
$$\frac{20.8}{0.4} = n$$
$$52 = n$$
20.8 is 40% of 52.

30. $\left(7-\sqrt{16}\right)^2 = (7-4)^2 = 3^2 = 9$

31. 31 is 4% of what number?
$$31 = 4\% \cdot n$$
$$31 = 0.04n$$
$$\frac{31}{0.04} = n$$
$$775 = n$$
There are 775 freshmen at Slidell High School.

32. The total area of a box of 40 tiles is $40 \cdot 1$ sq ft = 40 sq ft. So the unit price is
$$\frac{\$90}{40 \text{ sq ft}} = \$2.25 \text{ per sq ft.}$$

33. 1.5% of $\$214,000 = \frac{1.5}{100} \cdot \$214,000 = \$3210$

Her commission is $3210.

34. Let n be the number of exercises that the student can complete in 30 minutes.
$$\frac{n}{30}=\frac{7}{6}$$
$$6n = 30 \cdot 7$$
$$6n = 210$$
$$n = \frac{210}{6}$$
$$n = 35$$
The student can complete 35 exercises.

35. 8 feet $= \frac{8 \text{ feet}}{1} \cdot \frac{12 \text{ inches}}{1 \text{ foot}}$
$$= 8 \cdot 12 \text{ inches}$$
$$= 96 \text{ inches}$$

36. 100 in.
$$= 72 \text{ in.} + 24 \text{ in.} + 4 \text{ in.}$$
$$= \frac{72 \text{ in.}}{1} \cdot \frac{1 \text{ ft}}{12 \text{ in.}} \cdot \frac{1 \text{ yd}}{3 \text{ ft}} + \frac{24 \text{ in.}}{1} \cdot \frac{1 \text{ ft}}{12 \text{ in.}} + 4 \text{ in.}$$
$$= \frac{72}{36} \text{ yd} + \frac{24}{12} \text{ ft} + 4 \text{ in.}$$
$$= 2 \text{ yd } 2 \text{ ft } 4 \text{ in.}$$

37. $3.2 \text{ kg} = \dfrac{3.2 \text{ kg}}{1} \cdot \dfrac{1000 \text{ g}}{1 \text{ kg}} = 3.2 \cdot 1000 \text{ g} = 3200 \text{ g}$

38. $70 \text{ mm} = \dfrac{70 \text{ mm}}{1} \cdot \dfrac{1 \text{ m}}{1000 \text{ mm}} = \dfrac{70}{1000} \text{ m} = 0.07 \text{ m}$

39.
$$
\begin{array}{r}
4 \text{ gal } 2 \text{ qt} \\
- \quad\quad 3 \text{ qt} \\
\hline
\end{array}
\rightarrow
\begin{array}{r}
3 \text{ gal } 6 \text{ qt} \\
- \quad\quad 3 \text{ qt} \\
\hline
3 \text{ gal } 3 \text{ qt}
\end{array}
$$

40. Seventy thousand, fifty-two is 70,052.

41. $F = 1.8 \cdot C + 32$
$\quad\ = 1.8 \cdot 29 + 32$
$\quad\ = 52.2 + 32$
$\quad\ = 84.2$
$29°C = 84.2°F$

42. $\dfrac{1}{8} = \dfrac{1}{8} \cdot \dfrac{125}{125} = \dfrac{125}{1000} = 0.125 = 12.5\%$

43. $m\angle a = 180° - 95° - 35° = 50°$

44. $P = 8.1 \text{ m} + 14.1 \text{ m} + 10.8 \text{ m} = 33.0 \text{ m}$

45. $P = 2 \cdot l + 2 \cdot w$
$\quad\ = 2 \cdot 9 \text{ in.} + 2 \cdot 5 \text{ in.}$
$\quad\ = 18 \text{ in.} + 10 \text{ in.}$
$\quad\ = 28 \text{ in.}$

46. $A = l \cdot w = 9 \text{ in.} \cdot 5 \text{ in.} = 45 \text{ sq in.}$

47. $\sqrt{\dfrac{4}{25}} = \dfrac{2}{5}$ because $\dfrac{2}{5} \cdot \dfrac{2}{5} = \dfrac{4}{25}$.

48. $\sqrt{\dfrac{9}{16}} = \dfrac{3}{4}$ because $\dfrac{3}{4} \cdot \dfrac{3}{4} = \dfrac{9}{16}$.

49. The ratio is $\dfrac{12 \text{ ft}}{19 \text{ ft}} = \dfrac{12}{19}$.

50. $\dfrac{x}{10} = \dfrac{19}{12}$
$12x = 10 \cdot 19$
$12x = 190$
$x = \dfrac{190}{12}$
$x = \dfrac{95}{6}$
$x = 15\dfrac{5}{6}$

Chapter 9

Section 9.1 Practice

1. a. English has 6 symbols and each symbol represents 50 million speakers, so English is spoken by 6(50) = 300 million people.

b. Portuguese has 3.5 symbols, or 3.5(50) = 175 million speakers. So, 300 million − 175 million = 125 million more people speak English than Portuguese.

2. a. The European Space Agency corresponds to 1 symbol, and each symbol represents 7 spaceflights, so the European Space Agency undertook approximately 7 spaceflights.

b. The European Space Agency corresponds to 1 symbol, or approximately 7 spaceflights. Japan also corresponds to 1 symbol, or approximately 7 spaceflights. The total number is approximately 7 + 7 = 14 spaceflights.

3. a. The height of the bar for birds is 74, so approximately 74 endangered species are birds.

b. The shortest bar corresponds to arachnids, so arachnids have the fewest endangered species.

4.

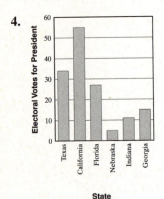

State

5. The height of the bar for 80–89 is 12, so 12 students scored 80–89 on the test.

6. The height of the bar for 40–49 is 1, for 50–59 is 3, for 60–69 is 2, for 70–79 is 10. So, 1 + 3 + 2 + 10 = 16 students scored less than 80 on the test.

7.

Class Interval (Credit Card Balances)	Tally	Class Frequency (Number of Months)
$0–$49	|||	3
$50–$99	||||	4
$100–$149	||	2
$150–$199	|	1
$200–$249	|	1
$250–$299	|	1

8.

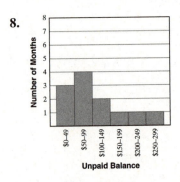

Unpaid Balance

9. a. The lowest point on the graph corresponds to January, so the average daily temperature is the lowest during January.

b. The point on the graph that corresponds to 25 is December, so the average daily temperature is 25°F in December.

c. The points on the graph that are greater than 70 are June, July, and August. So, the average daily temperature is greater than 70°F in June, July, and August.

Vocabulary and Readiness Check

1. A <u>bar</u> graph presents data using vertical or horizontal bars.

2. A <u>pictograph</u> is a graph in which pictures or symbols are used to visually present data.

3. A <u>line</u> graph displays information with a line that connects data points.

290

4. A <u>histogram</u> is a special bar graph in which the width of each bar represents a <u>class interval</u> and the height of each bar represents the <u>class frequency</u>.

Exercise Set 9.1

1. Kansas has the greatest number of wheat symbols, so the greatest quantity of acreage in wheat was planted by the state of Kansas.

3. Oklahoma is represented by 3.5 wheat symbols, and each symbol represents 1 million acres, so there were approximately 3.5(1 million) = 3.5 million or 3,500,000 acres of wheat planted.

5. Each wheat symbol represents 1 million acres, so find the state that has $\dfrac{5 \text{ million}}{1 \text{ million}} = 5$ wheat symbols. Montana has 5 wheat symbols, so it plants about 5,000,000 acres of wheat.

7. North Dakota is represented by 8 wheat symbols. From the pictograph, Montana, with 5 symbols, and South Dakota, with 3 symbols, together plant about the same acreage of wheat as North Dakota.

9. The year 2009 has 6.5 flames and each flame represents 12,000 wildfires, so there were approximately 6.5(12,000) = 78,000 wildfires in 2009.

11. The year with the most flames is 2006, so the most wildfires occurred in 2006.

13. 2004 has 5.5 flames and 2006 has 8 flames, which is 2.5 more. Thus, the increase in the number of wildfires from 2004 to 2006 was 2.5(12,000) or 30,000.

15. 2006 has 8 flames, 2007 has 7 flames, and 2008 has 6.5 flames. The average is $\dfrac{8 + 7 + 6.5}{3} = 7\dfrac{1}{6}$. Each flame represents 12,000 wildfires, so the average annual number of wildfires from 2006 to 2008 is $7\dfrac{1}{6}(12,000) = 86,000$.

17. The longest bar corresponds to September, so the month in which most hurricanes made landfall is September.

19. The length of the bar for August is 75, so approximately 75 hurricanes made landfall in August.

21. Two of the 76 hurricanes that made landfall in August did so in 2008. The fraction is $\dfrac{2}{76} = \dfrac{1}{38}$.

23. The longest bar corresponds to Tokyo, Japan, and the length of the bar is 33.8 million. So, the city with the largest population is Tokyo, Japan and its population is about 33.8 million or 33,800,000.

25. The longest bar corresponding to a city in the United States is the bar for New York. The population is approximately 21.9 million or 21,900,000.

27. The bar corresponding to Seoul, South Korea has length 23.8 and the bar corresponding to São Paolo, Brazil has length 20.9. Thus, Seoul, South Korea is about 23.8 − 20.9 = 2.9 ≈ 3 million larger than São Paolo, Brazil.

29.

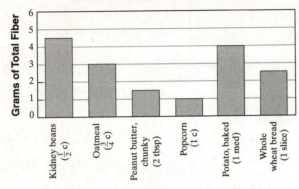

Fiber Content of Selected Foods

31.

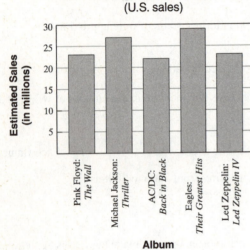

Best-selling Albums of All Time
(U.S. sales)

33. The height of the bar for 100–149 miles per week is 15, so 15 of the adults drive 100–149 miles per week.

35. 29 of the adults drive 0–49 miles per week, 17 of the adults drive 50–99 miles per week, and 15 of the adults drive 100–149 miles per week, so 29 + 17 + 15 = 61 of the adults drive fewer than 150 miles per week.

37. 15 of the adults drive 100–149 miles per week and 9 of the adults drive 150–199 miles per week, so 15 + 9 = 24 of the adults drive 100–199 miles per week.

39. 21 of the adults drive 250–299 miles per week and 9 of the adults drive 200–249 miles per week, so 21 − 9 = 12 more adults drive 250–299 miles per week than 200–249 miles per week.

41. 9 of the 100 adults surveyed drive 150–199 miles per week, so the ratio is $\dfrac{9}{100}$.

43. The tallest bar in the histogram represents the 20 to 44 age range, so the 20 to 44 age range will be the largest population group in 2020.

45. The population of 20-to-44 year olds is expected to be 109 million in 2020.

47. The population of those less than 4 years old is expected to be 23 million in 2020.

49. answers may vary

	Class Interval (Scores)	Tally	Class Frequency (Number of Games)
51.	70–79	|	1
53.	90–99	Ⅳ|||	8

	Class Interval (Account Balances)	Tally	Class Frequency (Number of People)
55.	$0–$99	Ⅳ|	6
57.	$200–$299	Ⅳ|	6
59.	$400–$499	||	2

61.

Golf Scores

63. The point on the line graph corresponding to 2003 is 69, so the total points scored in the Super Bowl in 2003 was 69.

65. The points on the line graph greater than 60 correspond to 2003 and 2004, so total points scored in the Super Bowl greater than 60 was in 2003 and 2004.

67. The lowest points on the line graph correspond to 2006 and 2008, so the lowest total points scored in the Super Bowl was in 2003.

69. The points on the line graph less than 50 correspond to 2005, 2006, 2007 and 2008, so total points scored in the Super Bowl less than 50 was in 2005, 2006, 2007, and 2008.

71. 30% of 12 is $0.30 \cdot 12 = 3.6$.

73. 10% of 62 is $0.10 \cdot 62 = 6.2$

75. $\frac{1}{4} = \frac{1}{4} \cdot 100\% = \frac{25 \cdot 4}{4}\% = 25\%$

77. $\frac{17}{50} = \frac{17}{50} \cdot 100\% = \frac{17 \cdot 2 \cdot 50}{50}\% = 34\%$

79. The point on the high temperature graph corresponding to Thursday is 83, so the high temperature reading on Thursday was 83°F.

81. The lowest point on the graph of low temperatures corresponds to Sunday. The low temperature on Sunday was 68°F.

83. The difference between the graphs is the greatest for Tuesday. The high temperature was 86°F and the low temperature was 73°F, so the difference is $86 - 73 = 13°F$.

85. answers may vary

Section 9.2 Practice

1. Eight of the 100 adults prefer golf. The ratio is $\frac{\text{adults preferring golf}}{\text{total adults}} = \frac{8}{100} = \frac{2}{25}$

2. Add the percents corresponding to Europe, Asia, and South America.
$25.3\% + 12.2\% + 5.0\% = 42.5\%$.

3. $\text{amount} = \text{percent} \cdot \text{base}$
$= 0.123 \cdot 61,000,000$
$= 7,503,000$
Thus, 7,503,000 tourists might come from Mexico in 2011.

4.

Year	Percent	Degrees in Sector
Freshmen	30%	30% of 360° = 0.30(360°) = 108°
Sophomores	27%	27% of 360° = 0.27(360°) = 97.2°
Juniors	25%	25% of 360° = 0.25(360°) = 90°
Seniors	18%	18% of 360° = 0.18(360°) = 64.8°

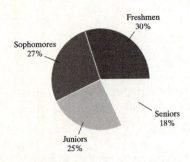

Freshmen 30%

Sophomores 27%

Seniors 18%

Juniors 25%

Vocabulary and Readiness Check

1. In a <u>circle</u> graph, each section (shaped like a piece of pie) shows a category and the relative size of the category.

2. A circle graph contains pie-shaped sections, each called a <u>sector</u>.

3. The number of degrees in a whole circle is <u>360</u>.

4. If a circle graph has percent labels, the percents should add up to <u>100</u>.

Exercise Set 9.2

1. The largest sector corresponds to the category "parent or guardian's home," thus most of the students live in a parent or guardian's home.

3. 180 of the 700 total students live in campus housing.
$$\frac{180}{700} = \frac{9}{35}$$
The ratio is $\frac{9}{35}$.

5. 180 of the students live in campus housing while 320 live in a parent or guardian's home.
$$\frac{180}{320} = \frac{9}{16}$$
The ratio is $\frac{9}{16}$.

7. The largest sector corresponds to Asia. Thus, the largest continent is Asia.

9. 30% + 7% = 37%
37% of the land on Earth is accounted for by Europe and Asia.

11. Asia accounts for 30% of the land on Earth.
$$30\% \text{ of } 57,000,000 = 0.30 \cdot 57,000,000$$
$$= 17,100,000$$
Asia is 17,100,000 square miles.

13. Australia accounts for 5% of the land on Earth.
$$5\% \text{ of } 57,000,000 = 0.05 \cdot 57,000,000$$
$$= 2,850,000$$
Australia is 2,850,000 square miles.

15. Add the percent for adult's fiction (33%) to the percent for children's fiction (22%).
33% + 22% = 55%
Thus, 55% of books are classified as some type of fiction.

17. The second-largest sector corresponds to nonfiction, so the second-largest category of books is nonfiction.

19. Nonfiction accounts for 25% of the books.
$$25\% \text{ of } 125,600 = 0.25 \cdot 125,600 = 31,400$$
The library has 31,400 nonfiction books.

21. Children's fiction accounts for 22% of the books.
$$22\% \text{ of } 125,600 = 0.22 \cdot 125,600 = 27,632$$
The library has 27,632 children's fiction books.

23. Reference or other accounts for 17% + 3% = 20% of the books.
$$20\% \text{ of } 125,600 = 0.20 \cdot 125,600 = 25,120$$
The library has 25,120 reference or other books.

25.

Type of Apple	Percent	Degrees in Sector
Red Delicious	37%	37% of 360° = 0.37(360°) ≈ 133°
Golden Delicious	13%	13% of 360° = 0.13(360°) ≈ 47°
Fuji	14%	14% of 360° = 0.14(360°) ≈ 50°
Gala	15%	15% of 360° = 0.15(360°) = 54°
Granny Smith	12%	12% of 360° = 0.12(360°) ≈ 43°
Other varieties	6%	6% of 360° = 0.06(360°) ≈ 22°
Braeburn	3%	3% of 360° = 0.03(360°) ≈ 11°

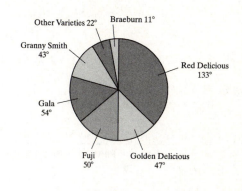

27.

Distribution of Large Dams by Continent		
Continent	Percent	Degrees in Sector
Europe	19%	19% of 360° = 0.19(360°) ≈ 68°
North America	32%	32% of 360° = 0.32(360°) ≈ 115°
South America	3%	3% of 360° = 0.03(360°) ≈ 11°
Asia	39%	39% of 360° = 0.39(360°) ≈ 140°
Africa	5%	5% of 360° = 0.05(360°) = 18°
Australia	2%	2% of 360° = 0.02(360°) ≈ 7°

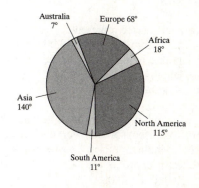

29. $20 = 2 \cdot 10 = 2 \cdot 2 \cdot 5 = 2^2 \cdot 5$

31. $40 = 2 \cdot 20 = 2 \cdot 2 \cdot 10 = 2 \cdot 2 \cdot 2 \cdot 5 = 2^3 \cdot 5$

33. $85 = 5 \cdot 17$

35. Pacific; answers may vary

37. Pacific Ocean:
$$49\% \cdot 264,489,800 = 0.49 \cdot 264,489,800$$
$$= 129,600,002 \text{ square}$$
$$\text{kilometers}$$

39. Indian Ocean:
$$21\% \cdot 264,489,800 = 0.21 \cdot 264,489,800$$
$$= 55,542,858 \text{ square}$$
$$\text{kilometers}$$

41. $21.5\% \cdot 2800 = 0.215 \cdot 2800 = 602$ respondents

43. $21.5\% + 59.8\% = 81.3\%$
$0.813 \cdot 2800 \approx 2276$ respondents

45. $\dfrac{\text{number of respondents who spend } \$0 - \$15}{\text{number of respondents who spend } \$15 - \$175}$
$$= \frac{602}{1674}$$
$$= \frac{2 \cdot 301}{2 \cdot 837}$$
$$= \frac{301}{837}$$

47. no; answers may vary

Integrated Review

1. Customer service representatives has 10 figures and each figure represents 50,000 workers. Thus, the increase in the number of customer service representatives is approximately $10 \cdot 50,000 = 500,000$.

2. Post-secondary teachers has 11 figures and each figure represents 50,000 workers. Thus, the increase in the number of post-secondary teachers is approximately $11 \cdot 50,000 = 550,000$.

3. Retail salespeople has the greatest number of figures, so the greatest increase is expected for retail salespeople.

4. Waitstaff has the least number of figures, so the least increase is expected for waitstaff.

5. The tallest bar corresponds to the Oroville dam. Its height is about 755 feet.

6. The only bar whose top is between the 625 and 650 foot levels is the one for the New Bullards Bar dam. Its height is about 635 feet.

7. The height of Hoover Dam is about 725 feet and the height of Glen Canyon Dam is about 710 feet. So Hoover Dam is about $725 - 710 = 15$ feet higher than Glen Canyon Dam.

8. There are 4 dams whose bars extend above the 700 foot level on the graph, namely Oroville, Hoover, Dworshak, and Glen Canyon.

9. The highest points on the graph correspond to Thursday and Saturday. The high temperature on those days was about 100°F.

10. The lowest point on the graph corresponds to Monday. The high temperature on Monday was about 82°F.

11. The high temperature was below 90°F on Sunday, Monday, and Tuesday.

12. The high temperature was above 90°F on Wednesday, Thursday, Friday, and Saturday.

13. Whole milk accounts for 35% of the milk consumed.
$$35\% \text{ of } 200 \text{ quarts} = 0.35 \cdot 200 \text{ quarts}$$
$$= 70 \text{ quarts}$$
About 70 quart containers of whole milk are sold per week.

14. Skim milk accounts for 26% of the milk consumed.
$$26\% \text{ of } 200 \text{ quarts} = 0.26 \cdot 200 \text{ quarts}$$
$$= 52 \text{ quarts}$$
About 52 quart containers of skim milk are sold per week.

15. Buttermilk accounts for 1% of the milk consumed.
$$1\% \text{ of } 200 \text{ quarts} = 0.01 \cdot 200 \text{ quarts}$$
$$= 2 \text{ quarts}$$
About 2 quart containers of buttermilk are sold per week.

16. Flavored reduced fat and skim milk accounts for 3% of the milk consumed.
$$3\% \text{ of } 200 \text{ quarts} = 0.03 \cdot 200 \text{ quarts}$$
$$= 6 \text{ quarts}$$
About 6 quart containers of flavored reduced fat and skim milk are sold per week.

17.

Class Interval (Scores)	Tally	Class Frequency (Number of Quizzes)
50–59	||	2

18.

Class Interval (Scores)	Tally	Class Frequency (Number of Quizzes)
60–69	|	1

19.

Class Interval (Scores)	Tally	Class Frequency (Number of Quizzes)
70–79	|||	3

20.

Class Interval (Scores)	Tally	Class Frequency (Number of Quizzes)
80–89	ⅢⅠ	6

21.

Class Interval (Scores)	Tally	Class Frequency (Number of Quizzes)
90–99	Ⅲ	5

22.

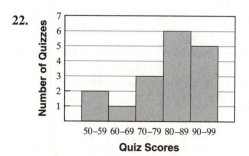

Quiz Scores

Section 9.3 Practice

1. Mean $= \dfrac{87+75+96+91+78}{5} = \dfrac{427}{5} = 85.4$

2. gpa $= \dfrac{4\cdot 2+3\cdot 4+2\cdot 5+1\cdot 2+4\cdot 2}{2+4+5+2+2} = \dfrac{40}{15} \approx 2.67$

3. Because the numbers are in numerical order, and there are an odd number of items, the median is the middle number, 24.

4. Write the numbers in numerical order:
   36, 65, 71, 78, 88, 91, 95, 95
   Since there are an even number of scores, the median is the mean of the two middle numbers.
   median $= \dfrac{78+88}{2} = 83$

5. Mode: 15 because it occurs most often, 3 times.

6. Median: Write the numbers in order.
   15, 15, 15, 16, 18, 26, 26, 30, 31, 35
   Median is mean of middle two numbers,
   $\dfrac{18+26}{2} = 22$.
   Mode: 15 because it occurs most often, 3 times.

Vocabulary and Readiness Check

1. Another word for "mean" is <u>average</u>.

2. The number that occurs most often in a set of numbers is called the <u>mode</u>.

3. The <u>mean (or average)</u> of a set of number items is $\dfrac{\text{sum of items}}{\text{number of items}}$.

4. The <u>median</u> of a set of numbers is the middle number. If the number of numbers is even, it is the <u>mean (or average)</u> of the two middle numbers.

5. An example of weighted mean is a calculation of <u>grade point average</u>.

Exercise Set 9.3

1. Mean: $\dfrac{15+23+24+18+25}{5} = \dfrac{105}{5} = 21$
   Median: Write the numbers in order:
   15, 18, 23, 24, 25
   The middle number is 23.
   Mode: There is no mode, since each number occurs once.

3. Mean:
$$\frac{7.6+8.2+8.2+9.6+5.7+9.1}{6}=\frac{48.4}{6}\approx 8.1$$
Median: Write the numbers in order:
5.7, 7.6, 8.2, 8.2, 9.1, 9.6

Median is mean of middle two: $\frac{8.2+8.2}{2}=8.2$

Mode: 8.2 since this number appears twice.

5. Mean:
$$\frac{0.5+0.2+0.2+0.6+0.3+1.3+0.8+0.1+0.5}{9}$$
$$=\frac{4.5}{9}$$
$$=0.5$$
Median: Write the numbers in order:
0.1, 0.2, 0.2, 0.3, 0.5, 0.5, 0.6, 0.8, 1.3
The middle number is 0.5.
Mode: Since 0.2 and 0.5 occur twice, there are two modes, 0.2 and 0.5.

7. Mean:
$$\frac{231+543+601+293+588+109+334+268}{8}$$
$$=\frac{2967}{8}$$
$$\approx 370.9$$
Median: Write the numbers in order:
109, 231, 268, 293, 334, 543, 588, 601

The mean of the middle two: $\frac{293+334}{2}=313.5$

Mode: There is no mode, since each number occurs once.

9. Mean:
$$\frac{1670+1614+1483+1483+1451}{5}=\frac{7701}{5}$$
$$=1540.2 \text{ feet}$$

11. Because the numbers are in numerical order, the median is mean of the middle two (of the top 8),
$$\frac{1483+1451}{2}=1467 \text{ feet.}$$

13. answers may vary

15. $\text{gpa}=\dfrac{3\cdot 3+2\cdot 3+4\cdot 4+2\cdot 4}{3+3+4+4}=\dfrac{39}{14}\approx 2.79$

17. $\text{gpa}=\dfrac{4\cdot 3+4\cdot 3+4\cdot 4+3\cdot 3+2\cdot 1}{3+3+4+3+1}$
$$=\frac{51}{14}$$
$$\approx 3.64$$

19. Mean:
$$\frac{7.8+6.9+7.5+4.7+6.9+7.0}{6}=\frac{40.8}{6}=6.8$$

21. Mode: 6.9, since this number appears twice.

23. Median: Write the numbers in order.
79, 85, 88, 89, 91, 93

The mean of the middle two: $\dfrac{88+89}{2}=88.5$

25. Mean: $\dfrac{\text{sum of 15 pulse rates}}{15}=\dfrac{1095}{15}=73$

27. Mode: Since 70 and 71 occur twice, there are two modes, 70 and 71.

29. There were 9 rates lower than the mean. They are 66, 68, 71, 64, 71, 70, 65, 70, and 72.

31. $\dfrac{6}{18}=\dfrac{1\cdot 6}{3\cdot 6}=\dfrac{1}{3}$

33. $\dfrac{18}{30}=\dfrac{6\cdot 3}{6\cdot 5}=\dfrac{3}{5}$

35. $\dfrac{55}{75}=\dfrac{5\cdot 11}{5\cdot 15}=\dfrac{11}{15}$

37. Since the mode is 35, 35 must occur at least twice in the set.
Since there is an odd number of numbers in the set, the median, 37 is in the set.
Let n be the remaining unknown number.

Mean: $\dfrac{35+35+37+40+n}{5}=38$
$$\frac{147+n}{5}=38$$
$$5\cdot \frac{147+n}{5}=5\cdot 38$$
$$147+n=190$$
$$147-147+n=190-147$$
$$n=43$$
The missing numbers are 35, 35, 37, and 43.

39. yes; answers may vary

Section 9.4 Practice

1.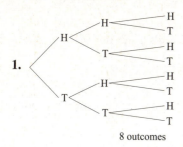

8 outcomes

2.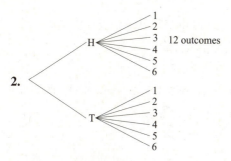

12 outcomes

3. The possibilities are:
   H, H, H,   H, H, T,   H, T, H,   H, T, T,
   T, H, H,   T, H, T,   T, T, H,   T, T, T
   T, H, T is one of the 8 possible outcomes, so the probability is $\frac{1}{8}$.

4. A 2 or a 5 are two of the six possible outcomes. The probability is $\frac{2}{6} = \frac{1}{3}$.

5. A blue is 2 out of the 4 possible marbles. The probability is $\frac{2}{4} = \frac{1}{2}$.

Vocabulary and Readiness Check

1. A possible result of an experiment is called an <u>outcome</u>.

2. A <u>tree diagram</u> shows each outcome of an experiment as a separate branch.

3. The <u>probability</u> of an event is a measure of the likelihood of it occurring.

4. <u>Probability</u> is calculated by number of ways that the event can occur divided by number of possible outcomes.

5. A probability of <u>0</u> means that an event won't occur.

6. A probability of <u>1</u> means that an event is certain to occur.

Exercise Set 9.4

1.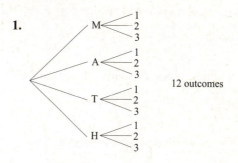

12 outcomes

3.

Red
Blue   3 outcomes
Yellow

5.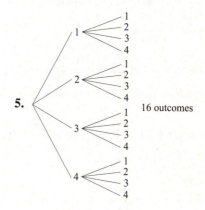

16 outcomes

7.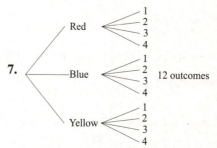

12 outcomes

9.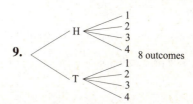

8 outcomes

11. A 5 is one of the six possible outcomes. The probability is $\frac{1}{6}$.

13. A 1 or a 6 are two of the six possible outcomes. The probability is $\frac{2}{6} = \frac{1}{3}$.

15. Three of the six possible outcomes are even. The probability is $\frac{3}{6} = \frac{1}{2}$.

17. Four of the six possible outcomes are numbers greater than 2. The probability is $\frac{4}{6} = \frac{2}{3}$.

19. A 2 is one of three possible outcomes. The probability is $\frac{1}{3}$.

21. A 1, a 2, or a 3 are three of three possible outcomes. The probability is $\frac{3}{3} = 1$.

23. An odd number is a 1 or a 3, which are two of three possible outcomes. The probability is $\frac{2}{3}$.

25. One of the seven marbles is red. The probability is $\frac{1}{7}$.

27. Two of the seven marbles are yellow. The probability is $\frac{2}{7}$.

29. Four of the seven marbles are either green or red. The probability is $\frac{4}{7}$.

31. The blood pressure was higher for 38 of the 200 people. The probability is $\frac{38}{200} = \frac{19}{100}$.

33. The blood pressure did not change for 10 of the 200 people. The probability is $\frac{10}{200} = \frac{1}{20}$.

35. $\frac{1}{2} + \frac{1}{3} = \frac{1}{2} \cdot \frac{3}{3} + \frac{1}{3} \cdot \frac{2}{2} = \frac{3}{6} + \frac{2}{6} = \frac{3+2}{6} = \frac{5}{6}$

37. $\frac{1}{2} \cdot \frac{1}{3} = \frac{1 \cdot 1}{2 \cdot 3} = \frac{1}{6}$

39. $5 \div \frac{3}{4} = \frac{5}{1} \div \frac{3}{4} = \frac{5}{1} \cdot \frac{4}{3} = \frac{5 \cdot 4}{1 \cdot 3} = \frac{20}{3}$ or $6\frac{2}{3}$

41. One of the 52 cards is the king of hearts. The probability is $\frac{1}{52}$.

43. Four of the 52 cards are kings. The probability is $\frac{4}{52} = \frac{1}{13}$.

45. Thirteen of the 52 cards are hearts. The probability is $\frac{13}{52} = \frac{1}{4}$.

47. Twenty six of the cards are in black ink. The probability is $\frac{26}{52} = \frac{1}{2}$.

Tree diagram for 49.–51.

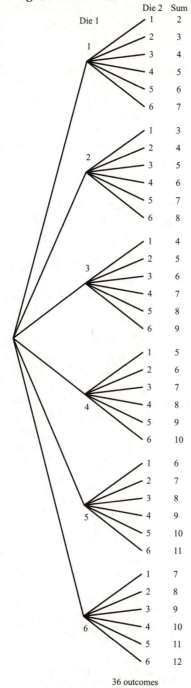

Die 1   Die 2   Sum

36 outcomes

49. Five of the 36 sums are 6. The probability is
$\frac{5}{36}$.

51. None of the 36 sums are 13. The probability is
$\frac{0}{36} = 0$.

53. answers may vary

Chapter 9 Vocabulary Check

1. A <u>bar</u> graph presents data using vertical or horizontal bars.

2. The <u>mean</u> of a set of number items is
$\frac{\text{sum of items}}{\text{number of items}}$.

3. The possible results of an experiment are the <u>outcomes</u>.

4. A <u>pictograph</u> is a graph in which pictures or symbols are used to visually present data.

5. The <u>mode</u> of a set of numbers is the number that occurs most often.

6. A <u>line</u> graph displays information with a line that connects data points.

7. The <u>median</u> of an ordered set of numbers is the middle number.

8. A <u>tree diagram</u> is one way to picture and count outcomes.

9. An <u>experiment</u> is an activity being considered, such as tossing a coin or rolling a die.

10. In a <u>circle</u> graph, each section (shaped like a piece of pie) shows a category and the relative size of the category.

11. The <u>probability</u> of an event is
$\frac{\text{number of ways that the event can occur}}{\text{number of possible outcomes}}$.

12. A <u>histogram</u> is a special bar graph in which the width of each bar represents a <u>class interval</u> and the height of each bar represents the <u>class frequency</u>.

Chapter 9 Review

1. Midwest has 3 houses, and each house represents 500,000 homes, so there were
3(500,000) = 1,500,000 new homes constructed in the Midwest.

2. Northeast has 4 houses, and each house represents 500,000 homes, so there were
4(500,000) = 2,000,000 new homes constructed in the Northeast.

3. South has the greatest number of houses, so the most new homes constructed were in the South.

4. Midwest has the least number of houses, so the fewest new homes constructed were in the Midwest.

5. Each house represents 500,000 homes, so look for the regions with $\dfrac{3,000,000}{500,000} = 6$ or more houses. The South had 3,000,000 or more new homes constructed.

6. Each house represents 500,000 homes, so look for the regions with fewer than $\dfrac{3,000,000}{500,000} = 6$ houses. The Northeast, West, and Midwest had fewer than 3,000,000 new homes constructed.

7. The height of the bar representing 1970 is 11. Thus, approximately 11% of persons completed four or more years of college in 1970.

8. The tallest bar corresponds to 2008. Thus, the greatest percent of persons completing four or more years of college was in 2008.

9. The bars whose height is at a level of 20 or more are 1990, 2000, and 2008. Thus, 20% or more persons completed four or more years of college in 1990, 2000, and 2008.

10. answers may vary

11. The point on the graph corresponding to 2008 is about 960. Thus, there were approximately 960 medals awarded at the Summer Olympics in 2008.

12. The point on the graph corresponding to 2000 is about 920. Thus, there were approximately 920 medals awarded at the Summer Olympics in 2000.

13. The point on the graph corresponding to 2004 is about 930. Thus, there were approximately 930 medals awarded at the Summer Olympics in 2004.

14. The point on the graph corresponding to 1992 is about 815. Thus, there were approximately 815 medals awarded at the Summer Olympics in 1992.

15. The points on the graph corresponding to 1996 and 1992 are 840 and 815, respectively. Thus, there were 840 − 815 = 25 more medals awarded in 1996 than in 1992.

16. The points on the graph corresponding to 2008 and 1992 are 960 and 815, respectively. Thus, there were 960 − 815 = 145 more medals awarded in 2008 than in 1992.

17. The height of the bar corresponding to 41−45 is 1. Thus, 1 employee works 41−45 hours per week.

18. The height of the bar corresponding to 21−25 is 4. Thus, 4 employees work 21−25 hours per week.

19. Add the heights of the bars corresponding to 16−20, 21−25, and 26−30. Thus, 6 + 4 + 8 = 18 employees work 30 hours or less per week.

20. Add the heights of the bars corresponding to 36−40 and 41−45. Thus, 8 + 1 = 9 employees work 36 or more hours per week.

	Class Interval (Temperatures)	**Tally**	**Class Frequency (Number of Months)**
21.	80°−89°	Ⱶ	5
22.	90°−99°	|||	3
23.	100°−109°	||||	4

24.

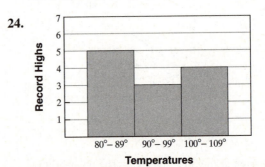

25. The largest sector corresponds to the category "Mortgage payment," thus the largest budget item is mortgage payment.

26. The smallest sector corresponds to the category "Utilities," thus the smallest budget item is utilities.

27. Add the amounts for mortgage payment and utilities. Thus, $975 + $250 = $1225 is budgeted for the mortgage payment and utilities.

28. Add the amounts for savings and contributions. Thus, $400 + $300 = $700 is budgeted for savings and contributions.

29. $\dfrac{\text{mortgage payment}}{\text{total}} = \dfrac{\$975}{\$4000} = \dfrac{39 \cdot 25}{160 \cdot 25} = \dfrac{39}{160}$

    The ratio is $\dfrac{39}{160}$.

30. $\dfrac{\text{food}}{\text{total}} = \dfrac{\$700}{\$4000} = \dfrac{7 \cdot 100}{40 \cdot 100} = \dfrac{7}{40}$

    The ratio is $\dfrac{7}{40}$.

31. The sector corresponding to Asia is 62%.
    62% of 61 = 0.62 · 61 ≈ 38
    Thus, 38 tall buildings are located in Asia.

32. The sector corresponding to North America is 29.5%.
    29.5% of 61 = 0.295 · 61 ≈ 18
    Thus, 18 tall buildings are located in North America.

33. The sector corresponding to Oceania is 1.6%.
    1.6% of 61 = 0.016 · 61 ≈ 1
    Thus, 1 tall building is located in Oceania.

34. The sector corresponding to Europe is 6.6%.
    6.6% of 61 = 0.066 · 61 ≈ 4
    Thus, 4 tall buildings are located in Europe.

35. Mean: $\dfrac{13 + 23 + 33 + 14 + 6}{5} = \dfrac{89}{5} = 17.8$

    Median: Write the numbers in order.
    6, 13, 14, 23, 33
    The middle number is 14.
    Mode: There is no mode, since each number occurs once.

36. Mean: $\dfrac{45 + 86 + 21 + 60 + 86 + 64 + 45}{7} = \dfrac{407}{7} \approx 58.1$

    Median: Write the numbers in order.
    21, 45, 45, 60, 64, 86, 86
    The middle number is 60.
    Mode: 45 and 86 are the mode, since they occur twice each.

37. Mean: $\dfrac{14{,}000 + 20{,}000 + 12{,}000 + 20{,}000 + 36{,}000 + 45{,}000}{6} = \dfrac{147{,}000}{6} = 24{,}500$

    Median: Write the numbers in order.
    12,000, 14,000, 20,000, 20,000, 36,000, 45,000

    The mean of the middle two numbers is $\dfrac{20{,}000 + 20{,}000}{2} = 20{,}000$.

    Mode: 20,000 is the mode, since it occurs twice.

38. Mean: $\dfrac{560+620+123+400+410+300+400+780+430+450}{10} = \dfrac{4473}{10} = 447.3$

Median: Write the numbers in order.
123, 300, 400, 400, 410, 430, 450, 560, 620, 780

The mean of the middle two numbers is $\dfrac{410+430}{2} = \dfrac{840}{2} = 420.$

Mode: 400 is the mode, since it occurs twice.

39. $\text{GPA} = \dfrac{4\cdot3+4\cdot3+2\cdot2+3\cdot3+2\cdot1}{3+3+2+3+1} = \dfrac{39}{12} = 3.25$

40. $\text{GPA} = \dfrac{3\cdot3+3\cdot4+2\cdot2+1\cdot2+3\cdot3}{3+4+2+2+3} = \dfrac{36}{14} \approx 2.57$

41.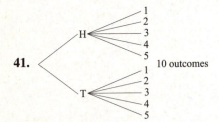

42.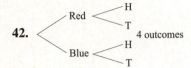

43.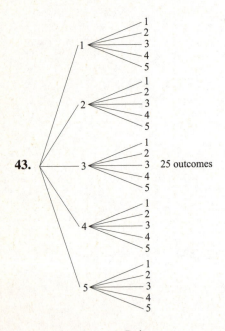

44.

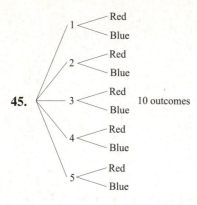

45. 10 outcomes

46. A 4 is one of the six possible outcomes. The probability is $\frac{1}{6}$.

47. A 3 is one of the six possible outcomes. The probability is $\frac{1}{6}$.

48. A 4 is one of the five possible outcomes. The probability is $\frac{1}{5}$.

49. A 3 is one of the five possible outcomes. The probability is $\frac{1}{5}$.

50. 1, 3, and 5 are three of the five possible outcomes. The probability of a 1, 3, or 5 is $\frac{3}{5}$.

51. 2 and 4 are two of the five possible outcomes. The probability of a 2 or 4 is $\frac{2}{5}$.

52. There are three even numbers on a standard die, 2, 4, and 6. They account for three of the six possible outcomes, so the probability of an even number is $\frac{3}{6} = \frac{1}{2}$.

53. There are three numbers greater than 3 on a standard die, 4, 5, and 6. They account for three of the six possible outcomes, so the probability of a number greater than 3 is $\frac{3}{6} = \frac{1}{2}$.

54. Mean: $\frac{73 + 82 + 95 + 68 + 54}{5} = \frac{372}{5} = 74.4$

Median: Write the numbers in order.
54, 68, 73, 82, 95
The middle number is 73.
Mode: There is no mode, since each number occurs once.

55. Mean: $\frac{25 + 27 + 32 + 98 + 62}{5} = \frac{244}{5} = 48.8$

Median: Write the numbers in order.
25, 27, 32, 62, 98
The middle number is 32.
Mode: There is no mode, since each number occurs once.

56. Mean:
$\frac{750 + 500 + 427 + 322 + 500 + 225}{6} = \frac{2724}{6}$
$= 454$

Median: Write the numbers in order.
225, 322, 427, 500, 500, 750
The mean of the middle two numbers is
$\frac{427 + 500}{2} = \frac{927}{2} = 463.5$
Mode: 500 is the mode, since it occurs twice.

57. Mean:
$\frac{952 + 327 + 566 + 814 + 327 + 729}{6} = \frac{3715}{6}$
≈ 619.17

Median: Write the numbers in order.
327, 327, 566, 729, 814, 952
The mean of the middle two numbers is
$\frac{566 + 729}{2} = \frac{1295}{2} = 647.5$.
Mode: 327 is the mode, since it occurs twice.

58. There are 2 blue marbles in a bag of 8 marbles, so the probability of choosing a blue marble is $\frac{2}{8} = \frac{1}{4}$.

59. There are 3 yellow marbles in a bag of 8 marbles, so the probability of choosing a yellow marble is $\frac{3}{8}$.

60. There are 2 red marbles in a bag of 8 marbles, so the probability of choosing a red marble is $\frac{2}{8} = \frac{1}{4}$.

61. There is 1 green marble in a bag of 8 marbles, so the probability of choosing a green marble is $\frac{1}{8}$.

Chapter 9 Test

1. There are $4\frac{1}{2}$ dollar symbols for the second week. Each dollar symbol corresponds to $50.

$$4\frac{1}{2} \cdot \$50 = \frac{9}{2} \cdot \$50 = \frac{\$450}{2} = \$225$$

$225 was collected during the second week.

2. Week 3 has more dollar symbols than any other week, so the most money was collected during week 3. There are 7 symbols for week 3, each representing $50, so the amount collected in week 3 was $7 \cdot \$50 = \$350$.

3. The number of dollar symbols for all 5 weeks is $3 + 4.5 + 7 + 5.5 + 2 = 22$. Each symbol represents $50, so the total amount collected was $22 \cdot \$50 = \$1100$.

4. The bars for June, August, and September extend above the 9 cm level, so the normal amount of precipitation is more than 9 cm during those months.

5. The shortest bar corresponds to February. The normal monthly precipitation in February in Chicago is 3 centimeters.

6. The tops of the bars for March and November are at the 7 cm level, so 7 cm of precipitation normally occurs in those months.

7.

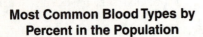

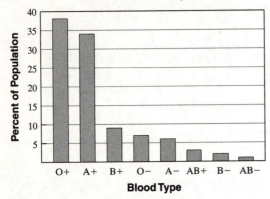

8. The point on the graph corresponding to 2003 is at about 2.25. Thus, the annual inflation rate in 2003 was about 2.25%.

9. The line graph is above the 3 level for 2000, 2005, 2006, and 2008. Thus the inflation rate was greater than 3% in 2000, 2005, 2006, and 2008.

10. Look for the sets of years where the line graph is increasing. During the sets 1998–1999, 1999–2000, 2002–2003, 2003–2004, 2004–2005, and 2007–2008, the inflation rate was increasing.

11. 85 of the 200 people prefer rock music, so the ratio is $\frac{85}{200} = \frac{17}{40}$.

12. 62 people prefer country music, and 44 prefer jazz, so the ratio is $\dfrac{62}{44} = \dfrac{31}{22}$.

13. The percent of visitors from Mexico was 12.3%.
$$12.3\% \text{ of } 50,500,000 = 0.123 \times 50,500,000$$
$$= 6,211,500$$
In 2008, approximately 6,211,500 foreign visitors were from Mexico.

14. The percent of visitors from Canada was 37.5%.
$$37.5\% \text{ of } 50,500,000 = 0.375 \times 50,500,000$$
$$= 18,937,500$$
In 2008, approximately 18,937,500 foreign visitors were from Canada.

15. The top of the bar marked $5'8'' - 5'11''$ is halfway between the levels marked 8 and 10, so 9 students are $5'8'' - 5'11''$ tall.

16. 5 students are $5'0'' - 5'3''$ tall and 6 students are $5'4'' - 5'7''$ tall, so $5 + 6 = 11$ students are $5'7''$ or shorter.

17.

Class Interval (Scores)	Tally	Class Frequency (Number of Students)
40–49	|	1
50–59	|||	3
60–69	||||	4
70–79	⫲	5
80–89	⫲ |||	8
90–99	||||	4

18.

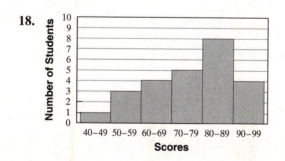

19. Mean: $\dfrac{26+32+42+43+49}{5} = \dfrac{192}{5} = 38.4$

Median: The numbers are already in order, and the middle number is 42.
Mode: There is no mode, since each number occurs once.

20. Mean:
$$\frac{8+10+16+16+14+12+12+13}{8} = \frac{101}{8} = 12.625$$
Median: Write the numbers in order.
8, 10, 12, 12, 13, 14, 16, 16
The mean of the middle two numbers is
$$\frac{12+13}{2} = \frac{25}{2} = 12.5.$$
Mode: 12 and 16 are the mode, since they occur twice.

21. $\text{GPA} = \dfrac{4 \cdot 3 + 3 \cdot 3 + 2 \cdot 3 + 3 \cdot 4 + 4 \cdot 1}{3+3+3+4+1} = \dfrac{43}{14} \approx 3.07$

22.

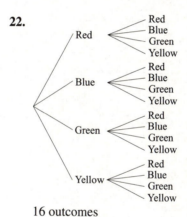

16 outcomes

23.

H —— H
H —— T
T —— H
T —— T

4 outcomes

24. 6 is one of the ten possible outcomes, so the probability of choosing a 6 is $\dfrac{1}{10}$.

25. A 3 or a 4 are two of the ten possible outcomes. The probability is $\dfrac{2}{10} = \dfrac{1}{5}$.

Cumulative Review Chapters 1–9

1. $(8-6)^2 + 2^3 \cdot 3 = 2^2 + 2^3 \cdot 3$
$$= 4 + 8 \cdot 3$$
$$= 4 + 24$$
$$= 28$$

2. $48 \div 8 \cdot 2 = 6 \cdot 2 = 12$

3. $\dfrac{30}{108} = \dfrac{6 \cdot 5}{6 \cdot 18} = \dfrac{5}{18}$

4. $\dfrac{19}{40} - \dfrac{3}{10} = \dfrac{19}{40} - \dfrac{12}{40} = \dfrac{7}{40}$

5. $\begin{array}{cc} 1\dfrac{4}{5} & 1\dfrac{8}{10} \\ & \\ 4 & 4 \\ +2\dfrac{1}{2} & +2\dfrac{5}{10} \\ \hline & 7\dfrac{13}{10} = 7 + 1\dfrac{3}{10} = 8\dfrac{3}{10} \end{array}$

6. $5\dfrac{1}{3} \cdot 2\dfrac{1}{8} = \dfrac{16}{3} \cdot \dfrac{17}{8} = \dfrac{2 \cdot 8 \cdot 17}{3 \cdot 8} = \dfrac{34}{3}$ or $11\dfrac{1}{3}$

7. $A = \dfrac{1}{2} \cdot b \cdot h = \dfrac{1}{2} \cdot 5.6 \text{ ft} \cdot 3 \text{ ft} = 8.4 \text{ sq ft}$

8. $P = 2 \cdot l + 2 \cdot w$
$$= 2 \cdot \left(3\dfrac{1}{2} \text{ m}\right) + 2 \cdot \left(1\dfrac{1}{2} \text{ m}\right)$$
$$= 7 \text{ m} + 3 \text{ m}$$
$$= 10 \text{ m}$$

9. $\dfrac{\$10}{\$15} = \dfrac{10}{15} = \dfrac{2}{3}$

10. 3 ft = 3 · 12 in. = 36 in., so the ratio is
$$\dfrac{14 \text{ in.}}{3 \text{ ft}} = \dfrac{14 \text{ in.}}{36 \text{ in.}} = \dfrac{14}{36} = \dfrac{7}{18}$$

11. $\dfrac{\$2160}{12 \text{ weeks}} = \dfrac{12 \cdot \$180}{12 \text{ weeks}} = \dfrac{\$180}{1 \text{ week}}$

12. $\dfrac{340 \text{ miles}}{5 \text{ hours}} = \dfrac{5 \cdot 68 \text{ miles}}{5 \text{ hours}} = \dfrac{68 \text{ miles}}{1 \text{ hour}}$

13. $\dfrac{360 \text{ miles}}{16 \text{ gallons}} = \dfrac{8 \cdot 45 \text{ miles}}{8 \cdot 2 \text{ gallons}} = \dfrac{45 \text{ miles}}{2 \text{ gallons}}$

14. $\dfrac{78 \text{ files}}{4 \text{ hours}} = \dfrac{2 \cdot 39 \text{ files}}{2 \cdot 2 \text{ hours}} = \dfrac{39 \text{ files}}{2 \text{ hours}}$

15. The cross products are $1\dfrac{1}{6} \cdot 4\dfrac{1}{2} = \dfrac{7}{6} \cdot \dfrac{9}{2} = \dfrac{63}{12} = \dfrac{21}{4}$
and $10\dfrac{1}{2} \cdot \dfrac{1}{2} = \dfrac{21}{2} \cdot \dfrac{1}{2} = \dfrac{21}{4}$. These are equal, so the proportion is true.

16. The cross products are $7.8 \cdot 2 = 15.6$ and $3 \cdot 5.2 = 15.6$. These are equal, so the proportion is true.

17. Let the standard dose for a 140-lb woman be n cubic centimeters.
$$\dfrac{n}{140} = \dfrac{4}{25}$$
$$25n = 4 \cdot 140$$
$$25n = 560$$
$$n = \dfrac{560}{25}$$
$$n = 22.5$$
The standard dose is 22.4 cc.

18. Let 7 inches on the map represent n miles in reality.
$$\dfrac{n}{7} = \dfrac{75}{2}$$
$$2n = 7 \cdot 75$$
$$2n = 525$$
$$n = \dfrac{525}{2}$$
$$n = 262.5$$
7 inches represents 262.5 miles.

19. $4.6\% = 4.6 \cdot 0.01 = 0.046$

20. $0.29\% = 0.29 \cdot 0.01 = 0.0029$

21. $190\% = 190 \cdot 0.01 = 1.9$

22. $452\% = 452 \cdot 0.01 = 4.52$

23. $40\% = 40 \cdot \dfrac{1}{100} = \dfrac{40}{100} = \dfrac{20 \cdot 2}{20 \cdot 5} = \dfrac{2}{5}$

24. $27\% = 27 \cdot \dfrac{1}{100} = \dfrac{27}{100}$

25. $33\dfrac{1}{3}\% = 33\dfrac{1}{3} \cdot \dfrac{1}{100} = \dfrac{100}{3} \cdot \dfrac{1}{100} = \dfrac{1}{3}$

26. $61\dfrac{1}{7}\% = 61\dfrac{1}{7} \cdot \dfrac{1}{100} = \dfrac{428}{7} \cdot \dfrac{1}{100} = \dfrac{428}{700} = \dfrac{107}{175}$

27. Let *n* be the unknown percent.
$5 = n \cdot 20$

28. The base is 20 and the amount is 5. The percent is unknown, call it *p*.
$$\dfrac{5}{20} = \dfrac{p}{100}$$

29. 7.5% of \$85.50 = 0.075 · \$85.50 = \$6.4125
The tax (rounded to the nearest cent) is \$6.41 and the total price is \$85.50 + \$6.41 = \$91.91.

30. 7% of \$23,000 = 0.07 · \$23,000 = \$1610
Her commission is \$1610.

31. simple interest = principal · rate · time
$$= \$2000 \cdot 0.10 \cdot 2$$
$$= \$400$$
total amount = principal + interest
$$= \$2000 + \$400$$
$$= \$2400$$
She will have \$2400 from her investment.

32. Mean $= \dfrac{28 + 35 + 40 + 32}{4} = \dfrac{135}{4} = 33.75$

33. 7 feet $= \dfrac{7 \text{ feet}}{1} \cdot \dfrac{1 \text{ yard}}{3 \text{ feet}} = \dfrac{7}{3}$ yards $= 2\dfrac{1}{3}$ yards

34. 2.5 tons $= \dfrac{2.5 \text{ tons}}{1} \cdot \dfrac{2000 \text{ lb}}{1 \text{ ton}}$
$$= 2.5 \cdot 2000 \text{ lb}$$
$$= 5000 \text{ lb}$$

35. 9 lb 6 oz ÷ 2 = 8 lb 22 oz ÷ 2 = 4 lb 11 oz

36. $1 + 3(10 - 8) = 1 + 3 \cdot 2 = 1 + 6 = 7$

37. 2.35 cg $= \dfrac{2.35 \text{ cg}}{1} \cdot \dfrac{1 \text{ g}}{100 \text{ cg}} = \dfrac{2.35}{100}$ g $= 0.0235$ g

38. 106 cm $= \dfrac{106 \text{ cm}}{1} \cdot \dfrac{10 \text{ mm}}{1 \text{ cm}}$
$$= 106 \cdot 10 \text{ mm}$$
$$= 1060 \text{ mm}$$

39. 3210 ml $= \dfrac{3210 \text{ ml}}{1} \cdot \dfrac{1 \text{ L}}{1000 \text{ ml}}$
$$= \dfrac{3210}{1000} \text{ L}$$
$$= 3.210 \text{ L}$$

40. 5 m $= \dfrac{5 \text{ m}}{1} \cdot \dfrac{100 \text{ cm}}{1 \text{ m}} = 5 \cdot 100 \text{ cm} = 500 \text{ cm}$

41. F = 1.8 · C + 32 = 1.8 · 15 + 32 = 27 + 32 = 59
15°C equals 59°F.

42. $C = \dfrac{5}{9} \cdot (F - 32) = \dfrac{5}{9} \cdot (41 - 32) = \dfrac{5}{9} \cdot 9 = 5$
41°F equals 5°C.

43. The complement of a 48° angle has measure
90° − 48° = 42°.

44. The supplement of a 48° angle has measure
180° − 48° = 132°.

45. $\sqrt{\dfrac{1}{36}} = \dfrac{1}{6}$ because $\dfrac{1}{6} \cdot \dfrac{1}{6} = \dfrac{1}{36}$.

46. $\sqrt{\dfrac{1}{25}} = \dfrac{1}{5}$ because $\dfrac{1}{5} \cdot \dfrac{1}{5} = \dfrac{1}{25}$.

47. 14 and 77 are the mode, since each of them occurs twice in the list, while the other numbers occur only once.

48. The numbers are already listed in order, and there are an odd number of them, so the median is the middle number, 56.

49. Tossing heads twice is one of the four equally likely outcomes of flipping a coin twice, so the probability is $\dfrac{1}{4}$.

50. Red marbles account for three of the five possible outcomes, so the probability of choosing a red marble is $\dfrac{3}{5}$.

Chapter 10

Section 10.1 Practice

1. a. If 0 represents the surface of the earth, then 3805 below the surface of the earth is −3805.

b. If zero degrees Fahrenheit is represented by 0°F, then 85 degrees below zero, Fahrenheit is represented by −85°F.

2.

3. 8 is to the right of −8, so 8 > −8.

4. −11 is to the left of 0, so −11 < 0.

5. −15 is to the left of −14, so −15 < −14.

6. 3 is to the right of −4.6, so 3 > −4.6.

7. 0 is to the right of −2, so 0 > −2.

8. −7.9 is to the left of 7.9, so −7.9 < 7.9.

9. $-\dfrac{3}{8}$ is to the right of $-1\dfrac{1}{7}$, so $-\dfrac{3}{8} > -1\dfrac{1}{7}$.

10. $|6| = 6$ because 6 is 6 units from 0.

11. $|-4| = 4$ because −4 is 4 units from 0.

12. $|0| = 0$ because 0 is 0 units from 0.

13. $\left|\dfrac{7}{8}\right| = \dfrac{7}{8}$ because $\dfrac{7}{8}$ is $\dfrac{7}{8}$ unit from 0.

14. $|-3.4| = 3.4$ because −3.4 is 3.4 units from 0.

15. The opposite of −7 is −(−7) or 7.

16. The opposite of 0 is −0 or 0.

17. The opposite of $\dfrac{11}{15}$ is $-\dfrac{11}{15}$.

18. The opposite of −9.6 is −(−9.6) or 9.6.

19. The opposite of 4 is −4.

20. −(−11) = 11

21. −|7| = −7

22. −|−2| = −2

23. The planet with the highest average temperature is the one that corresponds to the bar that extends the furthest in the positive direction (upward). Venus has the highest average temperature.

Vocabulary and Readiness Check

1. The numbers ...−3, −2, −1, 0, 1, 2, 3, ... are called <u>integers</u>.

2. Positive numbers, negative numbers, and zero, together are called <u>signed</u> numbers.

3. The symbols "<" and ">" are called <u>inequality</u> <u>symbols</u>.

4. Numbers greater than 0 are called <u>positive</u> numbers while numbers less than 0 are called <u>negative</u> numbers.

5. The sign "<" means <u>is less than</u> and ">" means <u>is greater than</u>.

6. On a number line, the greater number is to the <u>right</u> of the lesser number.

7. A number's distance from 0 on the number line is the number's <u>absolute value</u>.

8. The numbers −5 and 5 are called <u>opposites</u>.

Exercise Set 10.1

1. If 0 represents ground level, then 1235 feet underground is −1235.

3. If 0 represents sea level, then 14,433 feet above sea level is +14,433.

5. If 0 represents zero degrees Fahrenheit, then 118 degrees above zero is +118.

7. If 0 represents the surface of the ocean, then 13,000 feet below the surface of the ocean is −13,000.

9. If 0 represents a $0 change in annual sales, then an annual sales decrease of $49,584 is −49,584.

310

Copyright © 2011 Pearson Education, Inc. Publishing as Prentice Hall.

11. If 0 represents the surface of the ocean, then 160 feet below the surface is −160 and 147 feet below the surface is −147. Since −160 extends further in the negative direction, Guillermo is deeper.

13. If 0 represents a decrease of 0%, then a 9 percent decrease is −9.

15.

17.

19. The graph of 5 is to the left of 7 on a number line, so 5 is less than 7, or 5 < 7.

21. The graph of 4 is to the right of 0 on a number line so 4 is greater than 0, or 4 > 0.

23. The graph of −5 is to the right of −7 on a number line, so −5 is greater than −7, or −5 > −7.

25. The graph of 0 is to the right of −3 on a number line, so 0 is greater than −3, or 0 > −3.

27. The graph of −4.6 is to the left of −2.7 on a number line, so −4.6 is less than −2.7, or −4.6 < −2.7.

29. The graph of $-1\dfrac{3}{4}$ is to the left of 0 on a number line, so $-1\dfrac{3}{4}$ is less than 0, or $-1\dfrac{3}{4} < 0$.

31. The graph of $\dfrac{1}{4}$ is to the right of $-\dfrac{8}{11}$ on a number line, so $\dfrac{1}{4}$ is greater than $-\dfrac{8}{11}$, or $\dfrac{1}{4} > -\dfrac{8}{11}$.

33. The graph of $-2\dfrac{1}{2}$ is to the left of $-\dfrac{9}{10}$ on a number line, so $-2\dfrac{1}{2}$ is less than $-\dfrac{9}{10}$, or $-2\dfrac{1}{2} < -\dfrac{9}{10}$.

35. $|9| = 9$ since 9 is 9 units from 0.

37. $|-8| = 8$ since −8 is 8 units from 0.

39. $|0| = 0$ since 0 is 0 units from 0.

41. $|-5| = 5$ since −5 is 5 units from 0.

43. $|-8.1| = 8.1$ since −8.1 is 8.1 units from 0.

45. $\left|-\dfrac{1}{2}\right| = \dfrac{1}{2}$ since $-\dfrac{1}{2}$ is $\dfrac{1}{2}$ unit from 0.

47. $\left|-\dfrac{3}{8}\right| = \dfrac{3}{8}$ since $-\dfrac{3}{8}$ is $\dfrac{3}{8}$ unit from 0.

49. $|7.6| = 7.6$ since 7.6 is 7.6 units from 0.

51. The opposite of 5 is −5.

53. The opposite of −4 is −(−4) or 4.

55. The opposite of 23.6 is −23.6.

57. The opposite of $-\dfrac{9}{16}$ is $-\left(-\dfrac{9}{16}\right)$ or $\dfrac{9}{16}$.

59. The opposite of −0.7 is −(−0.7) or 0.7.

61. The opposite of $\dfrac{17}{18}$ is $-\dfrac{17}{18}$.

63. $|-7| = 7$

65. $-|20| = -20$

67. $-|-3| = -3$

69. $-(-8) = 8$

71. $|-14| = 14$

73. $-(-29) = 29$

75. If the number is 31, then the absolute value of 31 is 31 and the opposite of 31 is −31.

77. If the opposite of a number is 8.4, then the number is −8.4, and its absolute value is 8.4.

79. The bar that extends the farthest in the negative direction corresponds to the Caspian Sea, so the Caspian Sea has the lowest elevation.

81. The tallest bar on the graph corresponds to Lake Superior, so Lake Superior has the highest elevation.

83. The positive number on the graph closest to 100°C is 184°C, which corresponds to iodine.

85. The number on the graph closest to −200°C is −186°C, which corresponds to oxygen.

87. $0 + 13 = 13$

89.
$$\begin{array}{r} 15 \\ + 20 \\ \hline 35 \end{array}$$

91.
$$\begin{array}{r} 47 \\ 236 \\ + 77 \\ \hline 360 \end{array}$$

93. $2^2 = 2 \cdot 2 = 4$
$-|3| = -3$
$-(-5) = 5$
$-|-8| = -8$
In order from least to greatest:
$-|-8|, -|3|, 2^2, -(-5)$

95. $|-1| = 1$
$-|-6| = -6$
$-(-6) = 6$
$-|1| = -1$
In order from least to greatest:
$-|-6|, -|1|, |-1|, -(-6)$

97. $-(-2) = 2$
$5^2 = 5 \cdot 5 = 25$
$-10 = -10$
$-|-9| = -9$
$|-12| = 12$
In order from least to greatest:
$-10, -|-9|, -(-2), |-12|, 5^2$

99. a. $|0| = 0, 0 > 8$ false

    b. $|-5| = 5, 5 > 8$ false

    c. $|8| = 8, 8 > 8$ false

    d. $|-12| = 12, 12 > 8$ true

    Only d makes the statement true.

101. $-(-|-5|) = -(-5) = 5$

103. False; since $-1 > -2$ and -1 is not positive.

105. True; since every positive number is to the right of 0 on a number line and every negative number is to the left of 0, a positive number is always greater than a negative number.

107. False; since the opposite of a negative number is positive.

109. answers may vary

111. no; answers may vary.

Section 10.2 Practice

1.

$-5 + (-3) = -8$

2.

$5 + (-1) = 4$

3. $(-3) + (-9) = -12$

4. $-12 + (-3) = -15$

5. $9 + 5 = 14$

6. $-\dfrac{4}{15} + \left(-\dfrac{1}{15}\right) = -\dfrac{5}{15} = -\dfrac{1 \cdot 5}{3 \cdot 5} = -\dfrac{1}{3}$

7. $-8.3 + (-5.7) = -14$

8. $-3 + 9 = 6$

9. $2 + (-8) = -6$

10. $-46 + 20 = -26$

11. $8.6 + (-6.2) = 2.4$

12. $-\dfrac{3}{4} + \dfrac{1}{8} = -\dfrac{6}{8} + \dfrac{1}{8} = -\dfrac{5}{8}$

13. $-2 + 0 = -2 = -2$

14. $18 + (-18) = 0$

15. $-64 + 64 = 0$

16. $6 + (-2) + (-15) = 4 + (-15) = -11$

17. $5 + (-3) + 12 + (-14) = 2 + 12 + (-14)$
$$= 14 + (-14)$$
$$= 0$$

18. Temperature at 8 a.m. $= -7 + (+4) + (+7)$
$$= -3 + (+7)$$
$$= 4$$
The temperature was 4°F at 8 a.m.

Calculator Explorations

1. $-256 + 97 = -159$

2. $811 + (-1058) = -247$

3. $6(15) + (-46) = 44$

4. $-129 + 10(48) = 351$

5. $-108,650 + (-786,205) = -894,855$

6. $-196,662 + (-129,856) = -326,518$

Vocabulary and Readiness Check

1. If n is a number, then $-n + n = \underline{0}$.

2. Since $x + n = n + x$, we say that addition is <u>commutative</u>.

3. If a is a number, then $-(-a) = \underline{a}$.

4. Since $n + (x + a) = (n + x) + a$, we say that addition is <u>associative</u>.

Exercise Set 10.2

1.

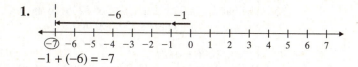

$-1 + (-6) = -7$

3.

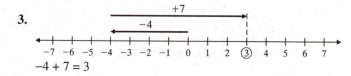

$-4 + 7 = 3$

5.

$-13 + 7 = -6$

7. $23 + 12 = 35$

9. $-6 + (-2) = -8$

11. $-43 + 43 = 0$

13. $6 + (-2) = 4$

15. $-6 + 8 = 2$

17. $3 + (-5) = -2$

19. $-2 + (-9) = -11$

21. $-7 + 7 = 0$

23. $12 + (-5) = 7$

25. $-6 + 3 = -3$

27. $-12 + 3 = -9$

29. $-12 + (-12) = -24$

31. $-25 + (-32) = -57$

33. $56 + (-26) = 30$

35. $-37 + 57 = 20$

37. $-123 + (-100) = -223$

39. $-42 + 193 = 151$

41. $-6.3 + (-2.2) = -8.5$

43. $-10.7 + 15.3 = 4.6$

45. $-\dfrac{7}{11} + \left(-\dfrac{1}{11}\right) = -\dfrac{8}{11}$

47. $-\dfrac{2}{3} + \left(-\dfrac{1}{6}\right) = -\dfrac{4}{6} + \left(-\dfrac{1}{6}\right) = -\dfrac{5}{6}$

49. $-\dfrac{4}{5} + \dfrac{1}{10} = -\dfrac{8}{10} + \dfrac{1}{10} = -\dfrac{7}{10}$

51. $-2\dfrac{1}{4} + 7\dfrac{7}{8} = -\dfrac{9}{4} + \dfrac{63}{8}$
$= -\dfrac{18}{8} + \dfrac{63}{8}$
$= \dfrac{45}{8}$ or $5\dfrac{5}{8}$

53. $-4 + 2 + (-5) = -2 + (-5) = -7$

55. $-5.2 + (-7.7) + (-11.7) = -12.9 + (-11.7)$
$= -24.6$

57. $12 + (-4) + (-4) + 12 = 8 + (-4) + 12$
$= 4 + 12$
$= 16$

59. $(-10) + 14 + 25 + (-16) = 4 + 25 + (-16)$
$= 29 + (-16)$
$= 13$

61. $-8 + 25 = 17$

63. $-31 + (-9) + 30 = -40 + 30 = -10$

65. $0 + (-165) + (-16) = -165 + (-16) = -181$
The diver is 181 feet below the surface.

67. Team 1:
$-2 + (-13) + 20 + 2 = -15 + 20 + 2 = 5 + 2 = 7$
Team 2: $5 + 11 + (-7) + (-3) = 16 + (-7) + (-3)$
$= 9 + (-3)$
$= 6$
Team 1 is the winning team.

69. Stanford:
$0 + 0 + 0 + (-1) + 0 + 0 + 0 + 1 + 0 = 0$
Wie:
$0 + 0 + 0 + (-1) + 0 + 0 + 0 + 0 + (-1) = -2$

71. Stanford:
$0 + 0 + 0 + (-1) + 0 + 0 + 0 + 1 + 0 + 1 + 0 + 0$
$+ (-1) + (-1) + (-1) + 0 + 0 + 0 = -2$
Wie:
$0 + 0 + 0 + (-1) + 0 + 0 + 0 + 0 + (-1) + 0 + 2$
$+ 0 + 0 + 0 + 0 + 0 + 1 + 0 = 1$

73. The bar for 2009 has a height of 8235, so the net income in 2009 was $8,235,000,000.

75. $1328 + 3496 = 4824$
The total net income for the years 2005 and 2007 was $4,824,000,000.

77. $-10 + 12 = 2$
The temperature at 11 p.m. was 2°C.

79. $-1786 + 15,395 = 13,609$
The sum of the net incomes for 2006 and 2007 is $13,609.

81. $-55 + 8 = -47$
West Virginia's record low temperature is -47°F.

83. $-10,924 + 3245 = -7679$
The depth of the Aleutian Trench is -7679 meters.

85. $44 - 0 = 44$

87. $76.1 - 4.09 = 72.01$

89. $\begin{array}{r} 200 \\ -\ 59 \\ \hline 141 \end{array}$

91. answers may vary

93. $7 + (-10) = -3$

95. $-10 + (-12) = -22$

97. True

99. False; for example, $4 + (-2) = 2 > 0$.

101. answers may vary

Section 10.3 Practice

1. $12 - 7 = 12 + (-7) = 5$

2. $-6 - 4 = -6 + (-4) = -10$

3. $11 - (-14) = 11 + 14 = 25$

4. $-9 - (-1) = -9 + 1 = -8$

5. $5 - 9 = 5 + (-9) = -4$

6. $-12 - 4 = -12 + (-4) = -16$

7. $-2 - (-7) = -2 + 7 = 5$

8. $-10.5 - 14.3 = -10.5 + (-14.3) = -24.8$

9. $\dfrac{5}{13} - \dfrac{12}{13} = \dfrac{5}{13} + \left(-\dfrac{12}{13}\right) = -\dfrac{7}{13}$

10. $-10 - 5 = -10 + (-5) = -15$

11. $\begin{aligned} -4 - 3 - 7 - (-5) &= -4 + (-3) + (-7) + 5 \\ &= -7 + (-7) + 5 \\ &= -14 + 5 \\ &= -9 \end{aligned}$

12. $\begin{aligned} 3 + (-5) - 6 - (-4) &= 3 + (-5) + (-6) + 4 \\ &= -2 + (-6) + 4 \\ &= -8 + 4 \\ &= -4 \end{aligned}$

13. $29,028 - (-1312) = 29,028 + 1312 = 30,340$
Mount Everest is 30,340 feet higher than Dead Sea.

Vocabulary and Readiness Check

1. It is true that $a - b = \underline{a + (-b)}$. b

2. The opposite of n is $\underline{-n}$. a

3. The expression $-10 - (-14)$ equals $\underline{-10 + 14}$. d

4. The expression $-5 - 10$ equals $\underline{-5 + (-10)}$. c

5. $5 - 5 = 0$

6. $7 - 7 = 0$

7. $6.2 - 6.2 = 0$

8. $1.9 - 1.9 = 0$

Exercise Set 10.3

1. $-5 - (-5) = -5 + 5 = 0$

3. $8 - 3 = 8 + (-3) = 5$

5. $3 - 8 = 3 + (-8) = -5$

7. $7 - (-7) = 7 + 7 = 14$

9. $-5 - (-8) = -5 + 8 = 3$

11. $-14 - 4 = -14 + (-4) = -18$

13. $2 - 16 = 2 + (-16) = -14$

15. $2.2 - 5.5 = 2.2 + (-5.5) = -3.3$

17. $3.62 - (-0.4) = 3.62 + 0.4 = 4.02$

19. $-\dfrac{3}{10} - \left(-\dfrac{7}{10}\right) = -\dfrac{3}{10} + \dfrac{7}{10} = \dfrac{4}{10} = \dfrac{2}{5}$

21. $\dfrac{2}{5} - \dfrac{7}{10} = \dfrac{2}{5} + \left(-\dfrac{7}{10}\right) = \dfrac{4}{10} + \left(-\dfrac{7}{10}\right) = -\dfrac{3}{10}$

23. $\dfrac{1}{2} - \left(-\dfrac{1}{3}\right) = \dfrac{1}{2} + \dfrac{1}{3} = \dfrac{3}{6} + \dfrac{2}{6} = \dfrac{5}{6}$

25. $-20 - 18 = -20 + (-18) = -38$

27. $-20 - (-3) = -20 + 3 = -17$

29. $2 - (-11) = 2 + 11 = 13$

31. $-21 + (-17) = -38$

33. $9 - 20 = 9 + (-20) = -11$

35. $-4.9 - 7.8 = -4.9 + (-7.8) = -12.7$

37. $\dfrac{4}{7} + \left(-\dfrac{1}{7}\right) = \dfrac{3}{7}$

39. $7 - 3 - 2 = 7 + (-3) + (-2) = 4 + (-2) = 2$

41. $\begin{aligned} 12 - 5 - 7 &= 12 + (-5) + (-7) \\ &= 7 + (-7) \\ &= 0 \end{aligned}$

43. $\begin{aligned} -5 - 8 - (-12) &= -5 + (-8) + 12 \\ &= -13 + 12 \\ &= -1 \end{aligned}$

45. $\begin{aligned} -10 + (-5) - 12 &= -10 + (-5) + (-12) \\ &= -15 + (-12) \\ &= -27 \end{aligned}$

47. $\begin{aligned} 12 - (-34) + (-6) &= 12 + 34 + (-6) \\ &= 46 + (-6) \\ &= 40 \end{aligned}$

49. $\begin{aligned} 19 &- 14 + (-6) + (-50) \\ &= 19 + (-14) + (-6) + (-50) \\ &= 5 + (-6) + (-50) \\ &= -1 + (-50) \\ &= -51 \end{aligned}$

51. The temperature in March is $11°F$ and in February is $-4°F$.
$11 - (-4) = 11 + 4 = 15$
The difference is $15°F$.

53. The two months with the lowest temperatures are January, $-10°F$, and December, $-6°F$.
$-6 - (-10) = -6 + 10 = 4$
The difference is $4°F$.

55. $136 - (-129) = 136 + 129 = 265$
Therefore, $136°F$ is $265°F$ warmer than $-129°F$.

57. Represent checks with negative numbers and deposits with positive numbers.
$\begin{aligned} 125 - 117 + 45 - 69 &= 125 + (-117) + 45 + (-69) \\ &= 8 + 45 + (-69) \\ &= 53 + (-69) \\ &= -16 \end{aligned}$
His balance is $-\$16$.

59. $\begin{aligned} -4 - 3 + 4 - 7 &= -4 + (-3) + 4 + (-7) \\ &= -7 + 4 + (-7) \\ &= -3 + (-7) \\ &= -10 \end{aligned}$
The temperature at 9 a.m. is $-10°C$.

61. $-282 - (-436) = -282 + 436 = 154$
The difference in elevation is 154 feet.

63. $-436 - (-505) = -436 + 505 = 69$
The difference in elevation is 69 feet.

65. $600 - (-52) = 600 + 52 = 652$
The difference in elevation is 652 feet.

67. $144 - 0 = 144$
The difference in elevation is 144 feet.

69. $867 - (-330) = 867 + 330 = 1197$
The difference in temperatures is $1197°F$.

71. $1287 - 2104 = 1287 + (-2104) = -817$
The U.S. trade balance in 2008 was $-\$817$ billion.

73. $8 \cdot 0 = 0$

75. $1 \cdot 8 = 8$

77.
$\begin{array}{r} 23 \\ \times\ 46 \\ \hline 138 \\ +\ 920 \\ \hline 1058 \end{array}$

79. answers may vary

81. $9 - (-7) = 9 + 7 = 16$

83. $10 - 30 = 10 + (-30) = -20$

85. $|-3| - |-7| = 3 - 7 = 3 + (-7) = -4$

87. $|-5| - |5| = 5 - 5 = 0$

89. $|-15| - |-29| = 15 - 29 = 15 + (-29) = -14$

91. $|-8 - 3| = |-8 + (-3)| = |-11| = 11$
$8 - 3 = 8 + (-3) = 5$
Since $11 \neq 5$, the statement is false.

93. answers may vary

Integrated Review

1. If 0 represents sea level, then 29,028 feet above sea level can be represented by +29,028.

2. If 0 represents sea level, then 35,840 feet below sea level can be represented by −35,840.

3. If 0 represents sea level, then 7 miles below sea level can be represented by −7.

4. If 0 represents sea level, then 28,250 feet above sea level can be represented by +28,250.

5.

6. The graph of 0 is to the right of −3 on a number line, so 0 is greater than −3, or $0 > -3$.

7. The graph of 7 is to the right of 0 on a number line, so 7 is greater than 0, or $7 > 0$.

8. The graph of $-1\frac{1}{2}$ is to the left of $1\frac{1}{4}$ on a number line, so $-1\frac{1}{2}$ is less than $1\frac{1}{4}$, or $-1\frac{1}{2} < 1\frac{1}{4}$.

9. The graph of −2 is to the right of −7.6 on a number line, so −2 is greater than −7.6, or $-2 > -7.6$.

10. The graph of −15 is to the left of −5 on a number line, so −15 is less than −5, or $-15 < -5$.

11. The graph of −4 is to the right of −40 on a number line, so −4 is greater than −40, or $-4 > -40$.

12. The graph of 3.9 is to the right of −3.9 on a number line, so 3.9 is greater than −3.9, or $3.9 > -3.9$.

13. The graph of $-\frac{1}{2}$ is to the left of $-\frac{1}{3}$ on a number line, so $-\frac{1}{2}$ is less than $-\frac{1}{3}$, or $-\frac{1}{2} < -\frac{1}{3}$.

14. $|-1| = 1$ since −1 is 1 unit from 0.

15. $|1| = 1$ since 1 is 1 unit from 0.

16. $|0| = 0$

17. $-|-4| = -4$

18. $|-8.6| = 8.6$

19. $|100.3| = 100.3$

20. $-\left(-\frac{3}{4}\right) = \frac{3}{4}$

21. $-\left|-\frac{3}{4}\right| = -\frac{3}{4}$

22. The opposite of 6 is −6.

23. The opposite of −13 is −(−13) or 13.

24. The opposite of 89.1 is −89.1.

25. The opposite of $-\frac{2}{9}$ is $-\left(-\frac{2}{9}\right)$ or $\frac{2}{9}$.

26. $-7 + 12 = 5$

27. $-9.2 + (-11.6) = -20.8$

28. $\frac{5}{9} + \left(-\frac{1}{3}\right) = \frac{5}{9} + \left(-\frac{3}{9}\right) = \frac{2}{9}$

29. $1 - 3 = 1 + (-3) = -2$

30. $\frac{3}{8} - \left(-\frac{3}{8}\right) = \frac{3}{8} + \frac{3}{8} = \frac{6}{8} = \frac{3}{4}$

31. $-2.6 - 1.4 = -2.6 + (-1.4) = -4.0$ or −4

32. $-7 + (-2.6) = -9.6$

33. $-14 + 8 = -6$

34. $-8 - (-20) = -8 + 20 = 12$

35. $18 - (-102) = 18 + 102 = 120$

36. $8.65 - 12.09 = 8.65 + (-12.09) = -3.44$

37. $\begin{aligned} -\frac{4}{5} - \frac{3}{10} &= -\frac{4}{5} + \left(-\frac{3}{10}\right) \\ &= -\frac{8}{10} + \left(-\frac{3}{10}\right) \\ &= -\frac{11}{10} \text{ or } -1\frac{1}{10} \end{aligned}$

38. $-8 + (-6) + 20 = -14 + 20 = 6$

39. $-11 - 7 - (-19) = -11 + (-7) + 19 = -18 + 19 = 1$

40. $\begin{aligned} -4 + (-8) - 16 - (-9) &= -4 + (-8) + (-16) + 9 \\ &= -12 + (-16) + 9 \\ &= -28 + 9 \\ &= -19 \end{aligned}$

41. $14 - 26 = 14 + (-26) = -12$

42. $-12 - (-8) = -12 + 8 = -4$

43. $-17 + (-27) = -44$

44. $-3.6 + 2.1 = -1.5$

45. $122 - (-67) = 122 + 67 = 189$
It is 189°F warmer.

46. $134 - (-80) = 134 + 80 = 214$
It is 214°F warmer.

Section 10.4 Practice

1. $-2 \cdot 6 = -12$

2. $-4(-3) = 12$

3. $0 \cdot (-10) = 0$

4. $\left(\frac{3}{7}\right)\left(-\frac{1}{3}\right) = -\frac{1}{7}$

5. $-4(-3.2) = 12.8$

6. $7(-2)(-4) = -14(-4) = 56$

7. $(-5)(-6)(-1) = 30(-1) = -30$

8. $(-2)(-5)(-6)(-1) = 10(-6)(-1) = -60(-1) = 60$

9. $\begin{aligned} (-3)^4 &= (-3)(-3)(-3)(-3) \\ &= 9(-3)(-3) \\ &= -27(-3) \\ &= 81 \end{aligned}$

10. $-9^2 = -(9 \cdot 9) = -81$

11. $\frac{28}{-7} = -4$

12. $-18 \div (-2) = 9$

13. $\frac{-4.6}{0.2} = -23$

14. $\frac{3}{5} \div \left(-\frac{2}{7}\right) = \frac{3}{5} \cdot \left(-\frac{7}{2}\right) = -\frac{3 \cdot 7}{5 \cdot 2} = -\frac{21}{10} \text{ or } -2\frac{1}{10}$

15. $\frac{-1}{0}$ is undefined because there is no number that gives a product of -1 when multiplied by 0.

16. $\frac{0}{-2} = 0$ because $0 \cdot (-2) = 0$.

17. total score $= 4 \cdot (-13) = -52$
The card player's total score was -52.

Vocabulary and Readiness Check

1. The product of a negative number and a positive number is a <u>negative</u> number.

2. The product of two negative numbers is a <u>positive</u> number.

3. The quotient of two negative numbers is a <u>positive</u> number.

4. The quotient of a negative number and a positive number is a <u>negative</u> number.

5. The product of a negative number and zero is <u>0</u>.

6. The quotient of 0 and a negative number is <u>0</u>.

7. The quotient of a negative number and 0 is <u>undefined</u>.

Exercise Set 10.4

1. $-2(-3) = 6$

3. $-4(9) = -36$

5. $(2.6)(-1.2) = -3.12$

7. $0(-14) = 0$

9. $-\dfrac{3}{5}\left(-\dfrac{2}{7}\right) = \dfrac{3 \cdot 2}{5 \cdot 7} = \dfrac{6}{35}$

11. $6(-4)(2) = -24(2) = -48$

13. $(-1)(-2)(-4) = 2(-4) = -8$

15. $-4(4)(-5) = -16(-5) = 80$

17. $\begin{aligned} 10(-5)(-1)(-3) &= -50(-1)(-3) \\ &= 50(-3) \\ &= -150 \end{aligned}$

19. $(-2)^2 = (-2)(-2) = 4$

21. $-5^2 = -(5 \cdot 5) = -25$

23. $(-3)^3 = (-3)(-3)(-3) = 9(-3) = -27$

25. $-2^3 = -(2 \cdot 2 \cdot 2) = -8$

27. $(-9)^2 = (-9)(-9) = 81$

29. $\left(-\dfrac{5}{11}\right)^2 = \left(-\dfrac{5}{11}\right)\left(-\dfrac{5}{11}\right) = \dfrac{25}{121}$

31. $-24 \div 6 = -4$ because $-4 \cdot 6 = -24$.

33. $\dfrac{-30}{6} = -5$ because $6 \cdot (-5) = -30$.

35. $\dfrac{-88}{-11} = 8$ because $8 \cdot (-11) = -88$.

37. $\dfrac{0}{14} = 0$ because $14 \cdot 0 = 0$.

39. $\dfrac{39}{-3} = -13$ because $-13 \cdot (-3) = 39$.

41. $\dfrac{7.8}{-0.3} = -26$ because $-0.3 \cdot (-26) = 7.8$.

43. $\begin{aligned} -\dfrac{5}{12} \div \left(-\dfrac{1}{5}\right) &= -\dfrac{5}{12} \cdot \left(-\dfrac{5}{1}\right) \\ &= \dfrac{5 \cdot 5}{12 \cdot 1} \\ &= \dfrac{25}{12} \text{ or } 2\dfrac{1}{12} \end{aligned}$

45. $\dfrac{100}{-20} = -5$ because $-20 \cdot (-5) = 100$

47. $450 \div (-9) = -50$ because $-50 \cdot (-9) = 450$.

49. $\dfrac{-12}{-4} = 3$ because $-4 \cdot 3 = -12$.

51. $\dfrac{-120}{0.4} = -300$ because $-300 \cdot 0.4 = -120$.

53. $-\dfrac{8}{15} \div \dfrac{2}{3} = -\dfrac{8}{15} \cdot \dfrac{3}{2} = -\dfrac{8 \cdot 3}{15 \cdot 2} = -\dfrac{24}{30} = -\dfrac{4}{5}$

55. $-12(0) = 0$

57. $(-5)^3 = (-5)(-5)(-5) = 25(-5) = -125$

59. $\dfrac{430}{-8.6} = -50$ because $-50 \cdot (-8.6) = 430$.

61. $\begin{aligned} -1(2)(7)(-3.1) &= -2(7)(-3.1) \\ &= -14(-3.1) \\ &= 43.4 \end{aligned}$

63. $\dfrac{4}{3}\left(-\dfrac{8}{9}\right) = -\dfrac{4 \cdot 8}{3 \cdot 9} = -\dfrac{32}{27}$

65. $\dfrac{4}{3} \div \left(-\dfrac{8}{9}\right) = \dfrac{4}{3} \cdot \left(-\dfrac{9}{8}\right) = -\dfrac{4 \cdot 9}{3 \cdot 8} = -\dfrac{36}{24} = -\dfrac{3}{2}$

67. $\dfrac{-54}{9} = -6$

The quotient of -54 and 9 is -6.

69. $\begin{array}{r} 42 \\ \times\ 6 \\ \hline 252 \end{array}$

$-42(-6) = 252$

The product of -42 and -6 is 252.

71. A loss of 4 yards is represented by –4.
$3 \cdot (-4) = -12$
The team had a total loss of 12 yards.

73. Each move of 20 feet down is represented by –20.
$5 \cdot (-20) = -100$
The diver is at a depth of 100 feet.

75. $3 \cdot (-70) = -210$
The melting point of nitrogen is –210°C.

77. $-3 \cdot 63 = -189$
The melting point of argon is –189°C.

79. $4 \cdot (-183) = -732$
Eastman Kodak's net income would have been
–$732 million after four years.

81. a. $275 - 27 = 248$
The number of California condors changed
by 248 from 1987 to 2009.

b. 1987 to 2009 is 22 years.
$\dfrac{248}{22} \approx 11$
The average change was 11 condors per
year.

83. $(11-8) \cdot (200-100) = (3) \cdot (100) = 300$

85. $(3 \cdot 5)^2 = (15)^2 = 15 \cdot 15 = 225$

87. $90 + 12^2 - 5^3 = 90 + 144 - 125$
$= 234 - 125$
$= 109$

89. $12 \div 4 - 2 + 7 = 3 - 2 + 7$
$= 1 + 7$
$= 8$

91. $-87 \div 3 = -29$

93. $-9 - 10 = -9 + (-10) = -19$

95. $-4 - 15 - (-11) = -4 + (-15) + 11$
$= -19 + 11$
$= -8$

97. False; since the product of two numbers having
the same sign is positive, the product of two
negative numbers is always a positive number.

99. True; since the product of a positive number and
a negative number is negative.

101. The product of an odd number of negative
numbers is negative, so the product of seven
negative numbers is negative.

103. $-\dfrac{20}{41} = \dfrac{-20}{41} = \dfrac{20}{-41}$

105. Note that a negative number raised to an even
power is positive, while a negative number
raised to an odd power is negative. In order from
least to greatest, the numbers are:
$(-5)^{17},\ (-2)^{17},\ (-2)^{12},\ (-5)^{12}$

107. answers may vary

Section 10.5 Practice

1. $20 + 50 + (-4)^3 = 20 + 50 + (-64)$
$= 70 + (-64)$
$= 6$

2. $\dfrac{25}{5(-1)} = \dfrac{25}{-5} = -5$

3. $\dfrac{-18 + 6}{-3 - 1} = \dfrac{-12}{-4} = 3$

4. $-2^3 + (-4)^2 + 1^5 = -8 + 16 + 1 = 8 + 1 = 9$

5. $2(2 - 8) + (-12) - \sqrt{9} = 2(-6) + (-12) - 3$
$= -12 + (-12) - 3$
$= -24 - 3$
$= -27$

6. $(-5) \cdot |-4| + (-3) + 2^3 = -5 \cdot 4 + (-3) + 2^3$
$= -5 \cdot 4 + (-3) + 8$
$= -20 + (-3) + 8$
$= -23 + 8$
$= -15$

7. $-4[-2 + 5(-3 + 5)] - 7.9 = -4[-2 + 5(2)] - 7.9$
$= -4[-2 + 10] - 7.9$
$= -4(8) - 7.9$
$= -32 - 7.9$
$= -39.9$

8.
$$\frac{5}{8} \div \left(\frac{1}{4} - \frac{3}{4} \right) = \frac{5}{8} \div \left(-\frac{2}{4} \right)$$
$$= \frac{5}{8} \div \left(-\frac{1}{2} \right)$$
$$= \frac{5}{8} \cdot \left(-\frac{2}{1} \right)$$
$$= -\frac{5 \cdot 2}{8 \cdot 1}$$
$$= -\frac{5}{4}$$

9. average

$$= \frac{\text{sum of numbers}}{\text{number of numbers}}$$
$$= \frac{15 + (-1) + (-11) + (-14) + (-16) + (-14) + (-1)}{7}$$
$$= \frac{-42}{7}$$
$$= -6$$

The average of the temperatures is –6°F.

Calculator Explorations

1. $\dfrac{-120 - 360}{-10} = 48$

2. $\dfrac{4750}{-2 + (-17)} = -250$

3. $\dfrac{-316 + (-458)}{28 + (-25)} = -258$

4. $\dfrac{-234 + 86}{-18 + 16} = 74$

Vocabulary and Readiness Check

1. To simplify $-2 \div 2 \cdot (3)$ which operation should be performed first? <u>division</u>

2. To simplify $-9 - 3 \cdot 4$, which operation should be performed first? <u>multiplication</u>

3. The <u>average</u> of a list of numbers is
$$\frac{\text{sum of numbers}}{\textit{number} \text{ of numbers}}.$$

4. To simplify $5[-9 + (-3)] \div 4$, which operation should be performed first? <u>addition</u>

5. To simplify $-2 + 3(10 - 12) \cdot (-8)$, which operation should be performed first? <u>subtraction</u>

Exercise Set 10.5

1. $(-1)(-2) + 1 = 2 + 1 = 3$

3. $\dfrac{-8(-5)}{10} = \dfrac{40}{10} = 4$

5. $9 - 12 - 4 = -3 - 4 = -7$

7. $4 + 3(-6) = 4 + (-18) = -14$

9.
$$\frac{4}{9} \left(\frac{2}{10} - \frac{7}{10} \right) = \frac{4}{9} \left(-\frac{5}{10} \right)$$
$$= \frac{4}{9} \left(-\frac{1}{2} \right)$$
$$= -\frac{4 \cdot 1}{9 \cdot 2}$$
$$= -\frac{2}{9}$$

11. $-10 + 4 \div 2 = -10 + 2 = -8$

13. $25 \div (-5) + \sqrt{81} = 25 \div (-5) + 9$
$$= -5 + 9$$
$$= 4$$

15. $\dfrac{-|-5| + 3}{7 - 9} = \dfrac{-5 + 3}{7 - 9} = \dfrac{-2}{-2} = 1$

17. $\dfrac{24}{10 + (-4)} = \dfrac{24}{6} = 4$

19. $5(-3) - (-12) = -15 - (-12)$
$$= -15 + 12$$
$$= -3$$

21. $(-19) - 12(3) = (-19) - 36 = -55$

23. $\dfrac{-23 + 7}{14 + (-10)} = \dfrac{-16}{4} = -4$

25. $[8 + (-4)]^2 = [4]^2 = 16$

27. $3^3 - 12 = 27 - 12 = 15$

29. $(3 - 12) \div 3 = (-9) \div 3 = -3$

31. $5 + 2^3 - 4^2 = 5 + 8 - 16 = 13 - 16 = -3$

33. $(5-9)^2 \div (4-2)^2 = (-4)^2 \div (2)^2$
$$= 16 \div 4$$
$$= 4$$

35. $|8 - 24| \cdot (-2) \div (-2) = |-16| \cdot (-2) \div (-2)$
$$= 16 \cdot (-2) \div (-2)$$
$$= -32 \div (-2)$$
$$= 16$$

37. $(-12 - 20) \div 16 - 25 = (-32) \div 16 - 25$
$$= -2 - 25$$
$$= -27$$

39. $5(5-2) + (-5)^2 - \sqrt{36} = 5(5-2) + 25 - 6$
$$= 5(3) + 25 - 6$$
$$= 15 + 25 - 6$$
$$= 40 - 6$$
$$= 34$$

41. $(0.2 - 0.7)(0.6 - 1.9) = (-0.5)(-1.3)$
$$= 0.65$$

43. $2 - 7 \cdot 6 - 19 = 2 - 42 - 19 = -40 - 19 = -59$

45. $(-36 \div 6) - (4 \div 4) = (-6) - (1) = -7$

47. $\left(\dfrac{1}{2}\right)^2 - \left(\dfrac{1}{3}\right)^2 = \dfrac{1}{4} - \dfrac{1}{9} = \dfrac{9}{36} - \dfrac{4}{36} = \dfrac{5}{36}$

49. $(-5)^2 - 6^2 = 25 - 36 = -11$

51. $(10 - 4^2)^2 = (10 - 16)^2 = (-6)^2 = 36$

53. $2(8 - 10)^2 - 5(1 - 6)^2 = 2(-2)^2 - 5(-5)^2$
$$= 2(4) - 5(25)$$
$$= 8 - 125$$
$$= -117$$

55. $3(-10) \div [5(-3) - 7(-2)]$
$$= 3(-10) \div [-15 + 14]$$
$$= 3(-10) \div (-1)$$
$$= -30 \div (-1)$$
$$= 30$$

57. $\dfrac{(-7)(-3) - 4(3)}{3[7 \div (3-10)]} = \dfrac{21 - 12}{3[7 \div (-7)]}$
$$= \dfrac{9}{3[-1]}$$
$$= \dfrac{9}{-3}$$
$$= -3$$

59. $(0.2)^2 - (1.5)^2 = 0.04 - 2.25 = -2.21$

61. $-3^2 + (-2)^3 - 4^2 = -9 + (-8) - 16$
$$= -17 - 16$$
$$= -33$$

63. $\dfrac{|8-13| + \sqrt{81}}{-5(3) - (-5)} = \dfrac{|-5| + 9}{-15 - (-5)}$
$$= \dfrac{5 + 9}{-15 + 5}$$
$$= \dfrac{14}{-10}$$
$$= -\dfrac{7}{5}$$

65. $\dfrac{-10 + 8 + (-4) + 2 + 7 + (-5) + (-12)}{7} = \dfrac{-14}{7} = -2$

The average is -2.

67. $-17 + (-26) + (-20) + (-13) = \dfrac{-76}{4} = -19$

The average is -19.

69. $\text{average} = \dfrac{-11 + (-14) + (-16) + (-14)}{4}$
$$= \dfrac{-55}{4}$$
$$= -13.75$$

The average temperature for the months December through March is $-13.75°F$.

71. no; answers may vary

73.
$$\begin{array}{r} 45 \\ \times\ 90 \\ \hline 4050 \end{array}$$

75.
$$\begin{array}{r} 90 \\ -\ 45 \\ \hline 45 \end{array}$$

77. $8 + 8 + 8 + 8 = 32$
The perimeter is 32 inches.

79. $9 + 6 + 9 + 6 = 30$
The perimeter is 30 feet.

81. $2 \cdot (7 - 5) \cdot 3 = 2 \cdot 2 \cdot 3 = 4 \cdot 3 = 12$

83. $-6 \cdot (10 - 4) = -6 \cdot 6 = -36$

85. answers may vary

87. answers may vary

89. $(-12)^4 = (-12)(-12)(-12)(-12) = 20,736$

Chapter 10 Vocabulary Check

1. Two numbers that are the same distance from 0 on the number line but are on opposite sides of 0 are called <u>opposites</u>.

2. Together, positive numbers, negative numbers, and 0 are called <u>signed</u> numbers.

3. The <u>absolute value</u> of a number is that number's distance from 0 on a number line.

4. The <u>integers</u> are ..., $-3, -2, -1, 0, 1, 2, 3, ...$

5. The opposite of a number is also called its <u>additive inverse</u>.

6. The <u>negative</u> numbers are numbers less than zero.

7. The <u>positive</u> numbers are numbers greater than zero.

8. The symbols "<" and ">" are called <u>inequality symbols</u>.

9. The <u>average</u> of a list of numbers is $\dfrac{\text{sum of numbers}}{\textit{number} \text{ of numbers}}$.

10. The sign "<" means <u>is less than</u> and ">" means <u>is greater than</u>.

Chapter 10 Review

1. If 0 represents the surface of Earth, then 1435 feet below the surface can be represented by -1435.

2. If 0 represents sea level, then 7562 meters above sea level can be represented by $+7562$.

3.

4.

5. The graph of -18 is to the right of -20 on a number line, so -18 is greater than -20, or $-18 > -20$.

6. The graph of $-5\frac{1}{3}$ is to the left of $-4\frac{7}{8}$ on a number line, so $-5\frac{1}{3}$ is less than $-4\frac{7}{8}$, or $-5\frac{1}{3} < -4\frac{7}{8}$.

7. The graph of -12.3 is to the right of -19.8 on a number line, so -12.3 is greater than -19.8, or $-12.3 > -19.8$.

8. $|-12| = 12$ because -12 is 12 units from 0.

9. $|0| = 0$ because 0 is 0 units from 0.

10. $\left|-\dfrac{7}{8}\right| = \dfrac{7}{8}$ because $-\dfrac{7}{8}$ is $\dfrac{7}{8}$ unit from 0.

11. The opposite of -12 is $-(-12)$ or 12.

12. The opposite of $\frac{1}{2}$ is $-\frac{1}{2}$.

13. $-(-3.9) = 3.9$

14. $-|-7| = -7$

15. True; a negative number is always to the left of a positive number on a number line.

16. True; the absolute value represents how many units the number is from 0.

17. $5 + (-3) = 2$

18. $18 + (-4) = 14$

19. $-12 + 16 = 4$

20. $-23 + 40 = 17$

21. $-8 + (-15) = -23$

22. $-5 + (-17)\ -22$

23. $-2.4 + 0.3 = -2.1$

24. $-8.9 + 1.9 = -7.0$ or -7

25. $\frac{2}{3} + \left(-\frac{2}{5}\right) = \frac{10}{15} + \left(-\frac{6}{15}\right) = \frac{4}{15}$

26. $-\frac{8}{9} + \frac{1}{3} = -\frac{8}{9} + \frac{3}{9} = -\frac{5}{9}$

27. $-43 + (-108) = -151$

28. $-100 + (-506) = -606$

29. Represent a fall in temperature by a negative number.
$-15 + (-5) = -20$
The temperature was $-20°C$ at 6 a.m.

30. Represent depths below the surface by negative numbers.
$-127 + (-23) = -150$
The diver's current depth is -150 feet.

31. $0 + 1 + (-1) + 0 = 0$
Her total score for the tournament was 0.

32. $-15 + 3 = -12$
Charles Howell III's score was -12.

33. $4 - 12 = 4 + (-12) = -8$

34. $-12 - 14 = -12 + (-14) = -26$

35. $-\frac{3}{4} - \frac{1}{2} = -\frac{3}{4} + \left(-\frac{2}{4}\right) = -\frac{5}{4}$ or $-1\frac{1}{4}$

36. $-\frac{2}{5} - \frac{7}{10} = -\frac{2}{5} + \left(-\frac{7}{10}\right)$
$= -\frac{4}{10} + \left(-\frac{7}{10}\right)$
$= -\frac{11}{10}$ or $-1\frac{1}{10}$

37. $7 - (-13) = 7 + 13 = 20$

38. $-6 - (-14) = -6 + 14 = 8$

39. $-(-5) - 12 - (-3) = 5 + (-12) + 3 = -7 + 3 = -4$

40. $-16 - 16 - (-4) - (-12) = -16 + (-16) + 4 + 12$
$= -32 + 4 + 12$
$= -28 + 12$
$= -16$

41. $-1.7 - (-2.9) = -1.7 + 2.9 = 1.2$

42. $-0.5 - (-1.2) = -0.5 + 1.2 = 0.7$

43. $\frac{4}{9} - \frac{7}{9} - \left(-\frac{1}{18}\right) = \frac{4}{9} + \left(-\frac{7}{9}\right) + \frac{1}{18}$
$= -\frac{3}{9} + \frac{1}{18}$
$= -\frac{6}{18} + \frac{1}{18}$
$= -\frac{5}{18}$

44. $\frac{3}{7} - \frac{5}{7} - \left(-\frac{1}{14}\right) = \frac{3}{7} + \left(-\frac{5}{7}\right) + \frac{1}{14}$
$= -\frac{2}{7} + \frac{1}{14}$
$= -\frac{4}{14} + \frac{1}{14}$
$= -\frac{3}{14}$

45. Represent checks as negative numbers and deposits as positive numbers.
$$142 - 125 + 43 - 85 = 142 + (-125) + 43 + (-85)$$
$$= 17 + 43 + (-85)$$
$$= 60 + (-85)$$
$$= -25$$
The balance is $-\$25$.

46. Lake Superior (600 ft) and Caspian Sea (−92).
$600 - (-92) = 600 + 92 = 692$
The difference in elevations is +692 feet.

47. $(-11)^2 = (-11)(-11) = 121$

48. $(-3)^4 = (-3)(-3)(-3)(-3)$
$$= 9(-3)(-3)$$
$$= -27(-3)$$
$$= 81$$

49. $-3^4 = -(3 \cdot 3 \cdot 3 \cdot 3) = -(9 \cdot 3 \cdot 3) = -(27 \cdot 3) = -81$

50. $-11^2 = -(11 \cdot 11) = -121$

51. $\left(-\dfrac{4}{5}\right)^3 = \left(-\dfrac{4}{5}\right)\left(-\dfrac{4}{5}\right)\left(-\dfrac{4}{5}\right) = -\dfrac{4 \cdot 4 \cdot 4}{5 \cdot 5 \cdot 5} = -\dfrac{64}{125}$

52. $\left(-\dfrac{2}{3}\right)^3 = \left(-\dfrac{2}{3}\right)\left(-\dfrac{2}{3}\right)\left(-\dfrac{2}{3}\right) = -\dfrac{2 \cdot 2 \cdot 2}{3 \cdot 3 \cdot 3} = -\dfrac{8}{27}$

53. $(-3)(-7) = 21$

54. $-6 \cdot 3 = -18$

55. $\dfrac{-78}{3} = -26$

56. $\dfrac{38}{-2} = -19$

57. $(-4)(-1)(-9) = 4(-9) = -36$

58. $(-2)(-2)(-10) = 4(-10) = -40$

59. $-\dfrac{2}{3} \cdot \dfrac{7}{8} = -\dfrac{14}{24} = -\dfrac{7}{12}$

60. $-0.5(-12) = 6$

61. $-2 \div 0$ is undefined.

62. $0 \div (-61) = 0$

63. $\dfrac{-0.072}{-0.06} = 1.2$

64. $-\dfrac{8}{25} \div \left(-\dfrac{2}{5}\right) = -\dfrac{8}{25} \cdot \left(-\dfrac{5}{2}\right) = \dfrac{8 \cdot 5}{25 \cdot 2} = \dfrac{4}{5}$

65. $(-5)(2) = -10$
The total loss is 10 yards.

66. $\dfrac{-20 + (-17) + 5 + 9 + (-12)}{5} = \dfrac{-35}{5} = -7$
The average is −7.

67. $-10 + 3(-2) = -10 + (-6) = -16$

68. $3(-12) - \sqrt{64} = -36 - 8 = -44$

69. $5.2 + \sqrt{36} \div (-0.3) = 5.2 + 6 \div (-0.3)$
$$= 5.2 + (-20)$$
$$= -14.8$$

70. $-6 + (-10) \div (-0.2) = -6 + 50 = 44$

71. $|-16| + (-3) \cdot 12 \div 4 = 16 + (-3) \cdot 12 \div 4$
$$= 16 + (-36) \div 4$$
$$= 16 + (-9)$$
$$= 7$$

72. $-(-4) \cdot |-3| - 5 = -(-4) \cdot 3 - 5$
$$= 4 \cdot 3 - 5$$
$$= 12 - 5$$
$$= 7$$

73. $4^3 - (8-3)^2 = 4^3 - 5^2 = 64 - 25 = 39$

74. $2^3 - (9-2)^2 = 2^3 - 7^2 = 8 - 49 = -41$

75. $\left(-\dfrac{1}{3}\right)^2 - \dfrac{8}{3} = \dfrac{1}{9} - \dfrac{8}{3} = \dfrac{1}{9} - \dfrac{24}{9} = -\dfrac{23}{9}$

76. $\left(-\dfrac{1}{10}\right)^2 - \dfrac{9}{10} = \dfrac{1}{100} - \dfrac{9}{10} = \dfrac{1}{100} - \dfrac{90}{100} = -\dfrac{89}{100}$

77. $\dfrac{(-4)(-3) - (-2)(-1)}{-10 + 5} = \dfrac{12 - 2}{-5} = \dfrac{10}{-5} = -2$

78. $\dfrac{4(12-18)}{-10 \div (-2-3)} = \dfrac{4(-6)}{-10 \div (-5)} = \dfrac{-24}{2} = -12$

79. $(-4)^2 = (-4)(-4) = 16$

80. $-4^2 = -(4 \cdot 4) = -16$

81. $-6 + (-9) = -15$

82. $-16 - 3 = -16 + (-3) = -19$

83. $-4(-12) = 48$

84. $\dfrac{84}{-4} = -21$

85. $-7.6 - (-9.7) = -7.6 + 9.7 = 2.1$

86. $-\dfrac{9}{20} + \dfrac{4}{20} = -\dfrac{5}{20} = -\dfrac{1}{4}$

87. $\begin{aligned} -\dfrac{15}{14} \div \left(-\dfrac{25}{28}\right) &= -\dfrac{15}{14} \cdot \left(-\dfrac{28}{25}\right) \\ &= \dfrac{15 \cdot 28}{14 \cdot 25} \\ &= \dfrac{3 \cdot 2}{1 \cdot 5} \\ &= \dfrac{6}{5} \end{aligned}$

88. $\dfrac{3}{11} \cdot \left(-\dfrac{22}{4}\right) = -\dfrac{3 \cdot 22}{11 \cdot 4} = -\dfrac{3 \cdot 2}{1 \cdot 4} = -\dfrac{3}{2}$

89. $\dfrac{7}{8} - \dfrac{9}{10} = \dfrac{35}{40} - \dfrac{36}{40} = -\dfrac{1}{40}$

90. $5(-0.27) = -1.35$

91. $-32 + 23 = -9$
His financial situation can be represented by $-\$9$.

92. $-11 + 17 = 6$
The temperature on Tuesday was 6°C.

93. $12,923 - (-195) = 12,923 + 195 = 13,118$
The difference in elevations is 13,118 feet.

94. $-18 + (-9) = -27$
The temperature on Friday was -27°C.

95. $(3 - 7)^2 \div (6 - 4)^3 = (-4)^2 \div (2)^3 = 16 \div 8 = 2$

96. $(4 + 6)^2 \div (2 - 7)^2 = 10^2 \div (-5)^2 = 100 \div 25 = 4$

97. $3(4+2)+(-6)-3^2 = 3(6)+(-6)-9$
$$= 18+(-6)+(-9)$$
$$= 12+(-9)$$
$$= 3$$

98. $4(5-3)-(-2)+3^3 = 4(2)-(-2)+27$
$$= 8+2+27$$
$$= 37$$

99. $2-4\cdot3+\sqrt{25} = 2-4\cdot3+5$
$$= 2-12+5$$
$$= -10+5$$
$$= -5$$

100. $4-6\cdot5+\sqrt{1} = 4-6\cdot5+1$
$$= 4-30+1$$
$$= -26+1$$
$$= -25$$

101. $\dfrac{-|-14|-6}{7+2(-3)} = \dfrac{-14-6}{7+(-6)} = \dfrac{-20}{1} = -20$

102. $5(7-6)^3-4(2-3)^2+2^4 = 5(1)^3-4(-1)^2+2^4$
$$= 5(1)-4(1)+16$$
$$= 5-4+16$$
$$= 1+16$$
$$= 17$$

Chapter 10 Test

1.

2. The opposite of $\dfrac{2}{3}$ is $-\dfrac{2}{3}$.

3. $|-2.9| = 2.9$

4. $-(-98) = 98$

5. $-|-98| = -98$

6. The graph of -17 is to the right of -19 on a number line, so -17 is greater than -19, or $-17 > -19$.

7. The graph of 0 is to the right of -2.5 on a number line, so 0 is greater than -2.5, or $0 > -2.5$.

8. $-5+8 = 3$

9. $18-24 = 18+(-24) = -6$

10. $0.5(-20) = -10$

11. $-16 \div (-4) = 4$

12. $-\frac{3}{11} + \left(-\frac{5}{22}\right) = -\frac{6}{22} + \left(-\frac{5}{22}\right) = -\frac{11}{22} = -\frac{1}{2}$

13. $-7 - (-19) = -7 + 19 = 12$

14. $-5(-13) = 65$

15. $\frac{-2.5}{-0.5} = 5$

16. $|-25| + (-13) = 25 + (-13) = 12$

17. $-8 + 9 \div (-3) = -8 + (-3) = -11$

18. $-7 + (-32) - 12 + \sqrt{25} = -7 + (-32) - 12 + 5$
$$= -39 - 12 + 5$$
$$= -51 + 5$$
$$= -46$$

19. $(-5)^3 - 24 \div (-3) = -125 - 24 \div (-3)$
$$= -125 - (-8)$$
$$= -125 + 8$$
$$= -117$$

20. $(5-9)^2 \cdot (8-6)^3 = (-4)^2 \cdot (2)^3 = 16 \cdot 8 = 128$

21. $\frac{-3(-2)+12}{-1(-4-5)} = \frac{6+12}{-1(-9)} = \frac{18}{9} = 2$

22. $\frac{|25-30|^2}{2(-6)+7} = \frac{|-5|^2}{-12+7} = \frac{5^2}{-5} = \frac{25}{-5} = -5$

23. $\left(\frac{5}{9} - \frac{7}{9}\right)^2 + \left(-\frac{2}{9}\right) = \left(-\frac{2}{9}\right)^2 + \left(-\frac{2}{9}\right)$
$$= \frac{4}{81} + \left(-\frac{2}{9}\right)$$
$$= \frac{4}{81} + \left(-\frac{18}{81}\right)$$
$$= -\frac{14}{81}$$

24. $-7 + (-9) + 35 = -16 + 35 = 19$

25. $40 - 43 = 40 + (-43) = -3$

26. $-\frac{1}{10} \cdot \frac{2}{9} = -\frac{1 \cdot 2}{10 \cdot 9} = -\frac{1}{45}$

27. $-\frac{1}{10} \div \frac{2}{9} = -\frac{1}{10} \cdot \frac{9}{2} = -\frac{1 \cdot 9}{10 \cdot 2} = -\frac{9}{20}$

28. Let 0 represent sea level.
$4(-22) = -88$
Her elevation is 88 feet below sea level.

29. Checks and withdrawals are represented by negative numbers, while deposits are represented by positive numbers.
$129 - 79 - 40 + 35 = 50 - 40 + 35$
$$= 10 + 35$$
$$= 45$$
His balance is $45, which can be represented by the number 45.

30. Let 0 represent sea level.
$6288 - (-25,354) = 6288 + 25,354 = 31,642$
The difference in elevations is 31,642 ft.

31. $1495 - 5315 = -3820$
The elevation of the deepest point in the lake is 3820 feet below sea level.

32. $\frac{-12 + (-13) + 0 + 9}{4} = \frac{-16}{4} = -4$
The average is −4.

Cumulative Review Chapters 1–10

1.
$$\begin{array}{r} 631 \\ \times 125 \\ \hline 3\,155 \\ 12\,620 \\ 63\,100 \\ \hline 78,875 \end{array}$$

2. $\frac{5}{8} \cdot \frac{10}{11} = \frac{5 \cdot 10}{8 \cdot 11} = \frac{5 \cdot 2 \cdot 5}{2 \cdot 4 \cdot 11} = \frac{5 \cdot 5}{4 \cdot 11} = \frac{25}{44}$

3. $\frac{2}{5} \div \frac{1}{2} = \frac{2}{5} \cdot \frac{2}{1} = \frac{2 \cdot 2}{5 \cdot 1} = \frac{4}{5}$

4.
$$\begin{array}{r} 236 \\ 9\overline{)2124} \\ \underline{-18} \\ 32 \\ \underline{-27} \\ 54 \\ \underline{-54} \\ 0 \end{array}$$
Thus $2124 \div 9 = 236$.

5. $\dfrac{2}{3} + \dfrac{1}{7} = \dfrac{2}{3} \cdot \dfrac{7}{7} + \dfrac{1}{7} \cdot \dfrac{3}{3} = \dfrac{14}{21} + \dfrac{3}{21} = \dfrac{17}{21}$

6. $9\dfrac{2}{7} - 7\dfrac{1}{2} = \dfrac{65}{7} - \dfrac{15}{2} = \dfrac{130}{14} - \dfrac{105}{14} = \dfrac{25}{14}$ or $1\dfrac{11}{14}$

   or

$$9\dfrac{2}{7} \qquad\qquad 9\dfrac{4}{14} \qquad\qquad 8\dfrac{18}{14}$$
$$\underline{-7\dfrac{1}{2}} \;\rightarrow\; \underline{-7\dfrac{7}{14}} \;\rightarrow\; \underline{-7\dfrac{7}{14}}$$
$$1\dfrac{11}{14}$$

7. Forty-eight and twenty-six hundredths is written as 48.26.

8. Eight hundredths is written as 0.08.

9. Six and ninety-five thousandths is written as 6.095.

10. $563.21 \times 100 = 56,321$

11. $$\begin{array}{r} 3.500 \\ -\,0.068 \\ \hline 3.432 \end{array}$$

12. $0.02\overline{)0.27}$ becomes $2\overline{)27.0}$

$$\begin{array}{r} 13.5 \\ 2\overline{)27.0} \\ \underline{-2} \\ 07 \\ \underline{-6} \\ 1\,0 \\ \underline{-1\,0} \\ 0 \end{array}$$

13. $\dfrac{5.68 + (0.9)^2 \div 100}{0.2} = \dfrac{5.68 + 0.81 \div 100}{0.2}$
$$= \dfrac{5.68 + 0.0081}{0.2}$$
$$= \dfrac{5.6881}{0.2}$$
$$= 28.4405$$

14. $50 \div 5 \cdot 2 = 10 \cdot 2 = 20$

15. $\dfrac{31,500 \text{ dollars}}{7 \text{ months}}$

$$\begin{array}{r} 4,500 \\ 7\overline{)31,500} \end{array}$$

The unit rate is $\dfrac{4500 \text{ dollars}}{1 \text{ month}}$ or 4500 dollars/month.

16. $\dfrac{300 \text{ miles}}{5 \text{ hours}}$

$$\begin{array}{r} 60 \\ 5\overline{)300} \end{array}$$

The unit rate is $\dfrac{60 \text{ miles}}{1 \text{ hour}}$ or 60 miles/hour.

17. $\dfrac{51}{34} = \dfrac{3}{n}$
$$51 \cdot n = 34 \cdot 3$$
$$51 \cdot n = 102$$
$$n = \dfrac{102}{51}$$
$$n = 2$$

18. $\dfrac{7}{8} = \dfrac{6}{n}$
$$7 \cdot n = 8 \cdot 6$$
$$7 \cdot n = 48$$
$$n = \dfrac{48}{7}$$

19. $\dfrac{46}{100} = 46\%$ of college students live at home.

20. $\dfrac{4}{5} = \dfrac{80}{100} = 80\%$ of free throws were made.

21. $\dfrac{9}{20} = \dfrac{9}{20} \cdot 100\% = \dfrac{9}{20} \cdot \dfrac{100}{1}\% = \dfrac{900}{20}\% = 45\%$

22. $\dfrac{53}{50} = \dfrac{53}{50} \cdot 100\% = \dfrac{53}{50} \cdot \dfrac{100}{1}\% = \dfrac{5300}{50}\% = 106\%$

23. $1\dfrac{1}{2} = \dfrac{3}{2} \cdot 100\% = \dfrac{3}{2} \cdot \dfrac{100}{1}\% = \dfrac{300}{2}\% = 150\%$

24. $5 = 5 \cdot 100\% = 500\%$

25.
$$13 = 6\frac{1}{2}\% \cdot n$$
$$13 = 0.065 \cdot n$$
$$\frac{13}{0.065} = n$$
$$200 = n$$

Thus, 13 is $6\frac{1}{2}\%$ of 200.

26. $n = 110\% \cdot 220$
$n = 1.10 \cdot 220$
$n = 242$
Thus, 242 is 110% of 220.

27. 101 is what percent of 200?
$$\frac{101}{200} = \frac{p}{100}$$

28. Let n be the percent.
$101 = n \cdot 200$

29. 8 months $= \dfrac{8}{12}$ year $= \dfrac{2}{3}$ year

Simple interest $=$ principal $\cdot$ rate $\cdot$ time
$$= \$2400 \cdot 10\% \cdot \frac{2}{3}$$
$$= \$2400 \cdot 0.10 \cdot \frac{2}{3}$$
$$= \$160$$

The interest on Ivan's loan is $160.

30. $762 - 237 = 525$
She will have $525 in her account.

31.
$$\begin{array}{r} 3 \text{ ft } 2 \text{ in.} \\ + 5 \text{ ft } 11 \text{ in.} \\ \hline 8 \text{ ft } 13 \text{ in.} \end{array}$$
Since 13 in. is the same as 1 ft 1 in., we have
8 ft 13 in. = 8 ft + 1 ft 1 in. = 9 ft 1 in.

32. $P = 4 \cdot s = 4 \cdot 17 \text{ m} = 68 \text{ m}$
The perimeter is 68 meters.

33. $C = \dfrac{5}{9} \cdot (F - 32) = \dfrac{5}{9}(98.6 - 32) = \dfrac{5}{9} \cdot (66.6) = 37$

Therefore, normal body temperature is 37°C.

34. $7.2 \text{ m} = \dfrac{7.2 \text{ m}}{1} \cdot \dfrac{100 \text{ cm}}{1 \text{ m}} = 720 \text{ cm}$

35. The supplement of a 107° angle is
$180° - 107° = 73°$.

36. The complement of a 34° angle is
$90° - 34° = 56°$.

37. The measure of $\angle b$ is $180° - 90° - 30° = 60°$.

38. The measure of $\angle a$ is $180° - 108° - 23° = 49°$.

39. $A = b \cdot h$
$= 3.4 \text{ miles} \cdot 1.5 \text{ miles}$
$= 5.1 \text{ square miles}$
The area is 5.1 square miles.

40. $A = \pi \cdot r^2 = \pi \cdot 5^2 = 25\pi \approx 25(3.14) = 78.5$
The area is $25\pi \approx 78.5$ sq m.

41. $V = \dfrac{4}{3} \cdot \pi \cdot r^3$
$= \dfrac{4}{3} \cdot \pi \cdot (3 \text{ in.})^2$
$= \dfrac{4}{3} \cdot \pi \cdot 27 \text{ cu in.}$
$= 36\pi \text{ cu in.}$

$V = 36\pi \text{ cu in.}$
$\approx 36 \cdot \dfrac{22}{7} \text{ cu in.}$
$= \dfrac{792}{7} \text{ or } 113\dfrac{1}{7} \text{ cu in.}$

The volume is $36\pi \approx 113\dfrac{1}{7}$ cu in.

42. $V = l \cdot w \cdot h = 2 \text{ cm} \cdot 4 \text{ cm} \cdot 3 \text{ cm} = 24 \text{ cu cm}$
The volume is 24 cu cm.

43. 25, 54, 56, 57, 60, 71, 98
Because this list is in numerical order, the median is the middle number, 57.

44. mean $= \dfrac{36 + 25 + 18 + 19}{4} = \dfrac{98}{4} = 24.5$

45. possible outcomes: 1, 2, 3, 4, 5, 6
probability of a 3 or a 4 $= \dfrac{2}{6} = \dfrac{1}{3}$

46. $-9 - (-4.1) = -9 + 4.1 = -4.9$

47. $|-2| = 2, |-21| = 21$, and $2 + 21 = 23$
Their common sign is negative, so the sum is negative.
$-2 + (-21) = -23$

48. a. $(-10)^3 = (-10) \cdot (-10) \cdot (-10) = -1000$

b. $-10^3 = -(10 \cdot 10 \cdot 10) = -1000$

49. $60 + 30 + (-2)^3 = 60 + 30 + (-8) = 90 + (-8) = 82$

50. $\dfrac{\sqrt{16} - (-8)}{9 - 12} = \dfrac{4 - (-8)}{9 - 12} = \dfrac{4 + 8}{9 - 12} = \dfrac{12}{-3} = -4$

Chapter 11

1. $5x - 12 = 5 \cdot 2 - 12 = 10 - 12 = -2$

2. $5x - y = 5 \cdot 2 - (-3) = 10 - (-3) = 10 + 3 = 13$

3. $\dfrac{5r - 2s}{-3q} = \dfrac{5 \cdot 3 - 2 \cdot 3}{-3 \cdot 1} = \dfrac{15 - 6}{-3} = \dfrac{9}{-3} = -3$

4. $13 - (3a + 8) = 13 - (3 \cdot (-2) + 8$
   $= 13 - (-6 + 8)$
   $= 13 - 2$
   $= 11$

5. $a^2 - 0.7b = 6^2 - 0.7(-2)$
   $= 36 - 0.7(-2)$
   $= 36 + 1.4$
   $= 37.4$

6. $P = 2l + 2w = 2(40) + 2(25) = 80 + 50 = 130$
   The perimeter is 130 meters

7. **a.** $8m - 11m = (8 - 11)m = -3m$

   b. $5a + a = 5a + 1a = (5 + 1)a = 6a$

8. $8m + 5 + m - 4 = 8m + 5 + m + (-4)$
   $= 8m + m + 5 + (-4)$
   $= (8 + 1)m + 5 + (-4)$
   $= 9m + 1$

9. $7y + 11y = 18y - 8$

10. $2y - 6 + y + 7y = 2y + y + 7y - 6 = 10y - 6$

11. $3.7x + 5 - 4.2x + 15 = 3.7x - 4.2x + 5 + 15$
    $= -0.5x + 20$

12. $-9y + 2 - 4y - 8x + 12 - x = -9x - 13y + 14$

13. $7(8a) = (7 \cdot 8)a = 56a$

14. $-5(9x) = (-5 \cdot 9)x = -45x$

15. $7(y + 2) = 7 \cdot y + 7 \cdot 2 = 7y + 14$

16. $4(7a - 5) = 4(7a) + 4(-5)$
    $= (4 \cdot 7)a - 20$
    $= 28a - 20$

17. $5(2y - 3) - 8 = 5(2y) + 5(-3) - 8$
    $= 10y - 15 - 8$
    $= 10y - 23$

18. $-7(x - 1) + 5(2x + 3)$
    $= -7(x) - 7(-1) + 5(2x) + 5(3)$
    $= -7x + 7 + 10x + 15$
    $= 3x + 22$

19. $4(2x) = (4 \cdot 2)x = 8x$
    The perimeter is $8x$ centimeters.

20. $3(12y + 9) = 3 \cdot 12y + 3 \cdot 9 = 36y + 27$
    The area of the garden is $(36y + 27)$ square yards.

Vocabulary and Readiness Check

1. $14y^2 + 2x - 23$ is called an <u>expression</u> while $14y^2$, $2x$, and -23 are each called a <u>term</u>.

2. To multiply $3(-7x + 1)$, we use the <u>distributive</u> property.

3. To simplify an expression like $y + 7y$, we <u>combine like terms</u>.

4. By the <u>commutative</u> properties, the *order* of adding or multiplying two numbers can be changed without changing their sum or product.

5. The term $5x$ is called a <u>variable</u> term while the term 7 is called a <u>constant</u> term.

6. The term z has an understood <u>numerical coefficient</u> of 1.

7. By the <u>associative</u> properties, the grouping of adding or multiplying numbers can be changed without changing their sum or product.

8. The terms $-x$ and $5x$ are <u>like</u> terms.

9. For the term $-3x^2y$, -3 is called the <u>numerical coefficient</u>.

10. The terms $5x$ and $5y$ are <u>unlike</u> terms.

11. $5x$ and $5y$ are unlike terms.

12. $-3a$ and $-3b$ are unlike terms.

13. x and $-2x$ are like terms.

14. $7y$ and y are like terms.

15. $-5n$ and $6n^2$ are unlike terms.

16. $4m^2$ and $2m$ are unlike terms.

17. $8b$ and $-6b$ are like terms.

18. $12a$ and $-11a$ are like terms.

Exercise Set 11.1

1. $3 + 2z = 3 + 2(-3)$
$= 3 + (-6)$
$= -3$

3. $-y - z = -5 - (-3) = -5 + 3 = -2$

5. $z - x + y = -3 - (-2) + 5$
$= -3 + 2 + 5$
$= -1 + 5$
$= 4$

7. $3x - z = 3(-2) - (-3)$
$= -6 - (-3)$
$= -6 + 3$
$= -3$

9. $8 - (5y - 7) = 8 - (5 \cdot 5 - 7)$
$= 8 - (25 - 7)$
$= 8 - 18$
$= -10$

11. $y^3 - 4x = 5^3 - 4(-2)$
$= 125 - 4(-2)$
$= 125 + 8$
$= 133$

13. $\dfrac{6xy}{4} = \dfrac{6(-2)(5)}{4}$
$= \dfrac{-12(5)}{4}$
$= \dfrac{-60}{4}$
$= -15$

15. $\dfrac{2y - 2}{x} = \dfrac{2(5) - 2}{-2} = \dfrac{10 - 2}{-2} = \dfrac{8}{-2} = -4$

17. $\dfrac{x + 2y}{2z} = \dfrac{-2 + 2 \cdot 5}{2(-3)}$
$= \dfrac{-2 + 10}{-6}$
$= \dfrac{8}{-6}$
$= -\dfrac{4}{3}$ or $-1\dfrac{1}{3}$

19. $\dfrac{5x}{y} - 10 = \dfrac{5(-2)}{5} - 10$
$= \dfrac{-10}{5} - 10$
$= -2 - 10$
$= -12$

21. $\dfrac{xz}{y} + \dfrac{3}{10} = \dfrac{-2(-3)}{5} + \dfrac{3}{10}$
$= \dfrac{6}{5} + \dfrac{3}{10}$
$= \dfrac{12}{10} + \dfrac{3}{10}$
$= \dfrac{15}{10}$
$= \dfrac{3}{2}$ or $1\dfrac{1}{2}$

23. $|x| - |y| - 7.6 = |-2| - |5| - 7.6$
$= 2 - 5 - 7.6$
$= -3 - 7.6$
$= -10.6$

25. $3x + 5x = (3 + 5)x = 8x$

27. $5n - 9n = (5 - 9)n = -4n$

29. $4c + c - 7c = (4 + 1 - 7)c$
$= (5 - 7)c$
$= -2c$

31. $5x - 7x + x - 3x = (5 - 7 + 1 - 3)x$
$= (-2 + 1 - 3)x$
$= (-1 - 3)x$
$= -4x$

33. $4a + 3a + 6a - 8 = (4 + 3 + 6)a - 8$
$= (7 + 6)a - 8$
$= 13a - 8$

35. $1.7x + 3.4 - 2.6x + 7.8$
$= 1.7x - 2.6x + 3.4 + 7.8$
$= (1.7 - 2.6)x + (3.4 + 7.8)$
$= -0.9x + 11.2$

37. $3x + 7 - x - 14 = 3x - x + 7 - 14$
$= (3 - 1)x + (7 - 14)$
$= 2x - 7$

39. $4x + 5y + 2 - y - 9x - 7$
$= 4x - 9x + 5y - y + 2 - 7$
$= (4 - 9)x + (5 - 1)y + (2 - 7)$
$= -5x + 4y - 5$

41. $\dfrac{5}{6} - \dfrac{7}{12}x - \dfrac{1}{3} - \dfrac{3}{10}x$
$= \dfrac{5}{6} - \dfrac{1}{3} - \dfrac{7}{12}x - \dfrac{3}{10}x$
$= \left(\dfrac{5}{6} - \dfrac{1}{3}\right) + \left(-\dfrac{7}{12} - \dfrac{3}{10}\right)x$
$= \left(\dfrac{5}{6} - \dfrac{2}{6}\right) + \left(-\dfrac{35}{60} - \dfrac{18}{60}\right)x$
$= \dfrac{3}{6} + \left(-\dfrac{53}{60}\right)x$
$= \dfrac{1}{2} - \dfrac{53}{60}x$

43. $-5m - 2.3m + 11 + 2.5m - 15.1$
$= -5m - 2.3m + 2.5m + 11 - 15.1$
$= (-5 - 2.3 + 2.5)m + (11 - 15.1)$
$= -4.8m - 4.1$

45. $6(5x) = (6 \cdot 5)x = 30x$

47. $-2(11y) = (-2 \cdot 11)y = -22y$

49. $-0.6(7a) = (-0.6 \cdot 7)a = -4.2a$

51. $\dfrac{2}{3}(-6a) = \left(\dfrac{2}{3} \cdot (-6)\right) \cdot a = -4a$

53. $2(y + 2) = 2 \cdot y + 2 \cdot 2 = 2y + 4$

55. $5(3a - 8) = 5 \cdot 3a + 5 \cdot (-8)$
$= 15a + (-40)$
$= 15a - 40$

57. $-4(3x + 7) = -4 \cdot 3x + (-4) \cdot 7$
$= (-12)x + (-28)$
$= -12x - 28$

59. $1.2(5x - 0.1) = 1.2 \cdot 5x + 1.2 \cdot (-0.1)$
$= 6x - 0.12$

61. $\dfrac{1}{2}(-8x - 3) = \dfrac{1}{2}(-8x) - \dfrac{1}{2}(3)$
$= \left(\dfrac{1}{2} \cdot (-8)\right)x - \dfrac{3}{2}$
$= -4x - \dfrac{3}{2}$

63. $2(x + 4) - 17 = 2 \cdot x + 2 \cdot 4 - 17$
$= 2x + 8 - 17$
$= 2x - 9$

65. $4(6n - 5) + 3n = 4 \cdot 6n - 4 \cdot 5 + 3n$
$= 24n - 20 + 3n$
$= 24n + 3n - 20$
$= (24 + 3)n - 20$
$= 27n - 20$

67. $3 + 6(w + 2) + w = 3 + 6 \cdot w + 6 \cdot 2 + w$
$= 3 + 6w + 12 + w$
$= 6w + w + 3 + 12$
$= 7w + 15$

69. $-2(3x + 1) - 5(x - 2)$
$= -2 \cdot 3x + (-2) \cdot 1 - 5 \cdot x - (-5) \cdot 2$
$= -6x - 2 - 5x + 10$
$= -6x - 5x - 2 + 10$
$= (-6 - 5)x + (-2 + 10)$
$= -11x + 8$

71. $3y + 4y + 2y + 6 + 5y + 16$
$= 3y + 4y + 2y + 5y + 6 + 16$
$= (3 + 4 + 2 + 5)y + 6 + 16$
$= 14y + 22$
The perimeter is $(14y + 22)$ meters.

73. $2a + 2a + 6 + 5a + 6 + 2a$
$= 2a + 2a + 5a + 2a + 6 + 6$
$= (2 + 2 + 5 + 2)a + 6 + 6$
$= 11a + 12$
The perimeter is $(11a + 12)$ feet.

75. $5(-5x + 11) = 5 \cdot (-5x) + 5 \cdot 11 = -25x + 55$
The perimeter is $(-25x + 55)$ inches.

77. Area $= (\text{length}) \cdot (\text{width})$
$= (4y) \cdot (9)$
$= (4 \cdot 9)y$
$= 36y$
The area is $36y$ square inches.

79. Area = (length) · (width)
$$= (x-2) \cdot (32)$$
$$= x \cdot 32 - 2 \cdot 32$$
$$= 32x - 64$$
The area is $(32x - 64)$ square kilometers.

81. Area = (length) · (width)
$$= (3y+1) \cdot (20)$$
$$= 3y \cdot 20 + 1 \cdot 20$$
$$= (3 \cdot 20)y + 20$$
$$= 60y + 20$$
The area is $(60y + 20)$ square miles.

83. Area = (length) · (width) = $94 \cdot 50 = 4700$
The area is 4700 square feet.

85. Perimeter = $2 \cdot$ (length) + $2 \cdot$ (width)
$$= 2 \cdot (18) + 2 \cdot (14)$$
$$= 36 + 28$$
$$= 64$$
The perimeter is 64 feet.

87. $5 + x + 2x + 1 = x + 2x + 5 + 1$
$$= (1+2)x + 5 + 1$$
$$= 3x + 6$$
The perimeter is $(3x + 6)$ feet.

89. Let $p = 3000$, $r = 0.06$, and $t = 2$.
$I = prt = 3000 \cdot (0.06) \cdot (2) = 180 \cdot 2 = 360$
The interest will be $360.

91. $A = \pi r^2 = \pi \cdot 5^2 \approx (3.14) \cdot 25 = 78.5$
The area is 78.5 square feet.

93. $F = \dfrac{9}{5}C + 32 = \dfrac{9}{5}(-5) + 32 = -9 + 32 = 23$
$-5°C$ is $23°F$.

95. Let $l = 12$, $w = 6$, and $h = 4$.
$V = lwh = 12 \cdot 6 \cdot 4 = 72 \cdot 4 = 288$
The volume is 288 cubic inches.

97. $V = lwh = 7.6(3)(4x) = 22.8(4x) = 91.2x$
The volume is $91.2x$ cubic inches.

99. $-13 + 10 = -3$

101. $-4 - (-12) = -4 + 12 = 8$

103. $-4 + 4 = 0$

105. The given result is incorrect.
$5(3x - 2) = 5 \cdot 3x - 5 \cdot 2 = 15x - 10$

107. The given result is incorrect.
$7x - (x + 2) = 7x - 1(x + 2)$
$$= 7x - 1 \cdot x + (-1)(2)$$
$$= 7x - x - 2$$
$$= 6x - 2$$

109. The 6 is multiplied by each term in the parentheses, $(2x - 3)$. Thus, $6(2x - 3) + 5 = 12x - 18 + 5$ demonstrates the distributive property.

111. The order of the terms is not changed, only the grouping within parentheses. Thus, $-7 + (4 + y) = (-7 + 4) + y$ demonstrates the associative property of addition.

113. answers may vary

115. Add the area of the left-hand rectangle and the right-hand rectangle.
$7(2x + 1) + 3(2x + 3)$
$$= 7 \cdot 2x + 7 \cdot 1 + 3 \cdot 2x + 3 \cdot 3$$
$$= 14x + 7 + 6x + 9$$
$$= 14x + 6x + 7 + 9$$
$$= 20x + 16$$
The area is $(20x + 16)$ square miles.

117. $A = 0.7854d^2$
$$= 0.7854(173)^2$$
$$= 0.7854(29,929)$$
$$\approx 23,506.2 \text{ sq in.}$$
The trunk area is 23,506.2 square inches.

119. $9684q - 686 - 4860q + 12,960$
$$= 9684q - 4860q - 686 + 12,960$$
$$= 4824q + 12,274$$

Section 11.2 Practice

1. $3(y - 6) = 6$
$3(4 - 6) \stackrel{?}{=} 6$
$3(-2) \stackrel{?}{=} 6$
$-6 \stackrel{?}{=} 6$ False
No, 4 is not a solution of the equation.

2. $-4x - 3 = 5$
$-4(-2) - 3 \stackrel{?}{=} 5$
$8 - 3 \stackrel{?}{=} 5$
$5 \stackrel{?}{=} 5$ True
Yes, -2 is a solution.

3.
$$y - 5 = -3$$
$$y - 5 + 5 = -3 + 5$$
$$y = 2$$
Check: $y - 5 = -3$
$$2 - 5 \stackrel{?}{=} -3$$
$$-3 \stackrel{?}{=} -3 \quad \text{True}$$
The solution of the equation is 2.

4.
$$-1 = z + 9$$
$$-1 - 9 = z + 9 - 9$$
$$-10 = z$$
Check: $-1 = z + 9$
$$-1 \stackrel{?}{=} -10 + 9$$
$$-1 \stackrel{?}{=} -1 \quad \text{True}$$
The solution is -10.

5.
$$x - 2.6 = -1.8 - 5.9$$
$$x - 2.6 = -7.7$$
$$x - 2.6 + 2.6 = -7.7 + 2.6$$
$$x = -5.1$$
Check: $x - 2.6 = -1.8 - 5.9$
$$-5.1 - 2.6 \stackrel{?}{=} -1.8 - 5.9$$
$$-7.7 \stackrel{?}{=} -7.7 \quad \text{True}$$
The solution is -5.1.

6.
$$-6y + 1 + 7y = 6 - 11$$
$$-6y + 7y + 1 = 6 - 11$$
$$y + 1 = -5$$
$$y + 1 - 1 = -5 - 1$$
$$y = -6$$
Check: $-6y + 1 + 7y = 6 - 11$
$$-6(-6) + 1 + 7(-6) \stackrel{?}{=} 6 - 11$$
$$36 + 1 - 42 \stackrel{?}{=} -5$$
$$-5 \stackrel{?}{=} -5 \quad \text{True}$$
The solution is -6.

7.
$$\frac{2}{3} = x - \frac{4}{9}$$
$$\frac{2}{3} + \frac{4}{9} = x - \frac{4}{9} + \frac{4}{9}$$
$$\frac{6}{9} + \frac{4}{9} = x$$
$$\frac{10}{9} = x$$

Check: $\dfrac{2}{3} = x - \dfrac{4}{9}$
$$\frac{2}{3} \stackrel{?}{=} \frac{10}{9} - \frac{4}{9}$$
$$\frac{2}{3} \stackrel{?}{=} \frac{6}{9}$$
$$\frac{2}{3} \stackrel{?}{=} \frac{2}{3} \quad \text{True}$$
The solution is $\dfrac{10}{9}$.

8.
$$13x = 4(3x - 1)$$
$$13x = 4 \cdot 3x - 4 \cdot 1$$
$$13x = 12x - 4$$
$$13x - 12x = 12x - 4 - 12x$$
$$x = -4$$

Vocabulary and Readiness Check

1. The equations $x + 6 = 10$ and $x + 6 - 6 = 10 - 6$ are called <u>equivalent</u> equations.

2. The difference between an equation and an expression is that an <u>equation</u> contains an equal sign, while an <u>expression</u> does not.

3. The process of writing $-3x + 10x$ as $7x$ is called <u>simplifying</u> the expression.

4. For the equation $-5x - 1 = -21$, the process of finding that 4 is the solution is called <u>solving</u> the equation.

5. By the <u>addition</u> property of equality, $x = -2$ and $x + 7 = -2 + 7$ are equivalent equations.

Exercise Set 11.2

1.
$$x - 8 = 2$$
$$10 - 8 \stackrel{?}{=} 2$$
$$2 \stackrel{?}{=} 2 \quad \text{True}$$
Yes, 10 is a solution of the equation.

3.
$$x + 12 = 17$$
$$-5 + 12 \stackrel{?}{=} 17$$
$$7 \stackrel{?}{=} 17 \quad \text{False}$$
No, -5 is not a solution of the equation.

5.
$$-9f = 64 - f$$
$$-9(-8) \stackrel{?}{=} 64 - (-8)$$
$$72 \stackrel{?}{=} 64 + 8$$
$$72 \stackrel{?}{=} 72 \quad \text{True}$$
Yes, -8 is a solution of the equation.

7. $5(c-5) = -10$

$5(3-5) \overset{?}{=} -10$

$5(-2) \overset{?}{=} -10$

$-10 \overset{?}{=} -10$ True

Yes, 3 is a solution of the equation.

9. $a + 5 = 23$

$a + 5 - 5 = 23 - 5$

$a = 18$

Check:

$a + 5 = 23$

$18 + 5 \overset{?}{=} 23$

$23 \overset{?}{=} 23$ True

The solution of the equation is 18.

11. $d - 9 = -17$

$d - 9 + 9 = -17 + 9$

$d = -8$

Check:

$d - 9 = -17$

$-8 - 9 \overset{?}{=} -17$

$-17 \overset{?}{=} -17$ True

The solution of the equation is -8.

13. $7 = y - 2$

$7 + 2 = y - 2 + 2$

$9 = y$

Check:

$7 = y - 2$

$7 \overset{?}{=} 9 - 2$

$7 \overset{?}{=} 7$ True

The solution of the equation is 9.

15. $-12 = x + 4$

$-12 - 4 = x + 4 - 4$

$-16 = x$

Check:

$-12 = x + 4$

$-12 \overset{?}{=} -16 + 4$

$-12 \overset{?}{=} -12$ True

The solution of the equation is -16.

17. $x + \dfrac{1}{2} = \dfrac{7}{2}$

$x + \dfrac{1}{2} - \dfrac{1}{2} = \dfrac{7}{2} - \dfrac{1}{2}$

$x = \dfrac{6}{2}$

$x = 3$

Check:

$x + \dfrac{1}{2} = \dfrac{7}{2}$

$3 + \dfrac{1}{2} \overset{?}{=} \dfrac{7}{2}$

$\dfrac{6}{2} + \dfrac{1}{2} \overset{?}{=} \dfrac{7}{2}$

$\dfrac{7}{2} \overset{?}{=} \dfrac{7}{2}$ True

The solution of the equation is 3.

19. $y - \dfrac{3}{4} = -\dfrac{5}{8}$

$y - \dfrac{3}{4} + \dfrac{3}{4} = -\dfrac{5}{8} + \dfrac{3}{4}$

$y = -\dfrac{5}{8} + \dfrac{6}{8}$

$y = \dfrac{1}{8}$

Check:

$y - \dfrac{3}{4} = -\dfrac{5}{8}$

$\dfrac{1}{8} - \dfrac{3}{4} \overset{?}{=} -\dfrac{5}{8}$

$\dfrac{1}{8} - \dfrac{6}{8} \overset{?}{=} -\dfrac{5}{8}$

$-\dfrac{5}{8} \overset{?}{=} -\dfrac{5}{8}$ True

The solution of the equation is $\dfrac{1}{8}$.

21. $x - 3 = -1 + 4$

$x - 3 = 3$

$x - 3 + 3 = 3 + 3$

$x = 6$

Check:

$x - 3 = -1 + 4$

$6 - 3 \overset{?}{=} -1 + 4$

$3 \overset{?}{=} 3$ True

The solution of the equation is 6.

23. $-7 + 10 = m - 5$

$3 = m - 5$

$3 + 5 = m - 5 + 5$

$8 = m$

Check: $-7 + 10 = m - 5$

$-7 + 10 \overset{?}{=} 8 - 5$

$3 \overset{?}{=} 3$ True

The solution of the equation is 8.

25.
$$x - 0.6 = 4.7$$
$$x - 0.6 + 0.6 = 4.7 + 0.6$$
$$x = 5.3$$
Check:
$$x - 0.6 = 4.7$$
$$5.3 - 0.6 \stackrel{?}{=} 4.7$$
$$4.7 \stackrel{?}{=} 4.7 \text{ True}$$
The solution of the equation is 5.3.

27.
$$-2 - 3 = -4 + x$$
$$-5 = -4 + x$$
$$4 - 5 = 4 - 4 + x$$
$$-1 = x$$
Check:
$$-2 - 3 = -4 + x$$
$$-2 - 3 \stackrel{?}{=} -4 - 1$$
$$-5 \stackrel{?}{=} -5 \text{ True}$$
The solution of the equation is -1.

29.
$$y + 2.3 = -9.2 - 8.6$$
$$y + 2.3 = -17.8$$
$$y + 2.3 - 2.3 = -17.8 - 2.3$$
$$y = -20.1$$
Check:
$$y + 2.3 = -9.2 - 8.6$$
$$-20.1 + 2.3 \stackrel{?}{=} -9.2 - 8.6$$
$$-17.8 \stackrel{?}{=} -17.8 \text{ True}$$
The solution of the equation is -20.1.

31.
$$-8x + 4 + 9x = -1 + 7$$
$$4 + x = 6$$
$$-4 + 4 + x = -4 + 6$$
$$x = 2$$
Check:
$$-8x + 4 + 9x = -1 + 7$$
$$-8(2) + 4 + 9(2) \stackrel{?}{=} -1 + 7$$
$$-16 + 4 + 18 \stackrel{?}{=} 6$$
$$6 \stackrel{?}{=} 6 \text{ True}$$
The solution of the equation is 2.

33.
$$5 + (-12) = 5x - 7 - 4x$$
$$-7 = x - 7$$
$$-7 + 7 = x - 7 + 7$$
$$0 = x$$
Check:
$$-5 + (-12) = 5x - 7 - 4x$$
$$-5 + (-12) \stackrel{?}{=} 5(0) - 7 - 4(0)$$
$$-7 \stackrel{?}{=} 0 - 7 - 0$$
$$-7 \stackrel{?}{=} -7 \text{ True}$$
The solution of the equation is 0.

35.
$$7x + 14 - 6x = -4 + (-10)$$
$$x + 14 = -14$$
$$x + 14 - 14 = -14 - 14$$
$$x = -28$$
Check:
$$7x + 14 - 6x = -4 + (-10)$$
$$7(-28) + 14 - 6(-28) \stackrel{?}{=} -4 + (-10)$$
$$-196 + 14 + 168 \stackrel{?}{=} -14$$
$$-14 \stackrel{?}{=} -14 \text{ True}$$
The solution of the equation is -28.

37.
$$2(5x - 3) = 11x$$
$$2 \cdot 5x - 2 \cdot 3 = 11x$$
$$10x - 6 = 11x$$
$$10x - 10x - 6 = 11x - 10x$$
$$-6 = x$$

39.
$$3y = 2(y + 12)$$
$$3y = 2 \cdot y + 2 \cdot 12$$
$$3y = 2y + 24$$
$$3y - 2y = 2y - 2y + 24$$
$$y = 24$$

41.
$$21y = 5(4y - 6)$$
$$21y = 5 \cdot 4y - 5 \cdot 6$$
$$21y = 20y - 30$$
$$21y - 20y = 20y - 20y - 30$$
$$y = -30$$

43.
$$-3(-4 - 2z) = 7z$$
$$-3 \cdot (-4) - (-3)(2z) = 7z$$
$$12 + 6z = 7z$$
$$12 + 6z - 6z = 7z - 6z$$
$$12 = z$$

45. $\dfrac{-7}{-7} = 1$ because $-7 \cdot 1 = -7$.

47. $\dfrac{1}{3} \cdot 3 = \dfrac{1}{3} \cdot \dfrac{3}{1} = \dfrac{1 \cdot 3}{3 \cdot 1} = 1$

49. $-\dfrac{2}{3} \cdot -\dfrac{3}{2} = \dfrac{2 \cdot 3}{3 \cdot 2} = \dfrac{6}{6} = 1$

51. To solve $\dfrac{2}{3} + x = \dfrac{1}{12}$, subtract $\dfrac{2}{3}$ from both sides of the equation.

53. To solve $-\dfrac{1}{7} = -\dfrac{4}{5} + x$, $\dfrac{4}{5}$ should be added to both sides of the equation.

55. answers may vary

57.
$$x - 76,862 = 86,102$$
$$x - 76,862 + 76,862 = 86,102 + 76,862$$
$$x = 162,964$$

59.
$$T = P + R$$
$$4990 = 3301 + R$$
$$-3301 + 4990 = -3301 + 3301 + R$$
$$1689 = R$$

The Steelers gained 1689 yards by rushing.

61.
$$I = R - E$$
$$885,000,000 = R - 15,504,000,000$$
$$885,000,000 + 15,504,000,000 = R - 15,504,000,000 + 15,504,000,000$$
$$R = 16,389,000,000$$

Kohl's total revenues were $16,389,000,000.

Section 11.3 Practice

1. $-3y = 18$
$$\frac{-3y}{-3} = \frac{18}{-3}$$
$$y = -6$$

Check: $-3y = 18$
$$-3(-6) \overset{?}{=} 18$$
$$18 \overset{?}{=} 18 \quad \text{True}$$

The solution is -6.

2. $-16 = 8x$
$$\frac{-16}{8} = \frac{8x}{8}$$
$$-2 = x \text{ or } x = -2$$

Check to see that -2 is the solution.

3. $-0.3y = -27$
$$\frac{-0.3y}{-0.3} = \frac{-27}{-0.3}$$
$$y = 90$$

Check to see that 90 is the solution.

4. $\dfrac{5}{7}b = 25$

$\dfrac{7}{5} \cdot \dfrac{5}{7}b = \dfrac{7}{5} \cdot 25$

$1 \cdot b = \dfrac{7 \cdot 25}{5 \cdot 1}$

$b = 35$

Check: $\dfrac{5}{7}b = 25$

$\dfrac{5}{7} \cdot 35 \overset{?}{=} 25$

$25 \overset{?}{=} 25$ True

The solution is 35.

5. $-\dfrac{7}{10}x = \dfrac{2}{5}$

$-\dfrac{10}{7} \cdot \left(-\dfrac{7}{10}x\right) = -\dfrac{10}{7} \cdot \dfrac{2}{5}$

$1 \cdot x = -\dfrac{10 \cdot 2}{7 \cdot 5}$

$x = -\dfrac{4}{7}$

Check to see that $-\dfrac{4}{7}$ is the solution.

6. $2m - 4m = 10$

$-2m = 10$

$\dfrac{-2m}{-2} = \dfrac{10}{-2}$

$m = -5$

Check: $2m - 4m = 10$

$2(-5) - 4(-5) \overset{?}{=} 10$

$-10 + 20 \overset{?}{=} 10$

$10 \overset{?}{=} 10$ True

The solution is −5.

7. $-3a + 2a = -8 + 6$

$-1a = -2$

$\dfrac{-1a}{-1} = \dfrac{-2}{-1}$

$a = 2$

Check to see that 2 is the solution.

8. $-8 + 6 = \dfrac{a}{3}$

$-2 = \dfrac{a}{3}$

$3 \cdot (-2) = 3 \cdot \dfrac{a}{3}$

$3 \cdot (-2) = \dfrac{3}{3} \cdot a$

$-6 = a$

Vocabulary and Readiness Check

1. The equations $-3x = 51$ and $\dfrac{-3x}{-3} = \dfrac{51}{-3}$ are called <u>equivalent</u> equations.

2. The difference between an equation and an expression is that an <u>equation</u> contains an equal sign, while an <u>expression</u> does not.

3. The process of writing $-3x + x$ as $-2x$ is called <u>simplifying</u> the expression.

4. For the equation $-5x - 1 = -21$, the process of finding that 4 is the solution is called <u>solving</u> the equation.

5. By the <u>multiplication</u> property of equality, $y = 8$ and $3 \cdot y = 3 \cdot 8$ are equivalent equations.

Exercise Set 11.3

1. $5x = 20$

$\dfrac{5x}{5} = \dfrac{20}{5}$

$x = 4$

3. $-3z = 12$

$\dfrac{-3z}{-3} = \dfrac{12}{-3}$

$z = -4$

5. $0.4y = -12$

$\dfrac{0.4y}{0.4} = \dfrac{-12}{0.4}$

$y = -30$

7. $2z = -34$

$\dfrac{2z}{2} = \dfrac{-34}{2}$

$z = -17$

9. $-0.3x = -15$

$$\frac{-0.3x}{-0.3} = \frac{-15}{-0.3}$$

$$x = 50$$

11. $10 = \frac{2}{5}x$

$$\frac{5}{2} \cdot 10 = \frac{5}{2} \cdot \frac{2}{5}x$$

$$\frac{5 \cdot 10}{2 \cdot 1} = x$$

$$25 = x \text{ or } x = 25$$

13. $\frac{1}{6}y = -5$

$$\frac{6}{1} \cdot \frac{1}{6}y = \frac{6}{1} \cdot -5$$

$$y = -30$$

15. $\frac{5}{6}x = \frac{5}{18}$

$$\frac{6}{5} \cdot \frac{5}{6}x = \frac{6}{5} \cdot \frac{5}{18}$$

$$x = \frac{6 \cdot 5}{5 \cdot 18}$$

$$x = \frac{1}{3}$$

17. $-\frac{2}{9}z = \frac{4}{27}$

$$-\frac{9}{2} \cdot -\frac{2}{9}z = -\frac{9}{2} \cdot \frac{4}{27}$$

$$z = -\frac{2}{3}$$

19. $2w - 12w = 40$

$$-10w = 40$$

$$\frac{-10w}{-10} = \frac{40}{-10}$$

$$w = -4$$

21. $16 = 10t - 8t$

$$16 = 2t$$

$$\frac{16}{2} = \frac{2t}{2}$$

$$8 = t$$

23. $2z = 1.2 + 1.4$

$$2z = 2.6$$

$$\frac{2z}{2} = \frac{2.6}{2}$$

$$z = 1.3$$

25. $4 - 10 = -3z$

$$-6 = -3z$$

$$\frac{-6}{-3} = \frac{-3z}{-3}$$

$$2 = z$$

27. $-7x = 0$

$$\frac{-7x}{-7} = \frac{0}{-7}$$

$$x = 0$$

29. $0.4 = -8z$

$$\frac{0.4}{-8} = \frac{-8z}{-8}$$

$$-0.05 = z$$

31. $\frac{8}{5}t = -\frac{3}{8}$

$$\frac{5}{8} \cdot \frac{8}{5}t = \frac{5}{8} \cdot \left(-\frac{3}{8}\right)$$

$$t = \frac{5 \cdot (-3)}{8 \cdot 8}$$

$$t = -\frac{15}{64}$$

33. $-\frac{3}{5}x = -\frac{6}{15}$

$$-\frac{5}{3} \cdot -\frac{3}{5}x = -\frac{5}{3} \cdot -\frac{6}{15}$$

$$x = \frac{2}{3}$$

35. $-3.6 = -0.9u + 0.3u$

$$-3.6 = -0.6u$$

$$\frac{-3.6}{-0.6} = \frac{-0.6u}{-0.6}$$

$$6 = u$$

37. $5 - 5 = 2x + 7x$

$$0 = 9x$$

$$\frac{0}{9} = \frac{9x}{9}$$

$$0 = x$$

39. $-42 + 20 = -2x + 13x$
$$-22 = 11x$$
$$\frac{-22}{11} = \frac{11x}{11}$$
$$-2 = x$$

41. $-3x - 3x = 50 - 2$
$$-6x = 48$$
$$\frac{-6x}{-6} = \frac{48}{-6}$$
$$x = -8$$

43. $23x - 25x = 7 - 9$
$$-2x = -2$$
$$\frac{-2x}{-2} = \frac{-2}{-2}$$
$$x = 1$$

45. $\dfrac{1}{4}x - \dfrac{5}{8}x = 20 - 47$
$$\frac{2}{8}x - \frac{5}{8}x = -27$$
$$-\frac{3}{8}x = -27$$
$$-\frac{8}{3} \cdot -\frac{3}{8}x = -\frac{8}{3} \cdot -27$$
$$x = 72$$

47. $18 - 11 = \dfrac{x}{5}$
$$7 = \frac{x}{5}$$
$$5 \cdot 7 = 5 \cdot \frac{x}{5}$$
$$35 = x$$

49. $\dfrac{x}{-4} = 1 - (-6)$
$$\frac{x}{-4} = 7$$
$$-4 \cdot \frac{x}{-4} = -4 \cdot 7$$
$$x = -28$$

51. $3x + 10 = 3(5) + 10 = 15 + 10 = 25$

53. $\dfrac{x-3}{2} = \dfrac{5-3}{2} = \dfrac{2}{2} = 1$

55. $\dfrac{3x+5}{x-7} = \dfrac{3 \cdot 5 + 5}{5 - 7} = \dfrac{15 + 5}{-2} = \dfrac{20}{-2} = -10$

57. The longest bar corresponds to the year in which the number of trumpeter swans was the greatest. The year is 2010.

59. From the length of the bar, there were approximately 35,000 trumpeter swans in 2005.

61. Addition should be used to solve the equation $12 = x - 5$. Specifically, 5 should be added to both sides of the equation.

63. Division should be used to solve the equation $-7x = 21$. Specifically, both sides of the equation should be divided by -7.

65. answers may vary

67. Use $d = r \cdot t$ where $d = 390$ and $r = 60$.
$$d = r \cdot t$$
$$390 = 60 \cdot t$$
$$\frac{390}{60} = \frac{60t}{60}$$
$$\frac{13}{2} = t$$
It will take $\dfrac{13}{2} = 6.5$ hours to make the drive.

69. Use $d = r \cdot t$ where $d = 294$ and $t = 5$.
$$d = r \cdot t$$
$$294 = r \cdot 5$$
$$\frac{294}{5} = \frac{5r}{5}$$
$$58.8 = r$$
The driver should travel at 58.8 miles per hour.

71. $-0.025x = 91.2$
$$\frac{-0.025x}{-0.025} = \frac{91.2}{-0.025}$$
$$x = -3648$$

73. $\dfrac{y}{72} = -86 - (-1029)$
$$\frac{y}{72} = -86 + 1029$$
$$\frac{y}{72} = 943$$
$$72 \cdot \frac{y}{72} = 72 \cdot 943$$
$$\frac{72}{72} \cdot y = 72 \cdot 943$$
$$y = 67,896$$

75.
$$\frac{x}{-2} = 5^2 - |-10| - (-9)$$
$$\frac{x}{-2} = 25 - |-10| - (-9)$$
$$\frac{x}{-2} = 25 - 10 - (-9)$$
$$\frac{x}{-2} = 25 - 10 + 9$$
$$\frac{x}{-2} = 24$$
$$-2 \cdot \frac{x}{-2} = -2 \cdot 24$$
$$\frac{-2}{-2} \cdot x = -2 \cdot 24$$
$$x = -48$$

Integrated Review

1. $7x - 5y + 14$ is an expression because it does not contain an equal sign.

2. $7x = 35 + 14$ is an equation because it contains an equal sign.

3. $3(x - 2) = 5(x + 1) - 17$ is an equation because it contains an equal sign.

4. $-9(2x + 1) - 4(x - 2) + 14$ is an expression because it does not contain an equal sign.

5. To <u>simplify</u> an expression, we combine any like terms.

6. To <u>solve</u> an equation, we use properties of equality to find any value of the variable that makes the equation a true statement.

7. $y - x = 3 - (-1) = 3 + 1 = 4$

8. $\dfrac{8y}{4x} = \dfrac{8(3)}{4(-1)} = \dfrac{24}{-4} = -6$

9. $5x + 2y = 5(-1) + 2(3) = -5 + 6 = 1$

10. $\dfrac{y^2 + x}{2x} = \dfrac{3^2 + (-1)}{2(-1)} = \dfrac{9 + (-1)}{2(-1)} = \dfrac{8}{-2} = -4$

11. $7x + x = (7 + 1)x = 8x$

12. $6y - 10y = (6 - 10)y = -4y$

13. $2a + 5a - 9a - 2 = (2 + 5 - 9)a - 2 = -2a - 2$

14.
$$3x - y + 4 - 5x + 4y - 11$$
$$= 3x - 5x - y + 4y + 4 - 11$$
$$= (3 - 5)x + (-1 + 4)y + (4 - 11)$$
$$= -2x + 3y - 7$$

15. $-2(4x + 7) = -2 \cdot 4x + (-2) \cdot 7 = -8x - 14$

16. $-3(2x - 10) = -3(2x) - (-3)(10) = -6x + 30$

17.
$$5(y + 2) - 20 = 5 \cdot y + 5 \cdot 2 - 20$$
$$= 5y + 10 - 20$$
$$= 5y - 10$$

18.
$$12x + 3(x - 6) - 13 = 12x + 3 \cdot x - 3 \cdot 6 - 13$$
$$= 12x + 3x - 18 - 13$$
$$= (12 + 3)x - 18 - 13$$
$$= 15x - 31$$

19.
$$A = lw$$
$$= (4x - 2) \cdot 3$$
$$= 3(4x - 2)$$
$$= 3 \cdot 4x - 3 \cdot 2$$
$$= 12x - 6$$
The area is $(12x - 6)$ square meters.

20. $A = s^2 = (5y)^2 = (5y)(5y) = 5 \cdot 5 \cdot y \cdot y = 25y^2$

The area is $25y^2$ square inches.

21.
$$x + 7 = 20$$
$$x + 7 - 7 = 20 - 7$$
$$x = 13$$
Check:
$$x + 7 = 20$$
$$13 + 7 \overset{?}{=} 20$$
$$20 \overset{?}{=} 20 \quad \text{True}$$
The solution of the equation is 13.

22.
$$-11 = x - 2$$
$$-11 + 2 = x - 2 + 2$$
$$-9 = x$$
Check:
$$-11 = x - 2$$
$$-11 \overset{?}{=} -9 - 2$$
$$-11 \overset{?}{=} -11 \quad \text{True}$$
The solution of the equation is -9.

23.
$$n - \frac{2}{5} = \frac{3}{10}$$
$$n - \frac{2}{5} + \frac{2}{5} = \frac{3}{10} + \frac{2}{5}$$
$$n = \frac{3}{10} + \frac{4}{10}$$
$$n = \frac{7}{10}$$

Check:
$$n - \frac{2}{5} = \frac{3}{10}$$
$$\frac{7}{10} - \frac{2}{5} \stackrel{?}{=} \frac{3}{10}$$
$$\frac{7}{10} - \frac{4}{10} \stackrel{?}{=} \frac{3}{10}$$
$$\frac{3}{10} \stackrel{?}{=} \frac{3}{10} \quad \text{True}$$

The solution of the equation is $\frac{7}{10}$.

24.
$$-7y = 0$$
$$\frac{-7y}{-7} = \frac{0}{-7}$$
$$y = 0$$

Check:
$$-7y = 0$$
$$-7 \cdot 0 \stackrel{?}{=} 0$$
$$0 \stackrel{?}{=} 0 \quad \text{True}$$

The solution of the equation is 0.

25.
$$12 = 11x - 14x$$
$$12 = -3x$$
$$\frac{12}{-3} = \frac{-3x}{-3}$$
$$-4 = x$$

Check:
$$12 = 11x - 14x$$
$$12 \stackrel{?}{=} 11(-4) - 14(-4)$$
$$12 \stackrel{?}{=} -44 + 56$$
$$12 \stackrel{?}{=} 12 \quad \text{True}$$

The solution of the equation is -4.

26.
$$\frac{3}{5}x = 15$$
$$\frac{5}{3} \cdot \frac{3}{5}x = \frac{5}{3} \cdot 15$$
$$x = 25$$

Check:
$$\frac{3}{5}x = 15$$
$$\frac{3}{5} \cdot 25 \stackrel{?}{=} 15$$
$$15 \stackrel{?}{=} 15 \quad \text{True}$$

The solution of the equation is 25.

27.
$$x - 1.2 = -4.5 + 2.3$$
$$x - 1.2 = -2.2$$
$$x - 1.2 + 1.2 = -2.2 + 1.2$$
$$x = -1$$

Check:
$$x - 1.2 = -4.5 + 2.3$$
$$-1 - 1.2 \stackrel{?}{=} -4.5 + 2.3$$
$$-2.2 \stackrel{?}{=} -2.2 \quad \text{True}$$

The solution of the equation is -1.

28.
$$8y + 7y = -45$$
$$15y = -45$$
$$\frac{15y}{15} = \frac{-45}{15}$$
$$y = -3$$

Check:
$$8y + 7y = -45$$
$$8(-3) + 7(-3) \stackrel{?}{=} -45$$
$$-24 - 21 \stackrel{?}{=} -45$$
$$-45 \stackrel{?}{=} -45 \quad \text{True}$$

The solution of the equation is -3.

29.
$$6 - (-5) = x + 5$$
$$6 + 5 = x + 5$$
$$11 = x + 5$$
$$11 - 5 = x + 5 - 5$$
$$6 = x$$

Check:
$$6 - (-5) = x + 5$$
$$6 - (-5) \stackrel{?}{=} 6 + 5$$
$$11 \stackrel{?}{=} 11 \quad \text{True}$$

The solution of the equation is 6.

30.
$$-0.2m = -1.6$$
$$\frac{-0.2m}{-0.2} = \frac{-1.6}{-0.2}$$
$$m = 8$$

Check:
$$-0.2m = -1.6$$
$$-0.2(8) \stackrel{?}{=} -1.6$$
$$-1.6 \stackrel{?}{=} -1.6 \quad \text{True}$$

The solution of the equation is 8.

31.

$$-\frac{2}{3}n = \frac{6}{11}$$

$$-\frac{3}{2} \cdot -\frac{2}{3}n = -\frac{3}{2} \cdot \frac{6}{11}$$

$$n = -\frac{9}{11}$$

Check:

$$-\frac{2}{3}n = \frac{6}{11}$$

$$-\frac{2}{3} \cdot \left(-\frac{9}{11}\right) \stackrel{?}{=} \frac{6}{11}$$

$$\frac{6}{11} \stackrel{?}{=} \frac{6}{11} \quad \text{True}$$

The solution of the equation is $-\dfrac{9}{11}$.

32. $11x = 55$

$$\frac{11x}{11} = \frac{55}{11}$$

$$x = 5$$

Check:

$$11x = 55$$

$$11 \cdot 5 \stackrel{?}{=} 55$$

$$55 \stackrel{?}{=} 55 \quad \text{True}$$

The solution of the equation is 5.

Section 11.4 Practice

1.

$$5y - 8 = 17$$

$$5y - 8 + 8 = 17 + 8$$

$$5y = 25$$

$$\frac{5y}{5} = \frac{25}{5}$$

$$y = 5$$

Check:

$$5y - 8 = 17$$

$$5(5) - 8 \stackrel{?}{=} 17$$

$$25 - 8 \stackrel{?}{=} 17$$

$$17 \stackrel{?}{=} 17 \quad \text{True}$$

The solution is 5.

2.

$$10 - y = 45$$

$$-10 + 10 - y = -10 + 45$$

$$-1y = 35$$

$$\frac{-1y}{-1} = \frac{35}{-1}$$

$$y = -35$$

Check:

$$10 - y = 45$$

$$10 - (-35) \stackrel{?}{=} 45$$

$$10 + 35 \stackrel{?}{=} 45$$

$$45 \stackrel{?}{=} 45 \quad \text{True}$$

The solution is -35.

3.

$$11 = \frac{3}{4}y + 20$$

$$11 - 20 = \frac{3}{4}y + 20 - 20$$

$$-9 = \frac{3}{4}y$$

$$\frac{4}{3} \cdot (-9) = \frac{4}{3} \cdot \frac{3}{4}y$$

$$-12 = y$$

Check to see that the solution is -12.

4.

$$9x - 12 = x + 4$$

$$9x - 12 + 12 = x + 4 + 12$$

$$9x = x + 16$$

$$9x - x = x + 16 - x$$

$$8x = 16$$

$$\frac{8x}{8} = \frac{16}{8}$$

$$x = 2$$

Check to see that the solution is 2.

5.

$$8x + 4.2 = 10x - 11.6$$

$$8x + 4.2 - 4.2 = 10x - 11.6 - 4.2$$

$$8x = 10x - 15.8$$

$$-10x + 8x = -10x + 10x - 15.8$$

$$-2x = -15.8$$

$$\frac{-2x}{-2} = \frac{-15.8}{-2}$$

$$x = 7.9$$

Check to see that 7.9 is the solution.

6.

$$6(a - 5) = 7a - 13$$

$$6a - 30 = 7a - 13$$

$$-6a + 6a - 30 = -6a + 7a - 13$$

$$-30 = a - 13$$

$$-30 + 13 = a - 13 + 13$$

$$-17 = a$$

Check to see that -17 is the solution.

7. $4(2x-3)+4=0$

$\quad 8x-12+4=0$

$\quad\quad 8x-8=0$

$\quad 8x-8+8=0+8$

$\quad\quad\quad 8x=8$

$\quad\quad\quad \dfrac{8x}{8}=\dfrac{8}{8}$

$\quad\quad\quad\quad x=1$

Check:

$\quad 4(2x-3)+4=0$

$\quad 4(2\cdot1-3)+4 \stackrel{?}{=} 0$

$\quad 4(2-3)+4 \stackrel{?}{=} 0$

$\quad 4(-1)+4 \stackrel{?}{=} 0$

$\quad -4+4 \stackrel{?}{=} 0$

$\quad\quad\quad 0 \stackrel{?}{=} 0$ True

The solution is 1.

8. a. The difference of 110 and 80 is 30 can be translated as $110-80=30$.

    b. The product of 3 and the sum of −9 and 11 amounts to 6 can be translated as $3(-9+11)=6$.

    c. The quotient of 24 and −6 yields −4 can be translated as $\dfrac{24}{-6}=-4$.

Calculator Explorations

1. Yes, 12 is a solution.

2. Yes, 35 is a solution.

3. No, −170 is not a solution.

4. Yes, −18 is a solution.

5. Yes, −21 is a solution.

6. No, 25 is not a solution.

Vocabulary and Readiness Check

1. An example of an expression is $\underline{3x-9+x-16}$ while an example of an equation is $\underline{5(2x+6)-1=39}$.

2. To solve $\dfrac{x}{-7}=-10$, we use the <u>multiplication</u> property of equality.

3. To solve $x-7=-10$, we use the <u>addition</u> property of equality.

4. To solve $9x-6x=10+6$, first <u>combine like terms</u>.

5. To solve $5(x-1)=25$, first use the <u>distributive</u> property.

6. To solve $4x+3=19$, first use the <u>addition</u> property of equality.

Exercise Set 11.4

1. $\quad 2x-6=0$

$\quad 2x-6+6=0+6$

$\quad\quad 2x=6$

$\quad\quad \dfrac{2x}{2}=\dfrac{6}{2}$

$\quad\quad\quad x=3$

3. $\quad 3n+3.6=9.3$

$\quad 3n+3.6-3.6=9.3-3.6$

$\quad\quad\quad 3n=5.7$

$\quad\quad\quad \dfrac{3n}{3}=\dfrac{5.7}{3}$

$\quad\quad\quad\quad n=1.9$

5. $\quad 6-n=10$

$\quad 6-6-n=10-6$

$\quad\quad -n=4$

$\quad\quad \dfrac{-n}{-1}=\dfrac{4}{-1}$

$\quad\quad\quad n=-4$

7. $\quad -\dfrac{2}{5}x+19=-21$

$\quad -\dfrac{2}{5}x+19-19=-21-19$

$\quad\quad -\dfrac{2}{5}x=-40$

$\quad -\dfrac{5}{2}\cdot-\dfrac{2}{5}x=-\dfrac{5}{2}\cdot-40$

$\quad\quad\quad x=100$

9. $\quad 1.7=2y+9.5$

$\quad 1.7-9.5=2y+9.5-9.5$

$\quad\quad -7.8=2y$

$\quad\quad \dfrac{-7.8}{2}=\dfrac{2y}{2}$

$\quad\quad -3.9=y$

11.
$$2n + 8 = 0$$
$$2n + 8 - 8 = 0 - 8$$
$$2n = -8$$
$$\frac{2n}{2} = \frac{-8}{2}$$
$$n = -4$$

13.
$$3x - 7 = 4x + 5$$
$$3x - 7 - 5 = 4x + 5 - 5$$
$$3x - 12 = 4x$$
$$3x - 3x - 12 = 4x - 3x$$
$$-12 = x$$

15.
$$10x + 15 = 6x + 3$$
$$10x + 15 - 15 = 6x + 3 - 15$$
$$10x = 6x - 12$$
$$10x - 6x = 6x - 12 - 6x$$
$$4x = -12$$
$$\frac{4x}{4} = \frac{-12}{4}$$
$$x = -3$$

17.
$$9 - 3x = 14 + 2x$$
$$9 - 14 - 3x = 14 - 14 + 2x$$
$$-5 - 3x = 2x$$
$$-5 - 3x + 3x = 2x + 3x$$
$$-5 = 5x$$
$$\frac{-5}{5} = \frac{5x}{5}$$
$$-1 = x$$

19.
$$-1.4x - 2 = -1.2x + 7$$
$$-1.4x - 2 + 2 = -1.2x + 7 + 2$$
$$-1.4x = -1.2x + 9$$
$$1.2x - 1.4x = 1.2x - 1.2x + 9$$
$$-0.2x = 9$$
$$\frac{-0.2x}{-0.2} = \frac{9}{-0.2}$$
$$x = -45$$

21.
$$x + 20 + 2x = -10 - 2x - 15$$
$$x + 2x + 20 = -10 - 15 - 2x$$
$$3x + 20 = -25 - 2x$$
$$3x + 2x + 20 = -25 - 2x + 2x$$
$$5x + 20 = -25$$
$$5x + 20 - 20 = -25 - 20$$
$$5x = -45$$
$$\frac{5x}{5} = \frac{-45}{5}$$
$$x = -9$$

23.
$$40 + 4y - 16 = 13y - 12 - 3y$$
$$40 - 16 + 4y = 13y - 3y - 12$$
$$24 + 4y = 10y - 12$$
$$24 + 4y - 4y = 10y - 4y - 12$$
$$24 = 6y - 12$$
$$24 + 12 = 6y - 12 + 12$$
$$36 = 6y$$
$$\frac{36}{6} = \frac{6y}{6}$$
$$6 = y$$

25.
$$-2(y + 4) = 2$$
$$-2y - 8 = 2$$
$$-2y - 8 + 8 = 2 + 8$$
$$-2y = 10$$
$$\frac{-2y}{-2} = \frac{10}{-2}$$
$$y = -5$$

27.
$$3(x - 1) - 12 = 0$$
$$3x - 3 - 12 = 0$$
$$3x - 15 = 0$$
$$3x - 15 + 15 = 0 + 15$$
$$3x = 15$$
$$\frac{3x}{3} = \frac{15}{3}$$
$$x = 5$$

29.
$$35 - 17 = 3(x - 2)$$
$$18 = 3x - 6$$
$$18 + 6 = 3x - 6 + 6$$
$$24 = 3x$$
$$\frac{24}{3} = \frac{3x}{3}$$
$$8 = x$$

31.
$$2(y - 3) = y - 6$$
$$2y - 6 = y - 6$$
$$2y - 6 + 6 = y - 6 + 6$$
$$2y = y$$
$$2y - y = y - y$$
$$y = 0$$

33.
$$2t - 1 = 3(t + 7)$$
$$2t - 1 = 3t + 21$$
$$2t - 1 - 21 = 3t + 21 - 21$$
$$2t - 22 = 3t$$
$$2t - 2t - 22 = 3t - 2t$$
$$-22 = t$$

35. $3(5c+1)-12=13c+3$
$15c+3-12=13c+3$
$15c-9=13c+3$
$15c-9+9=13c+3+9$
$15c=13c+12$
$15c-13c=13c+12-13c$
$2c=12$
$\dfrac{2c}{2}=\dfrac{12}{2}$
$c=6$

37. $-4x=44$
$\dfrac{-4x}{-4}=\dfrac{44}{-4}$
$x=-11$

39. $x+9=2$
$x+9-9=2-9$
$x=-7$

41. $8-b=13$
$8-8-b=13-8$
$-b=5$
$\dfrac{-b}{-1}=\dfrac{5}{-1}$
$b=-5$

43. $3r+4=19$
$3r+4-4=19-4$
$3r=15$
$\dfrac{3r}{3}=\dfrac{15}{3}$
$r=5$

45. $2x-1=-7$
$2x-1+1=-7+1$
$2x=-6$
$\dfrac{2x}{2}=\dfrac{-6}{2}$
$x=-3$

47. $7=4c-1$
$7+1=4c-1+1$
$8=4c$
$\dfrac{8}{4}=\dfrac{4c}{4}$
$2=c$

49. $9a+29=-7$
$9a+29-29=-7-29$
$9a=-36$
$\dfrac{9a}{9}=\dfrac{-36}{9}$
$a=-4$

51. $0=4x+4$
$0-4=4x+4-4$
$-4=4x$
$\dfrac{-4}{4}=\dfrac{4x}{4}$
$-1=x$

53. $11(x-2)=22$
$11x-22=22$
$11x-22+22=22+22$
$11x=44$
$\dfrac{11x}{11}=\dfrac{44}{11}$
$x=4$

55. $-7c+1=-20$
$-7c+1-1=-20-1$
$-7c=-21$
$\dfrac{-7c}{-7}=\dfrac{-21}{-7}$
$c=3$

57. $3(x-5)=-7-11$
$3x-15=-18$
$3x-15+15=-18+15$
$3x=-3$
$\dfrac{3x}{3}=\dfrac{-3}{3}$
$x=-1$

59. $-5+7k=-13+8k$
$-5+13+7k=-13+13+8k$
$8+7k=8k$
$8+7k-7k=8k-7k$
$8=k$

61. $4x+3=2x+11$
$4x+3-3=2x+11-3$
$4x=2x+8$
$4x-2x=2x+8-2x$
$2x=8$
$\dfrac{2x}{2}=\dfrac{8}{2}$
$x=4$

63. $-8(n+2)+17 = -6n-5$
$-8n-16+17 = -6n-5$
$-8n+1 = -6n-5$
$-8n+1+5 = -6n-5+5$
$-8n+6 = -6n$
$-8n+8n+6 = -6n+8n$
$6 = 2n$
$\dfrac{6}{2} = \dfrac{2n}{2}$
$3 = n$

65. $\dfrac{3}{8}x+14 = \dfrac{5}{8}x-2$

$\dfrac{3}{8}x+14+2 = \dfrac{5}{8}x-2+2$

$\dfrac{3}{8}x+16 = \dfrac{5}{8}x$

$-\dfrac{3}{8}x+\dfrac{3}{8}x+16 = -\dfrac{3}{8}x+\dfrac{5}{8}x$

$16 = \dfrac{2}{8}x$

$\dfrac{8}{2}\cdot16 = \dfrac{8}{2}\cdot\dfrac{2}{8}x$

$64 = x$

67. $10+5(z-2) = -4z+1$
$10+5z-10 = -4z+1$
$5z = -4z+1$
$5z+4z = -4z+4z+1$
$9z = 1$
$\dfrac{9z}{9} = \dfrac{1}{9}$
$z = \dfrac{1}{9}$

69. $\dfrac{5}{8}a = \dfrac{1}{8}a+\dfrac{3}{4}$

$-\dfrac{1}{8}a+\dfrac{5}{8}a = -\dfrac{1}{8}a+\dfrac{1}{8}a+\dfrac{3}{4}$

$\dfrac{4}{8}a = \dfrac{3}{4}$

$\dfrac{8}{4}\cdot\dfrac{4}{8}a = \dfrac{8}{4}\cdot\dfrac{3}{4}$

$a = \dfrac{3}{2}$

71. $7(6+w) = 6(w-2)$
$42+7w = 6w-12$
$42-42+7w = 6w-12-42$
$7w = 6w-54$
$7w-6w = 6w-6w-54$
$w = -54$

73. $3+2(2n-5) = 1$
$3+4n-10 = 1$
$4n-7 = 1$
$4n-7+7 = 1+7$
$4n = 8$
$\dfrac{4n}{4} = \dfrac{8}{4}$
$n = 2$

75. $2(3z-2)-2(5-2z) = 4$
$6z-4-10+4z = 4$
$10z-14 = 4$
$10z-14+14 = 4+14$
$10z = 18$
$\dfrac{10z}{10} = \dfrac{18}{10}$
$z = \dfrac{9}{5}$

77. $-20-(-50) = \dfrac{x}{9}$

$-20+50 = \dfrac{x}{9}$

$30 = \dfrac{x}{9}$

$9\cdot30 = 9\cdot\dfrac{x}{9}$

$270 = x$

79. $12+5t = 6(t+2)$
$12+5t = 6t+12$
$12+5t-5t = 6t-5t+12$
$12 = t+12$
$12-12 = t+12-12$
$0 = t$

81. $3(5c-1)-2=13c+3$
$15c-3-2=13c+3$
$15c-5=13c+3$
$15c-13c-5=13c-13c+3$
$2c-5=3$
$2c-5+5=3+5$
$2c=8$
$\dfrac{2c}{2}=\dfrac{8}{2}$
$c=4$

83. $10+5(z-2)=4z+1$
$10+5z-10=4z+1$
$5z=4z+1$
$5z-4z=4z-4z+1$
$z=1$

85. "The sum of" indicates addition.
$-42+16=-26$

87. "The product of" indicates multiplication.
$-5(-29)=145$

89. "Times" indicates multiplication and "difference" indicates subtraction.
$3(-14-2)=-48$

91. "The quotient of" indicates division.
$\dfrac{100}{2(50)}=1$

93. From the height of the bar, there were approximately 97 or 98 million electronically filed returns for 2010.

95. Subtract the number of projected returns for 2013 from the number of projected returns for 2016.
118 million – 109 million = 9 million
The number of electronically filed returns is projected to increase from 2013 to 2016 by 9 million.

97. The first step in solving $2x-5=-7$ is to add 5 to both sides, which is choice b.

99. The first step in solving $-3x=-12$ is to divide both sides by -3, which is choice a.

101. The error is in the second line.
$2(3x-5)=5x-7$
$6x-10=5x-7$
$6x-10+10=5x-7+10$
$6x=5x+3$
$6x-5x=5x+3-5x$
$x=3$

103. $(-8)^2+3x=5x+4^3$
$64+3x=5x+64$
$64+3x-3x=5x-3x+64$
$64=2x+64$
$64-64=2x+64-64$
$0=2x$
$\dfrac{0}{2}=\dfrac{2x}{2}$
$0=x$

105. $2^3(x+4)=3^2(x+4)$
$8(x+4)=9(x+4)$
$8x+32=9x+36$
$8x-8x+32=9x-8x+36$
$32=x+36$
$32-36=x+36-36$
$-4=x$

107. no; answers may vary

109. Use $C=\dfrac{5}{9}(F-32)$ with $C=50.7$.

$C=\dfrac{5}{9}(F-32)$

$50.7=\dfrac{5}{9}(F-32)$

$\dfrac{9}{5}\cdot 50.7=\dfrac{9}{5}\cdot\dfrac{5}{9}(F-32)$

$91.26=F-32$
$91.26+32=F-32+32$
$123.26=F$

The temperature was 123.26°F.

111. Use $C = \dfrac{5}{9}(F - 32)$ with $C = -23.0$.

$$C = \dfrac{5}{9}(F - 32)$$
$$-23.0 = \dfrac{5}{9}(F - 32)$$
$$\dfrac{9}{5} \cdot (-23.0) = \dfrac{9}{5} \cdot \dfrac{5}{9}(F - 32)$$
$$-41.4 = F - 32$$
$$-41.4 + 32 = F - 32 + 32$$
$$-9.4 = F$$

The temperature was $-9.4°F$.

Section 11.5 Practice

1. a. "Twice a number" is $2x$.

b. "8 increased by a number" is $8 + x$.

c. "10 minus a number" is $10 - x$.

d. "10 subtracted from a number" is $x - 10$.

e. "The quotient of 6 and a number" is $\dfrac{6}{x}$.

f. "The sum of 14 and triple a number" is $14 + 3x$.

2. a. "Five times a number is 20" is $5x = 20$.

b. "The sum of a number and -5 yields 14" is $x + (-5) = 14$.

c. "Ten subtracted from a number amounts to -23" is $x - 10 = -23$.

d. "Five times the difference of a number and 7 is equal to -8" is $5(x - 7) = -8$.

e. "The quotient of triple a number and 5 gives 1" is $\dfrac{3x}{5} = 1$ or $3x \div 5 = 1$.

3. "The difference of a number and 2 equals 6 added to three times the number" is $x - 2 = 6 + 3x$.

$$x - 2 = 6 + 3x$$
$$-x + x - 2 = -x + 6 + 3x$$
$$-2 = 6 + 2x$$
$$-6 - 2 = -6 + 6 + 2x$$
$$-8 = 2x$$
$$\dfrac{-8}{2} = \dfrac{2x}{2}$$
$$-4 = x$$

4. Let x be the distance from Denver to San Francisco. Since the distance from Cincinnati to Denver is 71 miles less than the distance from Denver to San Francisco, the distance from Cincinnati to Denver is $x - 71$. Since the total of the two distances is 2399, the sum of x and $x - 71$ is 2399.

$$x + x - 71 = 2399$$
$$2x - 71 = 2399$$
$$2x - 71 + 71 = 2399 + 71$$
$$2x = 2470$$
$$\dfrac{2x}{2} = \dfrac{2470}{2}$$
$$x = 1235$$

The distance from Denver to San Francisco is 1235 miles.

5. Let x be the amount her son receives. Since her husband receives twice as much as her son, her husband receives $2x$. Since the total estate is $\$57{,}000$, the sum of x and $2x$ is 57,000.

$$x + 2x = 57{,}000$$
$$3x = 57{,}000$$
$$\dfrac{3x}{3} = \dfrac{57{,}000}{3}$$
$$x = 19{,}000$$

$$2x = 2(19{,}000) = 38{,}000$$

Her husband will receive $\$38{,}000$ and her son will receive $\$19{,}000$.

Exercise Set 11.5

1. "The sum of a number and five" is $x + 5$.

3. "The total of a number and eight" is $x + 8$.

5. "Twenty decreased by a number" is $20 - x$.

7. "The product of 512 and a number" is $512x$.

9. "A number divided by 2" is $\dfrac{x}{2}$.

11. "The sum of seventeen, a number, and the product of five and the number" is
$17 + x + 5x$.

13. "A number added to -5 is -7" is
$-5 + x = -7$.

15. "Three times a number yields 27" is
$3x = 27$.

17. "A number subtracted from -20 amounts to 104" is $-20 - x = 104$.

19. "The product of five and a number" is $5x$.

21. "A number subtracted from 11" is $11 - x$.

23. "Twice a number gives 108" is $2x = 108$.

25. "Fifty decreased by eight times a number" is
$50 - 8x$.

27. "The product of 5 and the sum of -3 and a number is -20" is $5(-3 + x) = -20$.

29. "Three times a number, added to 9 is 33" is
$3x + 9 = 33$.
$$3x + 9 = 33$$
$$3x + 9 - 9 = 33 - 9$$
$$3x = 24$$
$$\frac{3x}{3} = \frac{24}{3}$$
$$x = 8$$

31. "The sum of 3, 4, and a number amounts to 16" is $3 + 4 + x = 16$.
$$3 + 4 + x = 16$$
$$7 + x = 16$$
$$-7 + 7 + x = -7 + 16$$
$$x = 9$$

33. "The difference of a number and 3 is equal to the quotient of 10 and 5" is $x - 3 = \dfrac{10}{5}$.
$$x - 3 = \frac{10}{5}$$
$$x - 3 = 2$$
$$x - 3 + 3 = 2 + 3$$
$$x = 5$$

35. "Thirty less a number is equal to the product of 3 and the sum of the number and 6" is
$30 - x = 3(x + 6)$.
$$30 - x = 3(x + 6)$$
$$30 - x = 3x + 18$$
$$-30 + 30 - x = 3x + 18 - 30$$
$$-x = 3x - 12$$
$$-3x - x = -3x + 3x - 12$$
$$-4x = -12$$
$$\frac{-4x}{-4} = \frac{-12}{-4}$$
$$x = 3$$

37. "40 subtracted from five times a number is 8 more than the number" is $5x - 40 = x + 8$.
$$5x - 40 = x + 8$$
$$5x - 40 + 40 = x + 8 + 40$$
$$5x = x + 48$$
$$5x - x = x + 48 - x$$
$$4x = 48$$
$$\frac{4x}{4} = \frac{48}{4}$$
$$x = 12$$

39. "Three times the difference of some number and 5 amounts to the quotient of 108 and 12" is
$3(x - 5) = \dfrac{108}{12}$.
$$3(x - 5) = \frac{108}{12}$$
$$3x - 15 = 9$$
$$3x - 15 + 15 = 9 + 15$$
$$3x = 24$$
$$\frac{3x}{3} = \frac{24}{3}$$
$$x = 8$$

41. "The product of 4 and a number is the same as 30 less twice that same number" is
$4x = 30 - 2x$.
$$4x = 30 - 2x$$
$$4x + 2x = 30 - 2x + 2x$$
$$6x = 30$$
$$\frac{6x}{6} = \frac{30}{6}$$
$$x = 5$$

43. The equation is $x + x - 28 = 82$.

$$x + x - 28 = 82$$
$$2x - 28 = 82$$
$$2x - 28 + 28 = 82 + 28$$
$$2x = 110$$
$$\frac{2x}{2} = \frac{110}{2}$$
$$x = 55$$

California has 55 votes and Florida has $55 - 28 = 27$ votes.

45. Let x be the fastest speed of the pheasant. Since a falcon's fastest speed is five times as fast as a pheasant, a falcon's fastest speed is $5x$. Since the total speeds for these two birds is 222 miles per hour, the sum of x and $5x$ is 222.

$$x + 5x = 222$$
$$6x = 222$$
$$\frac{6x}{6} = \frac{222}{6}$$
$$x = 37$$

The pheasant's fastest speed is 37 miles per hour and the falcon's fastest speed is $5(37) = 185$ miles per hour.

47. Let x be the enrollment (in thousands) at the largest university in India. Since the largest university in Pakistan has 306 thousand more students than the one in India, its enrollment is $x + 306$. Since the combined enrollment is 3306 thousand students, the sum of x and $x + 306$ is 3306.

$$x + x + 306 = 3306$$
$$2x + 306 = 3306$$
$$2x + 306 - 306 = 3306 - 306$$
$$2x = 3000$$
$$\frac{2x}{2} = \frac{3000}{2}$$
$$x = 1500$$

The enrollment at the largest university in India is 1500 thousand students, and the enrollment at the largest university in Pakistan is $1500 + 306 = 1806$ thousand students.

49. Let x be the cost of the games. Since the cost of the Xbox 360 is 3 times as much as the games, the cost of the Xbox 360 is $3x$. Since the total cost is \$560, the sum of x and $3x$ is 560.

$$x + 3x = 560$$
$$4x = 560$$
$$\frac{4x}{4} = \frac{560}{4}$$
$$x = 140$$

The games cost \$140 and the Xbox 360 costs $3(140) = \$420$.

51. Let x be the distance from Los Angeles to Tokyo. Since the distance from New York to London is 2001 miles less than the distance from Los Angeles to Tokyo, the distance from New York to London is $x - 2001$. Since the total of the two distances is 8939, the sum of x and $x - 2001$ is 8939.

$$x + x - 2001 = 8939$$
$$2x - 2001 = 8939$$
$$2x - 2001 + 2001 = 8939 + 2001$$
$$2x = 10,940$$
$$\frac{2x}{2} = \frac{10,940}{2}$$
$$x = 5470$$

The distance from Los Angeles to Tokyo is 5470 miles.

53. Let x be the capacity of Michigan Stadium. Since the capacity of Beaver Stadium is 1081 more than that of Michigan Stadium, the capacity of Beaver Stadium is $x + 1081$. Since the combined capacity is 213,483, the sum of x and $x + 1081$ is 213,483.

$$x + x + 1081 = 213,483$$
$$2x + 1081 = 213,483$$
$$2x + 1081 - 1081 = 213,483 - 1081$$
$$2x = 212,402$$
$$\frac{2x}{2} = \frac{212,402}{2}$$
$$x = 106,201$$

The capacity of Michigan Stadium is 106,201 and the capacity of Beaver Stadium is $106,201 + 1081 = 107,282$.

55. Let x be the number of tourists projected to visit Spain in 2020. Since the number of tourists projected to visit China in 2020 is twice the number projected for Spain, the number projected for China is $2x$. Since the total number of tourists projected for the two countries is 210 million, the sum of x and $2x$ is 210 million.

$$x + 2x = 210$$
$$3x = 210$$
$$\frac{3x}{3} = \frac{210}{3}$$
$$x = 70$$

$2x = 2(70) = 140$

70 million tourists are projected to visit Spain in 2020; 140 million are projected to visit China in 2020.

57. Let x be the number of points scored by the Arizona Cardinals. Since the Pittsburgh Steelers scored 4 more points than Arizona Cardinals, the number of points scored by the Pittsburgh Steels is $x + 4$. Since both teams together scored 50 points, the sum of x and $x + 4$ is 50.

$$x + x + 4 = 50$$
$$2x + 4 = 50$$
$$2x + 4 - 4 = 50 - 4$$
$$2x = 46$$
$$\frac{2x}{2} = \frac{56}{2}$$
$$x = 23$$

The Pittsburgh Steelers scored $23 + 4 = 27$ points.

59. Let x be the projected shortage of nurses in 2010. Then the projected shortage of nurses in 2020 is $x + 533{,}201$, and the sum of these two quantities is $1{,}083{,}631$.

$$x + x + 533{,}201 = 1{,}083{,}631$$
$$2x + 533{,}201 = 1{,}083{,}631$$
$$2x + 533{,}201 - 533{,}201 = 1{,}083{,}631 - 533{,}201$$
$$2x = 550{,}430$$
$$\frac{2x}{2} = \frac{550{,}430}{2}$$
$$x = 275{,}215$$

The shortage of nurses is projected to be 275,215 in 2010 and
$275{,}215 + 533{,}207 = 808{,}416$ in 2020.

61. Let x be the number of cars manufactured in Spain. Since the number of cars manufactured in Germany is twice as many as those manufactured in Spain, the number of cars manufactured in Germany is $2x$. Since the total number of these cars is 19,827, the sum of x and $2x$ is 19,827.

$$x + 2x = 19{,}827$$
$$3x = 19{,}827$$
$$\frac{3x}{3} = \frac{19{,}827}{3}$$
$$x = 6609$$

The number of cars manufactured in Spain is 6609 per day and the number in Germany is $2(6609) = 13{,}218$ per day.

63. Let x be the amount the biker received for the accessories. Since he received five times as much for the bike, he received $5x$ for the bike. Since he received a total of $270 for the bike and accessories, the sum of x and $5x$ is 270.

$$x + 5x = 270$$
$$6x = 270$$
$$\frac{6x}{6} = \frac{270}{6}$$
$$x = 45$$

Thus, the biker received $5 \cdot \$45 = \$225$ for the bike.

65. Let x be the number of computers in Japan. Since the USA has 162,550 million more computers than Japan, the number of computers in the USA is $x + 162{,}550$. Since the total number of computers in the two countries is 318,450 million, the sum of x and $x + 162{,}550$ is 318,450.

$$x + x + 162{,}550 = 318{,}450$$
$$2x + 162{,}550 = 318{,}450$$
$$2x + 162{,}550 - 162{,}550 = 318{,}450 - 162{,}550$$
$$2x = 155{,}900$$
$$\frac{2x}{2} = \frac{155{,}900}{2}$$
$$x = 77{,}950$$

Japan has 77,950 million computers, and the USA has $77{,}950 + 162{,}550 = 240{,}500$ million computers.

67. To round 586 to the nearest ten, observe that the digit in the ones place is 6. Since this digit is at least 5, we add 1 to the digit in the tens place. 586 rounded to the nearest ten is 590.

69. To round 1026 to the nearest hundred, observe that the digit in the tens place is 2. Since this digit is less than 5, we do not add 1 to the digit in the hundreds place. 1026 rounded to the nearest hundred is 1000.

71. To round 2986 to the nearest thousand, observe that the digit in the hundreds place is 9. Since this digit is at least 5, we add 1 to the digit in the thousands place. 2986 rounded to the nearest thousand is 3000.

73. Yes; answers may vary

75. Use $P = A + C$, where $P = 230{,}000$ and $C = 13{,}800$.

$$P = A + C$$
$$230{,}000 = A + 13{,}800$$
$$230{,}000 - 13{,}800 = A + 13{,}800 - 13{,}800$$
$$216{,}200 = A$$

The seller received $216,200.

77. Use $P = C + M$ where $P = 999$ and $M = 450$.

$$P = C + M$$
$$999 = C + 450$$
$$999 - 450 = C + 450 - 450$$
$$549 = C$$

The wholesale cost is \$549.

Chapter 11 Vocabulary Check

1. An algebraic expression is <u>simplified</u> when all like terms have been <u>combined</u>.

2. Terms that are exactly the same, except that they may have different numerical coefficients, are called <u>like</u> terms.

3. A letter used to represent a number is called a <u>variable</u>.

4. A combination of operations on variables and numbers is called an <u>algebraic expression</u>.

5. The addends of an algebraic expression are called the <u>terms</u> of the expression.

6. The number factor of a variable term is called the <u>numerical coefficient</u>.

7. Replacing a variable in an expression by a number and then finding the value of the expression is called <u>evaluating the expression</u> for the variable.

8. A term that is a number only is called a <u>constant</u>.

9. An <u>equation</u> is of the form expression = expression.

10. A <u>solution</u> of an equation is a value for the variable that makes an equation a true statement.

11. To multiply $-3(2x + 1)$, we use the <u>distributive</u> property.

12. By the <u>multiplication</u> property of equality, we may multiply or divide both sides of an equation by any nonzero number without changing the solution of the equation.

13. By the <u>addition</u> property of equality, the same number may be added to or subtracted from both sides of an equation without changing the solution of the equation.

Chapter 11 Review

1. $\dfrac{2x}{z} = \dfrac{2(5)}{-2} = \dfrac{10}{-2} = -5$

2. $4x - 3 = 4 \cdot 5 - 3 = 20 - 3 = 17$

3. $\dfrac{x+7}{y} = \dfrac{5+7}{0} = \dfrac{12}{0}$ is undefined.

4. $\dfrac{y}{5x} = \dfrac{0}{5 \cdot 5} = \dfrac{0}{25} = 0$

5. $x^3 - 2z = 5^3 - 2(-2) = 125 + 4 = 129$

6. $\dfrac{7+x}{3z} = \dfrac{7+5}{3(-2)} = \dfrac{12}{-6} = -2$

7. $V = s^3$
   $V = 2^3$
   $V = 8$
   The volume is 8 cubic feet.

8. $V = s^3$
   $V = 4^3$
   $V = 64$
   The volume is 64 cubic feet.

9. $I = prt$
   $I = 5000 \cdot 0.06 \cdot 6$
   $I = 300 \cdot 6$
   $I = 1800$
   The interest will be \$1800.

10. $I = prt$
    $I = 2000 \cdot 0.05 \cdot 3$
    $I = 100 \cdot 3$
    $I = 300$
    The interest will be \$300.

11. $-6x - 9x = (-6 - 9)x = -15x$

12. $\dfrac{2}{3}x - \dfrac{9}{10}x = \dfrac{20}{30}x - \dfrac{27}{30}x = \left(\dfrac{20}{30} - \dfrac{27}{30}\right)x = -\dfrac{7}{30}x$

13. $2y - 10 - 8y = 2y - 8y - 10$
    $= (2 - 8)y - 10$
    $= -6y - 10$

14. $8a + a - 7 - 15a = 8a + a - 15a - 7$
$\qquad\qquad\qquad = (8 + 1 - 15)a - 7$
$\qquad\qquad\qquad = -6a - 7$

15. $y + 3 - 9y - 1 = y - 9y + 3 - 1$
$\qquad\qquad\qquad = (1 - 9)y + 2$
$\qquad\qquad\qquad = -8y + 2$

16. $1.7x - 3.2 + 2.9x - 8.7 = 1.7x + 2.9x - 3.2 - 8.7$
$\qquad\qquad\qquad\qquad = (1.7 + 2.9)x - 11.9$
$\qquad\qquad\qquad\qquad = 4.6x - 11.9$

17. $-2(4y) = (-2 \cdot 4)y = -8y$

18. $3(5y - 8) = 3 \cdot 5y - 3 \cdot 8 = 15y - 24$

19. $7x + 3(x - 4) + x = 7x + 3 \cdot x - 3 \cdot 4 + x$
$\qquad\qquad\qquad\qquad = 7x + 3x - 12 + x$
$\qquad\qquad\qquad\qquad = 7x + 3x + x - 12$
$\qquad\qquad\qquad\qquad = (7 + 3 + 1)x - 12$
$\qquad\qquad\qquad\qquad = 11x - 12$

20. $4(x - 7) + 21 = 4 \cdot x - 4 \cdot 7 + 21$
$\qquad\qquad\qquad = 4x - 28 + 21$
$\qquad\qquad\qquad = 4x - 7$

21. $3(5a - 2) + 10(-2a + 1)$
$\quad = 3 \cdot 5a - 3 \cdot 2 + 10 \cdot (-2a) + 10$
$\quad = 15a - 6 - 20a + 10$
$\quad = 15a - 20a - 6 + 10$
$\quad = -5a + 4$

22. $6y + 3 + 2(3y - 6) = 6y + 3 + 2 \cdot 3y - 2 \cdot 6$
$\qquad\qquad\qquad\qquad = 6y + 3 + 6y - 12$
$\qquad\qquad\qquad\qquad = 6y + 6y + 3 - 12$
$\qquad\qquad\qquad\qquad = 12y - 9$

23. $A = lw$
$\quad = (2x - 1)(3)$
$\quad = 3(2x - 1)$
$\quad = 3 \cdot 2x - 3 \cdot 1$
$\quad = 6x - 3$
The area is $(6x - 3)$ square yards.

24. $A = s^2 = (5y)^2 = 5y \cdot 5y = 5 \cdot 5 \cdot y \cdot y = 25y^2$
The area is $25y^2$ square meters.

25. $5(2 - x) = -10$
$\quad 5(2 - 4) \overset{?}{=} -10$
$\quad\quad 5(-2) \overset{?}{=} -10$
$\quad\quad\quad -10 \overset{?}{=} -10$　True
Yes, 4 is a solution.

26. $6y + 2 = 23 + 4y$
$\quad 6 \cdot 0 + 2 \overset{?}{=} 23 + 4 \cdot 0$
$\quad\quad 0 + 2 \overset{?}{=} 23 + 0$
$\quad\quad\quad 2 \overset{?}{=} 23$　False
No, 0 is not a solution.

27. $z - 5 = -7$
$\quad z - 5 + 5 = -7 + 5$
$\quad\quad z = -2$

28. $x + 1 = 8$
$\quad x + 1 - 1 = 8 - 1$
$\quad\quad x = 7$

29. $x + \dfrac{7}{8} = \dfrac{3}{8}$
$\quad x + \dfrac{7}{8} - \dfrac{7}{8} = \dfrac{3}{8} - \dfrac{7}{8}$
$\quad\quad x = -\dfrac{4}{8}$
$\quad\quad x = -\dfrac{1}{2}$

30. $y + \dfrac{4}{11} = -\dfrac{2}{11}$
$\quad y + \dfrac{4}{11} - \dfrac{4}{11} = -\dfrac{2}{11} - \dfrac{4}{11}$
$\quad\quad y = -\dfrac{6}{11}$

31. $n + 18 = 10 - (-2)$
$\quad n + 18 = 10 + 2$
$\quad n + 18 = 12$
$\quad n + 18 - 18 = 12 - 18$
$\quad\quad n = -6$

32. $15 = 8x + 35 - 7x$
$\quad 15 = x + 35$
$\quad 15 - 35 = x + 35 - 35$
$\quad\quad -20 = x$

33. $m - 3.9 = -2.6$
$\quad m - 3.9 + 3.9 = -2.6 + 3.9$
$\quad\quad m = 1.3$

34.
$$z - 4.6 = -2.2$$
$$z - 4.6 + 4.6 = -2.2 + 4.6$$
$$z = 2.4$$

35.
$$-3y = -21$$
$$\frac{-3y}{-3} = \frac{-21}{-3}$$
$$y = 7$$

36.
$$-8x = 72$$
$$\frac{-8x}{-8} = \frac{72}{-8}$$
$$x = -9$$

37.
$$-5n = -5$$
$$\frac{-5n}{-5} = \frac{-5}{-5}$$
$$n = 1$$

38.
$$-3a = 15$$
$$\frac{-3a}{-3} = \frac{15}{-3}$$
$$a = -5$$

39.
$$\frac{2}{3}x = -\frac{8}{15}$$
$$\frac{3}{2} \cdot \frac{2}{3}x = \frac{3}{2} \cdot \left(-\frac{8}{15}\right)$$
$$x = -\frac{4}{5}$$

40.
$$-\frac{7}{8}y = 21$$
$$-\frac{8}{7}\left(-\frac{7}{8}y\right) = -\frac{8}{7} \cdot 21$$
$$y = -24$$

41.
$$-1.2x = 144$$
$$\frac{-1.2x}{-1.2} = \frac{144}{-1.2}$$
$$x = -120$$

42.
$$-0.8y = -10.4$$
$$\frac{-0.8y}{-0.8} = \frac{-10.4}{-0.8}$$
$$y = 13$$

43.
$$-5x = 100 - 120$$
$$-5x = -20$$
$$\frac{-5x}{-5} = \frac{-20}{-5}$$
$$x = 4$$

44.
$$18 - 30 = -4x$$
$$-12 = -4x$$
$$\frac{-12}{-4} = \frac{-4x}{-4}$$
$$3 = x$$

45.
$$3x - 4 = 11$$
$$3x - 4 + 4 = 11 + 4$$
$$3x = 15$$
$$\frac{3x}{3} = \frac{15}{3}$$
$$x = 5$$

46.
$$6y + 1 = 73$$
$$6y + 1 - 1 = 73 - 1$$
$$6y = 72$$
$$\frac{6y}{6} = \frac{72}{6}$$
$$y = 12$$

47.
$$-\frac{5}{9}x + 23 = -12$$
$$-\frac{5}{9}x + 23 - 23 = -12 - 23$$
$$-\frac{5}{9}x = -35$$
$$-\frac{9}{5} \cdot \left(-\frac{5}{9}x\right) = -\frac{9}{5} \cdot (-35)$$
$$x = 63$$

48.
$$-\frac{2}{3}x - 11 = \frac{2}{3}x - 55$$
$$-\frac{2}{3}x - 11 + 11 = \frac{2}{3}x - 55 + 11$$
$$-\frac{2}{3}x = \frac{2}{3}x - 44$$
$$-\frac{2}{3}x - \frac{2}{3}x = -\frac{2}{3}x + \frac{2}{3}x - 44$$
$$-\frac{4}{3}x = -44$$
$$-\frac{3}{4} \cdot \left(-\frac{4}{3}x\right) = -\frac{3}{4} \cdot (-44)$$
$$x = 33$$

49.
$$6.8 + 4y = -2.2$$
$$-6.8 + 6.8 + 4y = -2.2 - 6.8$$
$$4y = -9$$
$$\frac{4y}{4} = \frac{-9}{4}$$
$$y = -2.25$$

50.
$$-9.6 + 5y = -3.1$$
$$9.6 - 9.6 + 5y = 9.6 - 3.1$$
$$5y = 6.5$$
$$\frac{5y}{5} = \frac{6.5}{5}$$
$$y = 1.3$$

51.
$$2x + 7 = 6x - 1$$
$$2x + 7 - 7 = 6x - 1 - 7$$
$$2x = 6x - 8$$
$$-6x + 2x = -6x + 6x - 8$$
$$-4x = -8$$
$$\frac{-4x}{-4} = \frac{-8}{-4}$$
$$x = 2$$

52.
$$5x - 18 = -4x + 36$$
$$5x - 18 + 18 = -4x + 36 + 18$$
$$5x = -4x + 54$$
$$5x + 4x = -4x + 54 + 4x$$
$$9x = 54$$
$$\frac{9x}{9} = \frac{54}{9}$$
$$x = 6$$

53.
$$5(n - 3) = 7 + 3n$$
$$5n - 15 = 7 + 3n$$
$$5n - 15 + 15 = 7 + 3n + 15$$
$$5n = 22 + 3n$$
$$5n - 3n = 22 + 3n - 3n$$
$$2n = 22$$
$$\frac{2n}{2} = \frac{22}{2}$$
$$n = 11$$

54.
$$7(2 + x) = 4x - 1$$
$$14 + 7x = 4x - 1$$
$$-14 + 14 + 7x = 4x - 1 - 14$$
$$7x = 4x - 15$$
$$7x - 4x = 4x - 15 - 4x$$
$$3x = -15$$
$$\frac{3x}{3} = \frac{-15}{3}$$
$$x = -5$$

55.
$$2(4n - 11) + 8 = 5n + 4$$
$$8n - 22 + 8 = 5n + 4$$
$$8n - 14 = 5n + 4$$
$$8n - 14 + 14 = 5n + 4 + 14$$
$$8n = 5n + 18$$
$$-5n + 8n = -5n + 5n + 18$$
$$3n = 18$$
$$\frac{3n}{3} = \frac{18}{3}$$
$$n = 6$$

56.
$$3(5x - 6) + 9 = 13x + 7$$
$$15x - 18 + 9 = 13x + 7$$
$$15x - 9 = 13x + 7$$
$$15x - 9 + 9 = 13x + 7 + 9$$
$$15x = 13x + 16$$
$$15x - 13x = 13x + 16 - 13x$$
$$2x = 16$$
$$\frac{2x}{2} = \frac{16}{2}$$
$$x = 8$$

57. "The difference of 20 and −8 is 28" is
$20 - (-8) = 28$.

58. "Nineteen subtracted from −2 amounts to −21" is
$-2 - 19 = -21$.

59. "The quotient of −75 and the sum of 5 and 20 is
equal to −3" is $\dfrac{-75}{5 + 20} = -3$.

60. "The product of −5 and the sum of −2 and 6
yields −20" is $-5(-2 + 6) = -20$.

61. "The quotient of 70 and a number" is $\dfrac{70}{x}$.

62. "The difference of a number and 13" is $x - 13$.

63. "A number subtracted from 85" is $85 - x$.

64. "Eleven added to twice a number" is $2x + 11$.

65. "A number increased by 8 is 40" is $x + 8 = 40$.

66. 'Twelve subtracted from twice a number is 10"
is $2x - 12 = 10$.

67. "Five times a number subtracted from 40 is the same as three times the number" is $40 - 5x = 3x$.
$$40 - 5x = 3x$$
$$40 - 5x + 5x = 3x + 5x$$
$$40 = 8x$$
$$\frac{40}{8} = \frac{8x}{8}$$
$$5 = x$$

68. "The product of a number and 3 is twice the difference of that number and 8" is
$3x = 2(x - 8)$.
$$3x = 2(x - 8)$$
$$3x = 2x - 16$$
$$3x - 2x = 2x - 16 - 2x$$
$$x = -16$$

69. Let x be the number of inches that kelp can grow in one day. Since bamboo grows twice as fast, bamboo can grow $2x$ inches in one day. Since both can grow a total of 54 inches in one day, the sum of x and $2x$ is 54.
$$x + 2x = 54$$
$$3x = 54$$
$$\frac{3x}{3} = \frac{54}{3}$$
$$x = 18$$
Thus, kelp can grow 18 inches in one day and bamboo can grow $2 \cdot 18 = 36$ inches in one day.

70. Let x be the number of votes received by the challenger. Since the incumbent received 11,206 more votes than the challenger, the incumbent received $x + 11,206$. The total number, or sum, is 18,298.
$$x + x + 11,206 = 18,298$$
$$2x + 11,206 = 18,298$$
$$2x + 11,206 - 11,206 = 18,298 - 11,206$$
$$2x = 7092$$
$$\frac{2x}{2} = \frac{7092}{2}$$
$$x = 3546$$
Thus, the challenger received 3546 votes and the incumbent received
$3546 + 11,206 = 14,752$ votes.

71. $18 - (9 - 5x) = 18 - (9 - 5 \cdot 4)$
$$= 18 - (9 - 20)$$
$$= 18 - (-11)$$
$$= 18 + 11$$
$$= 29$$

72. $\dfrac{z}{100} + \dfrac{y}{10} = \dfrac{5}{100} + \dfrac{-3}{10}$
$$= \frac{5}{100} + \frac{-30}{100}$$
$$= \frac{5 - 30}{100}$$
$$= \frac{-25}{100}$$
$$= -\frac{1}{4}$$

73. $9x - 20x = (9 - 20)x = -11x$

74. $-5(7x) = (-5 \cdot 7)x = -35x$

75. $12x + 5(2x - 3) - 4 = 12x + 10x - 15 - 4$
$$= (12 + 10)x + (-15 - 4)$$
$$= 22x - 19$$

76. $-7(x + 6) - 2(x - 5) = -7x - 42 - 2x + 10$
$$= -7x - 2x - 42 + 10$$
$$= -9x - 32$$

77. "Seventeen less than a number" is $x - 17$.

78. "Three times the sum of a number and five" is $3(x + 5)$.

79. "The difference of a number and 3 is the quotient of the number and 4" is $x - 3 = \dfrac{x}{4}$.

80. "The product of a number and 6 is equal to the sum of the number and 2" is $6x = x + 2$.

81. $4y + 2 - 6y = 5 + 7$
$$4 \cdot 3 + 2 - 6 \cdot 3 \stackrel{?}{=} 5 + 7$$
$$12 + 2 - 18 \stackrel{?}{=} 12$$
$$14 - 18 \stackrel{?}{=} 12$$
$$-4 \stackrel{?}{=} 12 \quad \text{False}$$
No, 3 is not a solution.

82. $4(z - 8) + 12 = 8$
$$4(7 - 8) + 12 \stackrel{?}{=} 8$$
$$4(-1) + 12 \stackrel{?}{=} 8$$
$$-4 + 12 \stackrel{?}{=} 8$$
$$8 \stackrel{?}{=} 8 \quad \text{True}$$
Yes, 7 is a solution.

83.
$$c - 5 = -13 + 7$$
$$c - 5 = -6$$
$$c - 5 + 5 = -6 + 5$$
$$c = -1$$

84.
$$7x + 5 - 6x = -20$$
$$5 + x = -20$$
$$-5 + 5 + x = -5 - 20$$
$$x = -25$$

85.
$$-7x + 3x = -50 - 2$$
$$-4x = -52$$
$$\frac{-4x}{-4} = \frac{-52}{-4}$$
$$x = 13$$

86.
$$-x + 8x = -38 - 4$$
$$7x = -42$$
$$\frac{7x}{7} = \frac{-42}{7}$$
$$x = -6$$

87.
$$14 - y = -3$$
$$-14 + 14 - y = -14 - 3$$
$$-y = -17$$
$$\frac{-y}{-1} = \frac{-17}{-1}$$
$$y = 17$$

88.
$$7 - z = 0$$
$$-7 + 7 - z = -7 + 0$$
$$-z = -7$$
$$\frac{-z}{-1} = \frac{-7}{-1}$$
$$z = 7$$

89.
$$9x + 12 - 8x = -6 + (-4)$$
$$x + 12 = -10$$
$$x + 12 - 12 = -10 - 12$$
$$x = -22$$

90.
$$-17x + 14 + 20x - 2x = 5 - (-3)$$
$$x + 14 = 5 + 3$$
$$x + 14 = 8$$
$$x + 14 - 14 = 8 - 14$$
$$x = -6$$

91.
$$\frac{4}{9}x = -\frac{1}{3}$$
$$\frac{9}{4} \cdot \frac{4}{9}x = \frac{9}{4} \cdot \left(-\frac{1}{3}\right)$$
$$x = -\frac{3}{4}$$

92.
$$-\frac{5}{24}x = \frac{5}{6}$$
$$-\frac{24}{5} \cdot \left(-\frac{5}{24}x\right) = -\frac{24}{5} \cdot \frac{5}{6}$$
$$x = -4$$

93.
$$2y + 6y = 24 - 8$$
$$8y = 16$$
$$\frac{8y}{8} = \frac{16}{8}$$
$$y = 2$$

94.
$$13x - 7x = -4 - 12$$
$$6x = -16$$
$$\frac{6x}{6} = \frac{-16}{6}$$
$$x = -\frac{8}{3}$$

95.
$$\frac{2}{3}x - 12 = -4$$
$$\frac{2}{3}x - 12 + 12 = -4 + 12$$
$$\frac{2}{3}x = 8$$
$$\frac{3}{2} \cdot \frac{2}{3}x = \frac{3}{2} \cdot 8$$
$$x = 12$$

96.
$$\frac{7}{8}x + 5 = -2$$
$$\frac{7}{8}x + 5 - 5 = -2 - 5$$
$$\frac{7}{8}x = -7$$
$$\frac{8}{7} \cdot \frac{7}{8}x = \frac{8}{7} \cdot (-7)$$
$$x = -8$$

97.
$$-5z + 3z - 7 = 8z - 7$$
$$-2z - 7 = 8z - 7$$
$$-2z - 7 + 7 = 8z - 7 + 7$$
$$-2z = 8z$$
$$-2z - 8z = 8z - 8z$$
$$-10z = 0$$
$$\frac{-10z}{-10} = \frac{0}{-10}$$
$$z = 0$$

98.
$$4x - 3 + 6x = 5x - 3$$
$$-3 + 10x = 5x - 3$$
$$3 - 3 + 10x = 5x - 3 + 3$$
$$10x = 5x$$
$$10x - 5x = 5x - 5x$$
$$5x = 0$$
$$\frac{5x}{5} = \frac{0}{5}$$
$$x = 0$$

99. "Three times a number added to twelve is 27" is
$12 + 3x = 27$.
$$12 + 3x = 27$$
$$-12 + 12 + 3x = -12 + 27$$
$$3x = 15$$
$$\frac{3x}{3} = \frac{15}{3}$$
$$x = 5$$

100. "Twice the sum of a number and four is ten" is
$2(x + 4) = 10$.
$$2(x + 4) = 10$$
$$2x + 8 = 10$$
$$2x + 8 - 8 = 10 - 8$$
$$2x = 2$$
$$\frac{2x}{2} = \frac{2}{2}$$
$$x = 1$$

Chapter 11 Test

1.
$$\frac{3x - 5}{2y} = \frac{3 \cdot 7 - 5}{2(-8)}$$
$$= \frac{21 - 5}{-16}$$
$$= \frac{16}{-16}$$
$$= -1$$

2.
$$7x - 5 - 12x + 10 = 7x - 12x - 5 + 10$$
$$= (7 - 12)x + (-5 + 10)$$
$$= -5x + 5$$

3. $-2(3y + 7) = -2 \cdot 3y - 2 \cdot 7 = -6y - 14$

4.
$$5(3z + 2) - z - 18 = 5 \cdot 3z + 5 \cdot 2 - z - 18$$
$$= 15z + 10 - z - 18$$
$$= 15z - z + 10 - 18$$
$$= 14z - 8$$

5. Area = length $\cdot$ width
$$A = 4 \cdot (3x - 1)$$
$$= 4 \cdot 3x - 4 \cdot 1$$
$$= 12x - 4$$
The area is $(12x - 4)$ square meters.

6.
$$x - 17 = -10$$
$$x - 17 + 17 = -10 + 17$$
$$x = 7$$

7.
$$y + \frac{3}{4} = \frac{1}{4}$$
$$y + \frac{3}{4} - \frac{3}{4} = \frac{1}{4} - \frac{3}{4}$$
$$y = -\frac{2}{4}$$
$$y = -\frac{1}{2}$$

8.
$$-4x = 48$$
$$\frac{-4x}{-4} = \frac{48}{-4}$$
$$x = -12$$

9.
$$-\frac{5}{8}x = -25$$
$$-\frac{8}{5} \cdot -\frac{5}{8}x = -\frac{8}{5} \cdot -25$$
$$x = 40$$

10.
$$5x + 12 - 4x - 14 = 22$$
$$x - 2 = 22$$
$$x - 2 + 2 = 22 + 2$$
$$x = 24$$

11.
$$2 - c + 2c = 5$$
$$2 + c = 5$$
$$-2 + 2 + c = -2 + 5$$
$$c = 3$$

12.
$$3x - 5 = -11$$
$$3x - 5 + 5 = -11 + 5$$
$$3x = -6$$
$$\frac{3x}{3} = \frac{-6}{3}$$
$$x = -2$$

13.
$$-4x + 7 = 15$$
$$-4x + 7 - 7 = 15 - 7$$
$$-4x = 8$$
$$\frac{-4x}{-4} = \frac{8}{-4}$$
$$x = -2$$

14.
$$3.6 - 2x = -5.4$$
$$-3.6 + 3.6 - 2x = -5.4 - 3.6$$
$$-2x = -9$$
$$\frac{-2x}{-2} = \frac{-9}{-2}$$
$$x = 4.5$$

15.
$$12 = 3(4 + 2y)$$
$$12 = 3 \cdot 4 + 3 \cdot 2y$$
$$12 = 12 + 6y$$
$$-12 + 12 = -12 + 12 + 6y$$
$$0 = 6y$$
$$\frac{0}{6} = \frac{6y}{6}$$
$$0 = y$$

16.
$$5x - 2 = x - 10$$
$$5x - 2 + 2 = x - 10 + 2$$
$$5x = x - 8$$
$$5x - x = x - 8 - x$$
$$4x = -8$$
$$\frac{4x}{4} = \frac{-8}{4}$$
$$x = -2$$

17.
$$10y - 1 = 7y + 21$$
$$10y - 1 + 1 = 7y + 21 + 1$$
$$10y = 7y + 22$$
$$10y - 7y = 7y - 7y + 22$$
$$3y = 22$$
$$\frac{3y}{3} = \frac{22}{3}$$
$$y = \frac{22}{3}$$

18.
$$6 + 2(3n - 1) = 28$$
$$6 + 6n - 2 = 28$$
$$6n + 4 = 28$$
$$6n + 4 - 4 = 28 - 4$$
$$6n = 24$$
$$\frac{6n}{6} = \frac{24}{6}$$
$$n = 4$$

19.
$$4(5x + 3) = 2(7x + 6)$$
$$20x + 12 = 14x + 12$$
$$20x + 12 - 12 = 14x + 12 - 12$$
$$20x = 14x$$
$$20x - 14x = 14x - 14x$$
$$6x = 0$$
$$\frac{6x}{6} = \frac{0}{6}$$
$$x = 0$$

20.
$$A = \frac{1}{2} \cdot h \cdot (B + b)$$
$$= \frac{1}{2} \cdot 60 \cdot (70 + 130)$$
$$= \frac{1}{2} \cdot 60 \cdot (200)$$
$$= 30 \cdot (200)$$
$$= 6000$$
The area is 6000 square feet.

21. $A = \dfrac{1}{2} b \cdot h = \dfrac{1}{2} \cdot 5 \cdot 12 = 30$
The area is 30 square feet.

22. a. "The product of a number and 17" is $17x$.

b. "Twice a number subtracted from 20" is $20 - 2x$.

23. "The difference of three times a number and five times the same number is 4" is $3x - 5x = 4$.
$$3x - 5x = 4$$
$$-2x = 4$$
$$\frac{-2x}{-2} = \frac{4}{-2}$$
$$x = -2$$

24. Let x be the number of points scored by Maria. Then the number of points scored by Paula is $2x$. The total, or sum, is 51.

$$x + 2x = 51$$
$$3x = 51$$
$$\frac{3x}{3} = \frac{51}{3}$$
$$x = 17$$

Maria scored 17 points and Paula scored $2(17) = 34$ points.

25. Let x be the number of women runners entered. Then the number of men runners entered is $x + 112$, and the sum of these two quantities is 600.

$$x + x + 112 = 600$$
$$2x + 112 = 600$$
$$2x + 112 - 112 = 600 - 112$$
$$2x = 488$$
$$\frac{2x}{2} = \frac{488}{2}$$
$$x = 244$$

There were 244 women runners entered.

Cumulative Review Chapters 1–11

1.
$$\begin{array}{r} 0.0531 \\ \times\ \ \ \ 16 \\ \hline 3186 \\ 5310 \\ \hline 0.8496 \end{array}$$

2. $0.0531 \times 1000 = 53.1$

3. a. $\dfrac{\text{width}}{\text{length}} = \dfrac{5\ \text{feet}}{7\ \text{feet}} = \dfrac{5}{7}$

b. $\dfrac{\text{length}}{\text{perimeter}} = \dfrac{7\ \text{feet}}{7+5+7+5\ \text{feet}}$
$$= \frac{7\ \text{feet}}{24\ \text{feet}}$$
$$= \frac{7}{24}$$

4. $\dfrac{5}{12} + \dfrac{2}{9} = \dfrac{5}{12} \cdot \dfrac{3}{3} + \dfrac{2}{9} \cdot \dfrac{4}{4} = \dfrac{15}{36} + \dfrac{8}{36} = \dfrac{15+8}{36} = \dfrac{23}{36}$

5. $12\% \cdot n = 0.6$
$$0.12 \cdot n = 0.6$$
$$n = \frac{0.6}{0.12}$$
$$n = 5$$
Thus, 12% of 5 is 0.6.

6. $\dfrac{7}{8} \cdot \dfrac{2}{3} = \dfrac{7 \cdot 2}{8 \cdot 3} = \dfrac{7 \cdot 2}{2 \cdot 4 \cdot 3} = \dfrac{7}{4 \cdot 3} = \dfrac{7}{12}$

7. $n \cdot 12 = 9$
$$n = \frac{9}{12}$$
$$n = 0.75 = 75\%$$
Thus, 75% of 12 is 9.

8. $1\dfrac{4}{5} \div 2\dfrac{3}{10} = \dfrac{9}{5} \div \dfrac{23}{10} = \dfrac{9}{5} \cdot \dfrac{10}{23} = \dfrac{9 \cdot 10}{5 \cdot 23} = \dfrac{9 \cdot 2}{23} = \dfrac{18}{23}$

9. $3\ \text{lb} = \dfrac{3\ \text{lb}}{1} \cdot 1 = \dfrac{3\ \text{lb}}{1} \cdot \dfrac{16\ \text{oz}}{1\ \text{lb}} = 3 \cdot 16\ \text{oz} = 48\ \text{oz}$

10. 23,781 rounded to the nearest thousand is 24,000.

11. 2400 ml = 2.4 L or 8.9 L = 8900 ml
$$\begin{array}{r} 8.9\ \text{L} \\ +\ 2400\ \text{ml} \end{array} \quad \begin{array}{r} 8.9\ \text{L} \\ +\ 2.4\ \text{L} \\ \hline 11.3\ \text{L} \end{array} \text{ or } \begin{array}{r} 8900\ \text{ml} \\ +\ 2400\ \text{ml} \\ \hline 11,300\ \text{ml} \end{array}$$

12. 0.02351 rounded to the nearest thousandth is 0.024.

13. Figure (a) extends indefinitely in two directions. It is line CD or $\overleftrightarrow{CD}$.
Figure (b) has two endpoints. It is line segment EF or $\overline{EF}$.
Figure (c) has two rays with a common endpoint. It is $\angle MNO$, $\angle ONM$, or $\angle N$.
Figure (d) is part of a line with one endpoint. It is ray PT or $\overrightarrow{PT}$.

14. Supplementary angles sum to 180°.
$$180° - 12° = 168°$$
The supplement of a 12° angle is 168°.

15. The diameter is twice the radius.
$$d = 2 \cdot r$$
$$d = 2 \cdot 5\ \text{cm} = 10\ \text{cm}$$
The diameter is 10 cm.

16. The angles sum to 180°.
$180° − (54° + 92°) = 180° − 146° = 34°$
The unknown angle is 34°.

17. $\text{perimeter} = 10 \text{ ft} + 9 \text{ ft} + 3 \text{ ft} + 6 \text{ ft} + 7 \text{ ft} + 15 \text{ ft}$
$= 50 \text{ ft}$
The perimeter is 50 feet.

18. $\text{Area} = (10 \cdot 9) + (7 \cdot 6) = 90 + 42 = 132$
The area is 132 square feet.

19. $A = \dfrac{1}{2} \cdot b \cdot h$
$= \dfrac{1}{2} \cdot 14 \text{ cm} \cdot 8 \text{ cm}$
$= \dfrac{2 \cdot 7 \cdot 8}{2} \text{ sq cm}$
$= 56 \text{ sq cm}$
The area is 56 square centimeters.

20. $\dfrac{\left(4 + \sqrt{4}\right)^2}{\sqrt{100} - \sqrt{64}} = \dfrac{(4 + 2)^2}{10 - 8} = \dfrac{6^2}{2} = \dfrac{36}{2} = 18$

21. a. 64 and 81 are perfect squares. $\sqrt{64} = 8$ and $\sqrt{81} = 9$. Since 78 is between 64 and 81, $\sqrt{78}$ is between $\sqrt{64} = 8$ and $\sqrt{81} = 9$.

b. Since 78 is closer to 81, $\sqrt{78}$ is closer to $\sqrt{81} = 9$.

22. $0.02\overline{)0.1156}$ becomes

$$
\begin{array}{r}
5.78 \\
2\overline{)11.56} \\
-10 \\
\hline
1\,5 \\
-1\,4 \\
\hline
16 \\
-16 \\
\hline
0
\end{array}
$$

Thus, $0.1156 \div 0.02 = 5.78$.

23. $\dfrac{6}{n} = \dfrac{9}{69}$
or
$\dfrac{6}{n} = \dfrac{3}{23}$
$6 \cdot 23 = n \cdot 3$
$138 = n \cdot 3$
$\dfrac{138}{3} = n$
$46 = n$
The height of the tree is 46 feet.

24. $n \cdot 120 = 28.8$
$n = \dfrac{28.8}{120}$
$n = 0.24 = 24\%$
Thus, 24% of 120 is 28.8.

25. a. To approximate the number of endangered species that are clams, we go to the top of the bar that represents clams. From the top of this bar, we move horizontally to the left until the scale is reached. We read the height of the bar on the scale as approximately 62. There are approximately 62 species of clams that are endangered.

b. The most endangered species is represented by the tallest (longest) bar. There are two tallest bars. They correspond to birds and fishes.

26. $\text{mean} = \dfrac{1 + 7 + 8 + 10 + 11 + 11}{6} = \dfrac{48}{6} = 8$
median is the average of the two middle values:
$\dfrac{8 + 10}{2} = \dfrac{18}{2} = 9$
mode is the value that occurs most often: 11

27. −2.9 is to the left of −1, so $-2.9 < -1$.

28. $\dfrac{2}{3} = 0.6666... \approx 0.067$

$\dfrac{2}{3}$ is to the right of 0.66, so $\dfrac{2}{3} > 0.66$.

29. −9 is to the right of −11, so $-9 > -11$.

30. $\dfrac{4}{5} = 0.8, \dfrac{14}{17} \approx 0.82$

$\dfrac{4}{5}$ is to the left of $\dfrac{14}{17}$, so $\dfrac{4}{5} < \dfrac{14}{17}$.

31. $|-2| = 2$ because -2 is 2 units from 0.

32. $\left|\dfrac{5}{6}\right| = \dfrac{5}{6}$ because $\dfrac{5}{6}$ is $\dfrac{5}{6}$ unit from 0.

33. $|1.2| = 1.2$ because 1.2 is 1.2 units from 0.

34. $|0| = 0$ because 0 is 0 units from 0.

35. $-5 + (-11) = -16$

36. $-8.2 + 4.6 = -3.6$

37. $-\dfrac{3}{8} + \left(-\dfrac{1}{8}\right) = -\dfrac{4}{8} = -\dfrac{4}{2 \cdot 4} = -\dfrac{1}{2}$

38. $\dfrac{2}{5} + \left(-\dfrac{3}{10}\right) = \dfrac{4}{10} + \left(-\dfrac{3}{10}\right) = \dfrac{1}{10}$

39. $8 - 15 = 8 + (-15) = -7$

40. $4.6 - (-1.2) = 4.6 + 1.2 = 5.8$

41. $-4 - (-5) = -4 + 5 = 1$

42.
$$\begin{aligned}
\frac{7}{10} - \frac{23}{24} &= \frac{7}{10} \cdot \frac{12}{12} - \frac{23}{24} \cdot \frac{5}{5} \\
&= \frac{84}{120} - \frac{115}{120} \\
&= \frac{84 - 115}{120} \\
&= \frac{-31}{120} \\
&= -\frac{31}{120}
\end{aligned}$$

43. $-2(-5) = 10$

44. $-8(1.2) = -9.6$

45. $\left(-\dfrac{1}{2}\right)\left(-\dfrac{2}{3}\right) = \dfrac{1 \cdot 2}{2 \cdot 3} = \dfrac{1}{3}$

46. $-2\dfrac{2}{9}\left(1\dfrac{4}{5}\right) = -\dfrac{20}{9}\left(\dfrac{9}{5}\right) = -4$

47.
$$\begin{aligned}
(-3) \cdot |-5| - (-2) + 4^2 &= (-3) \cdot 5 - (-2) + 4^2 \\
&= (-3) \cdot 5 - (-2) + 16 \\
&= -15 - (-2) + 16 \\
&= -15 + 2 + 16 \\
&= -13 + 16 \\
&= 3
\end{aligned}$$

48.
$$\begin{aligned}
4x - 7.1 &= 3x + 2.6 \\
4x - 7.1 + 7.1 &= 3x + 2.6 + 7.1 \\
4x &= 3x + 9.7 \\
-3x + 4x &= -3x + 3x + 9.7 \\
x &= 9.7
\end{aligned}$$

49.
$$\begin{aligned}
3(2x - 6) + 6 &= 0 \\
6x - 18 + 6 &= 0 \\
6x - 12 &= 0 \\
6x - 12 + 12 &= 0 + 12 \\
6x &= 12 \\
\frac{6x}{6} &= \frac{12}{6} \\
x &= 2
\end{aligned}$$

50.
$$\begin{aligned}
6(x - 5) &= 4(x + 4) - 6 \\
6x - 30 &= 4x + 16 - 6 \\
6x - 30 &= 4x + 10 \\
6x - 30 + 30 &= 4x + 10 + 30 \\
6x &= 4x + 40 \\
-4x + 6x &= -4x + 4x + 40 \\
2x &= 40 \\
\frac{2x}{2} &= \frac{40}{2} \\
x &= 20
\end{aligned}$$

Appendix A

Exercise Set A.1

1.
```
    1
  + 4
  ―――
    5
```

3.
```
    2
  + 3
  ―――
    5
```

5.
```
    3
  + 9
  ―――
   12
```

7.
```
    4
  + 4
  ―――
    8
```

9.
```
    9
  + 5
  ―――
   14
```

11.
```
    5
  + 7
  ―――
   12
```

13.
```
    5
  + 5
  ―――
   10
```

15.
```
    8
  + 1
  ―――
    9
```

17.
```
    2
  + 9
  ―――
   11
```

19.
```
    9
  + 9
  ―――
   18
```

21.
```
    6
  + 4
  ―――
   10
```

23.
```
    1
  + 9
  ―――
   10
```

25.
```
    9
  + 8
  ―――
   17
```

27.
```
    4
  + 9
  ―――
   13
```

29.
```
    7
  + 5
  ―――
   12
```

31.
```
    9
  + 7
  ―――
   16
```

33.
```
    4
  + 3
  ―――
    7
```

35.
```
    3
  + 1
  ―――
    4
```

37.
```
    7
  + 1
  ―――
    8
```

39.
```
    8
  + 0
  ―――
    8
```

41.
```
    2
  + 4
  ―――
    6
```

43.
```
    8
  + 8
  ―――
   16
```

45.
```
    3
  + 6
  ―――
    9
```

47.
```
    4
  + 8
  ―――
   12
```

49.
```
    2
  + 5
  ―――
    7
```

51.
$$\begin{array}{r} 2 \\ + 0 \\ \hline 2 \end{array}$$

53.
$$\begin{array}{r} 8 \\ + 3 \\ \hline 11 \end{array}$$

55.
$$\begin{array}{r} 1 \\ + 7 \\ \hline 8 \end{array}$$

57.
$$\begin{array}{r} 0 \\ + 5 \\ \hline 5 \end{array}$$

59.
$$\begin{array}{r} 8 \\ + 6 \\ \hline 14 \end{array}$$

61.
$$\begin{array}{r} 6 \\ + 7 \\ \hline 13 \end{array}$$

63.
$$\begin{array}{r} 1 \\ + 6 \\ \hline 7 \end{array}$$

65.
$$\begin{array}{r} 0 \\ + 7 \\ \hline 7 \end{array}$$

67.
$$\begin{array}{r} 7 \\ + 6 \\ \hline 13 \end{array}$$

69.
$$\begin{array}{r} 4 \\ + 1 \\ \hline 5 \end{array}$$

71.
$$\begin{array}{r} 0 \\ + 4 \\ \hline 4 \end{array}$$

73.
$$\begin{array}{r} 7 \\ + 9 \\ \hline 16 \end{array}$$

75.
$$\begin{array}{r} 7 \\ + 7 \\ \hline 14 \end{array}$$

77.
$$\begin{array}{r} 1 \\ + 0 \\ \hline 1 \end{array}$$

79.
$$\begin{array}{r} 2 \\ + 2 \\ \hline 4 \end{array}$$

81.
$$\begin{array}{r} 2 \\ + 8 \\ \hline 10 \end{array}$$

83.
$$\begin{array}{r} 6 \\ + 2 \\ \hline 8 \end{array}$$

85.
$$\begin{array}{r} 5 \\ + 0 \\ \hline 5 \end{array}$$

87.
$$\begin{array}{r} 7 \\ + 3 \\ \hline 10 \end{array}$$

89.
$$\begin{array}{r} 9 \\ + 2 \\ \hline 11 \end{array}$$

91.
$$\begin{array}{r} 9 \\ + 3 \\ \hline 12 \end{array}$$

93.
$$\begin{array}{r} 2 \\ + 7 \\ \hline 9 \end{array}$$

95.
$$\begin{array}{r} 7 \\ + 2 \\ \hline 9 \end{array}$$

97.
$$\begin{array}{r} 6 \\ + 8 \\ \hline 14 \end{array}$$

99.
$$\begin{array}{r} 9 \\ + 0 \\ \hline 9 \end{array}$$

Exercise Set A.2

1.
$$\begin{array}{r} 1 \\ \times 1 \\ \hline 1 \end{array}$$

3.　　7
　　　×8
　　　‾‾‾
　　　56

5.　　8
　　　×4
　　　‾‾‾
　　　32

7.　　4
　　　×7
　　　‾‾‾
　　　28

9.　　2
　　　×2
　　　‾‾‾
　　　4

11.　9
　　　×7
　　　‾‾‾
　　　63

13.　3
　　　×2
　　　‾‾‾
　　　6

15.　5
　　　×6
　　　‾‾‾
　　　30

17.　4
　　　×6
　　　‾‾‾
　　　24

19.　6
　　　×3
　　　‾‾‾
　　　18

21.　5
　　　×8
　　　‾‾‾
　　　40

23.　4
　　　×8
　　　‾‾‾
　　　32

25.　9
　　　×6
　　　‾‾‾
　　　54

27.　8
　　　×7
　　　‾‾‾
　　　56

29.　6
　　　×9
　　　‾‾‾
　　　54

31.　2
　　　×1
　　　‾‾‾
　　　2

33.　4
　　　×9
　　　‾‾‾
　　　36

35.　6
　　　×2
　　　‾‾‾
　　　12

37.　9
　　　×4
　　　‾‾‾
　　　36

39.　3
　　　×4
　　　‾‾‾
　　　12

41.　8
　　　×6
　　　‾‾‾
　　　48

43.　1
　　　×8
　　　‾‾‾
　　　8

45.　9
　　　×0
　　　‾‾‾
　　　0

47.　9
　　　×3
　　　‾‾‾
　　　27

49.　3
　　　×5
　　　‾‾‾
　　　15

51.　5
　　　×9
　　　‾‾‾
　　　45

53.　1
　　　×0
　　　‾‾‾
　　　0

55.
$$\begin{array}{r} 9 \\ \times 9 \\ \hline 81 \end{array}$$

57.
$$\begin{array}{r} 0 \\ \times 6 \\ \hline 6 \end{array}$$

59.
$$\begin{array}{r} 5 \\ \times 0 \\ \hline 0 \end{array}$$

61.
$$\begin{array}{r} 9 \\ \times 2 \\ \hline 18 \end{array}$$

63.
$$\begin{array}{r} 1 \\ \times 3 \\ \hline 3 \end{array}$$

65.
$$\begin{array}{r} 6 \\ \times 6 \\ \hline 36 \end{array}$$

67.
$$\begin{array}{r} 7 \\ \times 9 \\ \hline 63 \end{array}$$

69.
$$\begin{array}{r} 7 \\ \times 5 \\ \hline 35 \end{array}$$

71.
$$\begin{array}{r} 6 \\ \times 7 \\ \hline 42 \end{array}$$

73.
$$\begin{array}{r} 8 \\ \times 5 \\ \hline 40 \end{array}$$

75.
$$\begin{array}{r} 0 \\ \times 1 \\ \hline 0 \end{array}$$

77.
$$\begin{array}{r} 9 \\ \times 1 \\ \hline 9 \end{array}$$

79.
$$\begin{array}{r} 5 \\ \times 3 \\ \hline 15 \end{array}$$

81.
$$\begin{array}{r} 1 \\ \times 5 \\ \hline 5 \end{array}$$

83.
$$\begin{array}{r} 3 \\ \times 0 \\ \hline 0 \end{array}$$

85.
$$\begin{array}{r} 3 \\ \times 7 \\ \hline 21 \end{array}$$

87.
$$\begin{array}{r} 0 \\ \times 2 \\ \hline 0 \end{array}$$

89.
$$\begin{array}{r} 8 \\ \times 2 \\ \hline 16 \end{array}$$

91.
$$\begin{array}{r} 0 \\ \times 0 \\ \hline 0 \end{array}$$

93.
$$\begin{array}{r} 4 \\ \times 1 \\ \hline 4 \end{array}$$

95.
$$\begin{array}{r} 2 \\ \times 3 \\ \hline 6 \end{array}$$

97.
$$\begin{array}{r} 3 \\ \times 6 \\ \hline 18 \end{array}$$

99.
$$\begin{array}{r} 0 \\ \times 9 \\ \hline 0 \end{array}$$

Appendix B

Exercise Set B

1. a. $\dfrac{\$16}{1 \text{ hr}} \cdot \dfrac{40 \text{ hr}}{1} = \dfrac{\$16 \cdot 40}{1 \cdot 1} = \$640$

Before deductions, your earnings for 40 hours are $640

b. $\dfrac{\$240}{1} \cdot \dfrac{1 \text{ hr}}{\$16} = \dfrac{240 \cdot 1 \text{ hr}}{1 \cdot 16} = 15 \text{ hr}$

It takes you 15 hours to earn $240 before deductions.

3. a. beats in 14 seconds

$= \dfrac{14 \text{ seconds}}{1} \cdot \dfrac{90 \text{ beats}}{1 \text{ second}}$

$= \dfrac{14 \cdot 90}{1 \cdot 1} \text{ beats}$

$= 1260 \text{ beats}$

A hummingbird's wings beat 1260 times in 14 seconds.

b. time for 5400 beats

$= \dfrac{5400 \text{ beats}}{1} \cdot \dfrac{1 \text{ second}}{90 \text{ beats}}$

$= \dfrac{5400 \cdot 1}{1 \cdot 90} \text{ seconds}$

$= 60 \text{ seconds}$

$= \dfrac{60 \text{ seconds}}{1} \cdot \dfrac{1 \text{ minute}}{60 \text{ seconds}}$

$= \dfrac{60 \cdot 1}{1 \cdot 60} \text{ minute}$

$= 1 \text{ minute}$

60 seconds or 1 minute has elapsed after 5400 wing beats.

c. time for 8100 beats

$= \dfrac{8100 \text{ beats}}{1} \cdot \dfrac{1 \text{ second}}{90 \text{ beats}}$

$= \dfrac{8100 \cdot 1}{1 \cdot 90} \text{ seconds}$

$= 90 \text{ seconds}$

$= \dfrac{90 \text{ seconds}}{1} \cdot \dfrac{1 \text{ minute}}{60 \text{ seconds}}$

$= \dfrac{90 \cdot 1}{1 \cdot 60} \text{ minutes}$

$= 1\dfrac{1}{2} \text{ minutes}$

90 seconds or $1\dfrac{1}{2}$ minutes has elapsed after 8100 wing beats.

5. a. $\dfrac{1 \text{ bed}}{155 \text{ people}} \cdot \dfrac{31{,}000 \text{ people}}{1}$

$= \dfrac{1 \text{ bed} \cdot 31{,}000}{155 \cdot 1}$

$= 200 \text{ beds}$

There would be 200 hospital beds in the city.

b. $\dfrac{155 \text{ people}}{1 \text{ bed}} \cdot \dfrac{73 \text{ beds}}{1} = \dfrac{155 \text{ people} \cdot 73}{1 \cdot 1}$

$= 11{,}315 \text{ people}$

The population is about 11,315 people.

7. a. flour for 9 cups water

$= \dfrac{9 \text{ c water}}{1} \cdot \dfrac{2 \text{ c flour}}{3 \text{ c water}}$

$= \dfrac{9 \cdot 2}{1 \cdot 3} \text{ c flour}$

$= 6 \text{ c flour}$

6 cups of flour are needed if 9 cups of water are used.

b. flour for 20 cups water

$= \dfrac{20 \text{ c water}}{1} \cdot \dfrac{2 \text{ c flour}}{3 \text{ c water}}$

$= \dfrac{20 \cdot 2}{1 \cdot 3} \text{ c flour}$

$= \dfrac{40}{3} \text{ c flour}$

$= 13\dfrac{1}{3} \text{ c flour}$

$13\dfrac{1}{3}$ cups of flour are needed if 20 cups of water are used.

c. water for 15 cups flour

$= \dfrac{15 \text{ c flour}}{1} \cdot \dfrac{3 \text{ c water}}{2 \text{ c flour}}$

$= \dfrac{15 \cdot 3}{1 \cdot 2} \text{ c water}$

$= \dfrac{45}{2} \text{ c water}$

$= 22\dfrac{1}{2} \text{ c water}$

$22\dfrac{1}{2}$ cups of water are needed if 15 cups of flour are used.

9. $\dfrac{\$0.75}{1 \text{ component}} \cdot \dfrac{26 \text{ components}}{1 \text{ hr}} \cdot \dfrac{33 \text{ hr}}{1}$

$= \dfrac{\$0.75 \cdot 26 \cdot 33}{1 \cdot 1 \cdot 1}$

$= \$643.50$

Her pay before deductions was \$643.50

11. **a.** drops for 17.5 liters

$= \dfrac{17.5 \text{ liters}}{1} \cdot \dfrac{15 \text{ drops}}{1 \text{ liter}}$

$= \dfrac{17.5 \cdot 15}{1 \cdot 1} \text{ drops}$

$= 262.5 \text{ drops}$

262.5 drops are needed for 17.5 liters.

b. liters for 300 drops

$= \dfrac{300 \text{ drops}}{1} \cdot \dfrac{1 \text{ liter}}{15 \text{ drops}}$

$= \dfrac{300 \cdot 1}{1 \cdot 15} \text{ liters}$

$= 20 \text{ liters}$

20 liters of saline solution are needed for 300 drops of medicine.

13. **a.** $\dfrac{16 \text{ tortillas}}{1 \text{ min}} \cdot \dfrac{60 \text{ min}}{1 \text{ hr}} \cdot \dfrac{7.5 \text{ hr}}{1}$

$= \dfrac{16 \text{ tortillas} \cdot 60 \cdot 7.5}{1 \cdot 1 \cdot 1}$

$= 7200 \text{ tortillas}$

The machine would make 7200 tortillas in 7.5 hours.

b. $\dfrac{200 \text{ tortillas}}{1} \cdot \dfrac{1 \text{ min}}{16 \text{ tortillas}} = \dfrac{200 \cdot 1 \text{ min}}{1 \cdot 16}$

$= 12.5 \text{ min}$

It takes the machine 12.5 minutes to make 200 tortillas.

15. **a.** amount for 6 episodes

$= \dfrac{6 \text{ episodes}}{1} \cdot \dfrac{\$1.8 \text{ million}}{1 \text{ episode}}$

$= \dfrac{6 \cdot \$1.8 \text{ million}}{1 \cdot 1}$

$= \$10.8 \text{ million}$

He earns \$10.8 million for 6 episodes.

b. episodes for \$8.1 million

$= \dfrac{\$8.1 \text{ million}}{1} \cdot \dfrac{1 \text{ episode}}{\$1.8 \text{ million}}$

$= \dfrac{8.1 \cdot 1}{1 \cdot 1.8} \text{ episodes}$

$= 4.5 \text{ episodes}$

It takes 4.5 episodes for him to earn \$8.1 million.

17. **a.** $\dfrac{45 \text{ miles}}{1 \text{ hr}} \cdot \dfrac{\frac{1}{4} \text{ hr}}{1} = \dfrac{45 \text{ miles} \cdot \frac{1}{4}}{1 \cdot 1}$

$= 11\dfrac{1}{4} \text{ miles}$

The cheetah has traveled $11\dfrac{1}{4}$ miles in $\dfrac{1}{4}$ hour.

b. $\dfrac{5 \text{ miles}}{1} \cdot \dfrac{1 \text{ hr}}{45 \text{ miles}} = \dfrac{5 \cdot 1 \text{ hr}}{1 \cdot 45} = \dfrac{1}{9} \text{ hr}$

$\dfrac{\frac{1}{9} \text{ hr}}{1} \cdot \dfrac{60 \text{ min}}{1 \text{ hr}} = \dfrac{\frac{1}{9} \cdot 60 \text{ min}}{1 \cdot 1}$

$= \dfrac{60}{9} \text{ min}$

$= 6\dfrac{2}{3} \text{ min}$

It takes the cheetah $\dfrac{1}{9}$ hour or $6\dfrac{2}{3}$ minutes to run 5 miles.

Practice Final Exam

1.
$$\begin{array}{r} 600 \\ -\ 487 \\ \hline 113 \end{array}$$

2. $(2^4 - 5) \cdot 3 = (16 - 5) \cdot 3 = (11) \cdot 3 = 33$

3.
$$\begin{aligned}
\frac{16}{3} \div \frac{3}{12} &= \frac{16}{3} \cdot \frac{12}{3} \\
&= \frac{16 \cdot 12}{3 \cdot 3} \\
&= \frac{16 \cdot 3 \cdot 4}{3 \cdot 3} \\
&= \frac{16 \cdot 4}{3} \\
&= \frac{64}{3} \text{ or } 21\frac{1}{3}
\end{aligned}$$

4. The LCD of 12, 8, and 24 is 24.
$$\begin{aligned}
\frac{11}{12} + \frac{3}{8} + \frac{5}{24} &= \frac{11}{12} \cdot \frac{2}{2} + \frac{3}{8} \cdot \frac{3}{3} + \frac{5}{24} \\
&= \frac{22}{24} + \frac{9}{24} + \frac{5}{24} \\
&= \frac{36}{24} \\
&= \frac{12 \cdot 3}{12 \cdot 2} \\
&= \frac{3}{2} \text{ or } 1\frac{1}{2}
\end{aligned}$$

5.
$$\begin{aligned}
\frac{64 \div 8 \cdot 2}{\left(\sqrt{9} - \sqrt{4}\right)^2 + 1} &= \frac{8 \cdot 2}{(3-2)^2 + 1} \\
&= \frac{16}{1^2 + 1} \\
&= \frac{16}{1 + 1} \\
&= \frac{16}{2} \\
&= 8
\end{aligned}$$

6.
$$\begin{array}{r} 10.2 \\ \times\ 4.3 \\ \hline 3\ 06 \\ 40\ 80 \\ \hline 43.86 \end{array}$$

7. $\dfrac{0.23 + 1.63}{0.3} = \dfrac{1.86}{0.3}$

$$0.3\overline{)1.86} \text{ becomes } 3\overline{\smash{)}\begin{array}{r} 6.2 \\ 18.6 \\ -18 \\ \hline 0\ 6 \\ -6 \\ \hline 0 \end{array}}$$

$$\frac{0.23 + 1.63}{0.3} = 6.2$$

8. The LCD of 6 and 8 is 24.
$$\begin{array}{ccc}
5\dfrac{1}{6} & 5\dfrac{4}{24} & 4\dfrac{28}{24} \\[2mm]
-\ 3\dfrac{7}{8} & -\ 3\dfrac{21}{24} & -\ 3\dfrac{21}{24} \\[2mm]
& & \hline \\[-2mm]
& & 1\dfrac{7}{24}
\end{array}$$

9.
$$\begin{aligned}
3\frac{1}{3} \cdot 6\frac{3}{4} &= \frac{10}{3} \cdot \frac{27}{4} \\
&= \frac{10 \cdot 27}{3 \cdot 4} \\
&= \frac{2 \cdot 5 \cdot 3 \cdot 9}{3 \cdot 2 \cdot 2} \\
&= \frac{5 \cdot 9}{2} \\
&= \frac{45}{2} \text{ or } 22\frac{1}{2}
\end{aligned}$$

10. $126.9 \times 100 = 12{,}690$

11.
$$\begin{aligned}
\left(\frac{3}{4}\right)^2 \div \left(\frac{2}{3} + \frac{5}{6}\right) &= \left(\frac{3}{4} \cdot \frac{3}{4}\right) \div \left(\frac{2}{3} \cdot \frac{2}{2} + \frac{5}{6}\right) \\
&= \frac{9}{16} \div \left(\frac{4}{6} + \frac{5}{6}\right) \\
&= \frac{9}{16} \div \frac{9}{6} \\
&= \frac{9}{16} \cdot \frac{6}{9} \\
&= \frac{9 \cdot 6}{16 \cdot 9} \\
&= \frac{9 \cdot 2 \cdot 3}{2 \cdot 8 \cdot 9} \\
&= \frac{3}{8}
\end{aligned}$$

12. To round 52,369 to the nearest thousand, observe that the digit in the hundreds place is 3. Since this digit is less than 5, we do not add 1 to the digit in the thousands place. The number 52,369 rounded to the nearest thousand is 52,000.

13. To round 34.8923 to the nearest tenth, observe that the digit in the hundredths place is 9. Since this digit is at least 5, we add 1 to the digit in the tenths place. The number 34.8923 rounded to the nearest tenth is 34.9.

14.
$$
\begin{array}{r}
0.9411 \approx 0.941 \\
17 \overline{)\ 16.0000} \\
-15\ 3 \\
\hline
70 \\
-68 \\
\hline
20 \\
-17 \\
\hline
30 \\
-17 \\
\hline
13
\end{array}
$$

$\dfrac{16}{17} \approx 0.941$

15. $85\% = 85(0.01) = 0.85$

16. $6.1 = 6.1 \cdot 100\% = 610\%$

17. $\dfrac{3}{8} = \dfrac{3}{8} \cdot 100\% = \dfrac{300}{8}\% = 37.5\%$

18. $0.2\% = 0.2 \cdot \dfrac{1}{100} = \dfrac{0.2}{100} = \dfrac{2}{1000} = \dfrac{1}{500}$

19. Perimeter $= 2 \cdot$ length $+ 2 \cdot$ width

$= 2 \cdot 1 + 2 \cdot \dfrac{2}{3}$

$= 2 + \dfrac{2}{1} \cdot \dfrac{2}{3}$

$= 2 + \dfrac{4}{3}$

$= 2 + 1 + \dfrac{1}{3}$

$= 3\dfrac{1}{3}$

The perimeter is $3\dfrac{1}{3}$ feet.

Area $=$ length $\cdot$ width $= 1\dfrac{2}{3} = \dfrac{2}{3}$

The area is $\dfrac{2}{3}$ square foot.

20. $\dfrac{4500 \text{ trees}}{6500 \text{ trees}} = \dfrac{4500}{6500} = \dfrac{500 \cdot 9}{500 \cdot 13} = \dfrac{9}{13}$

21. $\dfrac{9 \text{ inches}}{30 \text{ days}} = \dfrac{3 \cdot 3 \text{ inches}}{3 \cdot 10 \text{ days}} = \dfrac{3 \text{ inches}}{10 \text{ days}}$

22. $\dfrac{650 \text{ kilometers}}{8 \text{ hours}} = \dfrac{650}{8} \text{ km/hr} = 81.25 \text{ km/hr}$

23.
$$
\begin{array}{r}
0.0931 \approx 0.093 \\
16 \overline{)\ 1.4900} \\
-1\ 44 \\
\hline
50 \\
-48 \\
\hline
20 \\
-16 \\
\hline
4
\end{array}
$$

The 16-ounce size costs approximately $0.093/ounce.

$$
\begin{array}{r}
0.0995 \approx 0.100 \\
24 \overline{)\ 2.3900} \\
-2\ 16 \\
\hline
230 \\
-216 \\
\hline
140 \\
-120 \\
\hline
20
\end{array}
$$

The 24-ounce size costs approximately $0.100/ounce.
The 16-ounce size is the better buy.

24. $\dfrac{8}{n} = \dfrac{11}{6}$

$11n = 8 \cdot 6$

$11n = 48$

$n = \dfrac{48}{11} = 4\dfrac{4}{11}$

25.
$$
\begin{array}{r}
20.0 \\
-\ 8.6 \\
\hline
11.4
\end{array}
$$

26. Divide the number of gallons of fuel by the number of hours.

$$58\frac{3}{4} \div 7\frac{1}{2} = \frac{235}{4} \div \frac{15}{2}$$
$$= \frac{235}{4} \cdot \frac{2}{15}$$
$$= \frac{235 \cdot 2}{4 \cdot 15}$$
$$= \frac{5 \cdot 47 \cdot 2}{2 \cdot 2 \cdot 5 \cdot 3}$$
$$= \frac{47}{2 \cdot 3}$$
$$= \frac{47}{6} \text{ or } 7\frac{5}{6}$$

$7\frac{5}{6}$ gallons of fuel were used each hour.

27. Let n = the number of cartons.

$$\begin{array}{l} \text{cartons} \to \\ \text{hours} \to \end{array} \frac{n}{8} = \frac{86}{6} \begin{array}{l} \leftarrow \text{cartons} \\ \leftarrow \text{hours} \end{array}$$
$$6n = 8 \cdot 86$$
$$6n = 688$$
$$n = \frac{688}{6}$$
$$n = 114\frac{2}{3}$$

He can pack $114\frac{2}{3}$ cartons in 8 hours.

28.
$$\begin{array}{r} 17 \\ 29\overline{)493} \\ \underline{-29} \\ 203 \\ \underline{-203} \\ 0 \end{array}$$

Each can of paint was $17.

29. $7.5 = 0.6\% \cdot n$
$7.5 = 0.006 \cdot n$
$$n = \frac{7.5}{0.006}$$
$n = 1250$
0.6% of 1250 is 7.5.

30. $567 = n \cdot 756$
$$\frac{567}{756} = n$$
$0.75 = n$
$75\% = n$
567 is 75% of 756.

31. $12\% \cdot 320 = 0.12 \cdot 320 = 38.4$
The alloy contains 38.4 pounds of copper.

32. discount $= 15\% \cdot \$120 = 0.15 \cdot \$120 = \$18$
sale price $= \$120 - \$18 = \$102$

33. $2.4 \text{ kg} = \frac{2.4 \text{ kg}}{1} \cdot \frac{1000 \text{ g}}{1 \text{ kg}} = 2400 \text{ g}$

34. $2\frac{1}{2} \text{ gal} = \frac{2\frac{1}{2} \text{ gal}}{1} \cdot \frac{4 \text{ qt}}{1 \text{ gal}} = 10 \text{ qt}$

35.
$$\begin{array}{r} 2 \text{ ft } 9 \text{ in.} \\ \times \qquad 6 \\ \hline 12 \text{ ft } 54 \text{ in.} \end{array} = 12 \text{ ft} + 4 \text{ ft } 6 \text{ in.} = 16 \text{ ft } 6 \text{ in.}$$
Thus, 16 ft 6 in. of material is needed.

36. The complement of an angle that measures 78° is an angle that measures $90° - 78° = 12°$.

37. $\angle x$ and the angle marked 62° are adjacent angles, so $m\angle x = 180° - 62° = 118°$. $\angle y$ and the angle marked 62° are vertical angles, so $m\angle y = 62°$. $\angle x$ and $\angle z$ are vertical angles, so $m\angle z = m\angle x = 118°$.

38. $m\angle x = 180° - 92° - 62° = 26°$

39. $\frac{n}{12} = \frac{5}{8}$
$8n = 12 \cdot 5$
$8n = 60$
$$n = \frac{60}{8}$$
$n = 7.5 \text{ or } 7\frac{1}{2}$

40. Mean: $\frac{26 + 32 + 42 + 43 + 49}{5} = \frac{192}{5} = 38.4$
Median: The numbers are already in order, and the middle number is 42.
Mode: There is no mode, since each number occurs once.

41. 5 students are $5'0'' - 5'3''$ tall and 6 students are $5'4'' - 5'7''$ tall, so $5 + 6 = 11$ students are $5'7''$ or shorter.

42. $(-5)^3 - 24 \div (-3) = -125 - 24 \div (-3)$
$$= -125 - (-8)$$
$$= -125 + 8$$
$$= -117$$

43. $-7 - (-19) = -7 + 19 = 12$

44. $\dfrac{-3(-2) + 12}{-1(-4-5)} = \dfrac{6+12}{-1(-9)} = \dfrac{18}{9} = 2$

45. $\left(\dfrac{5}{9} - \dfrac{7}{9}\right)^2 + \left(-\dfrac{2}{9}\right) = \left(-\dfrac{2}{9}\right)^2 + \left(-\dfrac{2}{9}\right)$
$$= \dfrac{4}{81} + \left(-\dfrac{2}{9}\right)$$
$$= \dfrac{4}{81} + \left(-\dfrac{18}{81}\right)$$
$$= -\dfrac{14}{81}$$

46. $\dfrac{3x-5}{2y} = \dfrac{3 \cdot 7 - 5}{2(-8)}$
$$= \dfrac{21-5}{-16}$$
$$= \dfrac{16}{-16}$$
$$= -1$$

47. $5(3z+2) - z - 18 = 5 \cdot 3z + 5 \cdot 2 - z - 18$
$$= 15z + 10 - z - 18$$
$$= 15z - z + 10 - 18$$
$$= 14z - 8$$

48. $-\dfrac{5}{8}x = -25$
$$-\dfrac{8}{5} \cdot -\dfrac{5}{8}x = -\dfrac{8}{5} \cdot -25$$
$$x = 40$$

49. $5x + 12 - 4x - 14 = 22$
$$x - 2 = 22$$
$$x - 2 + 2 = 22 + 2$$
$$x = 24$$

50. $3x - 5 = -11$
$$3x - 5 + 5 = -11 + 5$$
$$3x = -6$$
$$\dfrac{3x}{3} = \dfrac{-6}{3}$$
$$x = -2$$

51. Let x be the number of women runners entered. Then the number of men runners entered is $x + 112$, and the sum of these two quantities is 600.
$$x + x + 112 = 600$$
$$2x + 112 = 600$$
$$2x + 112 - 112 = 600 - 112$$
$$2x = 488$$
$$\dfrac{2x}{2} = \dfrac{488}{2}$$
$$x = 244$$
There were 244 women runners entered.